COMPREHENSIVE SERIES IN PHOTOCHEMICAL AND PHOTOBIOLOGICAL SCIENCE

COMPREHENSIVE SERIES IN PHOTOCHEMISTRY AND PHOTOBIOLOGY

Series editors: Evelyne Sage, Lesley Rhodes and Massimo Trotta

Titles in this Series:

Volume 1	UV Effects in Aquatic Organisms and Ecosystems Edited by E.W. Helbling and H. Zagarese
Volume 2	Photodynamic Therapy Edited by T. Patrice
Volume 3	Photoreceptors and Light Signalling Edited by A. Batschauer
Volume 4	Lasers and Current Optical Techniques in Biology Edited by G. Palumbo and R. Pratesi
Volume 5	From DNA Photolesions to Mutations, Skin Cancer and Cell Death Edited by É. Sage, R. Drouin and M. Rouabhia
Volume 6	Flavins: Photochemistry and Photobiology Edited by E. Silva and A.M. Edwards
Volume 7	Photodynamic Therapy with ALA: A Clinical Handbook Edited by R. Pottier, B. Krammer, R. Baumgartner, H. Stepp
Volume 8	Primary Processes of Photosynthesis, Part 1: Principles and Apparatus Edited by G. Renger
Volume 9	Primary Processes of Photosynthesis, Part 2: Principles and Apparatus Edited by G. Renger
Volume 10	Biophysical and Physiological Effects of Solar Radiation on Human Skin Edited by Paolo U. Giacomoni
Volume 11	Photodynamic Inactivation of Microbial Pathogens: Medical and Environmental Applications Edited by Michael R. Hamblin and Giulio Jori

Volume 12 Surface Water Photochemistry
Edited by Paola Calza and Davide Vione

Volume 13 Singlet Oxygen: Applications in Biosciences and Nanosciences, Volume 1
Edited by Santi Nonell and Cristina Flors

Volume 14 Singlet Oxygen: Applications in Biosciences and Nanosciences, Volume 2
Edited by Santi Nonell and Cristina Flors

Volume 15 Photodynamic Medicine: From Bench to Clinic
Edited by Herwig Kostron and Tayyaba Hasan

Volume 16 Microalgal Hydrogen Production: Achievements and Perspectives
Edited by Michael Seibert and Giuseppe Torzillo

Volume 17 Light in Forensic Science: Issues and Applications
Edited by Giorgia Miolo, Jacqueline L. Stair and Mire Zloh

Visit our website at www.rsc.org/books

COMPREHENSIVE SERIES IN PHOTOCHEMISTRY AND PHOTOBIOLOGY – VOLUME 17

Light in Forensic Science
Issues and Applications

Editors

Giorgia Miolo
University of Padova
Italy
Email: giorgia.miolo@unipd.it

Jacqueline L. Stair
University of Hertfordshire
UK
Email: j.stair@herts.ac.uk

and

Mire Zloh
University of Hertfordshire
UK
Email: m.zloh@herts.ac.uk

ROYAL SOCIETY OF CHEMISTRY

THE QUEEN'S AWARDS FOR ENTERPRISE: INTERNATIONAL TRADE 2013

ISBN: 978-1-78262-768-5
PDF ISBN: 978-1-78801-034-4
EPUB ISBN: 978-1-78801-398-7
ISSN: 2041-9716

A catalogue record for this book is available from the British Library

The Royal Society of Chemistry is a charity, registered in England and Wales, Number 207890, and a company incorporated in England by Royal Charter (Registered No. RC000524), registered office: Burlington House, Piccadilly, London W1J 0BA, UK, Telephone: +44 (0) 207 4378 6556.

For further information see our web site at www.rsc.org

Printed in the United Kingdom by CPI Group (UK) Ltd, Croydon, CR0 4YY, UK

Preface

Many textbooks are currently available concerning crime scene examination and the forensic analyses of the material recovered from the crime scene, covering a large range of techniques and approaches.

The peculiarity of this book is that it is specifically targeted at light-based instrumental methods and at light of specific wavelengths for gathering evidence in the crime scene, assessing a number molecules useful in the crime scene solving and, at the same time, it presents the adverse photodegrading effects of light on materials of interest.

Since many traces are invisible to the naked eye, various types of illumination and visualisation techniques are devised in order to aid the activity of the examination teams.

The book covers the analysis of a range of molecular systems for forensic examination purposes, with a detailed review of different analytical methods that use light sources for either illumination or detection of trace samples, including illicit drugs and novel psychoactive substances (NPS), biological agents that may be used in bioterrorism, polymeric material, *i.e.* fibres, reflectors, adhesive tapes, textiles, and glass materials with a particular emphasis on detecting, analysing and interpreting evidence from crime scenes.

This book contains chapters with a review of such approaches, based on absorption, scattering and reflection of light of different wavelengths, and on the exploitation of fluorescence and chemiluminescence.

Moreover, the accurate and rapid identification of bacteria using laser irradiation *via* MALDI-TOF MS is widely discussed in a chapter dedicated to this methodology for detection of biological warfare agents.

Spectroscopic techniques are ideal tools for detecting analytes at trace levels *i.e.*, low concentrations. Microspectrophotometry, Raman spectroscopy, FTIR spectroscopy, IR-chemical imaging and X-ray fluorescence spectroscopy

Light in Forensic Science: Issues and Applications
Edited by Giorgia Miolo, Jacqueline L. Stair and Mire Zloh

Published by the Royal Society of Chemistry, www.rsc.org

are also discussed as recommended tools to determine the nature, class and subclass of polymers, such as fibers.

In regards to glass evidence, this has the ability to provide information not only to link individuals to broken glass items found at a crime scene but also to help establish what, when and how these were transferred. In this context, light plays a fundamental role in obtaining particularly source information, because the morphology, optical and chemical properties of glass can be characterised by different regions of both the visible and near visible electromagnetic spectrum.

In contextual chapters, NPS and illicit drugs' detection by light based techniques and the possibility of photodegradation in different matrices (both in water from various sources for epidemiological studies and in *in vivo* samples, *i.e.* hair of potential consumers) are extensively discussed and will provide the reader with a solid background in this area of crime investigation.

Normally, at a crime scene the evidence needed is often obtained by collecting the analytes of interest and the samples are often taken back to the lab for analysis. Nevertheless, there has been a recent surge in portable methods that can be used at the site of the crime or in emergency rooms. Supramolecular host-guest systems, providing the necessary architecture for portable systems while being cost effective, selective and sensitive, are described in a specific chapter of this book, focusing on the recognition of drugs of abuse that utilise optical changes *i.e.*, changes in colour or fluorescence response.

Furthermore, a chapter is dedicated to Raman spectroscopy which has been shown to decipher NPS chemical analogues as well as the NPS from cutting agents and adulterants present in associated products. Analysis can be done in the laboratory or in-field using handheld versions, in which the latter often have in-build matching algorithms for a quick response for non-experts.

Additionally, many synthetic novel psychoactive compounds contain a stereogenic centre leading to two possible enantiomers, whose pharmacological potency differs as it is the case for many chiral active pharmaceutical ingredients. Therefore, the analytical methods for enantioseparation of NPS by high performance separation techniques (HPLC, capillary electrophoresis, capillary electrochromatography and supercritical fluid chromatography) with UV absorption detection is of great forensic interest.

Another chapter is dedicated to hair analysis as a valuable tool in forensic toxicology to demonstrate drug exposure in the context of facilitated crimes. Since scalp hair is exposed to sunlight and/or artificial light for many hours per day; the action of light on hair could alter the content of drugs/illicit drugs and/or metabolites and the xenobiotics can gradually disappear or be transformed into other compounds. Thus, light exposure should be considered as a potential confounder in studies investigating xenobiotics in hair giving rise to reduced drug concentrations or even false negative results.

Indeed, the second aspect of the light discussed in this book concerns the degrading effects of light, both induced by the analytical methods employed (*i.e.* UV light breakage of DNA during DNA profiling) or by sunlight when outdoor samples are collected and then subjected to forensic analyses.

Moreover, the current processes used in the forensic analysis of DNA and RNA are covered in a selected chapter detailing why light is essential for detection but also destructive to the template DNA collected during a criminal investigation.

When traces have been exposed for long periods of time to weathering, degradation, especially by light, can modify their structure and complicate the work of the analyst. The issues related to degradation of material evidence found at the crime scene are discussed, focusing also on polymeric evidence, because polymers are the traces most sensitive to degradation, whereas glass or soil are not significantly changed by exposure to light for the time durations usually involved in normal casework.

At the same time the adverse effects of light on materials of interest as evidence could open opportunities for development of additional markers not yet explored or yield an enrichment of libraries for dedicated instruments by adding degradation products to reduce false negative or positive results.

In conclusion, this book aims to cover the latest research and development based on light methodologies in the crime scene investigation by specialists from forensic institutions and departments worldwide. Indeed, as light can play a powerful role in this environment, the presented material will allow reader to gain in depth understanding on how the light affects the evidence and when to apply techniques that employ light for crime scene investigations.

Giorgia Miolo, Jacqueline L. Stair and Mire Zloh

Contents

Light in Forensic Science: Issues and Applications
Edited by Giorgia Miolo, Jacqueline L. Stair and Mire Zloh

Published by the Royal Society of Chemistry, www.rsc.org

Chapter 1

Light for Crime Scene Examination

Valerio Causin*[a] and Giuseppe Guzzini[b]

[a]Università di Padova, Dipartimento di Scienze Chimiche, Via Marzolo 1, 35131 Padova, Italy; [b]Sezione Investigazioni Scientifiche Carabinieri, Via Salvo d'Acquisto 6, 37122 Verona, Italy

*E-mail: valerio.causin@unipd.it

Table of Contents

Light in Forensic Science: Issues and Applications
Edited by Giorgia Miolo, Jacqueline L. Stair and Mire Zloh

Published by the Royal Society of Chemistry, www.rsc.org

1.1. Introduction

Forensic science is a very powerful investigative tool, irreplaceable in many instances for the elucidation of complex cases and for an objective understanding of the dynamics of criminal acts. Looking back at the history of forensic science, every time a new technique became available for acquiring data on the crime scene, a disruptive step forward was introduced in the ability of police forces to identify and prosecute criminals and eventually to fight crime. This was especially true at the end of the 19th century when the first studies on fingerprints as means for the identification of individuals were published by Faulds and Galton.[1,2] Just a few years had elapsed from these seminal works when, in Argentine in 1892, Juan Vucetich was the first to solve a criminal case using fingerprints for the identification of a felon. This started the era of modern forensic science. An equally revolutionary advancement came with the development of DNA typing, in 1985.[3] Since then, more and more sensitive techniques have been devised, decreasing the minimum sample size for obtaining a reliable DNA profile. Less than 30 years later, it is almost impossible to imagine investigation without DNA. Technology and science are shaping and enhancing the ability of forensic science to achieve its purpose, *i.e.* the study of traces related to crimes.[4] Traces can be defined as the remnants of an activity and forensic science endeavours to deduce from the traces left on the crime scene as much information as possible on the crime itself. This concept is very well synthesised by the well known Locard's principle, which is often defined as 'every contact leaves a trace' even though Locard himself never formulated such an expression. Locard's words are very effective in stating this basic concept:

it is impossible for a criminal to act, and especially to act with the intensity that a crime requires, without leaving traces of his presence.[5] This was later elaborated introducing the notion that traces can be evidence left by the felon on the crime scene, but also, for a reverse action, they can be items collected from the crime scene and transferred to the felon.[6]

Kirk very fittingly formulated the definition of traces as mute witnesses:

wherever he steps, whatever he touches, whatever he leaves, even unconsciously, will serve as a silent witness against him. Not only his fingerprints or his footprints, but his hair, the fibers from his clothes, the glass he breaks, the tool mark he leaves, the paint he scratches, the blood or semen he deposits or collects. All of these and more, bear mute witness against him. This is evidence that does not forget. It is not confused by the excitement of the moment. It is not absent because human witnesses are. It is factual evidence. Physical evidence cannot be wrong, it cannot perjure itself, it cannot be wholly absent. Only human failure to find it, study and understand it, can diminish its value.[7]

From this short historical introduction, it should clearly emerge that traces and the Locard's principle are foundation stones without which forensic science would not exist. Acknowledging that the purpose of forensic science is interrogating material remnants of a criminal activity provides a theoretical and philosophical framework for implementing science into the administration of justice in the most effective way. Differently to what appears in fiction, the role of the forensic scientist, in fact, is not to determine if the suspect is guilty or not, but it is to reconstruct as precisely as possible the chain of events associated to a crime, giving to the Court, to the investigators or to the lawyers reliable information to properly do their job.

In such a context, if contact is always accompanied by the transfer of some material, then the analysis and characterisation of such material can allow the forensic scientist to describe, prove or confirm the contact that originated it. Of course, such a logical path will have a successful outcome depending on a number of non-negligible factors. Transfer, persistence and recovery are the three main processes that, if successful, allow the trace, and especially the information associated with it, to reach the laboratory and eventually the Courtroom. In other words, traces must be transferred onto the crime scene or to the felon, they must remain on the crime scene or on the felon, and they must be found and retrieved from the crime scene or from the felon. This latter step is a considerable bottleneck in the process. Transfer and persistence depend on the dynamics of the crime, they are not related to the training or ability of investigators or scientists. However, if suitable procedures are not applied in the search and recovery of the items from the crime scene, there is a severe risk that some important information is lost or that contamination is introduced. In both cases, the work of investigators would be hindered rather than aided by forensic science. The fragility, the lability, the latency and the corruptibility of traces calls for highly qualified personnel operating on the crime scene, because any mistake made in this phase will jeopardise all the subsequent analyses and interpretation.

Crime scene investigation started as the set of procedures aimed at crystallising the crime scene and at describing it. Ottolenghi, a pioneer in forensic sciences in Italy, in the early 1900's, extended the concept of Bertillon's portrait parlé, which was used for giving an objective description of individuals, to the crime scene.[8] Just as a detailed description of the physical features of a person can bring to a non-ambiguous identification, a careful depiction of the crime scene can give investigators and all those involved in the judicial process a solid foundation on which the verification of crimes and the search for the perpetrators can be developed.

If on one hand crime scene examination started as a mere descriptive activity, the modern implementation of such a critical step of forensic science includes the proactive search for items, the screening of the traces and the application of field tests.

Even though the number of texts dedicated to crime scene examination is much lower than forensic science books, most police forces and supernational bodies, such as European Network of Forensic Science Institutes, publish guidelines or best practice manuals, sometimes available on their websites.[9–11] It is not the purpose of this chapter to describe the technicalities of crime scene investigation, such as how to approach the crime scene, how to move around it and how to collect and store evidence. For such details, the interested reader is referred to the specialised literature.[12,13] and to the aforementioned guidelines and best practice manuals.

As anticipated above, crime scene examination is a set of analytical activities aimed at searching, collecting and preserving all the elements which, either *per se* or, even more importantly, due to their spatial location, can be considered evidence useful for the reconstruction for the dynamics of a crime and for identification of the perpetrators. The contextualisation of the items is therefore a fundamental element for attributing evidential value to a trace. The operator is not a mere gatherer of items, but is rather a specialist with a strong forensic background which can guide him towards an educated evaluation of traces and of their interrelationship. This is especially relevant in equivocal cases of death, in crimes perpetrated in a domestic context, or in cases of staged or simulated crime scenes, in which the significance of each trace is not due to the nature of the trace itself, but on its coherence with the possible hypotheses which can be set forth on the dynamics of the event.

Differently from what TV shows and fictional literature suggests, crime scene investigators are humans, and as such rely on their senses for searching and examining traces. Our eyes, though, have a limited sensitivity and much information would be lost both because it is too small to be detected in the chaos of a crime scene and because it is latent and invisible to the naked eye. The purpose of this chapter is to review the technical approaches which can be followed for widening the human senses and thus make the search for traces on a crime scene more effective and more productive.

1.2. A Brief Summary on the Theory of Light

As will emerge more clearly later in this chapter, exploiting the interactions between light and matter is a very effective method to detect latent traces and to find information on a crime scene.

Light is electromagnetic radiation, *i.e.* it is radiant energy which propagates as a wave. The features which define a wave are the wavelength, λ, *i.e.* the distance between adjacent crests or troughs, or the frequency, ν, *i.e.* the number of cycles passing by a fixed point per unit time (Figure 1.1).

Frequency is also a very important parameter because it is related to the energy of the electromagnetic radiation by the well known relationship $E = h\nu$, where E is the energy of the photon and h is the Planck constant, 6.62×10^{-34} J s. In other words, the higher the frequency, the greater the energy of the radiation. For practical and historical reasons, the electromagnetic spectrum, *i.e.* the set of all the possible radiations, has been broken down in several regions according to wavelength (Figure 1.2), even though the physical behaviour of electromagnetic radiation does not change as a function of frequency or energy.

The human eye is only sensitive to the visible range, a quite small portion of the whole electromagnetic spectrum with wavelengths comprised

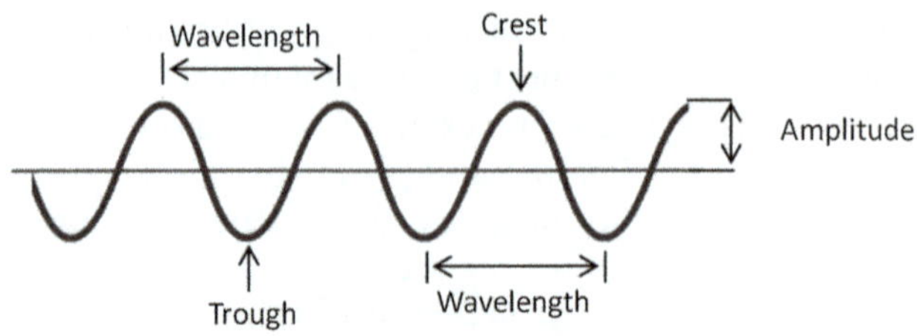

Figure 1.1. Describing the features of a wave. Polymers on the crime scene, 2015, V. Causin, © Springer International Publishing Switzerland 2015. With permission of Springer.

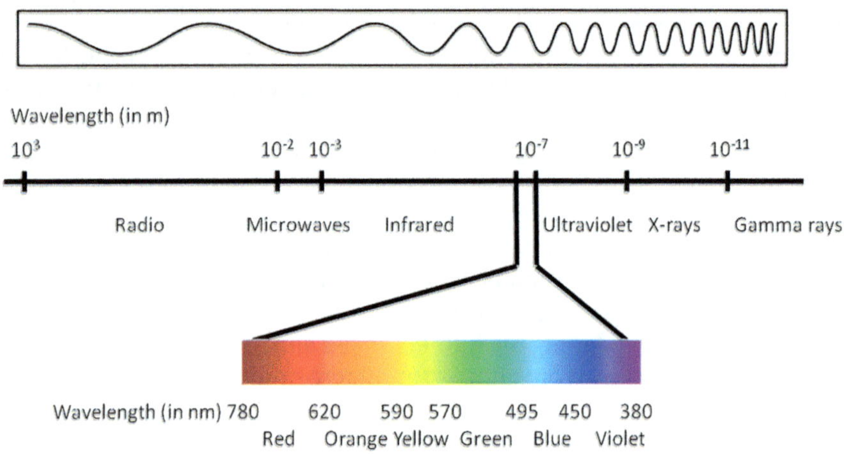

Figure 1.2. The electromagnetic spectrum. Polymers on the crime scene, 2015, V. Causin, © Springer International Publishing Switzerland 2015. With permission of Springer.

between 380 and 780 nm, which is a significant limit in the field of crime scene examination, where a number of traces remain latent when examined with this light. It should also be kept in mind that the sensitivity of the human eye is not equal for all the wavelengths of the visible range, but has a maximum for green, and decreases significantly towards red and blue/violet. The use of detectors rather than the naked eye, when working with these wavelengths, can significantly improve the chances of success in the search for tiny traces. Other useful portions of the electromagnetic spectrum, inaccessible to the human eye, but easily detectable with suitable technologies, are the near ultraviolet radiation, which is more energetic than visible light and has wavelengths from 200 to 380 nm, and the near infrared region, with a lower energy than the visible light and wavelengths comprised between 0.78 and 2.5 μm.

Before proceeding with the various observation techniques which may be useful during the examination of a crime scene, it may be useful to

Figure 1.3. Schematic of the absorption-reflection phenomenon.

summarise the different phenomena which may happen when light interacts with matter. The discussion will focus on a reflection geometry of illumination, because in most of the practical instances the traces observed are opaque, and not transparent.

When light containing all the visible wavelengths (white light) impinges an object, the object will absorb only certain wavelengths, whereas unabsorbed wavelengths will be reflected. These latter wavelengths will be perceived as a colour (Figure 1.3).

The colour seen will be the complementary colours to the absorbed colours. Figure 1.4 shows the "colour wheel", colours that are opposite on the wheel are complementary.

In other words, the object will absorb some of the colours contained in white light, allowing only the unabsorbed hues to be reflected back and to reach the eye of the observer. In our daily life we are very familiar with the absorption of visible light and the vision of colours, but the same physical phenomenon happens when the impinging light is from other regions of the electromagnetic spectrum, *e.g.* ultraviolet or infrared light.

Scattering is another phenomenon which may happen when an object is illuminated with light. When the wavelength of incoming light is comparable with the size of the illuminated objects, these start vibrating with the same frequency of light and they themselves become spherical sources of radiation. When this happens, some of the light is diffused at 360° around the object, blurring the purely geometrical propagation of light. In fact, when light interacts with smooth and shiny surfaces the radiation is reflected, with respect to the perpendicular to the surface, with an angle which equals the angle of incidence (Figure 1.5a). This mode is called specular reflection. In contrast, if light encounters a rough surface, scattering

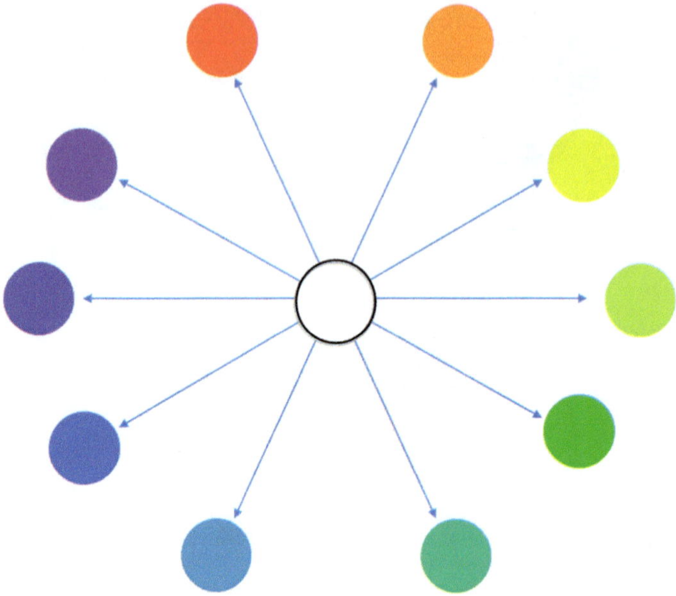

Figure 1.4. The colour wheel. Colours on the opposite sides of the wheel are complementary.

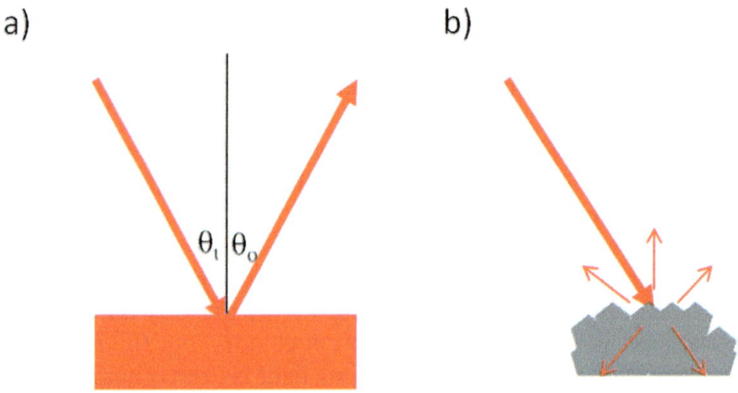

Figure 1.5. Schematic of light reflection. Incoming light (for clarity just monochromatic light is depicted here) a) can be reflected at an angle determined by classical optics equations (specular reflection), and/or b) it can partly or completely scattered at 360° (diffuse reflection).

occurs, and it is reflected in all directions of space, diffusing the radiation in the surrounding space (Figure 1.5b). In such case, the phenomenon is called diffuse reflection.

Fluorescence is a further phenomenon which is very useful for detecting traces. When a molecule absorbs light, it absorbs energy which is used to promote electrons to an excited state. Disexcitation may happen by the re-emission of light of the same wavelength, with a concurrent return of the electrons to their ground state. However, an alternative mechanism, called fluorescence, exists, in which some non-radiative decays, *i.e.* processes in which energy is lost without emission of radiation, accompany the return to the ground state by emission of light. In other words, when fluorescence happens, a compound absorbs light of a particular wavelength and energy, and re-emits radiation with a longer wavelength and lower energy. Many biological traces display fluorescence, which therefore is a very efficient method for their detection.

1.3. Imaging on the Crime Scene: Finding Traces

1.3.1. *Observation in the Absorption/Reflection Mode*

A crime scene is a very complex environment, in which traces with an evidential value coexist with a large number of items which do not carry any significant information of the event. Moreover, traces are often tiny, sometimes the perpetrator tries to erase them, and thus their detection is not easy. Finally, many kinds of traces are not directly visible to the naked eye, because they are not coloured and/or because they have a similar colour to the background surface where they lay. The purpose of the crime scene investigator is therefore to enhance, as much as possible, the contrast between the trace and the background, in order to clearly visualise it, aiding both its detection, documentation and retrieval.[14]

As mentioned above, the human eye has a sensitivity limited to the visible range, and within this wavelength region it is more effective in detecting green rather than red or violet.

A white light source is certainly suitable for a first survey of the scene, for detecting the most evident traces. However, the diversity of materials, in different colours and with different surface roughness, which can be encountered on crime scenes, calls for the application of more sophisticated approaches, if one wants to avoid missing important pieces of information.

In the first instance, absorption can be exploited to improve contrast. Using an illumination source of the same colour as the surface, the background will appear much lighter, and stains or traces on it will be dark features. An alternative to this could be illuminating with a colour complementary to that of the surface, which will appear dark, whereas traces should stand out as lighter features.

The choice of the illuminating wavelength also depends on the particular type of trace which the investigator is looking for. For example, the UV-visible spectrum of blood shows a prominent absorption peak at 415 nm.[15] If a blood spattered surface is observed with monochromatic light with this wavelength, the blood spots will appear as dark regions, because they will absorb all the radiation impinging upon them. If the surface does not absorb at 415 nm, or even if it reflects a portion of it, the substrate will lighten up, improving the contrast of the image. This can be practically performed with commercial tunable-wavelength light sources if working in dark conditions, or by illuminating with white light and observing through a filter at 415 nm.

Illuminating the sample with UV light and observing it with a UV sensitive camera, *i.e.* performing reflective ultraviolet photography, can be an effective way to improve contrast.[16] In some cases, traces which reflect UV light well stand on a UV opaque surface. Alternatively, it can happen that the substrate is reflective, whereas the trace is not. In both cases, reflective UV photography will catch a very good contrast between the reflective and non-reflective particulars of the object observed. Generally, if some pattern of a substance on a surface can be sensed, there are high chances that the pattern will stand out much more clearly if imaged in the near-UV. Short wavelength UV illumination is usually even more efficient, but at the same time it is a more aggressive kind of radiation, which could damage the most degradable and light sensitive materials.[17] Great care must be exercised when working with short wavelength UV light, because it can also be harmful for the operator: suitable protection for eyes and skin must be worn. Observation in the UV range with a reflection geometry is particularly suitable for observing bite marks and bruises on victims of aggression, because the penetration depth of UV light into the skin is deeper than that of visible light. In any case, good results can be obtained illuminating with the visible wavelengths of 415, 455 or 535 nm, associated to yellow, orange and red observation filters, respectively.

Observation with oblique or grazing illumination will help detecting traces on smooth surfaces, or particles, grooves or other 3D features which lay on a surface, such as footprints or blood prints.

1.3.2. *Fluorescence*

If on one hand these are quite practical approaches, on the other hand they are not very efficient in detecting fingerprints or other kinds of important biological fluids, such as semen, vaginal fluids, or saliva. In these cases, fluorescence is a very powerful tool. Semen, for example, is known to fluoresce in different conditions.[18]

Figure 1.6 shows, for example, that if semen is excited with light at 450 nm, it will emit fluorescent radiation with a maximum at about 520 nm. Practically, this operation is carried out using a tunable-wavelength light source set at 450 nm, and observing through a yellow/orange filter (*e.g.* 590 nm).[19] If the surface is not very reflective at this wavelength, the semen stains should appear as bright yellow/orange spots on a dark background. If the substrate were luminescent in these illumination conditions, the contrast between the stain and the background would not be ideal. In such cases, other wavelengths should be tried, because semen, as many other biological fluids, is fluorescent in a wide range of conditions. It should be born in mind

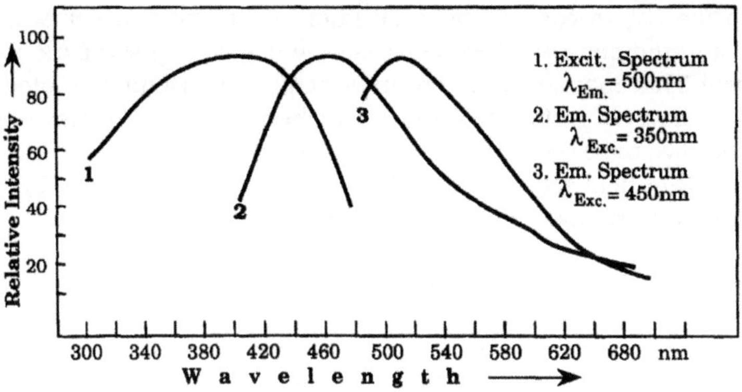

Figure 1.6. Fluorescence spectra of semen with different excitation wavelengths. Reprinted from *Forensic Science International*, **51**, M. Stoilovic, Detection of Semen and Blood Stains Using Polilight as a Light Source, Pages 289–296, Copyright (1991), with permission from Elsevier.

that as the wavelength of illuminating radiation is increased, the emission maximum is subsequently increased. Therefore, if for semen an illumination wavelength of 500 or 550 nm is used, the observation filter must have a colour more shifted to the red, orange and red, respectively.

Fluorescence can be used also to detect gunshot residues on the hands or clothes of a suspected shooter.[20–22] A simple set up can be illumination with 455 nm light and observation with an orange filter.

In field work, commercial light sources are employed, which allow the tuning of the wavelength of the lamp used for illuminating the scene. These are coupled to filters for the camera objectives or goggles to be worn by the operator with differing colours, in order to enhance the detection of the fluorescence luminescence. Each manufacturer will provide a handy table which indicates the combinations of illumination wavelength and observation filter most suitable for each class of trace, *e.g.* semen, vaginal fluid, saliva, blood, sweat, *etc.*

So far, only fluorescence excited by visible radiation and emitted in the visible range has been discussed. However, expanding the investigation beyond the visible limit, especially in the ultraviolet (UV) region, widens the possibility of detecting traces on a crime scene. Several substances fluoresce when excited by UV light, emitting either in the visible or in the UV. As mentioned before, this phenomenon, during investigation, is exploited by illuminating the crime scene by UV lamps, which allows one to choose the desired wavelength, and then observe it through a filter which selects the wavelength of maximum emission expected from the type of trace which is searched for. UV detectors should be used in the case of UV-UV fluorescence observations. In any case, since the visible emission due to UV excitation is usually located in the blue-violet region of the visible spectrum, where our eye is not very sensitive, the use of cameras or visors will help enhancing weaker traces. UV irradiation can help because many substrates are more reflective in the visible than in the UV, so contrast between fluorescent traces, which will appear bright, and the non-UV-reflective surface, which will appear dark, is greatly improved. This approach would be in principle very useful for biological fluids, which have a significant fluorescence when excited by UV. However, since DNA is known to be severely degraded by UV irradiation[23,24] and since the UV lamps used in field activity have a high intensity, the use of UV light, especially with a short wavelength, should be avoided for the search for biological traces.

On the other hand, UV irradiation can be very useful for the detection of many other traces, fingerprints in particular.[25]

Visible fingerprints are left either by fingers touching malleable materials, such as clay or mud, or by fingers stained by coloured substances, like paint or blood. They can be detected by illuminating them with just visible light, or by taking pictures in the absorption mode for enhancing contrast with respect to the substrate. These are rather rare cases, though, because the majority of fingerprints are latent and invisible to the naked eye. Fingerprints are composed of a dilute aqueous solution with about 2% proteins and lipids exuded by the sudoriparous pores in the fingertips or palms of the hand. Therefore, these traces are very hard to detect. When illuminated with UV light, fingerprints emit at 330 and a 440 nm, the emission at 330 nm being the strongest. It has been reported that 280 nm is the optimal excitation wavelength for maximising fluorescence.[26] It is therefore feasible to detect fingerprints by illuminating them with UV light and observing their blue visible emission. However, problems may arise when it is necessary to examine surfaces with a strong luminescence in the visible, due to optical brighteners. This is the case of some architectural paints, paper, or textiles.

Figure 1.7 shows some examples of untreated fingerprints on white paper. Since this substrate strongly fluoresces between 400 and 500 nm, in a spectral range which overlaps with the blue emission of fingerprints at 440 nm, it is necessary to perform imaging using the emission peak at 330 nm, which does not overlap with the fluorescence of paper. Being the light with a wavelength of 330 nm in the UV region, it will be necessary to use cameras with suitable detectors which, however, are quite common and available at affordable prices. For imaging a 360 nm band-pass will be suitable. Figure 1.7a–c show fluorescence images with excitation at 230, 280 and 300 nm, respectively. It may be seen that the clearest fluorescence image of a fingerprint was obtained with excitation at 280 nm.

As a further complication, it must be added that fluorescence emission from fingerprints tends to fade and/or change colour as the fingermark ages.[27]

Of course, it is not always possible to obtain an image of a fingerprint which is clear enough for a dactyloscopic assessment to be carried out, especially in difficult cases, such as weak prints or prints on surfaces that produce highly luminescent backgrounds. For this purpose, several chemical development methods have been devised for improving the level of detail

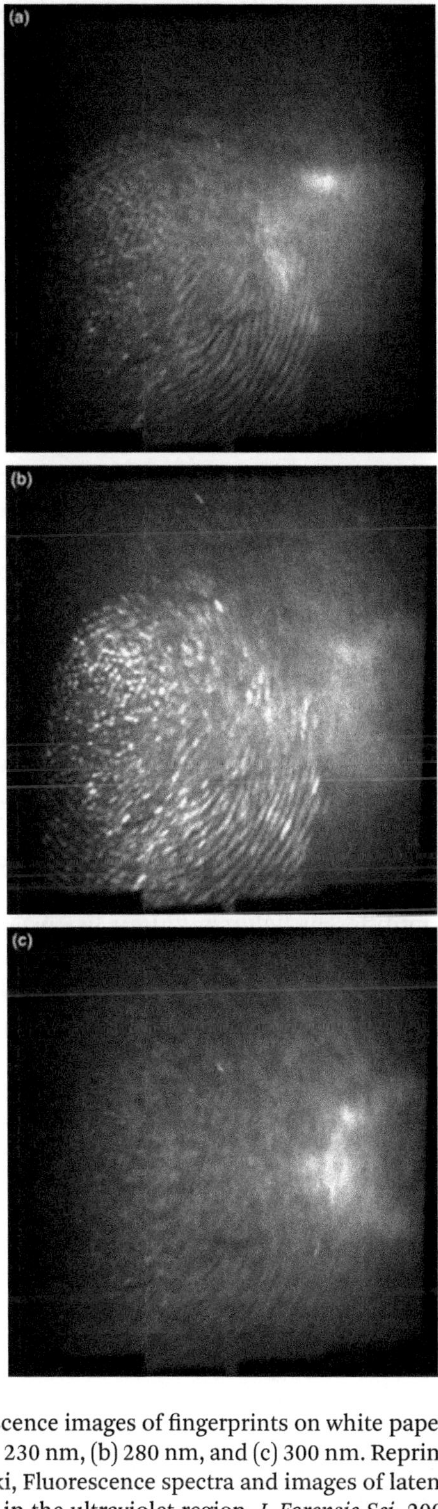

Figure 1.7. Fluorescence images of fingerprints on white paper observed at 330 nm with excitation at (a) 230 nm, (b) 280 nm, and (c) 300 nm. Reprinted from N. Akiba, N. Saitoh, and K. Kuroki, Fluorescence spectra and images of latent fingerprints excited with a tunable laser in the ultraviolet region. *J. Forensic Sci.* 2007, **52**, 1103 with permission from John Wiley and Sons. © 2007 American Academy of Forensic Sciences.

and contrast, in order to obtain the largest amount of information on this trace.[28] This topic will be further discussed and detailed in another chapter of this book.

The blue fluorescence emission due to optical brighteners can be very useful for detecting trace evidence such as textile fibres or paint chips or smears. Moreover, the lack of blue fluorescence emission can also yield useful information, such as in the case of counterfeit documents, where erasures or abrasions can alter the optical response of the paper substrate. In such cases, the paper illuminated by UV light will fluoresce emitting blue light, and the alterations will appear as dark areas.

Near-infrared is a suitable type of light as well,[21] for example for the detection of bone fragments on the ground or on complex surfaces. A near-infrared hyperspectral camera, associated to a data treatment based on chemometry, proved capable of distinguishing a wide variety of bones, including aged and weathered bone from stone, sand, leaves, moss, and wood.[29] Alsberg *et al.*[30] developed a specific instrument, PryJector, for highlighting *in situ* details which are otherwise invisible or difficult for human eyes to detect. By a combination of hyperspectral imaging, chemometric modelling, and back-projection by a computer projector, it is possible to emphasize the presence of particular items by creating a false coloured chemical image. Figure 1.8 shows the potential of this technique. A hyperspectral camera working in the 930–1670 nm range records a full spectrum in each pixel of a scene. Chemical information is extracted by multivariate data treatment, and each pixel is associated to a particular class, for example bone or non-bone in the current example. This chemical image is then projected back on the surface of interest, colouring the items of possible interest, in this case the bones in a heterogeneous mixture of bones, leaves, and soil.

The observation in different light conditions has been also proposed for distinguishing between fresh and old bruises, because it allows the detection of the yellow chromophores of bilirubin. Recent bruises in fact contain mainly blood, whereas as time goes by bilirubin progressively accumulates in the wounded region.[31]

Near IR radiation is particularly useful in the questioned document field. Many inks, with similar colours in the visible range, can be differentiated according to how they absorb or reflect IR. Figure 1.9 shows an application of this concept to obliterated writing. As evident, illumination by IR light succeeds in revealing the writing hidden beneath an obliteration only if there is a wavelength where the obliteration itself becomes transparent to IR, whereas the text remains opaque. If the situation is the opposite, *i.e.* the obliteration is opaque to IR and the text is transparent, it will not be possible to reconstruct what was originally written in the document, before the cancellation.[32]

IR illumination is very well suited also for detecting document counterfeiting by the addition of text. Figure 1.10 reports how, exploiting the different response of different inks to IR light, it is possible to clearly highlight that some alteration to the original text happened.

1.3.3. Chemiluminescence

A last approach worth mentioning is the exploitation of chemiluminescence. Different from fluorescence, which is an intrinsic characteristic of the target molecule (or of some of the molecules contained in the target trace), chemiluminescence consists of the emission of light triggered by a chemical reaction. In other words, if fluorescence can be activated just by illumination with light of the proper wavelength, for chemiluminescence interaction with a suitable reactant is needed. Luminol (5-Amino-2,3-dihydrophthalazine-1,4-dione) is the most renowned reagent for the detection of blood on the crime scene. Luminol and hydrogen peroxide are sprayed on the surface where blood traces are being searched for. If blood is present, the iron present in haemoglobin will act as a catalyst in Scheme 1.1. The first product is in the excited state, and then it loses the extra energy by emission of blue light.

From the mechanism of action, it is clear that any iron-containing substrate will not be suitable for treatment with Luminol, because it will give false positive results.

Latent blood traces, when treated with Luminol, will glow blue light which can be photographed. It should be noted that the luminescence lasts for about 30 seconds and then fades. Moreover, especially if the blood traces are small, the emitted intensity is quite low, and so the use of Luminol requires that the camera is operated in a darkened room, and with long exposure.

Despite Luminol being a reagent involved in a chemical reaction with blood components, its use does not jeopardise further treatments of the stains which are found. In particular, DNA typing remains feasible, because treatment with Luminol does not degrade it.

1.4. Photographic Techniques: Documenting Traces

In the previous section, a number of approaches have been described for detecting traces at a crime scene. Documentation is as important as finding traces, because, as introduced in the first paragraph of this chapter, most often the information related to the items depends on their location or on their relationship with other objects on the crime scene.

The modes of light-matter interaction described in paragraph 1.2 can be exploited to photograph traces on the crime scene.[33]

1.4.1. Absorption Mode

A coloured object or stain can be visualised emphasising it against the substrate where it stands. If there is a chromatic difference between the trace and its substrate, then it is possible to enhance it with photographs in the absorption mode. The best approach is illuminating the surface with

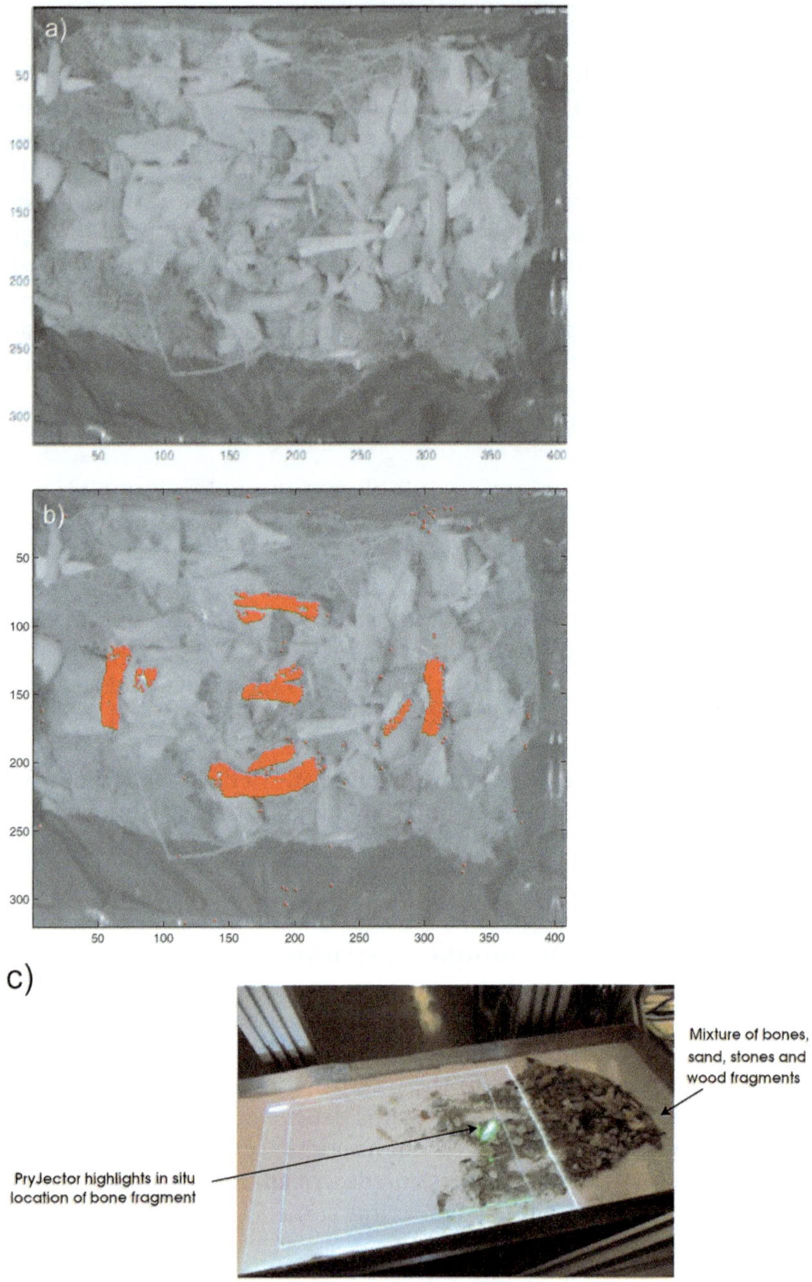

Figure 1.8. (a) The box containing the simulated forest ground with leaves, moss, plants, stones, sand, soil and bone fragments which were used to test the optimal classification model; (b) the result, after application of the optimal classification model on the simulated forest ground scene. Red is used to indicate predictions of bone. The location of the bone fragments can be clearly seen. False positives were removed

monochromatic light, obtained by a suitable filter. Working in the dark is suggested for the best results. The colour of light to shine on the surface must be complementary to the colour of the trace, and as close as possible to that of the substrate. For example, to enhance an orange stain on a blue substrate, blue light will be used. In this case, the light will be reflected by

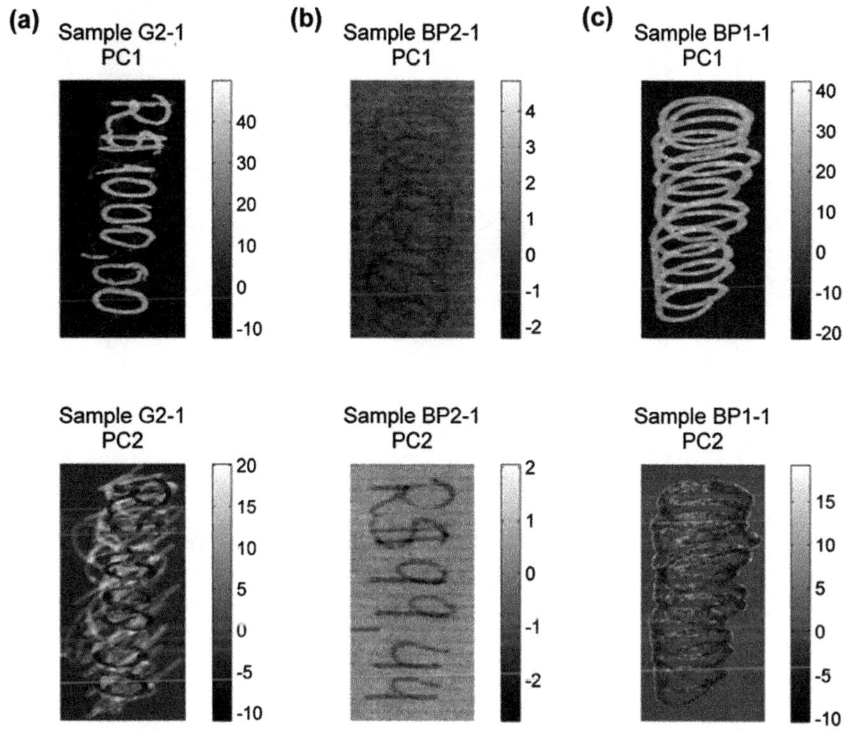

Figure 1.9. Images of three samples of obliterated text, obtained by acquiring a hyperspectral infrared image in the spectral range 930–2520 nm, followed by the application of principal component analysis. The images are produced by the scores of the two first principal components. Each sample, designated by the letters a), b) and c), was produced using different pens to write the text and to draw the obliteration. Reproduced from ref. 32 with permission from The Royal Society of Chemistry.

---

using a Mahalanobis-distance-based thresholding. (c) How the bone/non-bone classification model can be used by the PryJector system. Bone fragments are highlighted by a projector which continuously updates the chemical image of a surface using a line scanning hyperspectral camera. Here pixels belonging to the bone class are given a green colour and guide the localisation of bone fragments in a complex mixture. Reprinted from B. K. Alsberg, and J. Rosvold, Rapid localization of bone fragments on surfaces using back-projection and hyperspectral imaging. *J. Forensic Sci.* 2014, **59**, 474 with permission from John Wiley and Sons. © 2013 American Academy of Forensic Sciences.

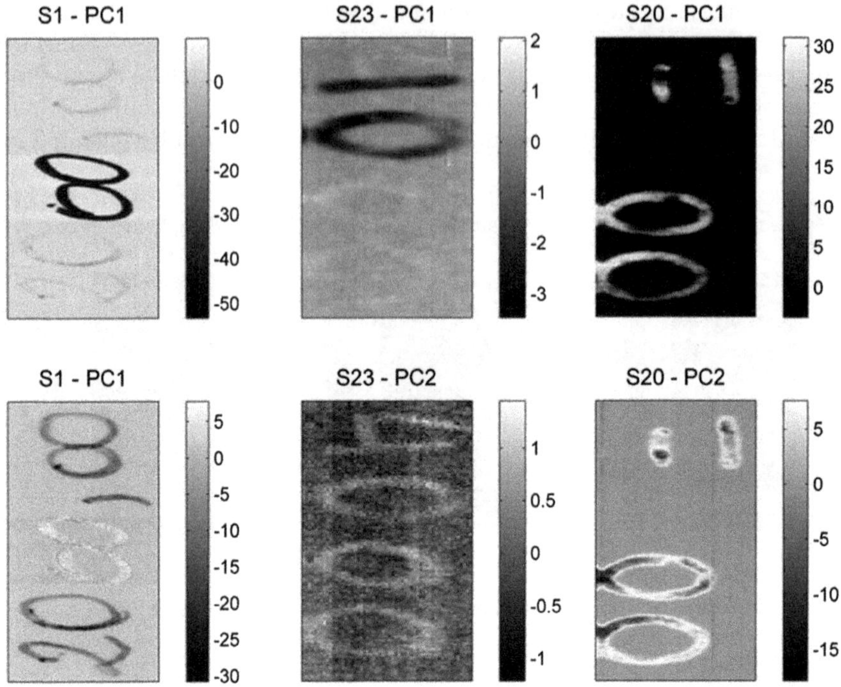

Figure 1.10. Images of three samples of text partly written with different pens. The mode of imaging and the data treatment by multivariate analysis is described in Figure 1.9. Reproduced from ref. 32 with permission from The Royal Society of Chemistry.

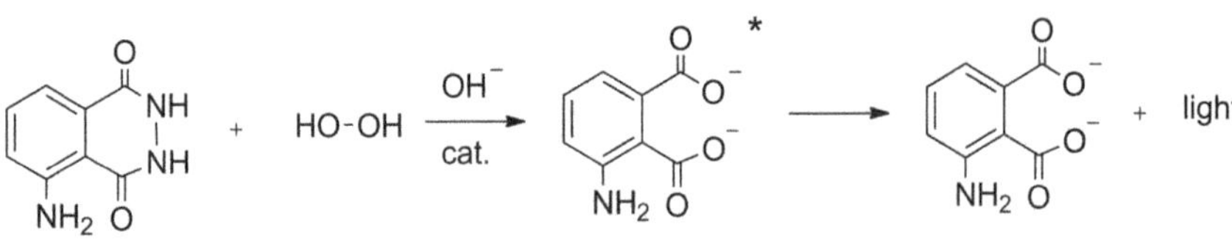

Scheme 1.1. Mechanism of chemiluminescence of Luminol.

the surface, whereas it will be absorbed by the stain. Therefore, just the blue light reflected by the surface will reach the detector of the camera, and as a result the stain will stand out as dark against a light background. The technique can be applied using a filter to monochromatise illuminating light and observing with an unfiltered camera or by the naked eye (Figure 1.11a). As an alternative, a filter can be applied to the objective of the camera, or the object can be observed through a filter, while illuminating it with white light (Figure 1.11b). The former method is more efficient, but must be performed

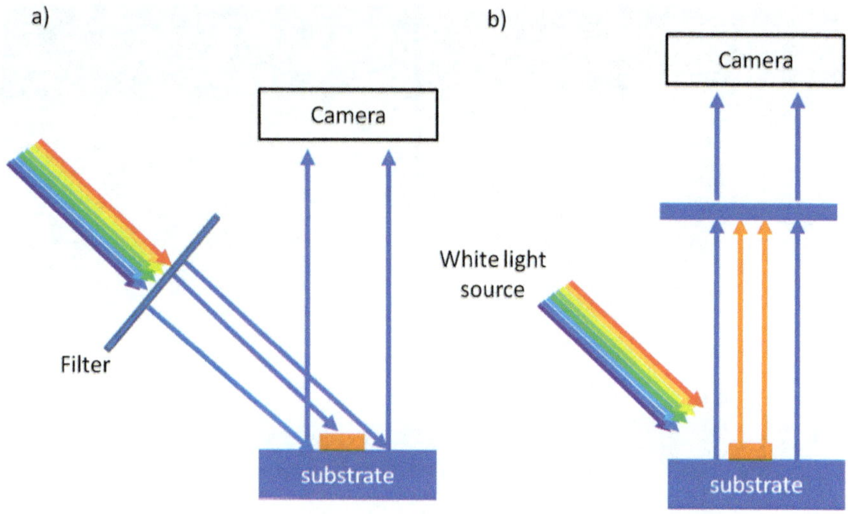

Figure 1.11. Imaging in the absorption mode of an orange stain on a blue substrate. (a) Light, monochromatised by a filter, is shone on the object. Just the substrate reflects light, whereas the stain absorbs it and thus no light is reflected by it and will not reach the camera. (b) White light is shone on the object. The substrate reflects blue light, the stain reflects orange light. A filter applied to the camera will let just blue light pass, and thus the image will show a black stain in an intense blue background.

in the dark. The latter has the advantage that can be applied in sunlight or in conditions of artificial lighting.

Figure 1.12 shows an example of the potential of absorption photography in enhancing the contrast of stains or traces.

As an alternative, the trace can be imaged as a light feature on a dark background illuminating with light which will be absorbed by the substrate and reflected by the trace. For the example examined before of an orange stain on a blue surface, orange light will work well: this radiation will be reflected by the stain, but it will be absorbed by the substrate. Orange is in fact a complementary colour to blue, being opposite to it in the colour wheel of Figure 1.4.

Imaging in the absorption mode shows its limits when the colour of the trace is very similar to that of the substrate.

1.4.2. Diffuse Reflection Mode

Imaging techniques which exploit the diffusion of light can be very useful for observing traces which lie on a smooth or very reflecting surface, or for enhancing their 3D structure. Examples include greasy fingerprints on a shiny glass surface or latent grooves on a piece of paper. This technique can be applied either with monochromatic or with white light. In the case where

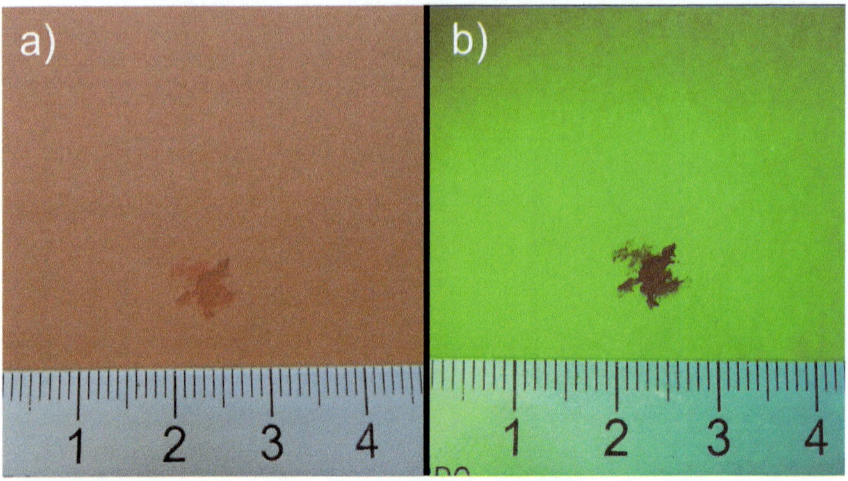

Figure 1.12. Red stain on an orange cardboard background. Imaging mode (a) white light (b) green light (500 nm).

monochromatisation was available, a colour as similar as possible to the colour of the trace should be chosen, or at least a complementary colour to that of the surface. The method works best, in fact, on smooth and dark surfaces which will absorb most of the radiation impinging on them, reflecting the rest. The rationale of this technique can be understood if the two main mechanisms of reflection of light from a surface are acknowledged, *i.e.* specular reflection and diffuse reflection. When light impinges the surface, it will be specularly reflected by its smooth regions, and it will be diffused by the rough regions. Illumination is oriented at an angle such that specular reflection does not reach the camera, whereas diffused light, which is directed in all the directions of space, can be acquired by the detector. This way, the trace appears as a light feature on a dark surface. The technique yields even better results if some of the radiation is absorbed from the substrate, making the background of the image even darker and thus the contrast even higher. In this case also, it is suggested that one should work in the dark, for better results.

As evident from Figure 1.13, illumination should be angled, especially if it is of interest to enhance the 3D structure of the observed object. Oblique or incident light should be used in this case, and there is no particular need for monochromatisation: white light can be used. The physical phenomenon which is exploited is the same as above: when light almost parallel to the surface impinges some protruding feature, it is scattered towards the camera, whereas the rest of light does not reach the detector.

Figure 1.14 shows how the diffuse reflectance mode allows to image a fingerprint left on a dark and smooth plastic surface.

An alternative set up for taking pictures in reflection mode is by an episcopic coaxial illumination set up. In this case, through a semitransparent

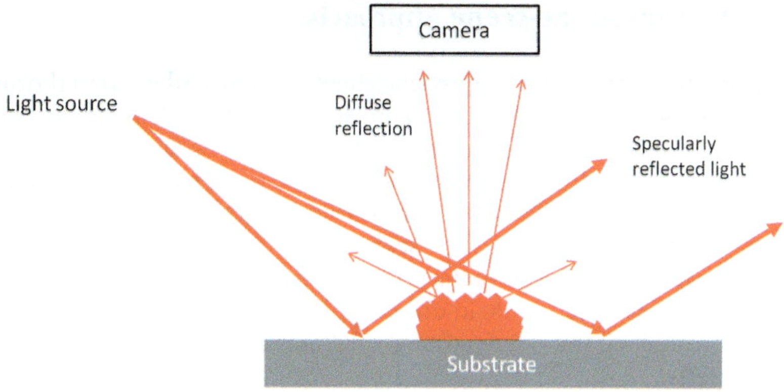

Figure 1.13. Imaging in the diffuse reflectance mode of a red, rough trace on a smooth grey substrate. Illumination by light of a colour as similar to that of the trace as possible (red in this case) is performed. The substrate will reflect the incoming light by specular reflection, away from the camera, whereas the diffuse reflection originated by the rough trace will reach the camera and will be detected as a light feature on a dark background.

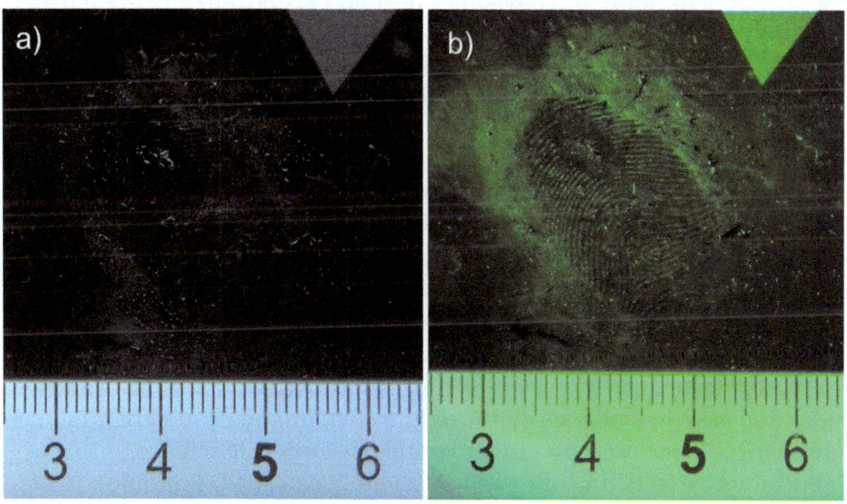

Figure 1.14. Image of an untreated fingerprint left with blue ink on a black, smooth plastic surface. Imaging modes are (a) reflected white light (b) 45° illumination with green light (540 nm).

mirror, incoming light is directed towards the object perpendicularly to its surface. Reflected light follows the same path and reaches the camera. In this case, the smooth part of the sample will be more efficient in reflecting light, whereas the rough trace will diffuse some of the radiation away from the camera. As a consequence, with this geometry the trace will appear dark against a light background.

1.5. The Lab-on-the-scene Approach

In this chapter, a number of approaches have been described for a thorough examination of the crime scene. They are all aimed at the detection of traces, which are subsequently documented, collected, and brought to the laboratory. In other words, traditionally the laboratory is the place where the analysis, identification and interpretation of evidence is carried out. There is a growing trend to bring to the crime scene some of the instrumentations and techniques applied in the laboratory. Under this perspective, the investigator can start screening the items, in order to pick those which are more significant and to reduce the overload of the laboratory and the amount of collected material.

Among the techniques which fit the scope of this book, *i.e.* the relationship between light and the crime scene, two can be cited: miniaturised spectroscopy and presumptive tests.

Given the small size that traces often have, forensic science exerted a strong driving force towards miniaturisation of analytical instruments. This is true not only for the miniaturisation aimed at analysing very small items, but also aimed at making instruments portable for allowing their use in the field. Raman, UV-visible, IR, mass spectrometers are currently available in portable, and also hand-held versions, by which first instance analyses can be performed directly on the crime scene to identify a material and decide whether or not it is worth collecting it. The performance of such instruments is obviously lower than their benchtop counterparts in the laboratory, but they still can give useful data in some cases. Needless to say, it is of the utmost importance that the operator is fully trained not just on the practical use of the instrument, but on the interpretation of the data. Otherwise, the risk is that decisions are taken on the strategy of crime scene examination based on wrongly interpreted data, jeopardising the whole investigation.

Presumptive tests have been in the tool box of the crime scene investigator for quite a long time. These are mainly colorimetric tests, which can give a quick and easy general idea on the nature of an unknown material. Table 1.1 summarises some examples in several fields of forensic science.

Presumptive tests must be used with care. Their main limitation is a lack of selectivity. For example, amylase is found in a variety of body fluids, such as saliva, blood, urine, sweat, tears, semen, breast milk, feces, and vaginal secretions. A positive test to the amylase test may support the hypothesis that the tested sample is indeed saliva, but chances are that it could be another kind of biological fluid. False positives are always an option. For instance, aspirin gives a positive reaction to the Marquis test, even though it is not an illicit drug. Many tests for blood involve a redox step catalysed by haemoglobin present in blood, which however can be replicated also by the peroxidase enzymes which can be found in many kinds of fruit. Any presumptive test must be followed by a confirmatory analysis, carried out with established protocols in a proper laboratory.

Table 1.1. Some examples of presumptive tests for drugs of abuse, biological specimens, explosives.

Test name	Analyte	Mode of action
Marquis	Amphetamines, opioids	The active principles form a coloured complex with the reagent
Nitric acid test	Opioids	A yellow-orange-red complex is formed as a consequence of the reaction of the opium derivative with HNO_3
Duquenois-Levine	Cannabinoids	A purple complex is formed between THC and the reagents
Young	Cocaine	A blue complex is formed between cocaine and the reagents
Nessler	Ammonium	A brown colour or precipitate forms in the presence of ammonium, an ion commonly present in inorganic explosives
Precipitin or other antigen/antibody type of tests	Blood	Combining human antiserum and human blood gives agglutination
Jaffe test	Urine	A brown/orange dye is formed by reaction of creatinine and picric acid
Amylase test	Saliva	Immunochemical test based on antibodies for α-amylase
Acid phosphatase test	Semen	Acid phosphatase present in sperm forms purple complexes with the reagent
Kastle-Meyer Colour Test	Blood	Phenolphthalein and hydrogen peroxide give a pink coloured compound with blood

In the case of tests of the antigen-antibody type, the selectivity is extremely high, and the risk of false positives is negligible. However, laboratory tests are needed to extract the most useful information from the traces, *i.e.* DNA.

Another drawback of presumptive colorimetric tests is that they are destructive. In fact, they require a small aliquot of the unknown sample to be treated with the reagent, which after the test will no longer be suitable for further analyses.

1.6. Conclusion

In this chapter, a number of approaches has been described for a thorough examination of the crime scene. Focus has been mainly on the physical bases of each approach. There is currently a large number of commercial instruments allowing the investigation of the crime scene with many different types and geometries of illumination. Hyperspectral instruments are being manufactured, and the fast advancement of software and hardware allowed the introduction to the market spherical cameras or 3D scanners which automatically acquire images from their surroundings and which can merge this data array to obtain a 3D picture of the crime scene. Software allows also one to take measurements within the crime scene, without actually entering it. This is an interesting technology which can greatly aid the assessment of the crime scene. Systems like these are useful on one hand, but they should be used with great care. They can be employed as an effective way for documenting the location and relative position of all the items in the crime scene, but they cannot substitute an assessment in person. Even though it is an exquisitely technical activity, crime scene investigation retains a strong dependence on the ability of a trained examiner to catch the subtle details which lead to the discovery of traces and of patterns. Tools such as these avoid the need to enter the crime scene for operations such as planimetry drawing or measuring steps. However, an excessive virtualisation of the crime scene should be avoided, because there is the risk of focusing too much on the most evident traces, losing the tiny particulars which only a hands-on investigation of the crime scene can give.

As in other branches of forensic science, there is not, among the tools which have been described, a silver bullet which solves every situation. The variability encountered in crime scenes just allows for the drafting of guidelines and best practices, no rigid rule can be established. The most fitting approach for each different circumstance must be identified, and this is dependent on the training and capabilities of the personnel operating on the crime scene. Choices must be made on every crime scene, about which items should be collected and which traces should be considered useful. In the opinion of the authors, these are best left to an expert investigator, rather than to an instrument. Competence, more than tools or technologies, is the most powerful weapon that can be used for solving crimes.

References

1. H. Faulds, *Nature*, 1880, **22**, 605.
2. F. Galton, *Fingerprints*, MacMillian, London, 1892.
3. P. Gill, A. J. Jeffreys and D. J. Werrett, *Nature*, 1985, **318**, 577.
4. O. Ribaux, A. Baylon, E. Lock, O. Delémont, C. Roux, C. Zingg and P. Margot, *Forensic Sci. Int.*, 2010, **199**, 63.
5. E. Locard, *L'enquète criminelle et les méthodes scientifiques*, Flammarion, Paris, 1920.
6. E. Locard, *Manuel de technique policière*, Payot, Paris, 1923.
7. P. L. Kirk, *Crime Investigation: Physical Evidence and the Police Laboratory*, Interscience, New York, 1953.
8. S. Ottolenghi, *Trattato di Polizia Scientifica*, Società Editrice Libraria, Milano, 1910.
9. https://archives.fbi.gov/archives/about-us/lab/forensic-science-communications/fsc/oct1999/trace.htm, last accessed, December 16, 2016.
10. http://enfsi.eu/documents/best-practice-manuals/, last accessed, December 16, 2016.
11. http://www.crime-scene-investigator.net/PDF/crime-scene-investigation-a-guide-to-law-enforcement.pdf, last accessed, December 16, 2016.
12. B. A. J. Fisher and D. R. Fisher, *Techniques of Crime Scene Investigation*, CRC Press, Boca Raton, 2012.
13. J. Horswell, *The Practice of Crime Scene Investigation*, CRC Press, Boca Raton, 2004.

14. V. Causin, *Polymers on the Crime Scene*, Springer, New York, 2015.
15. M. Friebel, A. André Roggan, G. Müller and M. Meinke, *J. Biomed. Opt.*, 2006, **11**, 034021.
16. T. C. Krauss and S. C. Warlen, *J. Forensic Sci.*, 1985, **30**, 262.
17. A. Richards and R. Leintz, *J. Forensic Identif.*, 2013, **63**, 46.
18. M. Stoilovic, *Forensic Sci. Int.*, 1991, **51**, 289.
19. K. Virkler and I. K. Lednev, *Forensic Sci. Int.*, 2009, **188**, 1.
20. F. S. Romolo and P. Margot, *Forensic Sci. Int.*, 2001, **119**, 195.
21. G. Payne, C. Wallace, B. Reedy, C. Lennard, R. Schuler, D. Exline and C. Roux, *Talanta*, 2005, **67**, 334.
22. I. Elworthy, *Detection of Gunshot Residues Using the Polilight, 16th International Symposium on the Forensic Sciences*, National Convention Centre, Canberra, ACT, Australia, 2002.
23. P. J. Gomes, P. A. Ribeiro, D. Shaw, N. J. Mason and M. Raposo, *Polym. Degrad. Stab.*, 2009, **94**, 2134.
24. R. V. Ballari and A. Martin, *Food Chem.*, 2013, **141**, 2130.
25. D. L. Exline, C. Wallace, C. Roux, C. Lennard, M. P. Nelson and P. J. Treado, *J Forensic Sci.*, 2003, **48**, 1047.
26. N. Akiba, N. Saitoh and K. Kuroki, *J Forensic Sci.*, 2007, **52**, 1103.
27. A. van Dam, M. C. G. Aalders, T. Todorovski, T. G. van Leeuwen and S. A. G. Lambrechts, *Forensic Sci. Int.*, 2016, **258**, 19.
28. G. Payne, B. Reedy, C. Lennard, B. Comber, D. Exline and C. Roux, *Forensic Sci. Int.*, 2005, **150**, 33.
29. B. K. Alsberg and J. Rosvold, *J Forensic Sci.*, 2014, **59**, 474.
30. B. K. Alsberg, T. Løke and I. Baarstad, *J Forensic Sci.*, 2011, **56**, 976.
31. G. Payne, N. Langlois, C. Lennard and C. Roux, *Med., Sci. Law*, 2007, **47**, 225.
32. C. S. Silva, M. F. Pimentel, R. S. Honorato, C. Pasquini, J. M. Prats-Montalbán and A. Ferrer, *Analyst*, 2014, **139**, 5176.
33. C. Lennard and M. Stoilovic, Application of Forensic Light Sources at the Crime Scene, in *The Practice of Crime Scene Investigation*, ed. J. Horswell, CRC Press, Boca Raton, 2004.

Chapter 2

Degradation and Damage due to Exposure to Light in Trace Evidence

Valerio Causin

Università di Padova, Dipartimento di Scienze Chimiche, Via Marzolo 1, 35131 Padova, Italy

*E-mail: valerio.causin@unipd.it

Table of Contents

Light in Forensic Science: Issues and Applications
Edited by Giorgia Miolo, Jacqueline L. Stair and Mire Zloh

Published by the Royal Society of Chemistry, www.rsc.org

2.1. Introduction

As it was more thoroughly discussed in Chapter 1, dedicated to crime scene investigation, forensic science consists of the study of traces, *i.e.* of the remnants of an activity, specifically of a criminal activity.[1]

As such, this discipline can be applied in two ways: justice driven forensic science and intelligence driven forensic science. In the justice driven approach, forensic data are used to evaluate evidence in the context of the court trial. In other words, analyses are performed to verify or substantiate a particular reconstruction of the criminal event.

On the other hand, forensic science is applied in an intelligence driven context when analytical data are part of the investigative process and are used as one of several inputs necessary for taking decisions or to select strategies of action. In this case, the task is identifying recurring traces, which can refer to a single individual or organisation, which are responsible of a reiteration of a particular crime or illegal activity.

If several cocaine bricks which share consistent packaging mode and materials are seized, in different locations and on different days, it could be deduced that the same organisation or individual is responsible for their preparation. The ability to describe the *modus operandi* of felons allows the connection of a series of crimes, even if they are far away in time and space.

Normally the vast majority of casework in a forensic laboratory is in the justice driven sense, because it is the most established and traditional mode of action. Intelligence driven forensic science, being much more complex and requiring the merging of different kinds of expertise and scientific or technical background, is much less common, but shows huge potential to become a powerful weapon in investigations on serial and organised crime.[1–8]

Irrespective of the purpose, the analytical strategies in a forensic case can be divided into two classes: comparison or identification.

In the first case, traces found on or related to a crime scene must be compared with an alleged source, in order to assess their compatibility. For example, if a burglar, while breaking into a house, uses a tool which leaves a paint smear on the door frame, and a suspect is apprehended which owns a utensil similar to that used by such burglar, it will be necessary to determine how likely it is that the traces on the door frame came from the paint on the suspect's tool.

The analyses aimed at identification are applied for gathering intelligence on the dynamics of the events and to suggest possible leads to investigators. If for example particular pollen is found on a dead body, the identification of the botanical origin of such pollen could bring about a description of the trees and plants present on the premises where the homicide was committed.

Scrupulous and targeted characterisation are fundamental in any case, because often the analyst has to deal with mass produced items, which differ by very subtle features. This is in fact what delayed the development of the contact traces branch of forensic science. According to a false perception, many lawyers or judges think that the evidential value of textile fibres, paint chips, glass fragments, *etc.* is low because "all red fibres are the same, so everybody who has a red T-shirt in his closet can be the killer". Indeed, there is a high degree of variability also in mass produced items, which, if correctly caught and measured, allows the class of possible sources for a trace to be narrowed to just a very small number of objects on the market. In other words, if correctly interrogated, trace evidence can be used as a mute witness able to describe with a surprising level of detail what happened on the crime scene. For example, several studies showed that the probability that two fibres coming from different sources match just by chance is extremely low, being the frequency of most fibres far below 0.5%.[9,10]

Among contact traces, polymers stand as prominent materials, ubiquitous on many crime scenes. Polymers are in fact everywhere in our daily life, and the chances are high that some polymer trace may be left as a consequence of a criminal action. However, among the several possible cases that may happen in actual casework, it is not infrequent that the traces on the crime scene were exposed to the elements, whereas their possible source materials were not. A typical example is that of dead victims, left in an outdoor location and found several days or weeks after the homicide. If textile fibres are found on them, and compared to those of the garments of a suspect, care must be taken to guarantee that the comparison is significant. The molecular weight, the degree of crystallinity or the chemical structure of polymers are quite sensitive to heat and interaction with moisture or light, so the exposed traces could in fact be altered by exposure to the elements. Comparison between a trace and its possible source is surely possible and meaningful if they did not experience radically different weathering conditions; otherwise, the outcome of the analysis must be critically weighted and interpreted to assess its significance.

The purpose of this chapter is to discuss these issues, firstly summarising the topic of photodegradation of polymers, and then contextualising it in the field of polymeric contact traces of forensic interest. Other contact traces, such as minerals in soils, or glass, will not be covered in this contribution because they are not appreciably degraded by light.

2.2. The Degradation of Polymers

Like many organic materials, polymers are reactive to oxygen, moisture or light. Especially in outdoor use, polymers tend to become opaque and brittle, their colour fades, and their mechanical properties dramatically decrease. This process is called degradation, weathering or ageing, and it is an industrially very relevant problem, particularly when the material is used for structural purposes. Degradation is a general label for all the chemical or physical processes associated to a deterioration of the properties of a material. Weathering is a term used to define the negative effects of exposure to an outdoor environment on the properties and service life of a polymeric material.[11,12] Ageing is generally applied to the effects of weathering, extended for long periods of time.

Polymer degradation can be originated by heat (thermal degradation), light (photodegradation), ionising radiations (radio degradation), mechanical action or by the action of microorganisms (biodegradation). However, the deterioration of plastics in an outdoor environment is primarily due to photodegradation caused by the ultraviolet portion of sunlight, combined with atmospheric oxygen and moisture. Therefore, the overall process is termed oxidative photodegradation or photooxidation.[13] Even though the atmosphere acts as a filter for sunlight, removing radiation with wavelengths shorter than 280–310 nm (UV-B) which are very energetic and hazardous for living tissues, about 5% of the total solar spectrum reaching our planet is still composed of light with wavelengths in the near ultraviolet range, *i.e.* between 300 and 380 nm (UV-A).[14]

Many organic functional groups absorb ultraviolet light, for example the carbonyl C=O group (270–360 nm), the aromatic ring (up to 350 nm, but at higher wavelengths if associated with a C=O chromophore) or the carbon–carbon C=C double bond, which alone absorbs below 200–250 nm, but in case of an extended conjugation can absorb at higher wavelengths, eventually even reaching the visible range.[15]

When a molecule absorbs UV light, its energy increases and an unstable excited state is reached.

It is worth remarking that such chromophores can be internally present in the macromolecule, or they can appear within the formulation of the polymeric item or in the environment surrounding it in normal use.[16] Among the examples of possible UV-absorbing molecules in the formulation of polymeric objects are traces of solvents, catalysts, initiators, *etc*, or a variety of additives such as pigments, dyes, thermal stabilisers, photostabilisers, *etc*. Contaminants of the item from the environment are those derived from a polluted atmosphere or smog (*e.g.* polycyclic hydrocarbons) or traces of metals and metal oxides from processing equipment and containers.[13]

The excess of energy in excited molecules can be dissipated in a variety of ways, some of them able to initiate degradative processes:

1. Excitation can be transferred to another molecule. This is the mechanism exploited for initiating chain reactions, such as for example chain radical polymerisations.
2. Re-emission of the excess energy as radiation with a lower energy than the exciting UV light associated to heat: this phenomenon is called photoluminescence. When this process is simultaneous to excitation it is defined as fluorescence, if it continues also after illumination by the exciting light has finished, it is phosphorescence.
3. Excess energy can be dissipated as heat.
4. The excited intermediate can rearrange, releasing the excess energy slowly as heat.
5. Raman vibrations can be triggered.
6. If none of the dissipation modes listed above are possible, the excess energy is used to break bonds, therefore degrading the molecule.

Some of these processes may modify the structure of the molecules composing the material, with a subsequent decrease or change in its mechanical properties, exacerbated if illumination by UV light is associated to an elevated temperature.

As far as polymers are concerned, two groups can be identified according to their behaviour in response to UV irradiation. In the first category are materials which are not extremely light sensitive and which are able to retain most of their physical–mechanical properties, for example polyvinyl chloride and polyacrylonitrile. In these polymers, the mode of disexcitation involves the rearrangement of the chemical structure, without breaking the macromolecular backbone. If rearrangement involves chromophoric groups, discolouration can occur. The second group includes materials which become sensibly more brittle as a consequence of UV illumination. In this category are polyolefins such as polyethylene, polypropylene or polystyrene. This behaviour can be ascribed to the following three reasons, singularly or happening simultaneously: scission of the main chain, photoinduced crystallisation and/or crosslinking. Crystallisation and crosslinking may at first improve some of the mechanical properties, but eventually will bring about embrittlement and failure of the material.

In particular, when UV radiation results in the breaking of the polymer chains, radicals are produced, and a reduction of molecular weight or an extensive crosslinking occurs, deteriorating the mechanical properties and ultimately causing the failure of the material.

Polymer degradation can occur in several ways, which all involve bond scission as a starting step. This happens when energy is transferred to a polymer chain, with the formation of a couple of radicals:

$$\sim CH_2CH_2 \sim \xrightarrow{\text{Energy}} \sim H_2C^{\bullet} + {}^{\bullet}CH_2 \sim$$

In general, when a free radical is formed from the dissociation of a polymer molecule, the following processes can happen:[15]

1. reduction of molecular weight, by cleavage of the chains into smaller fragments;
2. elimination of small molecules;
3. formation of unsaturations along the polymer chains;
4. depolymerisation;
5. rearrangement of the molecular structure;
6. cross-linking between adjacent polymer molecules;
7. oxidation of the polymer, especially at the surface exposed to atmospheric oxygen.

Of course, all these reactions can occur simultaneously, widening the range of possible products that can appear as a result of the degradation of polymers.

According to the mode of energy transfer to the polymer, chemical, thermal, mechanical, radiolytic degradation, biodegradation and photodegradation can be mentioned.[16]

Chemical degradation is started by the reaction with a chemical such as an acid, a base, a solvent, *etc.* In thermal degradation, cleavage of bonds is obtained solely by transferring heat to the material, without a reaction with any other compound. In mechanical and radiolytic degradation, the energy necessary to break the bonds is administered through application of a mechanical force or by electromagnetic radiation. Biodegradation could be considered a special case of chemical degradation, because the scission of the bonds of the substrate material is obtained through the action of the enzymes of the microbes involved.

Finally, photodegradation consists of the degradation of a molecule, in which chain scission is initiated by the absorption of photons, especially from sunlight. In other words, photodegradation includes photodissociation: the breakup of molecules into smaller pieces by photons. The distribution of the energy from the sun as a function of wavelength has a maximum at 500 nm, corresponding to an energy sufficient to break just weak bonds such as the O–O bond in a peroxide or an N–N bond. To break the much stronger carbon–carbon or carbon—hydrogen bonds, a wavelength shorter than about 380 and 300 nm, respectively, is required. However, the proportion of radiation in this wavelength range which reaches the surface of the Earth is quite low, and none is available to break bonds such as C=O or C=C, which have energies greater than 100 kcal mole^{-1}.[17] Since the chemical bonds involved in the backbone of the most common polymers have strengths comparable to those of the carbon–carbon bond, we can conclude that visible radiation will be ineffective in bond-breaking processes. As anticipated before, then, just UV-A light will be effective in the photodissociation of polymers.

In practical applications, it is probably more correct to talk about photooxidation, rather than of photodegradation, because, in the presence of air, the effect of light is accompanied by the reactive action of oxygen.

Photooxidation can happen according to two main mechanisms. The first involves direct reaction of singlet oxygen with the substrate, whereas, in the one other one, radicals are produced, which subsequently react with oxygen.[18,19]

In the singlet oxygen mechanism of oxidation, the most important reaction is the formation of a hydroperoxide by oxidation of an olefin containing an allylic hydrogen, which can further decompose towards carbonyl-terminated chains.

The radical mechanism of photooxidation of polymers is a chain reaction, with initiation, propagation and termination steps.[20–23] Initiation consists of the light-induced homolytic scission of a bond in a polymer chain, with the formation of radicals (R•) (Scheme 2.1).

As said above, bond scission can be initiated by a variety of physical factors, such as light, heat, ionising radiation, *etc.*

The presence of singlet oxygen produces hydroperoxides, which are very efficient initiators in this process, because they are very photolabile and reactive compounds. Other reactions which are main contributors to the photodegradation of polymers is ketone photolysis, which proceeds through the

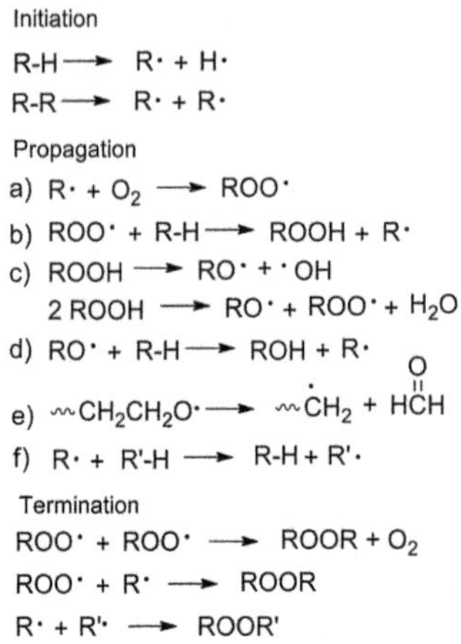

Scheme 2.1. Reactions in the radical chain mechanism of photooxidation.

Norrish I and Norrish II reactions, which will be described later. Ketones appear in the macromolecular chains as a result of photooxidation, and they are very strong chromophores for UV light.

Once the radicals are formed in the initiation step, six different propagation steps can occur:

(a) Reactions of the radicals with oxygen, with the formation of alkylperoxy radicals (ROO•).
(b) Abstraction of hydrogen from the same or another polymer molecule by the alkylperoxy radical, with the formation of a hydroperoxide and another radical.
(c) Photodecomposition of hydroperoxide groups with the formation of alkyloxy (RO•), alkylperoxy and hydroxyl (HO•) radicals.
(d) Abstraction of hydrogen by an alkyloxy radical, with the formation of hydroxyl functional groups in the polymer.
(e) Disproportionation reaction (scission process) of alkoxy radicals, with the formation of an aldehyde and an alkyl radical.
(f) Hydrogen abstraction from a polymer chain by a radical.

The relative stabilities of the radicals that can possibly form, the tendency to recombination of alkyl radicals before reaction with oxygen, and the energy of sunlight irradiation which will preferentially cleave RO–OH and R–OOH, rather than ROO–H bonds, will favour the formation of RO• and •OH radicals, which will predominate during light irradiation.

Termination steps happen when two radicals react together, coupling their lone electrons by formation of a bond. Examples of possible termination reactions are shown in Scheme 2.1.

Polymers containing a carbonyl group can undergo photolysis according to either of the two Norrish mechanisms (Scheme 2.2). The Norrish I reaction

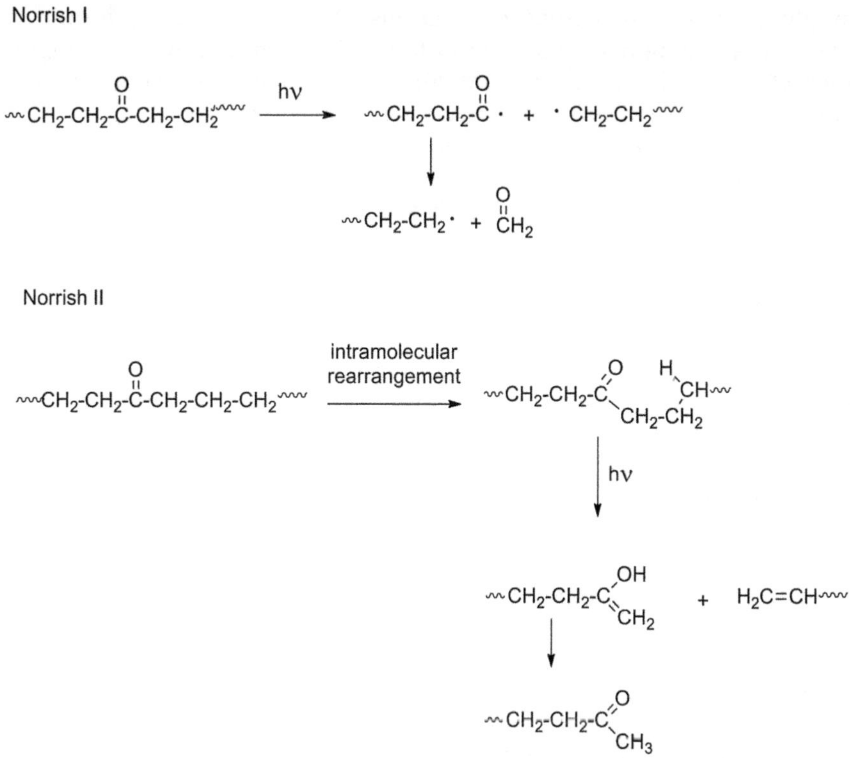

Scheme 2.2. Norrish I and Norrish II photolysis reactions.

leads to chain cleavage and to the formation of radicals which may initiate the chain photooxidation just described. Norris II reaction is an intramolecular rearrangement, followed by chain cleavage.

If on one hand these are the most recurrent mechanisms in commercial polymers, each material has its own preferred pathway of degradation in response to light irradiation.

However, as has been detailed above, exposure to UV light most often originates significant degradation of many polymeric materials. Damage by UV radiation is the most common reason for discolouration, weathering, yellowing, loss of gloss and deterioration of the mechanical properties. The logical consequence of this is that there is a great interest in the industry for prolonging the service life of polymeric materials under conditions of light exposure. Thus, ultraviolet stabilisers are among the most common additives used in the formulation of plastics and resins.[24] These products act according to a variety of mechanisms.

Pigments that are opaque to UV can screen the material from this kind of radiation and protect it. They reflect UV light before it can considerably penetrate into the polymer, thus protecting all but the extreme surface layer. Among the most common examples of such additives, titanium dioxide is amply inserted into coating formulations. Dark pigments, *e.g.* brown and black ones, are more effective than white or light ones. In order to improve their efficacy, it is suggested, if possible, to preferentially accumulate these additives on the surface of the object, rather than in the bulk.

In order to improve the reflection mechanism of defence against UV rays, a coating can be used, for example a metallic layer deposited by electroplating, which will effectively screen the bulk of the polymer from incoming UV rays.

Another approach for the protection of polymeric materials from photo-degradation consists of exploiting substances which absorb UV light and convert it into harmless species with a lower energy level before it can reach and attack the polymer itself. Carbon black is by far the most effective UV absorber, but it also has a very wide absorption of wavelengths in the visible range, therefore it confers to products a black colour, which could be aesthetically not pleasant for all applications. Zinc oxide is another example of this class of additives, widely used for this purpose in rubber, paints and plastics.

If a colourless material is sought after, then suitable additives, which absorb just in the UV and not in the visible, must be chosen. Derivates of benzotriazole are quite effective for this purpose, utilising UV energy for achieving an excited state, with a subsequent return to the ground state by emitting mainly heat.

This is the typical mode of action of deactivating antioxidants, which absorb UV light and then dissipate the energy in a variety of non-aggressive ways, usually by emission of fluorescent radiation or of heat. Free radical inhibitors, on the other hand, do not directly absorb light, but they scavenge the radicals formed within the system as a consequence of UV irradiation. The radicals thus formed are stable and non-reactive, so they stop any undesired reaction which can deteriorate the polymer. Typical examples of additives which act according to this mode are hindered amine light stabilisers (HALS).

It should be remembered that UV stabilising additives often function through a combination of the mechanisms cited above. For example, carbon black, zinc oxide, and organic ultraviolet light absorbers all work primarily by preferentially absorbing UV radiation, but then they quench excited polymer molecules or they interrupt the propagation of free-radical chain-reactions.

As in most other fields of polymer science, there are no UV-stabilisers which are universally better than others. Some will work well for one polymer, but degrade another; some polymers are easily stabilised, others are next to impossible to stabilise. The choice of an optimal additive for a polymeric item involves determining what wavelengths it absorbs and what wavelengths it re-emits, in the perspective of which radiation must be stopped for stabilising the particular matrix material which is being considered. Other features to be kept in mind are the concentration of the stabilising additive, which must be sufficient to properly screen all the damaging radiation, without being too high, with the effect of avoiding an unintended colouration of the item, often with tones of yellow. The duration of an ultraviolet stabiliser is also an issue, because after repeated exposition to UV light for a long period, it may degrade itself, becoming exhausted. Finally, synergistic effects must be considered, because combinations of several stabilisers can improve the protection, but in some cases the concurrent use of different additives at the same time may lead to a worse performance.

2.3. Polymer Degradation in a Forensic Context

As introduced above, an understanding of the effects of weathering and ageing on polymeric traces of forensic interest is a significant issue. In fact, currently, forensic analysis of these items is mainly of a comparative kind: a sample of unknown origin is examined in order to determine if it could share a common origin with similar objects from a known source. It is thus paramount that the results of these analyses are significant, to be used in a court of justice. Complete and comprehensive protocols are applied, by which a very detailed description of the trace is obtained, and subsequently compared to an analogously detailed picture of an object taken from where it possibly came from. If they match, a common origin can be concluded. These detailed protocols allow to discriminate between very similar mass-produced objects such as adhesive tapes, textile fibres or paint coatings, assessing their rarity and evidential value. However, trace materials may have acquired characteristics, for example those derived by some kind of degradation, and it is crucial to be able to infer an explanation for possible differences in a comparative analysis.

For example, textile fibres are quite common objects found on crime scenes, and thus many established protocols exist for their characterisation. In general terms, a typical analytical scheme starts with an observation of their morphology with an optical microscope, by qualitatively identifying the polymer composition by infrared (IR) spectroscopy and by a microspectrophotometric comparison of colour, if the items are dyed. Other techniques such as thin layer chromatography, high performance liquid chromatography or capillary electrophoresis may complement the comparisons but are applicable only if the fibres are coloured.[25]

It may happen in casework, though, that fibres coming from a crime scene have been exposed to the elements for a long time, even for many weeks.[26,27] Each time a forensic protocol is devised, then, it would be necessary to confirm that comparisons are significant, even when they are performed on samples that experienced very different histories. It may in fact happen that fibres that remained on open-air crime scenes, exposed for long periods to the elements, must be compared to items coming, for example, from the well protected closet of a suspect. For instance, in 2004 a method was proposed to differentiate morphologically similar colourless poly(ethylene terephthalate) (PET) single fibres, using a quantitative IR microspectroscopy method.[28] This was intended to provide a tool to increase the evidential value of colourless polyester fibres, which are often considered too common to be significant. In the absence of colour, in fact, a key feature for individualising the trace is lacking, so decreasing the significance of colourless evidence. This method was based on unpolarised IR spectroscopy. Four peaks in the IR spectrum, namely at 3440, 1370, 874 and 846 cm^{-1}, were integrated and two ratios of absorbancies were computed. The ratio between the peaks at 3440 cm^{-1} and at 874 cm^{-1} was proportional to the abundance of end groups in the chains, and consequently to the molecular weight of the polymer. The second ratio considered for differentiation, between 1370 and 846 cm^{-1}, was related to the conformation adopted by the chains, and thus to the crystallinity of the fibre.[28] Crystallinity is a very process-dependent property of polymers, and the molecular weight is the feature which is most starkly modified as a consequence of degradation processes. In other words, it was highly likely that exposure to sunlight, which may start degradation processes and may heat up the material, or immersion in water, which may swell the polymeric network, with a plasticising effect on the mobility of macromolecules, even in the solid state, could bring about a consequential modification of the semi-crystalline network or a decrease in the molecular weight of the polymer. So, it was considered necessary to validate this protocol by monitoring the effects of sunlight and water on PET fibres.[29] PET fibres specimens were exposed to sunlight or immersed in water for three months. After this period the IR spectra were gathered on single fibres, as described in the proposed protocol,[28] and integrated. The data were then compared to those obtained on the same fibre that had not undergone the treatment.

Neither the morphology of the fibres, *i.e.* their diameter and their microscopic appearance, nor the ratios between IR bands related to the molecular weight of the polymer and to the crystallinity degree of the material were changed by long exposures to strong sunlight or immersion in water. This means that, in the case of PET fibres, the material resists quite well in an outdoor environment, that it does not degrade, and so that characterisation procedures aimed at describing the structure and microstructure of the polymeric chains are suitable for obtaining significant forensic information.

Paint is a particularly valuable type of trace, because it may be transferred in a variety of ways and from a variety of objects. Paint smears or chips are practically always found on crime accident scenes, and are especially informative when vehicles hit walls or other objects during the perpetration of crimes such as robberies or thefts. Paint is also often transferred when tools are used, for example for breaking into edifices in order to commit burglary. As a consequence, the study of paint degradation and paint weathering is of particular interest to forensic scientists. The main purpose of every coating is indeed to protect the substrate from environmental conditions, and just as a secondary aim there are aesthetical functions such as colour or texture.

Exactly for this reason, paint is usually exposed to light, heat and moisture, which are likely, in the long run, to bring about chemical, physical and structural modifications to the film. These may be evidenced by macroscopical colour changes or deteriorations of the surface, *e.g.* peeling or detachment of the coating, but they may also be detected through more detailed chemical characterisation techniques, such as spectroscopical or chromatographic techniques. Understanding how photooxidation modifies the outcome of the characterisation of paint traces is necessary to be able to assess the significance of comparative analyses. Analogously to the previous example on PET fibres, it should be verified that the changes originated by exposure to light for a certain amount of time will not interfere in the analytical data so that erroneous conclusions are taken on the comparison between traces. It should be remarked that UV radiation is not the only factor responsible for degradation. Moisture and water uptake are also important agents which can modify materials. Unlike light, moisture affects both the coating and the support, and it often penetrates through the substrate, corroding it, and peeling off the paint layer. The main effect of a continuous contact with water is usually hydrolysis, and a consequent modification of the chemical structure of the paint coating.

Many papers exist on the degradation of automotive paints [ref. 30–34 and references within].

Automotive coatings are quite complex materials, composed of at least four layers, the two inner layers are mainly devoted to protecting the metal of the car body. On top of these, the basecoat is deposited, whose purpose is to give colour and texture to the vehicle. The outermost layer is the clearcoat, a transparent coating which protects the surface from scratches and deterioration by UV light. The typical characterisation protocols for paints is focused on the pigmented basecoat, because it offers the possibility of a rapid compatibility screening, based on the measurement of colour, when comparing traces. However, in terms of likelihood of transfer and also the probability of photooxidation, the very outer layer, the transparent clearcoat, is probably most interesting.[34] A number of databases, of which the best known is the PDQ, Paint Query Database, were developed, which allow identification of the make, model and production year of a car from a paint trace. However, the data contained in these valuable resources are collected on new vehicles, and thus they contain spectra and analytical data of pristine coatings, never subject to extended exposure to elements or illumination by sunlight. Sauzier *et al.*[35] were the first to test the effect that environmental exposure could have on the classification of clearcoat samples. A number of automotive paint samples was aged for 8 months, and then their clearcoats were compared with their pristine counterparts. No significant changes in their analytical features, compared through a chemiometric approach, were detected. However, for 12 samples an incorrect classification was obtained when the database was interrogated. For 11 of them, evidence was found that the car had been subject to repainting, therefore explaining why the composition of its coating did not match the one originally applied by the manufacturer. One last sample, though, was misclassified, even though the vehicle was not repainted. The mismatch was attributed to a possible degradation due to UV irradiation of the polyurethane basecoat. This finding motivated a further, more detailed study on the degradation of automotive clearcoats.[34] By synchrotron IR spectroscopy, van der Paul *et al.* investigated how the degradation progresses through the clearcoat layer, and if such degradation is indeed attributable to exposure to outdoor environmental conditions, or in contrast it is solely due to the effect of slow reactions which happen irrespectively of sunlight irradiation or of weathering.

In this study, paint samples were exposed to outdoor weathering conditions for up to 435 days, and their IR spectra were periodically taken. An aliquot of the same samples was stored in the dark in an office for one year, in order to verify whether degradation could occur away from weather exposure. As can be seen in Figure 2.1, some changes in the spectra were detected, in the areas denoted by letters a to e.

In particular, the increase in the absorptions at 3700 (area A) and 3450 cm^{-1} (Area B) could be attributed to a modification of the quantity and bonding of O–H groups and to the formation of methylol groups, respectively.

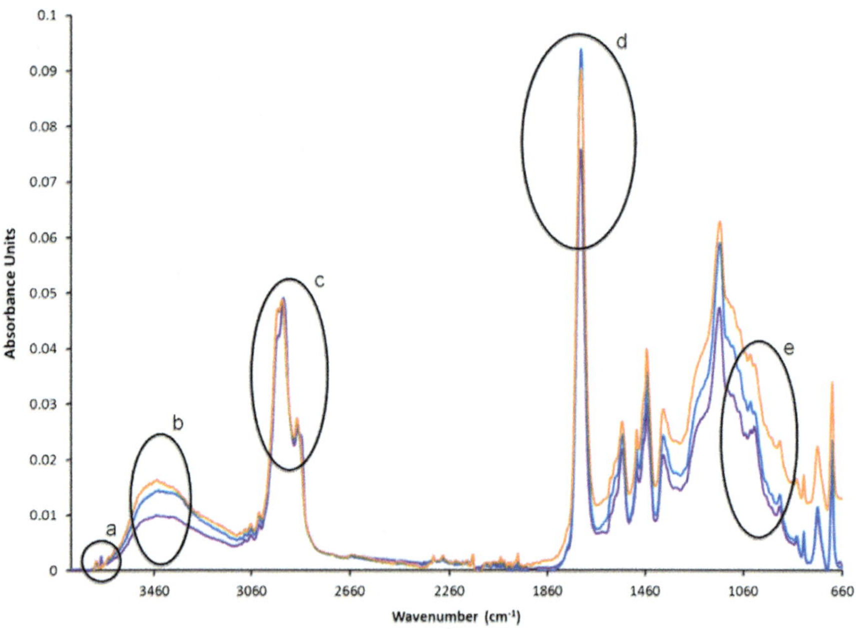

Figure 2.1. Normalised infrared spectra from an environmentally aged automotive clearcoat sample at day zero (purple), day 175 (blue) and day 435 (orange). Reprinted from *Talanta*, **148**, K. J. van der Pal, G. Sauzier, M. Maric, W. van Bronswijk, K. Pitts, and S. W. Lewis, The effect of environmental degradation on the characterisation of automotive clearcoats by infrared spectroscopy, 715–720, Copyright (2016) with permission from Elsevier.

The changes in the group of bands at about 2900 cm^{-1} (Area C) are due to a modification of the aliphatic chain backbone, the increase of the 1725 cm^{-1} peak (Area D) is the typical consequence of the formation of C=O bonds due to oxidation reactions and finally the decrease in the intensity of the group of bands around 1030 cm^{-1} (Area E) can be associated to the self-condensation of the OH and NH groups formed as a consequence of hydrolysis.[34] Due to these spectral changes, when databases were interrogated, the clearcoat samples aged for 435 days considered in this work were not correctly identified and were associated to an incorrect vehicle. In contrast, the classification method proved robust and insensitive to the physiological degradation of paint during service up to 175 days.[35,36] These results highlight a relevant limitation of paint databases, because they could fail in their main function, which is of identifying, from the trace found on the crime scene, the make and model of the vehicle which left it. The results of such a tool must therefore be viewed with great care, especially if the result of the database query is an old model, which has, since its production, experienced a significant exposure to a variety of weather conditions which reasonably brought about a modification of the chemical structure of its paint, especially of the outer layer such as the clearcoat. Moreover, the spectral changes reported in Figure 2.1 could lead to a negative result of a comparison of paint chips which experienced different exposure histories, and in this case also samples aged for 175 days show significant differences with respect to pristine samples. Comparative analyses are therefore much more sensitive to photooxidation and exposure to weathering than those aimed at identification. This should be always considered, even though the cases where significant paint traces are found months after the moment when they were transferred are quite rare.

Another result of the study by van der Pal *et al.*[34] is notable. By synchrotron IR microscopy it was verified that degradation processes proceed in a top-down manner, so they involve the inner part of the clearcoat much later than its outer layer (Figure 2.2). Therefore, when possible, spectra should be acquired from the middle of a paint cross-section, in order to minimise the effect of weathering on the outer part and migration of the coloured components from the basecoat into the clearcoat. This will improve the accuracy of the results, and will increase the chances of obtaining meaningful identifications of automotive paint samples.

Other paint types which attracted extensive interest from the scientific community are those used in the architectural, artistic and cultural heritage field. Also in these cases, evidence of degradation was found. For example, acrylic paint used for coating of concrete is quite sensitive to UV illumination, and it undergoes significant degradation.[37] In such conditions, the polymer hydrolyses and is subject to decarboxylation, with a concurrent formation within the material of radicals, which lead to disruption of the cooperation in the coating–substrate system, and eventually to cracking or pitting of the coating (Figure 2.3).

For these samples, evidence of chemical degradation can be found by IR spectroscopy. From panel C in Figure 2.4 the disappearance, in the aged

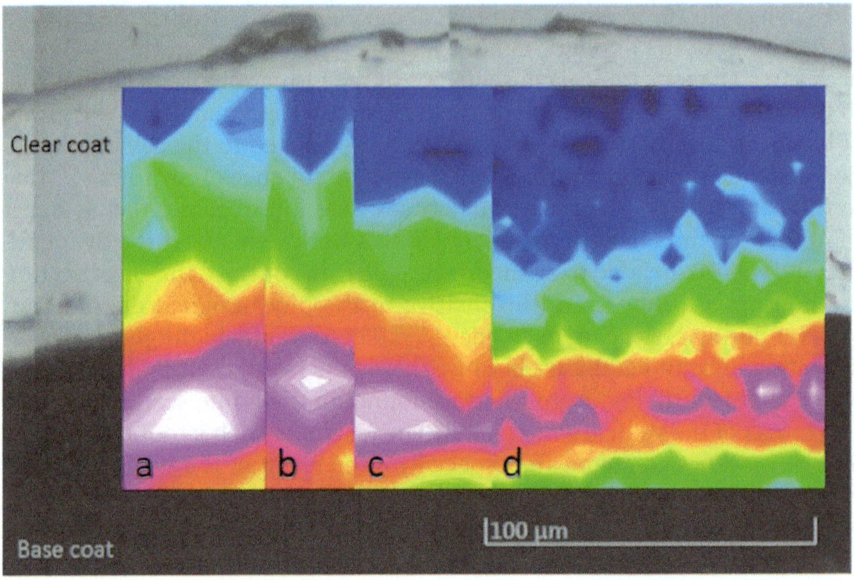

Figure 2.2. Composite 2-dimensional FTIR chemical image from paint cross-sections aged for (a) 2 months; (b) 4 months; (c) 8 months and (d) 25 months, following the integration of 1130–1065 cm^{-1} C–O vibrations. Decreasing intensity from dark blue to green. Reprinted from *Talanta*, **148**, K. J. van der Pal, G. Sauzier, M. Maric, W. van Bronswijk, K. Pitts, and S.W. Lewis, The effect of environmental degradation on the characterisation of automotive clearcoats by infrared spectroscopy, 715–720, Copyright (2016) with permission from Elsevier.

samples, of the absorption at 1650 cm^{-1} is evident, and it can be attributed to decarboxylation and to the hydrolysis of the esters groups present in acrylic polymers.

The relative instability of the paints described above can be quite surprising, because these materials are engineered for resistance to weathering, and they are normally protected using hindered amine light absorbers and ultra-violet absorbers. However, it should be kept in mind that the changes observed can be detected just by a detailed chemical analysis, and they are not reflected in the macroscopical performance of the coating. In other words, these paints work very well during their service life, even if their chemical structure evolves with time.

Many simpler paints, not specifically designed for outdoor use, such as for examples spray or artistic paints, exist, which in principle should be more prone to degradation or deterioration [ref. 38 and references within]. In the case of these materials, it is decisive to be able to assess if degradation has occurred. A due interpretation of possible photooxidation mechanisms in a paint trace can make a difference between concluding if the origin of two compared traces is common or not. Binders in paints devised for indoor use are often based on alkyd modified polyesters. The name 'alkyd' derives

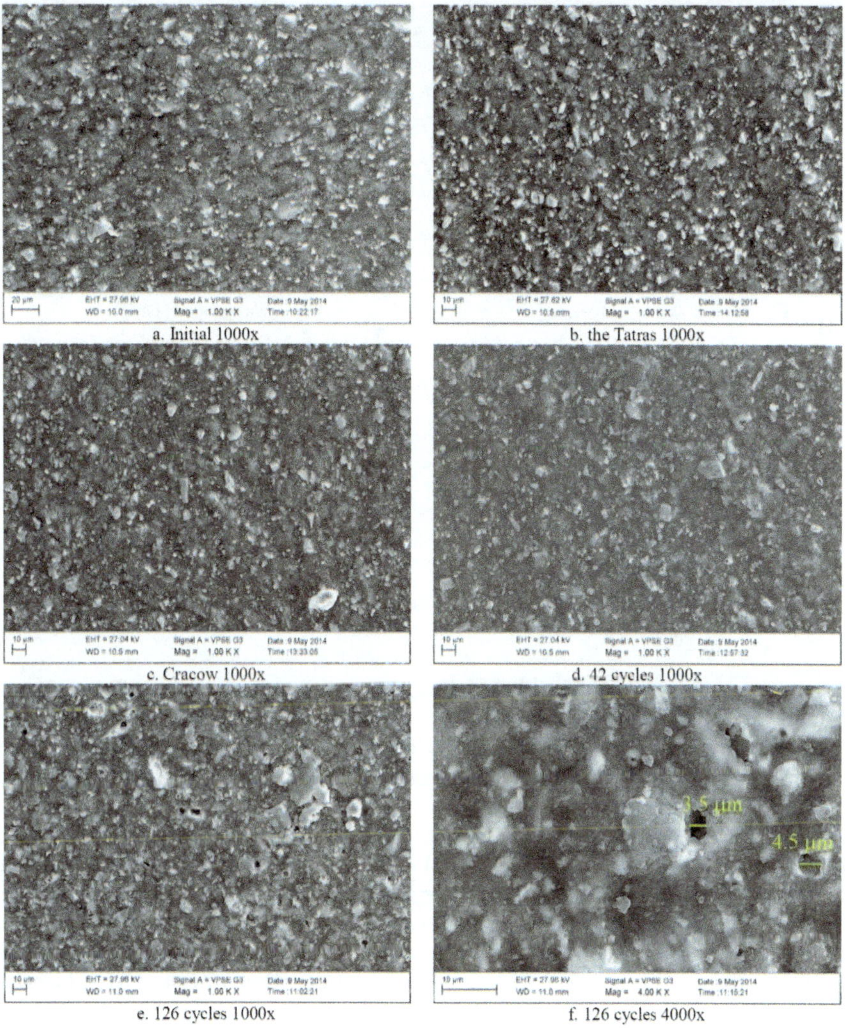

Figure 2.3. Scanning Electron Micrographs of acrylic coatings on concrete treated in different conditions. (a) pristine sample; (b) aged for one year in a mountain region (the Tatras), situated around 1000 m above the sea level; (c) aged for one year in an urban environment (Cracow), characterised by highly contaminated atmosphere and acid rains; (d–f) aged in a weathering chamber in compliance with ASTM G 154a for (d) 42 cycles, (e) and (f) 126 cycles. Pictures (e) and (f) differ by magnification. Each weathering cycle consisted of irradiation with the use of UV lamps at 60 ± 30 °C for 6 h and then sprinkling with distilled water at 40 ± 30 °C for 2 h. Reprinted from *Procedia Engineering*, **108**, A. Kozak, Multi-criteria assessment of an acrylic coating exposed to natural and artificial weathering, 664–672, Copyright (2015), with permission from Elsevier.

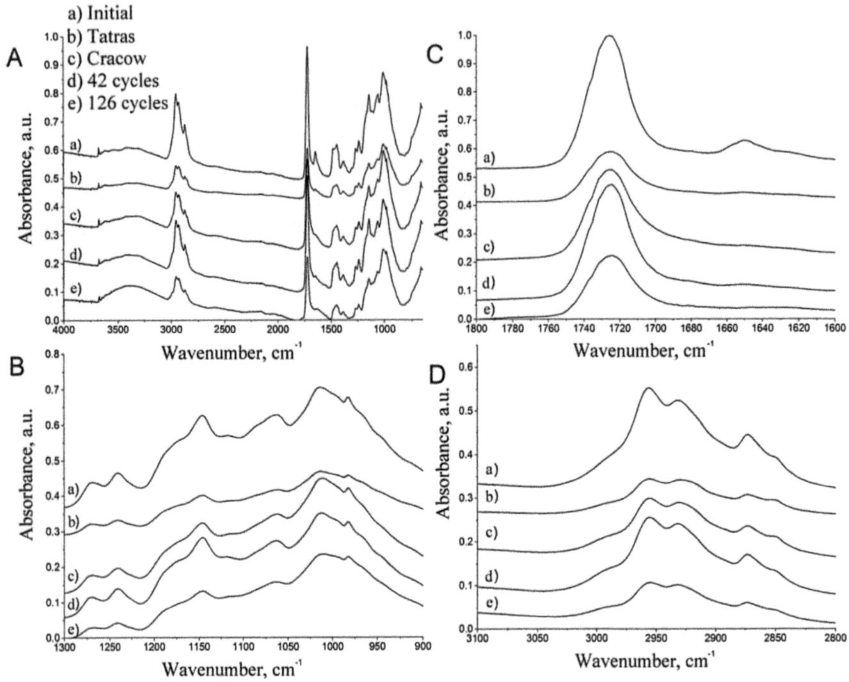

Figure 2.4. IR spectra of acrylic coatings on concrete, treated in different conditions. Panel A shows the entire spectral range, panels B to D are magnifications of spectral subregions. Curves are relative to (a) pristine coating; (b) coating aged for one year in a mountain region (the Tatras), situated around 1000 m above the sea level; (c) coating aged for one year in an urban environment (Cracow), characterized by highly contaminated atmosphere and acid rains; (d,e) coating aged in a weathering chamber in compliance with ASTM G 154a for (d) 42 cycles, (e) 126 cycles. Each weathering cycle consisted of irradiation with the use of UV lamps at 60 ± 30 °C for 6 h and then sprinkling with distilled water at 40 ± 30 °C for 2 h. Reprinted from *Procedia Engineering*, **108**, A. Kozak, Multi-criteria assessment of an acrylic coating exposed to natural and artificial weathering, 664–672, Copyright (2015), with permission from Elsevier.

from the al in alcohol and the cid in acid, and identifies polyesters modified with fatty acid polyesters. They are normally prepared by the condensation polymerisation of polyalcohols, polyprotic acids and fatty acids. Natural ageing of these resins involves an increase and broadening of the OH group band around 3300 cm^{-1} and an increase in bands in the carbonyl stretching region at about 1730–1680 cm^{-1}. These spectral changes are the result of autooxidation reactions during natural ageing and crosslinking. However, such a chemical drying and cross-linking process lasts about 250 days. Once it has ended, alkyd colours are very stable.[39] This was confirmed by Jost *et al.*, who investigated under a forensic perspective the UV-initiated degradation of spray paints.[38] Spectral variations due to oxidisation were detected by IR spectroscopy, but not by Raman spectroscopy, even though these are both techniques sensitive to vibrations of bonds in molecules. In the case of spray paints, the most significant changes in structure were observed up to 2 months after the deposition of the coating. Afterwards, they became more stable, as previously verified in other scientific contexts.[39] Spectral modifications brought about by photooxidation were concentrated in the 2800–3400 cm^{-1} region, quite minor, with respect to the whole spectrum, and thus unlikely to lead to a false exclusion of common source.

One of the requirements for a molecule to be a polymer is that it has a high molecular weight. The intrinsic heterogeneity of polymers, which are invariably composed by a variety of molecules with different molecular weight, imposes the introduction of concepts such as the average molecular weight and the distribution of molecular weights, which are unknown in the chemistry of small molecules, where every species is monodisperse as far as molecular weight is concerned. This offers scope for the characterisation of polymeric traces, because most often plastic items, similar by morphology but produced by a different manufacturer, were obtained from polymers with different molecular weight. For example, many pressure sensitive adhesives, used in the adhesive tape industry, are a mixture of similar acrylic polymers, but chances are high that different producers will use polymers with different molecular weight. Surprisingly, very few works exist in the literature in which molecular weight is used for characterising traces of forensic interest.[40] However, all the various degradation mechanisms of polymers, including those described in the previous paragraph, have the consequence of reducing the molecular weight of the macromolecules. If analytical protocols are designed for the forensic characterisation of polymeric traces, it is therefore suggested that they are accompanied with an evaluation of the effect of degradation induced by light or heat. This will be necessary to assess the limitations of the proposed approach, and also to correctly interpret differences in comparisons.

Figure 2.5 shows a comparison of size exclusion chromatograms obtained from samples of electric cables and of adhesive tapes, before and after irradiation with a xenon illuminator (Atlas Suntest CPS+, power 1500 W) for 40 hours.

As may be seen, in both these samples there is a multimodal distribution of molecular weights, indicating that there are two distinct populations of macromolecules, starkly different by size. This is not surprising, because the formulation of polymers can be quite complex. In this particular case, the insulant layer of the electric cable, whose chromatogram is depicted in Figure 2.5a, was composed of PVC plasticised with oligomeric plasticisers. The adhesive related to Figure 2.5b, on the other hand, was a mixture of acrylic polymers. It is worth recalling that in size exclusion chromatography (SEC) the lower the elution time, the higher the molecular weight of the corresponding fraction of macromolecules. In Figure 2.5, there are detectable and significant changes in the molecular weight distribution after the samples have been irradiated with UV/visible light. The maxima of the most prominent peaks in the chromatograms shift towards lower elution time, revealing

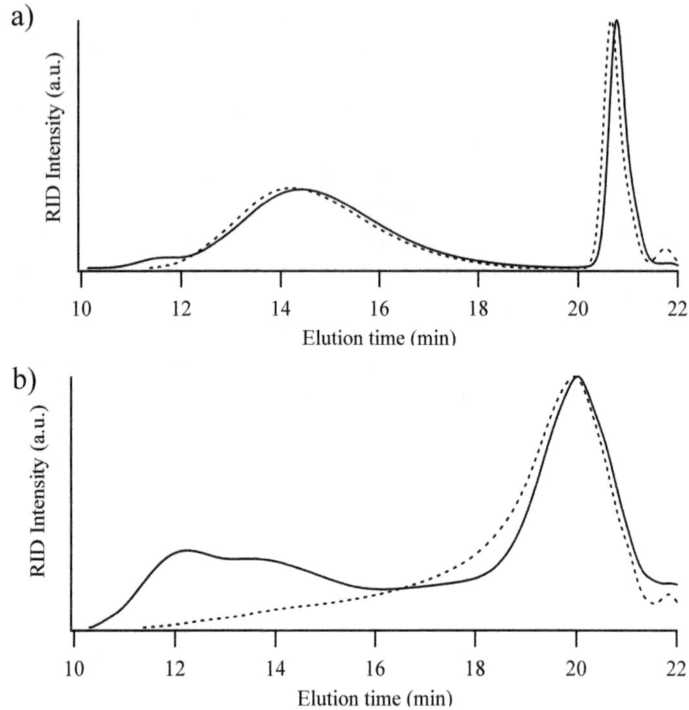

Figure 2.5. Size exclusion chromatogram of the tetrahydrofuran (THF) extracts (15 mg mL^{-1}) of (a) the insulant layer of an electric cable and of (b) the pressure sensitive adhesive of an adhesive tape. Solid lines are relative to the pristine sample, dashed lines are relative to the same sample after 40 h ageing with a xenon illuminator. Chromatographic conditions: Agilent 1260 Infinity GPC/SEC System, 1 ml min^{-1} mobile phase (THF) flow, injection volume 10 μL, Phenogel 10^4 and 10^5 columns, 30–32 bar pressure, 35 °C temperature. The signal was acquired with a refractive index detector.

an apparent increase in the molecular weight of the corresponding population of macromolecules. This is consistent with a crosslinking initiated by the appearance on the molecular chain of reactive radical sites. Moreover, in the adhesive sample (Figure 2.5b), the signal located between 10 and 16 minutes disappears, reflecting the degradation of the fraction with a higher molecular weight. The concurrent broadening of the peak located between 18 and 22 minutes is consistent with the decrease in size of the larger molecules, which after degradation become part of the low molecular weight fraction. Results such as those reported in Figure 2.5 show that an analytical approach based on the measurement of molecular weight should be used with great care on items which were exposed to weathering for a long time before being retrieved from the crime scene.

Explosive and fire investigation is not a typical context in which polymer analysis is preponderant. In this kind of casework, in fact, it is mostly of interest to determine what originated the fire or the explosion, and so the analysis is focused on the identification of accelerants or explosives from the remaining traces. Among the few polymers which are of interest to explosive investigators is nitrocellulose (NC). This is an important industrial material obtained by nitration of the hydroxyl groups of cellulose. As a function of the degree of nitration, defined as nitrogen content, this polymer can have several different uses. At low nitrogen content, *i.e.* below 12%, it is used in cosmetics, printing inks, paints and lacquers. If the nitrogen content is above 12%, it will be a component of gun and rocket propellants. NC is, of course, prone to thermal decomposition, but it is also subject to photo-decomposition. Due to this remarkable instability, and also to other features such as a high molecular weight and a high structural complexity, most of the characterisation techniques for NC yield qualitative data, far fewer methods are available for obtaining quantitative information.[41] Moniruzzaman *et al.* studied the changes in viscosity of NC solutions and of the molecular mass of NC irradiated by UV and visible light.[42] When exposed to UV light, the viscosity of the NC solutions noticeably decreases, due to photo-induced heating or de-aggregation/degradation of polymer chains. Wet NC tends to photodegrade faster than dry NC, suggesting that water has a critical role in catalysing degradation. The most evident effect was recorded on the molecular weight of NC: after less than 100 minutes of UV irradiation it was decreased by 15% in the most stable sample, and it was more than halved in the NC sample most sensitive to degradation.[42] Even though they are rare, cases may happen when an unexploded device is recovered, and the materials which compose it are compared with those found in the premises of the suspect bomber. In these instances, the natural degradation of the explosive, or of the additives of the explosive, must be taken into account. Another example of this issue was reported by Sáiz *et al.*,[43] who studied how the formulation of a dynamite changed with storage time.

As shown in Figure 2.6, nitroglycol is progressively lost from dynamite. After approximately one month of storage, nearly 30% of this dynamite's nitroglycol content had left the material, and in two months about 66% was gone. This is of course mainly due to the high vapour pressure of this component, but also to the modifications that the solid matrix of the dynamite, which is mostly polymeric in nature, undergo.

2.4. Conclusion

One of the reasons why contact trace analysis is considered a less prominent branch of forensic science, if compared to DNA typing or fingerprint analysis, is due to the fact that comparisons between items of material evidence can be affected by the conditions which such objects experienced. If on one hand much work has been done for advancing the level of detail by which contact traces can be described and individualised, much research has yet to be done on the effect of weathering and ageing. This knowledge is necessary for increasing the evidential value of contact traces, and to improve

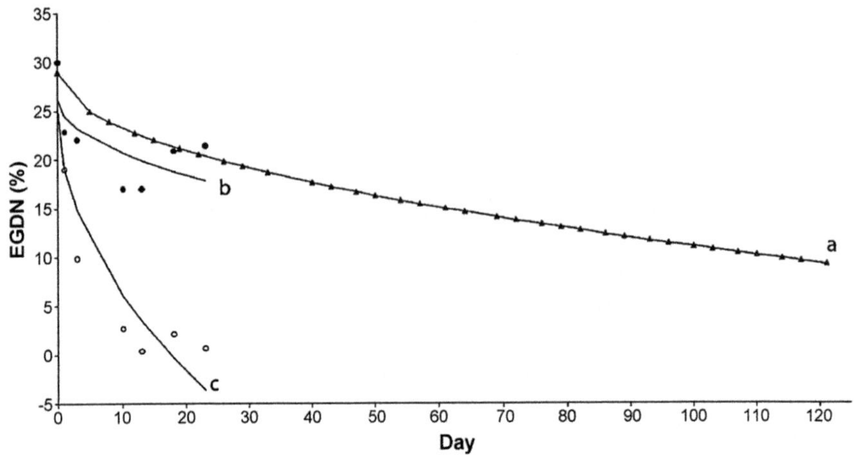

Figure 2.6. Evolution over time of nitroglycol in specimens of dynamite stored in polyethylene bags. (a) Long-term study during 121 days at room temperature, (b) short term study during 24 days at room temperature and (c) temperature study during 24 days at 60 °C. Reprinted from *Forensic Science International*, **211**, J. Sáiz, J. L. Ferrando, J. C. Atoche, M. Torre, and C. García-Ruiz, Study of losses of volatile compounds from dynamites. Investigation of cross-contamination between dynamites stored in polyethylene bags, 27–33, Copyright (2011) with permission from Elsevier.

the efficacy of the interpretation of the information carried by them. This is indeed a fertile avenue of research for all the forensic scientists involved in this field, and one which will undoubtedly enlarge and empower the toolbox of the investigators in the resolution of crimes.

References

1. O. Ribaux, A. Baylon, E. Lock, O. Delémont, C. Roux, C. Zingg and P. Margot, *Forensic Sci. Int.*, 2010, **199**, 63.
2. O. Ribaux and P. Margot, *Forensic Sci. Int.*, 1999, **100**, 193.
3. O. Ribaux, S. J. Walsh and P. Margot, *Forensic Sci. Int.*, 2006, **156**, 171.
4. O. Ribaux, A. Baylon, C. Roux, O. Delémont, E. Lock, C. Zingg and P. Margot, *Forensic Sci. Int.*, 2010, **195**, 10.
5. P. Esseiva, S. Ioset, F. Anglada, L. Gasté, O. Ribaux, P. Margot, A. Gallusser, A. Biedermann, Y. Specht and E. Ottinger, *Forensic Sci. Int.*, 2007, **167**, 247.
6. S. Lociciro, P. Esseiva, P. Hayoz, L. Dujourdy, F. Besacier and P. Margot, *Forensic Sci. Int.*, 2008, **177**, 199.
7. S. Lociciro, P. Hayoz, P. Esseiva, L. Dujourdy, F. Besacier and P. Margot, *Forensic Sci. Int.*, 2007, **167**, 220.
8. P. Esseiva, L. Dujourdy, F. Anglada, F. Taroni and P. Margot, *Forensic Sci. Int.*, 2003, **132**, 139.
9. C. Roux and P. Margot, *Sci. Justice*, 1997, **37**, 225.
10. W. Brüschweiler and M. C. Grieve, *Sci. Justice*, 1997, **37**, 85.
11. G. Wypych, *Handbook of Material Weathering*, Chemtec Publishing, Toronto, 2008.

12. H. Zweifel, *Stabilization of Polymeric Materials*, Springer, Berlin Heidelberg, 1998.
13. E. Yousif and R. Haddad, *SpringerPlus*, 2013, **2**, 398.
14. J. Pospíšil and S. Nešpurek, *Prog. Polym. Sci.*, 2000, **25**, 1261.
15. R. D. Deanin, A. Okroth, R. W. Eliasen and T. N. Greer, *Polym. Eng. Sci.*, 1970, **10**, 228.
16. W. Schnabel, *Polymer Degradation: Principle and Practical Applications*, Hanser, München, 1981.
17. J. E. Guillet, *Pure Appl. Chem.*, 1972, **30**, 135.
18. G. Geuskens and C. David, *Pure Appl. Chem.*, 1979, **51**, 233.
19. G. Geuskens and C. David, *Pure Appl. Chem.*, 1979, **51**, 233.
20. J. F. Rabek, *Polymer Degradation Mechanisms and Experimental Methods*, Springer, Berlin, 1994.
21. J. F. Rabek, in *Comprehensive Chemical Kinetics*, ed. C. H. Bamford and C. F. Tipper, Elsevier, Amsterdam, 1974.
22. J. E. Guillet, *Pure Appl. Chem.*, 1972, **30**, 135.
23. J. E. Guillet, *Polym. Prepr. (Am. Chem. Soc., Div. Polym. Chem.)*, 1972, **13**, 628.
24. M. Bolgar, J. Hubball, J. Groeger and S. Meronek, *Handbook for the Chemical Analysis of Plastic and Polymer Additives*, CRC Press, Boca Raton, 2016.
25. J. R. Robertson and M. Grieve, *Forensic Examination of Fibres*, Taylor and Francis, London, 1999.
26. R. Spencer, *J. Forensic Sci.*, 1994, **39**, 854.
27. V. Causin, S. Schiavone, A. Marigo and P. Carresi, *Forensic Sci. Int.*, 2004, **141**, 159.
28. V. Causin, C. Marega, G. Guzzini and A. Marigo, *Appl. Spectrosc.*, 2004, **58**, 1272.
29. V. Causin, C. Marega, G. Guzzini and A. Marigo, *J. Forensic Sci.*, 2005, **50**, 519.
30. K. Adamsons, *Prog. Polym. Sci.*, 2000, **25**, 1363.
31. J. L. Gerlock, *Polym. Degrad. Stab.*, 1998, **62**, 225.
32. C. M. Seubert, *Polym. Degrad. Stab.*, 2003, **81**, 103.
33. H. Yari, M. Mohseni and B. Ramezanzadeh, *J. Appl. Polym. Sci.*, 2009, **111**, 2946.
34. K. J. van der Pal, G. Sauzier, M. Maric, W. van Bronswijk, K. Pitts and S. W. Lewis, *Talanta*, 2016, **148**, 715.
35. G. Sauzier, M. Maric, W. van Bronswijk and S. W. Lewis, *Anal. Methods*, 2013, **5**, 4984.
36. M. Maric, W. van Bronswijk, S. W. Lewis and K. Pitts, *Anal. Methods*, 2012, **4**, 2687.
37. A. Kozak, *Procedia Eng.*, 2015, **108**, 664.
38. C. Jost, C. Muehlethaler and G. Massonnet, *Forensic Sci. Int.*, 2016, **258**, 32.
39. C. Duce, V. Della Porta, M. R. Tiné, A. Spepi, L. Ghezzi, M. P. Colombini and E. Bramanti, *Spectrochim. Acta, Part A*, 2014, **130**, 214.
40. V. Causin, *Polymers on the Crime Scene*, Springer, New York, 2015.
41. M. A. F. de la Ossa, M. López-López, M. Torre and C. Carmen García-Ruiz, *Trends Anal. Chem.*, 2011, **30**, 1740.
42. M. Moniruzzaman, J. M. Bellerby and N. Mai, *Polym. Degrad. Stab.*, 2011, **96**, 929.
43. J. Sáiz, J. L. Ferrando, J. C. Atoche, M. Torre and C. García-Ruiz, *Forensic Sci. Int.*, 2011, **211**, 27.

Chapter 3

Use of Light in the Detection of Genetic Variation of Forensic Importance

Nick Dawnay

Liverpool John Moores University, James Parsons Building, Byrom Street, Liverpool, L3 3AF, UK

*E-mail: n.m.dawnay@ljmu.ac.uk

Table of Contents

Light in Forensic Science: Issues and Applications
Edited by Giorgia Miolo, Jacqueline L. Stair and Mire Zloh

Published by the Royal Society of Chemistry, www.rsc.org

3.1. Introduction to DNA and RNA Structure, Function and Variation

DNA is common in its basic elements in all forms of biological life. The structure comprises a sugar-phosphate backbone whereby each individual deoxyribose sugar is linked to each other by phosphodiester bonds. These bonds form between the 5′ carbon of one deoxyribose sugar and the 3′ carbon of the next. It is for this reason DNA sequences are commonly reported in a 5′-3′ direction. Attached to each sugar is a nucleotide base which can either be a purine in the case with Adenine (A) and Guanine (G) or a pyrimidine as is the case with Cytosine (C) and Thymine (T). Hydrogen bonding can form between the A and T bases and the C and G bases (purine: pyrimidine) with two complementary strands of DNA forming the DNA double helix. Bonds can form and break naturally through processes such as cellular division and transcription and artificially in optimised biochemical reactions used to analyse DNA.

While the basic elements of DNA remain constant between organisms, the genome size, sequence, polymorphism and mode of inheritance can vary. Such differences allow comparison within and between groups of species, populations and individuals. Most DNA that exists in a cell is located within the cell's nucleolus and termed genomic DNA (gDNA). This DNA is tightly wound by histone proteins to form the chromosomes which are encased with the cell's nucleus protected from external factors that can promote decay and mutation.[1–3] Each autosomal cell contains approximately 6.6 pg of DNA with two copies of each nuclear gene (paternal and maternal) termed a locus. With only two locus copies per cell the use of nuclear DNA in forensic investigations relies on the ability of the forensic scientist to collect sufficient quantities of biological material before it decays and is lost.[4,5] DNA can also exist in the form of mitochondrial DNA (mtDNA), a simple circular genome existing within the cell's mitochondria. The mtDNA genome also has a high copy number per cell allowing it to be used in forensic investigation after the decay and loss of any nuclear DNA.[6,7]

Regions of DNA are often described as being either coding or non-coding, a simple reference to the fact that some regions of DNA will code for genes. Each gene is formed of domains called exons and introns. Exons form mature messenger RNA (mRNA) while the introns (non-coding regions of the gene) are removed during the process of transcription. RNA has a similar structure to DNA except it exists as a single transcribed sequence and there is an oxygen molecule located on the 2′ carbon in the sugar group. The base Thymine is also replaced by the base Uracil (U) during the transcription process. Base pairs in the mature mRNA form part of the triplet code with sets of three nucleotides coding each amino acid. The amino acid forms long chains and folds into proteins during the process of translation. These expressed proteins form the basis of phenotypic variation and like regions of DNA, RNA can be detected using laboratory instrumentation to aid forensic investigations.[8,9]

Nuclear DNA has two key features that make it a desirable tool in forensic investigations. First, the genetic material located on the autosomal chromosomes is inherited in a Mendelian fashion through the fertilization of gametes formed by meiosis. This process explains why individuals in our population are genetically and phenotypically different and explains why an individual's DNA profile is unique. The second key feature is that almost all autosomal DNA is identical as it replicates through the process of mitosis. This process explains why biological material recovered from a crime scene maintains a link to the individual who left it. Variation in the DNA sequence between individuals also occurs due to the process of mutation. Being located outside the nuclear casing, mtDNA has a higher mutation rate than nuclear DNA due to the absence of repair mechanisms that minimise errors during replication.[10,11] Despite the higher mutation rate the mtDNA genes remain less variable because of its mode of inheritance. The mtDNA genome is uni-parentally inherited from the mother and therefore, like genetic material on the Y chromosome (uni-parentally inherited from the father), shows little difference between close relatives.[12,13] The inheritance and exchange of genetic material within and between populations together with the process of mutation creates genetic polymorphism. These individual and population genetic differences become the relevant information the forensic geneticist collects to help answer the questions at the centre of the criminal investigation.

3.2. Enabling Technologies for DNA Analysis

The ability to detect and differentiate between regions of variable DNA whether it be single base pair differences, different numbers of DNA repeats, whole genes or expressed mRNA relies on instrumentation that use sensors that are capable of distinguishing different wavelengths and intensities of light. There are several assays and approaches that are commonly used to generate data in criminal and civil investigations. Before these are discussed it is necessary to identify and introduce two key enabling technologies that allow the process of DNA analysis, namely Polymerase Chain Reaction (PCR) and fluorescent dye labelling.

The process of PCR was pioneered in the mid 1980's and rapidly became a standard approach in forensic genetics in the early 1990's.[14–17] The success and universal adoption of the technique in molecular biology comes from the ability to use a DNA polymerase enzyme to make copies of the existing genetic material *in vitro* in the laboratory. This means that a single cell copy of a DNA sequence can be recovered from a crime scene sample and through the process of PCR create millions of copies of that sequence. All modern Forensic Genetic processes requiring detection of DNA use PCR in some shape or form and while there are a number of different applications, the stages PCR are relatively similar. The first stage called '**denaturation**' is the process whereby the hydrogen bonds between each strand of DNA in a double helix break and the DNA becomes single stranded. This process usually requires heating the

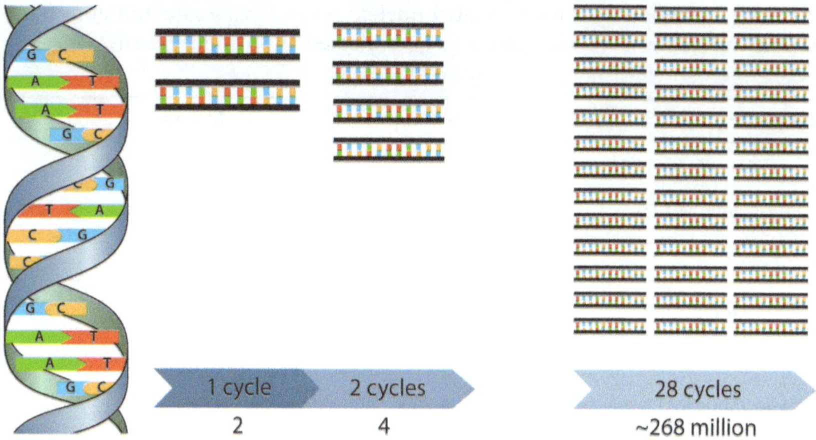

Figure 3.1. The Polymerase Chain Reaction (PCR) allows amplification of the specific DNA region under analysis. The method is common to all current forms of DNA analysis. Adapted from http://www.gmotesting.com/Testing-Options/Genetic-analysis.

DNA to ~95 °C for several seconds and in some cases minutes depending on the DNA polymerase being used. Once the DNA has become single stranded, the reaction is cooled to a temperature that allows short DNA oligonucleotides (termed primers) to bond to homologous DNA regions. This second stage is termed '**annealing**'. The temperature at which the primer anneals is specific and based on the primer sequence meaning that different primers can have different 'optimal' annealing temperatures. This stage is also where fluorescent probes can anneal to the target DNA strand. The last stage of the PCR is termed the '**elongation**' phase and occurs when the Polymerase enzyme starts forming a new complementary strand by adding free floating dNTPs onto the 3′ end of the newly annealed oligo primer. The temperature of elongation can vary, but is generally between 68–72 °C. By the end of this process two new complementary strands are formed, one from the forward primer and one for the reverse, effectively doubling the amount of starting material. The process is repeated 28–50 times depending on the application and method of detection. The speed of amplification is very rapid (1–2 hours) and has the impact of increasing the number of DNA targets from a single copy to in excess of 1 billion copies (Figure 3.1).[18]

The advent of PCR has had a significant impact on forensic science because prior to its use DNA analysis could only be performed on a large quantity of well preserved biological samples.[16,19] However, simply creating more DNA does not allow visualisation or identification of the polymorphism under study. Such variation between DNA sequences is detected using well characterised fluorescent dyes that excite and emit light at different wavelengths. The method of dye labelling and detection varies between techniques but they can be loosely grouped into those that use either (i) non-sequence specific intercalating dyes, (ii) sequence specific fluorophore labelled probes and primers, and (iii) di-deoxy terminal labelled nucleotide triphosphates.[20–25]

Table 3.1. Common dyes used to label nucleic acids. Dyes excite and emit light at different wavelengths allowing data to be collected from multiple DNA fragments simultaneously.

Fluorophore	Excitation	Emission
FAM	495	520
TET	521	536
CAL Fluor Gold 540	522	544
JOE	529	555
VIC	538	554
HEX	535	556
CAL Fluor Orange 560	538	559
Quasar 570	548	566
NED	546	575
TAMRA	557	583
CAL Fluor Red 590	569	591
ROX	586	610
CAL Fluor Red 610	590	610
Texas Red	597	616
CAL Fluor Red 635	618	637
Pulsar 650	460	660
Cy5	646	669
Quasar 670	647	670
Cy5.5	675	694
Quasar 705	690	705

Processes that use intercalating dyes are somewhat limited in their application as they offer single wavelength detection while the existence of different fluorophores allows the visualisation of multiple DNA targets at once (Table 3.1). These multiplex approaches bind the fluorophore to either a single di-nucleotide triphosphate in solution or a single stranded DNA construct. The advent of fluorescent labelling and detection has allowed the detection of multiple genetic markers including Short Tandem Repeats (STRs), Single Nucleotide Polymorphisms (SNPs) and whole gene sequences. However, to detect each dye signal the appropriate machinery and chemistry is required which can vary with application.

Common elements of the detection apparatus include a source of light designed to excite the fluorescent dye thereby causing it to emit light at a specific wavelength.[26–28] The light emitted is then passed through a lens and sometimes a prism filter which separates the light to allow detection by a photodiode (Figure 3.2). With the ability to detect the light emitted from the DNA target comes the necessity to differentiate between the signal (from dye bound to target marker) and noise (from unbound dye or dye bound to non-target DNA). The differentiation between signal and noise is essential to any analytical science where misidentification can lead to false positives and false negatives and ultimately low accuracy at identification.[29–31] The application of analysis thresholds to differentiate between signal and noise is common practice and recommendations and best practice guidelines are provided for each method.[32–35]

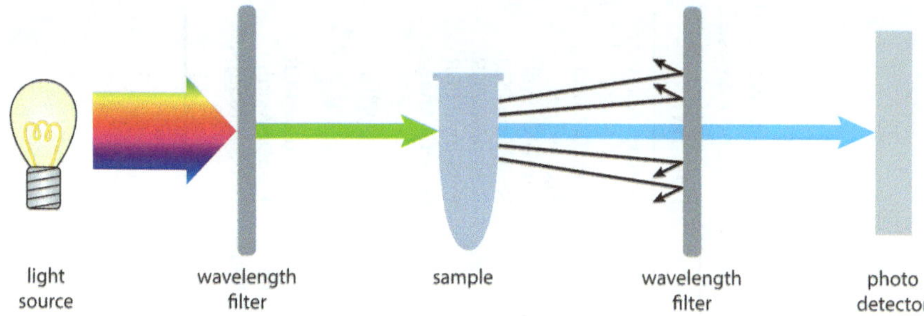

Figure 3.2. The basics of light detection. Light is separated and a specific wavelength is used to excite a fluorescent component causing it to emit light at a different wavelength. This is then passed through another filter and detected by a photodetector. Adapted from https://nanohub.org/resources/17359/supportingdocs.

3.3. Analysis of DNA in Forensic Casework

The combined approach of PCR and fluorescence detection makes DNA analysis a very sensitive technique in Forensic Science. Current applications of DNA in Forensic casework are shown in Table 3.2. Many forensic investigations involving DNA analysis use a few common approaches but occasionally specialist techniques and analysis may be required. The next section covers these strategies and groups the approaches in relation to the question under study, specifically DNA detection, DNA quantification, individual identification, mtDNA DNA sequencing, body-fluid detection and phenotypic trait inference.

3.3.1. DNA Detection

Forensic samples are variable in terms of the quality and quantity of genetic material they yield.[36–38] The ability to know that DNA has been successfully recovered from evidence is important in casework management and allows those concerned to make decisions on how to proceed with each case based on relevant data.[39–42] In its simplest form the question of whether you have recovered DNA from the evidence becomes a binary 'yes' or 'no' answer. If no DNA has been recovered then the sample will not undergo any further genetic profiling, if DNA has been recovered then further analysis can occur. This can be performed with a certain degree of success by simply observing the extracted total cellular DNA (comprised of nuclear and mitochondrial DNA) on a dye stained agarose gel using the process of electrophoresis. This approach sees little use in a modern forensic laboratory due to the development of new accurate quantification tests but is commonly used in research and development and validation activities and stands as the pre-cursor technology of modern approaches. The process separates DNA

Table 3.2. Common molecular approaches used in forensic genetic investigations. There are multiple methods and platforms used and different approaches can be used to ask the same question. The common element is that data in the form of light is collected and analysed.

Forensic Subject	Method	Common Markers	Labelling	Platform
Individual identification	Size separation of DNA fragments using capillary electrophoresis	Short Tandem Repeats	Four colour fluorescently labelled primers	Genetic Analyser
	End-point allele specific melt temperatures		Three colour HyBeacon Probe	ParaDNA unit
DNA detection	DNA extraction and agarose gel visualisation	Whole genome	Intercalating dyes	Agarose gel
	End-point analysis of PCR	Amel, D16, THO1	Single Colour HyBeacon Probe	ParaDNA unit
DNA quantification	Real time PCR quantification of unknown sample compared to standard curve	HTRT	TaqMan® Probe	Real-time PCR
		RNU2	Plexor® Probe	Real-time PCR
		Alu gene	SYBR green dye	Real-time PCR
mtDNA gene sequence identification	Size separation of terminally labelled DNA fragments	Cytochrome oxidase I, cytochrome b, HVI, HVII	Four colour dye terminating ddNTPs	Genetic Analyser
	Gene specific amplification and size separation and visualisation on agarose gel (species ID only)		Intercalating dye	Agarose gel
Body fluid identification	End-point analysis of gene specific melt temperatures	HTN3, MMP10, ALAS2, CYP2B7P1, SEMG1, IL19	SYBR green dye	Real-time PCR
			Three colour HyBeacon Probe	ParaDNA unit
	Size separation of DNA fragments using capillary electrophoresis		Four colour fluorescently labelled primers	Genetic Analyser

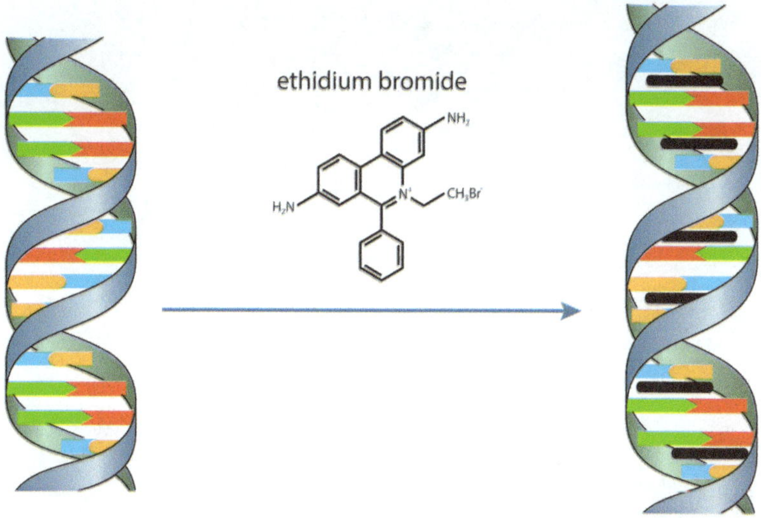

Figure 3.3. Intercalating dyes such as ethidium bromide become sequestered within the double helix and fluoresce when bound. Copyright © 2012. From *Introduction to genetics: A molecular approach* by T. A. Brown. Reproduced by permission of Garland Science/Taylor & Francis Group LLC.

fragments based on their relative size and mobility when an electrical current is applied. DNA has a total negative charge meaning the DNA migrates toward the positive electrode. Short and more mobile fragments moving at a quicker rate. To visualise the DNA an intercalating dye is used. Intercalation describes the insertion of molecules between adjacent nucleotide base pairs of a DNA duplex (Figure 3.3). Ethidium bromide has an excitation maximum of 525 nm and emission maximum at 600 nm in 10 mM TBE buffer.[43] The dye fluoresces at 25 times higher intensity when bound to double stranded DNA. Intercalative binding occurs at every second site along the helix.[44,45] The fluorescence yield of EtBr increases as solvent polarity decreases.[46] While ethidium bromide is a widely-used dye other less hazardous intercalating dyes are available. SYBR Safe™ is a cyanine dye that has excitation maxima at 280nnm and 502 nm and an emission maximum at 530 nm[47] or gel red which has excitation maxima at 285 and an emission maximum at 590 nm. Bands are visualised through exposure to UV (or blue light in the case of SYBR Safe) which allows the bands to be visualised by eye or using a camera.

When total genomic DNA is analysed the expectation is that the material should consist of large molecular weight molecules and therefore migrate slowly. If DNA is degraded and fragmented the band will appear to 'smear' into a single long product. While it becomes possible to visualise the DNA on the gel the analysis is very subjective in nature. The process is also laboratory based and has largely been replaced by the more accurate method of quantification discussed later.

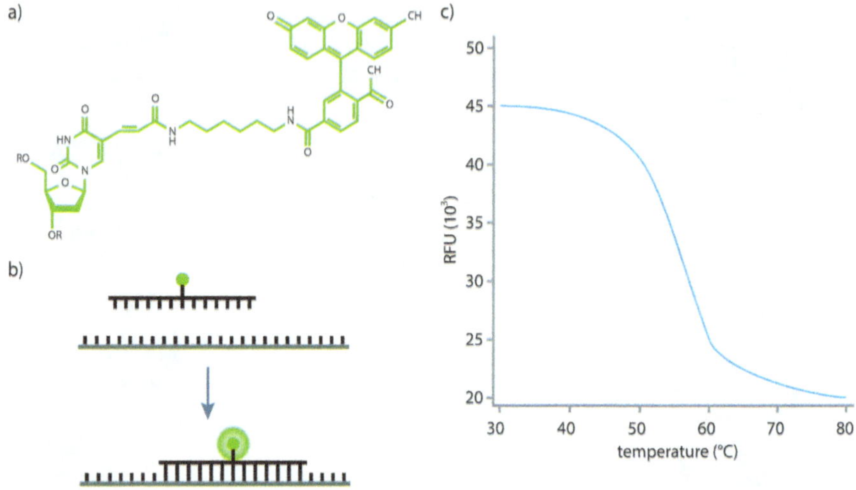

Figure 3.4. Mechanics of HyBeacon Fluorescence as used in the ParaDNA Screening System. Fluorescein label (a) is attached to a thymine base in a Hybeacon Probe. The Probe fluoresces at a greater intensity when bound to corresponding DNA (b). As the bound complex is heated the hydrogen bonds break releasing the probe resulting in a reduction in fluorescent signal (c). Parts (a) and (b) reproduced from ref. 23 with permission from the Royal Society of Chemistry.

A recent advance that offers to extend the DNA detection approach out of the laboratory and into the hands of the investigating authority uses a portable fluorescent PCR detection system and pre-made stabilised reagents.[48–50] Developed by LGC LTD, the ParaDNA Screening System™ offers a method to detect DNA on evidence items prior to submission to a Forensic Service Provider. The system uses HyBeacon Probe Technology to detect DNA amplified at the D16S359, THO1 STR loci as well as Amelogenin. The HyBeacon probe is comprised of a single stranded short oligonucleotide that possess two fluorescein labelled Thymine nucleotides (Figure 3.4a), a 3′ phosphate to prevent PCR extension and a 5′ Trimethoxystilbene cap. The Fluorescein HyBeacon Probe has excitation maxima at 495 nm and an emission maximum at 520 nm. Probes are included in the PCR reaction and emit greater amounts of fluorescence when hybridised to complementary target sequences than when single-stranded (Figure 3.4b). This system does not require a specific quencher molecule as fluorescence is reduced in the hybridised form due to the inherent quenching properties of DNA.[51,52] As this approach requires the probe to bind to single stranded DNA (ssDNA) target, an excess of ssDNA is generated using an asymmetric PCR approach. The PCR is typically 40–50 cycles long with ssDNA being generated after ~30 cycles as the limiting primer is exhausted.[53,54]

Utilising three different fluorescein probes in a bespoke four-well PCR plate the ParaDNA Screening system uses end-point analysis of a melt curve. After the 50 cycle asymmetric PCR, the reaction containing the excess ssDNA target and probe is cooled from 90 °C to 20 °C to allow hybridisation of probe and DNA. The sample is then heated to 70 °C until the bonds linking the probe and target DNA denature leading to a sharp reduction in fluorescence (Figure 3.4c). Summing the level of fluorescence observed between the four wells gives a rudimentary assessment of the amount of DNA that was recovered from the sample which is then graded as having a strong, medium or low chance of yielding a downstream STR profile.[55,56] While the novel approach offered by the non-expert user systems fulfils the requirements of informed decision making it still falls short of offering the accurate and precise DNA quantification data required for STR profiling.

3.3.2. *DNA Quantification*

DNA quantification is a key step in the process work flow. Performed after DNA extraction, it provides data on the amount of DNA present in the extract. It allows the forensic scientist to adjust the amount of input DNA entering the PCR reaction. The relationship between amount of input template into PCR and signal obtained is routinely documented and forms part of the validation experiments performed when developing and characterising new forensic assays. If too much DNA enters the PCR then it is highly likely that the signal is saturated, complicating the interpretation. If too little DNA enters the reaction the signal will be reduced and may result in stochastic artefacts including allele drop-in and drop-out in very low template samples.[35,57] To minimise this from happening developers of commercial assays recommend a certain amount of genomic DNA be used in the PCR reaction.[58,59] This ensures that the signal is within the bounds of interpretation. The two most common methods used to quantify DNA are UV-spectrophotometry and quantitative PCR (qPCR).

UV-spectrophotometric methods calculate the concentration of DNA based on the amount of UV-light that is detected after it is passed through the sample. Nucleic acids absorb light at 260 nm meaning that the more concentrated a sample is, the less light reaches the photodetector at this wavelength.[60,61] Accurate quantification of the unknown DNA sample can be achieved relative to a series of known standards after creating a standard curve. Another approach is to simply calculate concentration based on the observation that DNA at a concentration of 50 ng μl^{-1} gives an absorbance reading at 260 nm of 1.0. However, both methods are prone to some error given that DNA is not the only molecule to absorb light at 260 nm (RNA and aromatic acids in proteins can also contribute to the value).[62] Another disadvantage is that the approach cannot differentiate between different species' DNA meaning that it becomes impossible to determine whether the stated concentration is target human DNA or from any other biological contaminant.[63] The main limitation of spectrophotometric methods is its sensitivity.[64] Data demonstrates that below a concentration of approximately 2 ng μl^{-1} it becomes impossible to differentiate between the A260 of a true sample and a blank sample containing no DNA.[65] This is problematic in a field where DNA recovered from crime scenes can often be in the sub 1 ng μl^{-1} range.

In addition to the traditional UV approach there are other spectrophotometric methods based on fluorescence which are more sensitive.[63] However, in many applications they have largely been superseded by more sensitive methods of quantification using qPCR.[66] These approaches use the power of PCR to detect and amplify gene targets from single cells and compare them against standards of known concentration. One of the earliest uses in forensic science describes the use of SYBR Green to quantify the ALU gene relative to known standards.[67,68] This dye is commonly used in research and development activities as it has very little binding to ssDNA and a high quantum yield when bound to dsDNA.[69] It has an excitation maxima at 494 and an emission maximum at 521 nm. Data suggests multiple modes of interaction between SYBR Green and double stranded DNA and shows a 1000× enhancement in fluorescence between the bound and unbound states.[70] As with Ethidium Bromide, the intercalation between base pairs and dye contributes to increased fluorescence (Figure 3.5a). Quantification of the Alu gene using SYBR Green has shown to be a sensitive and cost effective approach although careful optimisation of the PCR conditions and primer designs is required.[71] This is because any non-specific amplification is also detected by the intercalating dye. Another limitation is that q-PCR detection of multiple gene products is not possible due to the inability to differentiate the signal from each PCR amplicon at a single wavelength and is only possible through the use of melt curve analysis post-PCR.[72] Sequence-specific probes offer a solution to these limitations. Two of the most common probe based systems used for DNA quantification for forensic applications are Life Technologies TaqMan® Probes[73] used in the Life Tech Quantifier Duo range[74] and dabcyl-iso labelled Primer Probes[75] used in the Promega Plexor® range.[76]

TaqMan® probes are short, single stranded oligonucleotides that are designed complementary to the target DNA sequence and internal to both the forward and reverse primers. The probe contains a donor fluorophore and an acceptor, termed a quencher. In solution the probes emit fluoresce at a low intensity due to the interaction between the fluorophore and quencher. During the PCR annealing phase the TaqMan® probe anneals to a region of DNA downstream of the forward primer. During extension when the DNA polymerase bonds other single nucleotides to the ssDNA the enzyme hydrolyses the quencher molecule through 3′ exonuclease activity and the probe dissociates from the ssDNA target as the strand elongates. The probe, now unquenched, fluoresces at a higher intensity than before in the solution which is observed as an increase in fluorescence signal (Figure 3.5b).[73,74] The Plexor® HY assay uses a system containing a primer labelled with a fluorescent 5′-methylisocytosine (iso-C) reside automatically fluorescing as ssDNA. The second primer is unlabelled. The PCR mix contains normal dNTPs (A, T, C, G) for sequence extension and also some isoguanine (iso-G) nucleotides

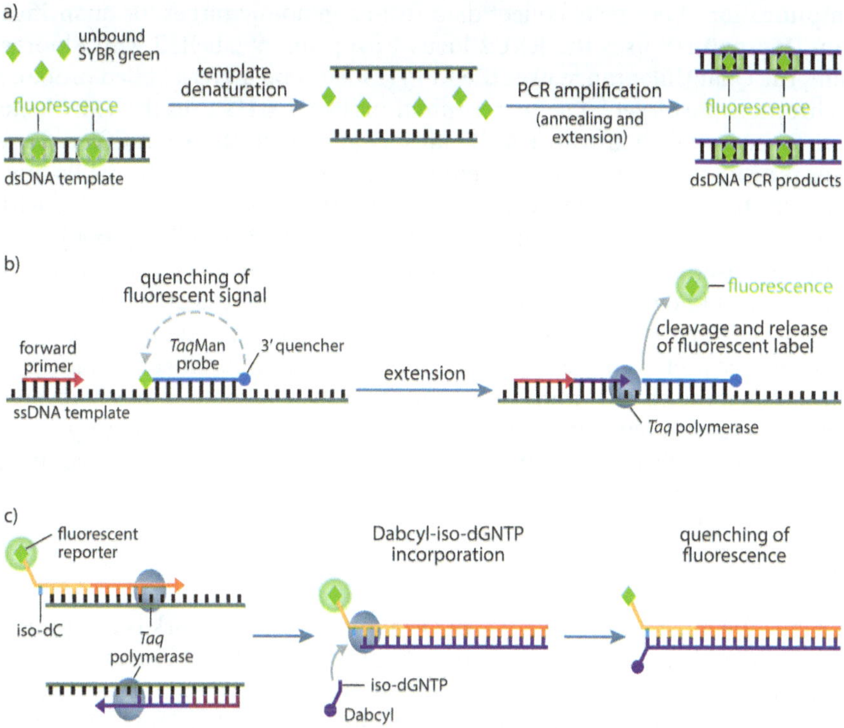

Figure 3.5. Common methods used in DNA Quantification. Intercalating dyes such as SYBR Green become sequestered within the double helix and fluoresce when bound (a). As more double stranded DNA is generated during PCR the fluorescence increases. TaqMan® Probes are quenched when bound (b). The action of the Taq polymerase releases the fluorophore resulting in an increase in fluorescence. The Plexor® HY approach uses a quenching dNTP which is bound to the fluorescent reporter during PCR extension (c). This results in a decrease in fluorescence as more PCR product is generated. Parts (a) and (b) reproduced from Cindy J. Smith and A. Mark Osborn, Advantages and limitations of quantitative PCR (Q-PCR)-based approaches in microbial ecology, *FEMS Microbiology Ecology*, 2008, **67**, Issue 1, by permission of Oxford University Press. Part (c) adapted with permission from Benjamin E. Krenke, *et al.* (2005) Development of a Novel, Fluorescent, Two-Primer Approach to Quantitative PCR, Promega Corporation.

modified to include a dabcyl quencher. During successive rounds of PCR as new complementary DNA strands are generated the iso-G is attached and the signal from the iso-C primer is quenched leading to a reduction in signal over time (Figure 3.5c).[77] As the fluorophore is attached to an oligo sequence these approaches can offer multi-target detection in a single tube with each fluorophore exciting and emitting data at different, non-overlapping wavelengths.

During successive cycles of q-PCR process the fluorescence is continuously recorded and the cycle threshold (ct) is established as a measure of DNA amplification. Both tests collect data from a genomic target for quantification (Plexor® HY uses the RNU2 locus with primers labelled with Fluorescein, the Quantifiler range uses the RPPH1 locus and a VIC labelled probe), a Y-chromosome target for sex determination (Plexor® HY uses the TSPY locus with primers labelled with CAL Fluor® Orange 560, the Quantifiler range uses the SRY locus and a Fluorescein labelled probe) and an Internal Positive Control (IPC) synthetic DNA target (Plexor® HY uses primers labelled with CAL Fluor® Red 610, the Quantifiler range uses a NED labelled probe). Data from the casework samples of unknown concentration are then compared to standards of known concentration and a standard curve is drawn and a concentration in ng μl^{-1} is provided. As these methods provide more accurate data and are preferred over gel base approaches this quantification data provides one of the earliest indications that a forensic laboratory has recovered DNA from an evidence item. Forming part of the normal laboratory workflow, the quantification result correlates well with the chance of generating downstream data.

3.3.3. STR Typing

Perhaps the most well-known use of DNA in forensic casework is its application in individual identification. The ability to match the DNA profile from a suspect with the profile collected at a crime scene offers unparalleled power to link an individual to a specific location. To generate this data PCR is used to amplify multiple loci from a group of molecular markers called Short Tandem Repeats (STRs). STRs are formed from repeating units of DNA either 2–6 base pairs in length. They are diploid (two copies of the marker exist in every autosomal cell), inherited in a Mendelian fashion and are located in the nucleus. The suite of STR loci selected for forensic identity testing are in non-coding regions of DNA. This ensures that the markers are neutral in response to selection and also preserves any novel length mutation as it has a non-deleterious effect on survival.[36,78] These different length DNA sequences termed alleles exist in multiple numbers in the population with each person inheriting one allele from each parent. An individual can be termed homozygote if both alleles are the same length or heterozygote if alleles are different length. Given their mode of inheritance STR profiles can also be used in paternity investigations, where the profiles of the putative fathers are compared to that of the child.[79]

Modern STR profiling kits consist of 17 + STR loci and the generation of a DNA profile requires two distinct steps, (i) the PCR amplification of target STRs and (ii) the separation and identification of the PCR product by capillary electrophoresis on a Genetic Analyser. The PCR amplification uses a traditional forward and reverse primer, one of which is labelled with a fluorescent dye bound to the 5′ end of the primer.[80] Through primer extension during PCR all amplified DNA target is automatically fluorescently labelled (Figure 3.6a). Allele ranges for each STR can span several hundred base pairs so differentiation between alleles at different loci is achieved by using

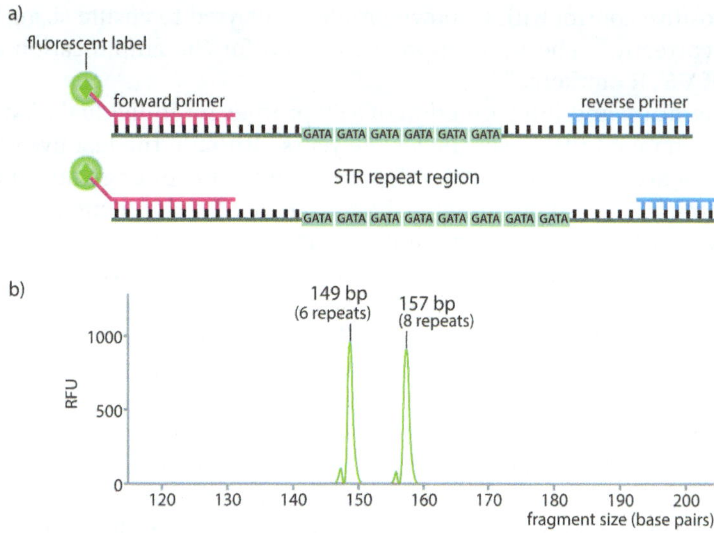

Figure 3.6. Mechanics of STR analysis. Fluorescently labelled primers that are unique to an STR locus result in the amplification of two STR alleles (a). Due to the presence of a variable number of repeats the resulting labelled PCR product is of different length which visualised after separation on a Genetic Analyser (b).

primer pairs that do not generate fragment sizes that overlap and using primer labelled with probes of different wavelength for any size ranges that do overlap.[81] DNA profiling kits are commercially produced and the dyes used to label the primers can vary.[82–84] The instrumentation used to generate a DNA profile is termed a Genetic Analyser.[85,86] The instrument is capable of detecting the fluorescence emitted from the labelled probes across five channels (Channel 1 = FAM 518 nm; Channel 2 = VIC 554 nm or JOE 548 nm; Channel 3 = NED 575 nm or TAMRA 580 nm; Channel 4 = PET595 nm or CXR-ET 602 nm; Channel 5 = LIZ655 nm). Channels 1–4 provide information specific to the sample being analysed. Channel 5 is allocated for a DNA size standard.

Size separation is achieved using *Capillary Electrophoresis* through a matrix of polyacrylamide.[87,88] The application of a positive charge will cause the negatively charged DNA to migrate through the matrix with smaller fragments moving faster due to their relative mobility. The resulting STR profile comprises several peaks detected by the instrument's photodetector as it passes across a filter. The signal is recorded as an increase in fluorescence relative to the background noise and appears as a peak on the electropherogram (Figure 3.6b). The sizes of the alleles are confirmed through three quality assurance steps. First, the DNA size standard (added to the PCR sample prior to separation) contains peaks of known size and allows the instrument software and analyst to confirm the sizes of the unknown alleles in the sample of interest.[86] Second, an allelic ladder containing all the alleles of interest is processed at the same time as the sample allowing any allele to be identified.[89]

Third, a positive control with a known profile is analysed to ensure sizing has occurred correctly.[86] The same approach is used for the amplification and analysis of Y-STR markers.[90,91]

The generation and interpretation of STR profiles for individual identification has changed little over the last 20 years although the last five years has seen research focussing on developing systems that offer greater speed and increased data. Such research has seen the development of Rapid DNA Systems which have allowed non-expert users the ability to generate STR profiles outside of the laboratory using miniaturised Genetic Analysers with expert software performing much of the analysis.[92–94] Another approach developed by LGC is the ParaDNA Intelligence Test which offers a field based system to generate mini-STR profiles. The system uses direct PCR amplification in a closed tube system and multiple HyBeacon Probes to collect end-point melt curve data to differentiate between alleles.[95,96] The test uses a custom designed 4-well plate pre-loaded with all the required reagents. The test amplifies and detects five STR loci: D3S1358, D16S539, D8S1179, D18S51 and TH01 and also the amelogenin gender marker. As with the ParaDNA Screening Test described above, the system requires the generation of an excess amount of ssDNA for the Probe to hybridise to. Allele sizes are differentiated due to the presence of a blocker oligo that only allows a portion of the HyBeacon probe to hybridise to the allele and also due to the different fluorophore dyes used to label the probes (FAM, JOE and TAMRA). How much of the probe hybridises to the allele will vary with its length (Figure 3.7). As the assay is heated after PCR the temperature at which the probe disassociates itself from the allele will vary as a function of how much is hybridised giving an allele specific melt temperature. Analysis is a fully automated process that involves quantifying the

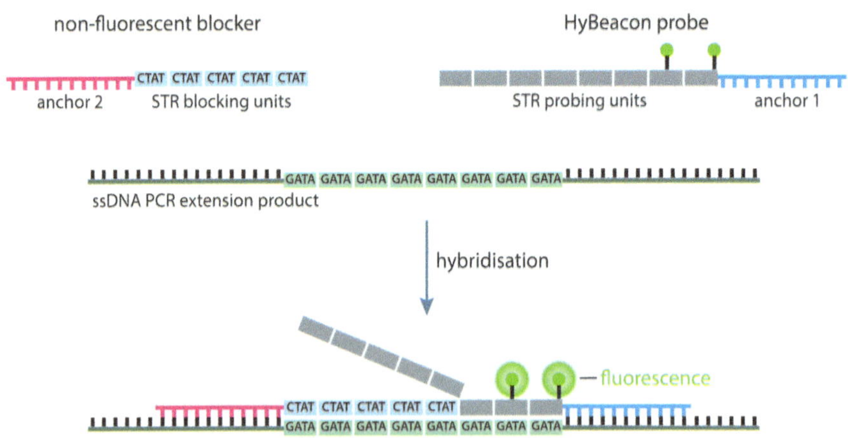

Figure 3.7. Mechanics of HyBeacon STR analysis. The presence of a blocker which attaches to the ssDNA first allows the partial hybridisation of the HyBeacon probe to the STR region. As the STR regions increase in length more of the probe is allowed to hybridise to the region giving alleles specific melt temperatures.

fluorescence change associated with each of the possible allele contributions. This analysis is based on an understanding of how the fluorescence of bound and unbound probe varies with temperature. The quantitative allelic data are compared to a series of thresholds that are designed to reject system noise and stutter and thereby only retain genuine allelic contributions.

3.3.4. mtDNA Gene Sequence Identification

The analysis of mtDNA genes in Forensic Science is used in instances where some information on human identity is required but the biological material to perform full nuclear STR typing is simply not present. The categorisation of human mtDNA haplotypes shows a relatively strong correlation with observed living population groups, although gene flow between these populations prevents complete separation of haplogroups.[97] The application of mtDNA typing in forensic science is most often used in a historical contexts such as the post-mortem identification of soldiers recovered from ancient battlefields, the identification of historically significant individuals and ancient DNA analysis.[98–102] The two most often used mtDNA genes used in human haplogroup typing are HVI and HVII (Hypervariable I and II).[103,104] These markers are non-coding so have an even higher mutation rate than the coding genes also located in the mitochondria.[105] DNA sequences are amplified *via* PCR and separated using capillary electrophoresis and analysed on a Genetic Analyser.

Mitochondrial DNA markers are also used for the identification of animal and wildlife species of forensic importance. In such instances the limitation, that mtDNA can only identify the taxonomic level of sub-population or population, is not necessarily problematic as much of the casework requires identification at the species level.[106] Casework involving the trade in endangered species, animal persecution and illegal food substitution all require identification at the species level.[107–109] The choice of gene region is important when performing work of this kind as the success of the approach relies on the use of public databases holding DNA sequence data of the species of interest.[110,111] Commonly used genes for animal identification are the Cytochrome b gene and Cytochrome Oxidase I.[112,113]

The most accurate and often used approach in forensic genetics is the generation and visualisation of the whole amplified gene sequence using capillary electrophoresis on a Genetic Analyser.[114,115] This requires two rounds of PCR. The first to amplify the original strand using a forward and reverse primer and deoxy nucleotide triphosphates (dNTPs). The second '*sequencing PCR*' uses a single primer for linear extension and both dNTPs and chain-terminating fluorescently labelled di-deoxy nucleotide triphosphates (ddNTPs). As strand extension occurs and a fluorescently labelled ddNTP is attached the strand ceases to elongate. Depending on which dideoxy base is attached a strand of a specific length labelled with a unique termination dye is produced.[116,117] Over multiple cycles there is a

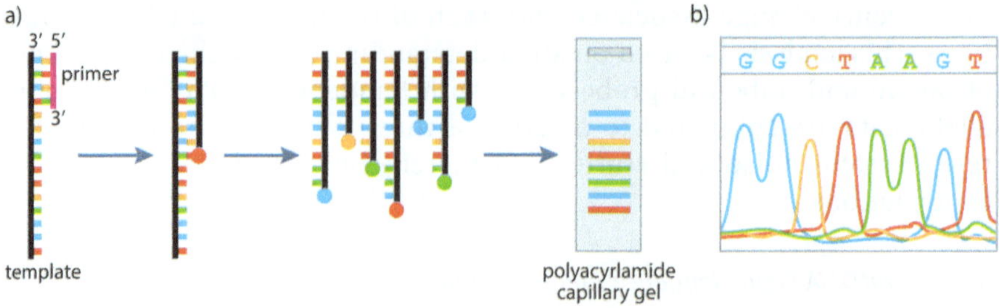

Figure 3.8. Traditional DNA Sequencing. The sequencing PCR uses fluorescently labelled ddNTPs resulting in chain terminating once incorporated. The resulting fragments are labelled uniquely according to the 3′ base (a). Size separation through a polyacrylamide capillary results in single base pair detection of each nucleotide (b).

build-up of fragments each differing by a single base pair with every terminal 3′ labelled with a fluorescent ddNTP (Figure 3.8a). Separating these different sized fragments using Capillary Electrophoresis produces the DNA data as a continual sequence of nucleotides as multiple strands of different lengths across four wavelengths are detected by the instrument's photodetector (Figure 3.8b).[114,115] Individual nucleotide differences occur as point mutations in the sequence. In this manner, unique DNA sequences for the HVI and HVII regions are collected with each sequence differing in as little as a single base pair from any of the others. Using information on the frequency of each SNP the frequency at which the haplotype is observed at in the population is known and the chance of observing matching haplotypes is determined.

Another cost effective and rapid method to detect and identify target DNA is agarose gel electrophoresis. These tests are often designed rapidly, validated and serve a single application. The DNA target under analysis is a single gene region amplified by PCR. The PCR amplicon transfers through the gel matrix towards the positive electrode at a rate based on its size. The size of the amplicon is known based on the position of the primers that are used during PCR and any bands amplified in the putative sample of interest are compared in size to those of a DNA size standard and also the use of a positive control. Visualisation of the band is through staining the gel with intercalating dye following the same mechanism as covered in Section 3.3.1. An issue with the use of agarose gel as a method of assessment is its sensitivity. Even with the amplifying effect of PCR power of PCR and a staining dye, detection of the DNA band is often subjective if left solely to the analyst. As such many of the traditional tests developed for use on agarose gel have now been re-optimised and re-validated for use with real time instrumentation although agarose gel approaches are often used as an initial stepping stone for further optimisation and development. Just as with any PCR based method mtDNA analysis is heavily dependent on primer design. There needs to be high intra-specific homology between the target DNA and the PCR primer and low inter-specific homology between the non-target DNA and the PCR primer.[109] This promotes amplification of the target DNA if it is present in the evidence sample while other species DNA should not be amplified or amplify and differentiate based on the size of the DNA fragment. Primer design requires access to public resources containing pre-published DNA sequence data *e.g.* (NCBI or BOLD databases)[110,111] and requires DNA sequence alignment between the target DNA sequence and any other non-target regions that may co-amplify. Even with high homology between target and non-target, primer design approaches such as ARMS can be attempted which together with careful optimisation can result in a powerful assay.[118,119]

3.3.5. *Body Fluid Identification*

The identification of body fluids in forensic casework provides activity level information that may either support or refute the defence/prosecution hypothesis. A number of different genes have been documented and characterised by researchers in the forensic community for a number of years and the adoption of marker panels by practitioners is slowly increasing.[120,121] This research has shown a number of genes and primer sequences that are highly specific to a body fluid. Approaches using mRNA and microRNA have been developed that allow the reverse transcription of the RNA sequence into complementary DNA (cDNA).[122–124] This ensures that only tissue that contains the expressed gene product are detected and not the genomic DNA. After the reverse transcription PCR (RT-PCR) the product can be analysed using the same techniques and instrumentation as DNA.

Identification by capillary electrophoresis uses the same principals as discussed for STRs above. One of the primers used to amplify the cDNA is fluorescently labelled and the amplicon size is designed to allow size specific peaks to be observed at the desired wavelength. Given the low number of genes identified that show specificity to the target body fluid and the standard size of the gene amplification product there is only a single peak observed at a specific size for each body fluid marker. While this makes it possible to design a primer set labelled with a single fluorescent dye most research has used a number of different labelled primers.[121,125] This provides a very effective method of identification and allows for multiple markers to be detected. A limitation is that it requires a multi process approach before identification can occur, namely extraction and purification of mRNA, generated of cDNA, PCR of gene markers and finally size separation. Generally speaking in forensic applications it is favourable to reduce the number of processing steps to their minimum as multiple handling steps represent an increased contamination risk. Strategies to reduce the steps are to perform a single RT-PCR step or to use a qRT-PCR approach instead of size separation allowing simultaneous amplification and detection in a closed tube.[126,127] The limitation of this approach is that fewer markers can be differentiated as the systems use q-RT-PCR amplification plot data. High Resolution Melting offers an analytical approach that allows differentiation of different product signal at end-point using which preferentially binds to double stranded DNA.[128] Under recommended reaction conditions the SYBR green dye will bind to and fluoresce with double stranded DNA. If the DNA is heated and becomes single stranded the level of fluorescence detected by the Real-Time PCR machine drops. HRM uses an end-point analysis of fluorescence. The temperature at which this occurs varies depending on how many bonds occur between the two strands of DNA which varies as a function of length allowing detection of different fragment lengths (Figure 3.9a). Recent advances have seen the development of an mRNA based profiling technique for non-expert users at crime scene.[129,130] This further reduces the number of handling steps as there is no requirement to perform an RNA extraction and isolation step which is required for both the capillary electrophoresis method and also the qRT-PCR methods. Instead this approachused direct PCR which sees the reverse transcription, PCR and melt curve detection all happening from the single crude sample. This approach again developed by

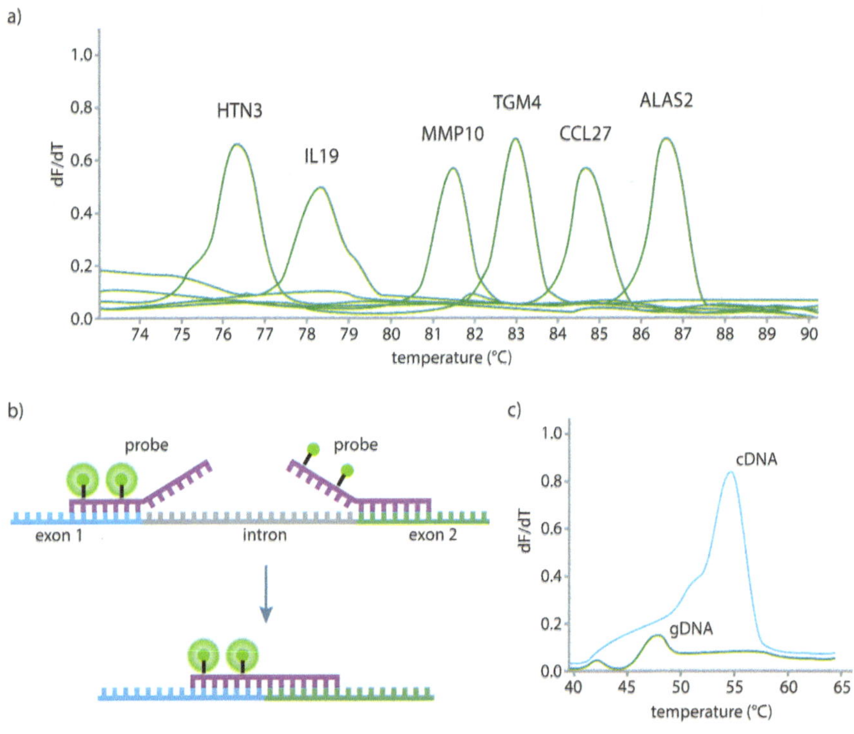

Figure 3.9. Body Fluid Identification. Approaches have been developed to allow melt curve analysis to differentiate between different mRNA markers either using intercalating dyes (a) or HyBeacon Probes (b) allowing clear differentiation between gDNA and cDNA products (c). Part (a) adapted from E. K. Hanson and J. Ballantyne, *F1000 Research*, 2014, **2**, 281. © 2014 E. K. Hanson and J. Ballantyne. Published under the terms of the Creative Commons CC BY license (https://creativecommons.org/licenses/by/3.0/).

LGC for the ParaDNA platform uses Hybeacon probes labelled with a FAM, JOE or TAMRA fluorophore. The system uses RT-PCR to generate an excess of ssDNA for the Probe to hybridise to and fluoresce. Using this system mRNA is differentiated from gDNA as each probe sits on the exon-intron boundary (Figure 3.9b) it has a different melt temperature depending on whether the probe sits on gDNA or cDNA (Figure 3.9c). Analysis is a fully automated process that involves quantifying the fluorescence change associated with each of the possible allele contributions. The exact method of analysing body fluid data remains the subject of debate in the forensic community with some groups calling for redundancy in the panel of markers chosen so that multiple markers per body fluid need amplifying to achieve confidence in the result, while another interesting approach is that markers are weighed and a probability is provided based on the chance that it would fail or amplify in another tissue.[131,132] The routine application of these approaches in forensic laboratories has undergone several assessments through international collaborative exercises.[133]

3.3.6. *Biogeographic Origin and Phenotypic Trait Inference*

It has long been noted that the high degree of genetic similarity between human populations reflects the recent divergence of the human species into continental groups. It is also suggested that the marked physical differences between individuals and populations have evolved due to local environmental pressures, random genetic drift, and sexual selection. The ability to identify genetic markers that code for such complex traits and use the data to infer the biogeographic origin and physical appearance of an individual has obvious application in forensic science.[134] In the absence of a suspect identified through either witness testimony or DNA database, or in a case with very few investigative leads reducing the pool of possible suspects through inference of physical traits is an attractive if not ethically problematic solution. Early research looking to use such genetic differences demonstrate that inferring ethnic identify from genetic data is possible.[135] While not an application that saw routine use in forensic laboratories it highlighted an area of research that has been slowing building its evidence base and is now beginning to become realised. Mapping the human genome throughout the 1990's provided geneticists with the basic elements to begin identifying possible gene regions that can be linked to physical traits.[136] The advent of massively parallel sequencing platforms (Next Generation Sequencing) in the last decade has allowed large amounts of data to be generated rapidly and has changed the way in which modern genetic laboratories collect data. In turn this has led to the development and discovery of the underlying Single Nucleotide Polymorphism (SNP) for key traits and the application of such approaches to the field of forensics.[137,138] Termed forensic genomics, these advances have made it possible to infer the eye colour, hair colour, skin colour and ancestral origin of the individual who left a biological sample at a crime scene.[139–142] For this purpose SNP markers have successfully been identified using both a Single Base Extension (SBE) approach where a fluorescently labelled ddNTP complementary to the SNP in question is directly added to an SBE primer and identified through capillary electrophoresis, and also using the TaqMan® probe detection approach described above. This research has laid the ground work and demonstrated the capabilities of this approach culminating in the development of a panel of markers for use on the Illumina MiSeq platform.[143]

Massively Parallel Sequencing (MPS), also commonly referred to as Next Generation Sequencing (NGS), can be performed using several different methods, each specific to the platform of use. Here we describe that used on the MiSeqFGx Forensic Genomics System which is at the most advanced stage of application in forensic genetics. This system detects several DNA target regions including 27 global autosomal STRs for individual identification, 24 Y-STRs for paternity and kinship analysis, 7 X-STRs, 94 identity SNPs, 22 Phenotypic SNPs for common physical characteristics and 56 biogeographical ancestry SNPs. The system works using sequencing by synthesis (SBS)[144] and involves a number of distinct steps; *Library Preparation* which involves the amplification and enrichment for the target DNA regions using adapters that are ligated to the target DNA strands, *Cluster Generation* whereby the enriched DNA targets are first bound to a surface holding complementary oligos and then amplified using an isothermal PCR approach called bridge amplification. The final stage is *sequencing* where a reversible terminator-based method detects the single bases as they are incorporated into the DNA template strand. Detection occurs after addition of each fluorescently labelled dNTP through excitation by a light source. The light source causes the fluorophore to cleave from the strand and fluoresce at a greater intensity which is detected by the instruments photodetector. In doing so another fluorescently labelled dNTP is free for incorporation in the DNA strand.[144] This occurs hundreds of times in millions of the amplified fragments which are separated by specific regions on the surface of the flow cell. The amount of information derived by MPS approaches is large and the systems implementation into forensic casework has been slow. There is some uncertainty in whether having 90% probability that the individual who left the sample had brown eyes may actually be helpful as there is still a 10% chance they do not. There are also concerns regarding the legality of collecting data that may be unique to racial and ethnic groups in some countries which are being openly discussed and appropriately considered in the forensic community.[134] While it is important to reach a consensus within the community prior to routine application there continues to be much evidence to support the accuracy, sensitivity and specificity of the system in preparation for its eventual use.[145–148]

3.4. Degradation of DNA by Light

While the detection of DNA using fluorescence at different wavelengths is used in many applications it is not the only way light impacts effects DNA analysis. Solar radiation in the form of ultraviolet (UV) light has a detrimental impact on the DNA in both fauna and flora, from single cell to multicellular organisms. The most damaging component of UV light is UV-B (280–315 nm) as it is absorbed directly by the DNA while UV-A (315–400 nm) causes secondary damage by indirect photosensitizing reactions. Least damaging is UV-C (<280 nm) which is mostly absorbed by oxygen and ozone in the atmosphere.[149] While many biological cells contain UV absorbing pigments as a method to counteract the damaging effect of UV light it cannot be completely avoided. Damage occurs through the development of DNA lesions in the double stranded helix which result in the breaking of the structure and the formation of DNA dimers. Lesions develop in different ways giving rise to characteristic shapes and forms that are symptomatic of the underlying mechanism for damage. Some of the more damaging types of DNA lesion are caused by photoproduct DNA.[150] Caused as a direct result of UV-B radiation this process results in the formation of cross-linked bonds between Thymine bases in the DNA such as cyclobutene pyrimidine dimers (CPDs), pyrimidine 6-4 pyrimidone photoproducts (6-4PPs) and their associated isomers (Figure 3.10a). Oxidative DNA damage can also occur through the creation of free radicals in the presence of ionizing and UV-radiation. Such reactive oxygen species interact directly with the DNA causing base changes which may also result in single and double stranded DNA breaks (Figure 3.10b). The creation of reactive oxygen species also damages cell structures, lipids and proteins interfering with the structural integrity and functioning of the cell. Additionally, the loss of the amine group termed deamination is caused by hydrolytic DNA damage and can result in the conversion of one base to another (Figure 3.10c). This process can also result in the complete removal of the purine and pyrimidine structures from the DNA double helix that may cause breakage to the DNA backbone. Together these mechanisms of damage result in double strand breaks in the DNA which are very deleterious as it affects both strands of DNA over a large region.

Damage to the DNA can be reduced and even repaired at a cellular level but only while the cell is alive. Cellular material left at a crime scene begins to decay and degrade through both biological and any damage through UV radiation goes unrepaired. The resulting double stranded DNA breaks reduce the chance of recovering high molecular weight DNA from the scene, which if left for too long will prevent the isolation and amplification of the target region of interest using the mechanisms described above. Therefore, the collection of material from crime scene needs to occur in as quick a time as possible without detriment to the normal casework and investigation flow. Once collected the biological material is dried if wet and stored appropriately to reduce any further biological or chemical damage. The main emphasis when collecting and storing evidence items is on the prevention of biological decay rather than the prevention of UV-induced DNA damage. This may be because the cellular material containing the DNA is still intact and therefore provides some form of continued protection. The provision of clear crime scene packaging for the collection of items underlines this and there is little research to suggest that using packaging with a UV-barrier would better preserve the DNA.

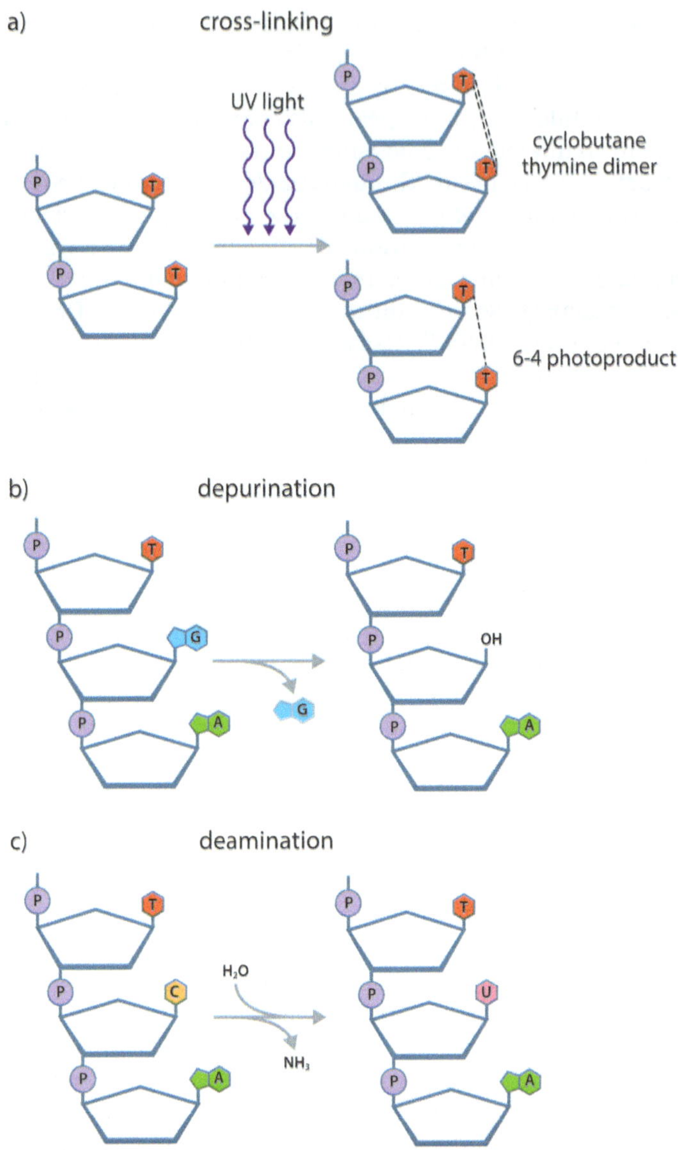

Figure 3.10. Mechanics of UV damage. Cross-linking and the creation of photo-products including cyclobutane thymine dimers and 6-4 photoproducts (a), depurination (b) and deamination (c) can weaken the DNA helix structure resulting in structural damage. Copyright © 2015 From *Molecular Biology: Structure and Dynamics of Genomes and Proteomes* by J. Zlatanova and K. E. van Holde. Reproduced by permission of Garland Science/Taylor & Francis Group LLC.

While UV-light can have a damaging effect on the biological material of interest, it can also be used to the forensic scientist's advantage as a sterilisation technique. The use of the PCR technique described in Section 3.2 above is a very powerful technique that is used to amplify the sample of interest. However, the amplification of non-target contamination product is not uncommon in the laboratory.[151–153] Laboratory workstations are routinely available with UV-lamps which are activated for a period prior to commencing work.[154,155] Certain laboratories have UV-lighting installed overhead which can be activated prior to entering the room whereby the work surfaces and laboratory equipment are sterilised.[153] The amount of time required to completely degrade the exogenous contaminating DNA has been experimentally tested a number of times with some researchers noting that profiles from extracted DNA exposed to UV light may still yield a DNA profile. One the other hand bloodstains containing DNA still presumably contained within the cells matrix still yielded PCR amplification product until 102 hours of UV exposure.[156]

3.5. Summary

DNA provides forensic scientists with a single analyte that can be used to answer, infer or provide critical evidence in forensic casework. It becomes the job of the forensic scientist to know which approach is best suited for each piece of evidence and in conjunction with the investigating authority decide on how best to proceed. The mechanism by which the data is generated has become appropriately standardised and commercial biotechnology groups develop assays for many application and release forensic specific software for analysis. At the same time, bespoke analysis is possible by specialist researchers providing appropriate quality assurance and control standards are followed. Throughout the development of the genetic processes used in molecular applications to forensic investigations light has been central to the detection of this analyte. As instruments become increasing powerful and test chemistries become more varied it is likely that light will continue to play a key role in forensic genetic science.

Acknowledgements

I would like to thank Dr Louise Dawnay for personally creating the illustrations that accompany this text.

References

1. T. Lindahl, *Nature*, 1993, **6422**, 709.
2. S. Pääbo, H. Poinar, D. Serre, V. Jaenicke-Després, J. Hebler, N. Rohland, M. Kuch, J. Krause, L. Vigilant and M. Hofreiter, *Annu. Rev. Genet.*, 2004, **38**, 645.
3. R. Bermejo, A. Kumar and M. Foiani, *Trends Cell Mol. Biol.*, 2012, **22**, 465.

4. R. Alaeddini, S. J. Walsh and A. Abbas, *Forensic Sci. Int.: Genet.*, 2010, **4**, 148.
5. W. Bär, A. Kratzer, M. Mächler and W. Schmid, *Forensic Sci. Int.*, 1988, **39**, 59.
6. M. M. Holland and T. J. Parsons, *Forensic Sci. Rev.*, 1999, **11**, 21.
7. B. Budowle, M. W. Allard, M. R. Wilson and R. Chakraborty, *Annu. Rev. Genomics Hum. Genet.*, 2003, **4**, 119.
8. M. Bauer, *Forensic Sci. Int.: Genet.*, 2007, **1**, 69.
9. M. Setzer, J. Juusola and J. Ballantyne, *J. Forensic Sci*, 2008, **53**, 296.
10. W. M. Brown, M. George and A. C. Wilson, *Proc. Natl. Acad. Sci. U. S. A.*, 1979, **76**, 1967.
11. F. M. Yakes and B. Van Houten, *Proc. Natl. Acad. Sci. U. S. A.*, 1997, **94**, 514.
12. G. S. Shadel and D. A. Clayton, *Annu. Rev. Biochem.*, 1997, **66**, 409.
13. S. Sigurðardóttir, A. Helgason, J. R. Gulcher, K. Stefansson and P. Donnelly, *Am. J. Hum. Genet.*, 2000, **66**, 1599.
14. R. K. Saiki, S. Scharf, F. Faloona, K. B. Mullis, G. T. Horn, H. A. Erlich and N. Arnheim, *Science*, 1985, **230**, 1350.
15. K. Mullis, F. Faloona, S. Scharf, R. Saik, G. Horn and H. Erlich, *Cold Spring Harbor Symp. Quant. Biol.*, 1986, **51**, 263.
16. R. Higuchi, C. H. von Beroldingen, G. F. Sensabaugh and H. A. Erlich, *Nature*, 1988, **332**, 543.
17. B. Budowle, R. Chakraborty, A. M. Giusti, A. J. Eisenberg and A. R. Allen, *Am. J. Hum. Genet.*, 1991, **48**, 137.
18. J. M. Butler, *Advanced Topics in Forensic DNA Typing: Methodology*, Academic Press, 2011, ch. 4, pp. 69–98.
19. R. Reynolds, G. Sensabaugh and E. Blake, *Anal. Chem.*, 1991, **63**, 2.
20. T. K. Christopoulos, *Anal. Chem.*, 1999, **71**, 425.
21. N. J. Walker, *Science*, 2002, **296**, 557.
22. H. Ihmels and D. Otto, *Topics in Current Chemistry*, Springer, Berlin, 2005, ch. 258, pp. 161–204.
23. R. T. Ranasinghe and T. Brown, *Chem. Commun.*, 2005, **44**, 5487.
24. J. G. Wetmur and J. Fresco, *Crit. Rev. Biochem. Mol. Biol.*, 2008, **26**, 227.
25. A. I. Dragan, R. Pavlovic, J. B. McGivney, J. R. Casas-Finet, E. S. Bishop, R. J. Strouse, M. A. Schenerman and C. D. Geddes, *J. Fluoresc.*, 2012, **22**, 1189.
26. A. Castro, F. R. Fairfield and E. B. Shera, *Anal. Chem.*, 1993, **65**, 849.
27. S. Nie and R. N. Zare, *Annu. Rev. Biophys. Biomol. Struct.*, 1997, **26**, 567.
28. R. A. Keller, W. P. Ambrose, P. M. Goodwin, J. H. Jett, J. C. Martin and M. Wu, *Appl. Spectrosc.*, 1996, **50**, 12.
29. D. G. Altman and J. M. Bland, *BMJ*, 1994, **308**, 1552.
30. A. Lario, A. González and G. Dorado, *Anal. Biochem.*, 1997, **247**, 30.
31. R. A. Mathies, K. Peck and L. Stryer, *Anal. Chem.*, 1990, **62**, 1786.
32. P. Gill, L. Gusmao, H. Haned, W. R. Mayr, N. Morling, W. Parson, L. Prieto, M. Prinz, H. Schneider, P. M. Schneider and B. S. Weir, *Forensic Sci. Int.: Genet.*, 2012, **6**, 679.
33. P. Gill, C. H. Brenner, J. S. Buckleton, A. Carracedo, M. Krawczak, W. R. Mayr, N. Morling, M. Prinz, P. M. Schneider and B. S. Weir, *Forensic Sci. Int.*, 2006, **160**, 90.
34. P. Gill, C. Brenner, B. Brinkmann, B. Budowle, A. Carracedo, M. A. Jobling, P. de Knijff, M. Kayser, M. Krawczak, W. R. Mayr and N. Morling, *Leg. Med.*, 2001, **3**, 252.
35. J. M. Butler, *Advanced Topics in Forensic DNA Typing: Interpretation*, Academic Press, 2014, ch. 2, pp. 25–46.
36. M. A. Jobling and P. Gill, *Nat. Rev. Genet.*, 2004, **5**, 739.

37. R. A. van Oorschot, K. N. Ballantyne and R. J. Mitchell, *Invest. Genet.*, 2010, **1**, 1.
38. J. J. Raymond, R. A. van Oorschot, P. R. Gunn, S. J. Walsh and C. Roux, *Forensic Sci. Int.: Genet.*, 2009, **4**, 26.
39. J. W. Bond and C. Hammond, *J. Forensic Sci.*, 2008, **53**, 797.
40. O. Ribaux, A. Baylon, C. Roux, O. Delémont, E. Lock, C. Zingg and P. Margot, *Forensic Sci. Int.*, 2010, **195**, 10.
41. O. Ribaux, A. Baylon, E. Lock, O. Delémont, C. Roux, C. Zingg and P. Margot, *Forensic Sci. Int.*, 2010, **199**, 63.
42. A. A. Mapes, A. D. Kloosterman, C. J. de Poot and V. van Marion, *Forensic Sci. Int.*, 2016, **264**, 28.
43. Sigma Aldridge, *Ethidium Bromide Product Information Sheet*, available at: http://www.sigmaaldrich.com/content/dam/sigma-aldrich/docs/Sigma/Product_Information_Sheet/e7637pis.pdf, accessed on 12.01.2017.
44. M. J. Waring, Complex formation between ethidium bromide and nucleic acids, *J. Mol. Biol.*, 1965, **13**, 269.
45. J. B. LePecq and C. Paoletti, *J. Mol. Biol.*, 1967, **27**, 87.
46. J. Olmsted III and D. R. Kearns, *Biochemistry*, 1977, **16**, 3647.
47. Invitrogen, *SYBR™ Safe DNA Gel Stain User Guide*, available at: https://tools.thermofisher.com/content/sfs/manuals/sybr_safe_dna_gel_stain_man.pdf, accessed on 12.01.2017.
48. N. Dawnay, B. Stafford-Allen, D. Moore, C. Blackman, P. Rendell, E. K. Hanson, J. Ballantyne, B. Kallifatidis, J. Mendel, D. K. Mills and R. Nagy, *Forensic Sci. Int.: Genet.*, 2014, **11**, 73.
49. N. Dawnay, R. Ahmed and S. Naif, *Sci. Justice*, 2014, **54**, 481.
50. N. D. Tribble, J. A. Miller, N. Dawnay and N. J. Duxbury, *J. Forensic Sci*, 2015, **60**, 690.
51. A. H. Marks, P. K. Bhadra, D. G. McDowell, D. J. French, K. T. Douglas, E. V. Bichenkova and R. A. Bryce, *J. Biomol. Struct. Dyn.*, 2005, **23**, 49.
52. D. J. French, C. L. Archard, T. Brown and D. G. McDowell, *Mol. Cell. Probes*, 2001, **15**, 363.
53. S. K. Poddar, *Mol. Cell. Probes*, 2000, **14**, 25.
54. J. A. Sanchez, K. E. Pierce, J. E. Rice and J. L. Wangh, *Proc. Natl. Acad. Sci. U. S. A.*, 2004, **101**, 1933.
55. M. Aufiero, Predicting the Quality of DNA Profiles through the Evaluation of the ParaDNA® Screening Instrument, MSc Thesis, Marshal University, 2013.
56. ParaDNA, *Screening System UK Police Pilot Study Application Note*, available at: https://www.lgcgroup.com/LGCGroup/media/PDFs/Products/paradna/App%20notes/ParaDNA-Screening-System-UK-Police-Pilot-Study_an.pdf, accessed on 12.01.2017.
57. C. A. Rakay, J. Bregu and C. M. Grgicak, *Forensic Sci. Int.: Genet.*, 2012, **6**, 723.
58. V. C. Tucker, A. J. Hopwood, C. J. Sprecher, R. S. McLaren, D. R. Rabbach, M. G. Ensenberger, J. M. Thompson and D. R. Storts, *Forensic Sci. Int.: Genet.*, 2011, **5**, 436.
59. R. L. Green, R. E. Lagacé, N. J. Oldroyd, L. K. Hennessy and J. J. Mulero, *Forensic Sci. Int.: Genet.*, 2013, **7**, 41.
60. J. Sambrook and D. W. Russell, *Molecular Cloning: A Laboratory Manual*, Cold Spring Harbor Laboratory Press, Cold Spring Habor, NY, USA, 3rd edn, 2001.
61. G. Ceriotti, *J. Biol. Chem.*, 1955, **214**, 59.
62. S. R. Gallagher and P. R. Desjardins, *Curr. Protoc. Hum. Genet.*, 2007, A-3D.
63. K. A. Haque, R. M. Pfeiffer, M. B. Beerman, J. P. Struewing, S. J. Chanock and A. W. Bergen, *BMC Biotechnol.*, 2003, **3**, 1.

64. K. Nielsen, H. S. Mogensen, J. Hedman, H. Niederstätter, W. Parson and N. Morling, *Forensic Sci. Int.: Genet.*, 2008, **2**, 226.
65. Thermo Scientific, *NanoDrop200 Specifications*, available at: http://www.nanodrop.com/Productnd2000specs.aspx, accessed on 12.01.2017.
66. H. D. VanGuilder, K. E. Vrana and W. M. Freeman, *BioTechniques*, 2008, **44**, 619.
67. J. A. Nicklas and E. Buel, *Anal. Bioanal. Chem.*, 2003, **376**, 1160.
68. J. A. Nicklas and E. Buel, *J. Forensic Sci*, 2005, **50**, 1081.
69. H. Zipper, H. Brunner, J. Bernhagen and F. Vitzthum, *Nucleic Acids Res.*, 2004, **32**, 103.
70. A. I. Dragan, R. Pavlovic, J. B. McGivney, J. R. Casas-Finet, E. S. Bishop, R. J. Strouse, M. A. Schenerman and C. D. Geddes, *J. Fluoresc.*, 2012, **22**, 1189.
71. C. T. Wittwer, M. G. Herrmann, A. A. Moss and R. P. Rasmussen, *BioTechniques*, 1997, **22**, 130.
72. S. Giglio, P. T. Monis and C. P. Saint, *Nucleic Acids Res.*, 2003, **31**, 136.
73. P. M. Holland, R. D. Abramson, R. Watson and D. H. Gelfand, *Proc. Natl. Acad. Sci. U. S. A.*, 1991, **88**, 7276.
74. R. L. Green, I. C. Roinestad, C. Boland and L. K. Hennessy, *J. Forensic Sci.*, 2005, **50**, 1.
75. C. B. Sherrill, D. J. Marshall, M. J. Moser, C. A. Larsen, L. Daudé-Snow and J. R. Prudent, *J. Am. Chem. Soc.*, 2004, **126**, 4550.
76. B. E. Krenke, N. Nassif, C. J. Sprecher, C. Knox, M. Schwandt and D. R. Storts, *Forensic Sci. Int.: Genet.*, 2008, **3**, 14.
77. Promega, *Plexor® qPCR and RT-qPCR Systems Applications Note*, available at: http://www.promega.co.uk/resources/pubhub/applications-notes/plexor-qpcr-and-rtqpcr-systems-applications/, accessed on 13.01.2017.
78. N. M. Silva, L. Pereira, E. S. Poloni and M. Currat, *PLoS One*, 2012, **7**, e49666.
79. N. Morling, R. W. Allen, A. Carracedo, H. Geada, F. Guidet, C. Hallenberg, W. Martin, W. R. Mayr, B. Olaisen, V. L. Pascali and P. M. Schneider, *Forensic Sci. Int.*, 2002, **129**, 148.
80. W. G. Giusti and T. Adriano, *Genome Res.*, 1993, **2**, 223.
81. P. Gill, C. P. Kimpton, A. Urquhart, N. Oldroyd, E. S. Millican, S. K. Watson and T. J. Downes, *Electrophoresis*, 1995, **16**, 1543.
82. Promega, *PowerPlex® 5-Dye Matrix Standards, 3100/3130*, available at: https://www.promega.com/-/media/files/resources/protocols/technical-bulletins/101/powerplex-5-dye-matrix-standards-3100-3130-protocol.pdf, accessed on 13.01.2017.
83. Promega, *PowerPlex® ESI 17 Pro System Technical Manual*, available at: https://www.promega.com/-/media/files/resources/protocols/technical-manuals/101/powerplex-esi-17-pro-system-protocol.pdf, accessed on 13.01.2017.
84. Applied Biosystems, *AmpFlSTR® NGM SElect™ PCR Amplification Kit User Guide*, available at: https://www3.appliedbiosystems.com/cms/groups/applied_markets_marketing/documents/generaldocuments/cms_089008.pdf, accessed on 13.01.2017.
85. Applied Biosystems, *3130/3130xl Genetic Analyzers User Guide*, available at: https://www3.appliedbiosystems.com/cms/groups/mcb_support/documents/generaldocuments/cms_041468.pdf, accessed on 13.01.2017.
86. J. M. Butler, E. Buel, F. Crivellente and B. R. McCord, *Electrophoresis*, 2004, **25**, 1397.
87. Y. F. Pariat, J. Berka, D. N. Heiger, T. Schmitt, M. Vilenchik, A. S. Cohen, F. Foret and B. L. Karger, *J. Chromatogr. A*, 1993, **652**, 57.
88. C. Heller, *Electrophoresis*, 2001, **22**, 629.

89. R. A. Griffiths, M. D. Barber, P. E. Johnson, S. M. Gillbard, M. D. Haywood, C. D. Smith, J. Arnold, T. Burke, A. J. Urquhart and P. Gill, *Int. J. Leg. Med.*, 1998, **111**, 267.
90. J. M. Butler, R. Schoske, P. M. Vallone, M. C. Kline, A. J. Redd and M. F. Hammer, *Forensic Sci. Int.*, 2002, **129**, 10.
91. J. J. Mulero, C. W. Chang, L. M. Calandro, R. L. Green, Y. Li, C. L. Johnson and L. K. Hennessy, *J. Forensic Sci.*, 2006, **51**, 64.
92. L. K. Hennessy, N. Mehendale, K. Chear, S. Jovanovich, S. Williams, C. Park and S. Gangano, *Forensic Sci. Int.: Genet.*, 2014, **13**, 247.
93. J. A. Lounsbury, A. Karlsson, D. C. Miranian, S. M. Cronk, D. A. Nelson, J. Li, D. M. Haverstick, P. Kinnon, D. J. Saul and J. P. Landers, *Lab Chip*, 2013, **13**, 1384.
94. E. L. Romsos, S. Lembirick and P. M. Vallone, *Forensic Sci. Int.: Genet. Suppl. Ser.*, 2015, **5**, e1–e2.
95. G. Ball, N. Dawnay, B. Stafford-Allen, M. Panasiuk, P. Rendell, S. Blackman, N. Duxbury and S. Wells, *Forensic Sci. Int.: Genet.*, 2015, **16**, 48.
96. S. Blackman, N. Dawnay, G. Ball, B. Stafford-Allen, N. Tribble, P. Rendell, K. Neary, E. K. Hanson, J. Ballantyne, B. Kallifatidis, J. Mendel, D. K. Mills and S. Wells, *Forensic Sci. Int.: Genet.*, 2015, **17**, 137.
97. B. Pakendorf and M. Stoneking, *Annu. Rev. Genomics Hum. Genet.*, 2005, **6**, 165.
98. M. R. Wilson, M. Stoneking, M. M. Holland, J. A. DiZinno and B. Budowle, *Crime Lab. Dig.*, 1993, **20**, 68.
99. B. Budowle, M. W. Allard, M. R. Wilson and R. Chakraborty, *Annu. Rev. Genomics Hum. Genet.*, 2003, **4**, 119.
100. M. M. Holland, D. L. Fisher, L. G. Mitchell, W. C. Rodriquez, J. J. Canik, C. R. Merril and V. W. Weedn, *J. Forensic Sci.*, 1993, **38**, 542.
101. T. E. King, G. G. Fortes, P. Balaresque, M. G. Thomas, D. Balding, P. M. Delser, R. Neumann, W. Parson, M. Knapp, S. Walsh and L. Tonasso, *Nat. Commun.*, 2014, **5**, 5631.
102. P. Gill, P. L. Ivanov, C. Kimpton, R. Piercy, N. Benson, G. Tully, I. Evett, E. Hagelberg and K. Sullivan, *Nat. Genet.*, 1994, **6**, 130.
103. B. Budowle, M. R. Wilson, J. A. DiZinno, C. Stauffer, M. A. Fasano, M. M. Holland and K. L. Monson, *Forensic Sci. Int.*, 1999, **103**, 23.
104. W. Bär, B. Brinkmann, B. Budowle, A. Carracedo, P. Gill, M. Holland, P. J. Lincoln, W. Mayr, N. Morling, B. Olaisen and P. M. Schneider, *Int. J. Leg. Med.*, 2000, **113**, 193.
105. M. Van Oven and M. Kayser, *Hum. Mutat.*, 2009, **30**, e386.
106. M. A. Cronin, D. A. Palmisciano, E. R. Vyse and D. G. Cameron, *Wildl. Soc. Bull.*, 1991, **19**, 94.
107. R. Ogden, N. Dawnay and R. McEwing, *Endangered Species Res.*, 2009, **9**, 179.
108. E. A. Alacs, A. Georges, N. N. FitzSimmons and J. Robertson, *Forensic Sci., Med., Pathol.*, 2010, **6**, 180.
109. R. N. Johnson, L. Wilson-Wilde and A. Linacre, *Forensic Sci. Int.: Genet.*, 2014, **10**, 1.
110. S. Ratnasingham and P. D. Hebert, *Mol. Ecol. Notes*, 2007, **7**, 355.
111. E. W. Sayers, T. Barrett, D. A. Benson, E. Bolton, S. H. Bryant, K. Canese, V. Chetvernin, D. M. Church, M. DiCuccio, S. Federhen and M. Feolo, *Nucleic Acids Res.*, 2011, **39**, D38.
112. R. Pawlowski, T. Kupiec and W. Branicki, *J. Forensic Sci.*, 2003, **48**, 1.
113. N. Dawnay, R. Ogden, R. McEwing, G. R. Carvalho and R. S. Thorpe, *Forensic Sci. Int.*, 2007, **173**, 1.

114. M. R. Wilson, J. A. DiZinno, D. Polanskey, J. Replogle and B. Budowle, *Int. J. Leg. Med.*, 1995, **108**, 68.
115. M. M. Holland and T. J. Parsons, *Forensic Sci. Rev.*, 1999, **11**, 21.
116. S. Tabor and C. C. Richardson, *Proc. Natl. Acad. Sci. U. S. A.*, 1987, **84**, 4767.
117. J. M. Prober, G. L. Trainor, R. J. Dam, F. W. Hobbs, C. W. Robertson, R. J. Zagursky, A. J. Cocuzza, M. A. Jensen and K. Baumeister, *Science*, 1987, **238**, 336.
118. S. Ye, S. Dhillon, X. Ke, A. R. Collins and I. N. M. Day, *Nucleic Acids Res.*, 2001, **29**, e88.
119. J. H. Wetton, C. S. Tsang, C. A. Roney and A. C. Spriggs, *Forensic Sci. Int.*, 2002, **126**, 137.
120. J. Juusola and J. Ballantyne, *Forensic Sci. Int.*, 2003, **135**, 85.
121. J. Juusola and J. Ballantyne, *Forensic Sci. Int.*, 2005, **152**, 1.
122. E. K. Hanson, H. Lubenow and J. Ballantyne, *Anal. Biochem.*, 2009, **387**, 303.
123. M. A. Cortez, C. Bueso-Ramos, J. Ferdin, G. Lopez-Berestein, A. K. Sood and G. A. Calin, *Nat. Rev. Clin. Oncol.*, 2011, **8**, 467.
124. A. Lindenbergh, P. Maaskant and T. Sijen, *Forensic Sci. Int.: Genet.*, 2013, **7**, 159.
125. M. L. Richard, K. A. Harpe, R. L. Craig, A. J. Onorato, J. M. Robertson and J. Donfack, *Forensic Sci. Int.: Genet.*, 2012, **6**, 452.
126. J. Juusola and J. Ballantyne, *J. Forensic Sci.*, 2007, **52**, 1252.
127. C. Nussbaumer, E. Gharehbaghi-Schnell and I. Korschineck, *Forensic Sci. Int.*, 2006, **157**, 181.
128. E. K. Hanson and J. Ballantyne, *F1000Research*, 2013, **2**, 281.
129. ParaDNA, *Body Fluid ID Test*, available at: https://www.lgcgroup.com/products/paradna-technology/product-overview-(1)/#.WH5UPk1vhQs, accessed on 17.01.2017.
130. A. Breddy, The Effect of mRNA Amplicon Length on Direct RT-PCR Efficiency of Inhibited and Degraded Samples, MSc Thesis, King's College London, 2015.
131. T. R. De Wolff, A. J. Kal, C. E. Berger and B. Kokshoorn, *Law Probab. Risk*, 2015, **14**, 323.
132. J. De Zoete, J. Curran and M. Sjerps, *Forensic Sci. Int.: Genet.*, 2016, **20**, 30.
133. C. Haas, E. Hanson, W. Bär, R. Banemann, A. M. Bento, A. Berti, E. Borges, C. Bouakaze, A. Carracedo, M. Carvalho and A. Choma, *Forensic Sci. Int.: Genet.*, 2011, **5**, 21.
134. M. Kayser and P. M. Schneider, *Forensic Sci. Int.: Genet.*, 2009, **3**, 154.
135. M. D. Shriver, M. W. Smith, L. Jin, A. Marcini, J. M. Akey, R. Deka and R. E. Ferrell, *Am. J. Hum. Genet.*, 1997, **60**, 957.
136. M. D. Adams, J. M. Kelley, J. D. Gocayne, M. Dubnick, M. H. Polymeropoulos, H. Xiao, C. R. Merril, A. Wu, B. Olde, R. F. Moreno and A. R. Kerlavage, *Science*, 1991, **252**, 1651.
137. E. C. Berglund, A. Kiialainen and A. C. Syvänen, *Invest. Genet.*, 2011, **2**, 1.
138. C. Børsting and N. Morling, *Forensic Sci. Int.: Genet.*, 2015, **18**, 78.
139. S. Walsh, F. Liu, K. N. Ballantyne, M. van Oven, O. Lao and M. Kayser, *Forensic Sci. Int.: Genet.*, 2011, **5**, 170.
140. W. Branicki, F. Liu, K. van Duijn, J. Draus-Barini, E. Pośpiech, S. Walsh, T. Kupiec, A. Wojas-Pelc and M. Kayser, *Hum. Genet.*, 2011, **129**, 443.
141. O. Spichenok, Z. M. Budimlija, A. A. Mitchell, A. Jenny, L. Kovacevic, D. Marjanovic, T. Caragine, M. Prinz and E. Wurmbach, *Forensic Sci. Int.: Genet.*, 2011, **5**, 472.
142. C. Phillips, A. Salas, J. J. Sanchez, M. Fondevila, A. Gomez-Tato, J. Alvarez-Dios, M. Calaza, M. C. de Cal, D. Ballard, M. V. Lareu and A. Carracedo, *Forensic Sci. Int.: Genet.*, 2007, **1**, 273.

143. S. Caratti, S. Turrina, M. Ferrian, E. Cosentino and D. De Leo, *Forensic Sci. Int.: Genet. Suppl. Ser.*, 2015, **5**, e98.
144. J. Ju, D. H. Kim, L. Bi, Q. Meng, X. Bai, Z. Li, X. Li, M. S. Marma, S. Shi, J. Wu and J. R. Edwards, *Proc. Natl. Acad. Sci. U. S. A.*, 2006, **103**, 19635.
145. J. D. Churchill, S. E. Schmedes, J. L. King and B. Budowle, *Forensic Sci. Int.: Genet.*, 2016, **20**, 20.
146. S. Caratti, S. Turrina, M. Ferrian, E. Cosentino and D. De Leo, *Forensic Sci. Int.: Genet. Suppl. Ser.*, 2015, **5**, e98.
147. N. Almalki, H. Y. Chow, V. Sharma, K. Hart, D. Siegel and E. Wurmbach, *Electrophoresis*, 2017, 846–854.
148. C. Hussing, C. Børsting, H. S. Mogensen and N. Morling, *Forensic Sci. Int.: Genet. Suppl. Ser.*, 2015, **5**, e449.
149. R. P. Sinha and D. P. Häder, *Photochem. Photobiol. Sci.*, 2002, **1**, 225.
150. R. P. Rastogi, A. Kumar, M. B. Tyagi and R. P. Sinha, *J. Nucleic Acids*, 2010, **16**, 1–32.
151. DNA Advisory Board, *Forensic Sci. Commun.*, 2000, **2**, 1–14.
152. T. E. Mifflin, *Cold Spring Harb. Protoc.*, 2007, **7**, 5.
153. K. Shaw, I. Sesardić, N. Bristol, C. Ames, K. Dagnall, C. Ellis, F. Whittaker and B. Daniel, *Int. J. Leg. Med.*, 2008, **122**, 29.
154. J. Tamariz, K. Voynarovska, M. Prinz and T. Caragine, *J. Forensic Sci.*, 2006, **51**, 790.
155. L. A. Gefrides, M. C. Powell, M. A. Donley and R. Kahn, *Forensic Sci. Int.: Genet.*, 2010, **4**, 89.
156. A. Hall and J. Ballantyne, *Anal. Bioanal. Chem.*, 2004, **380**, 72.

Chapter 4

Use of Light in Fingerprint Detection

Laurent Galmiche*, Clémence Allain and Gilles Clavier

PPSM, ENS Cachan, CNRS, Université Paris-Saclay, 94235 Cachan, France
*E-mail: laurent.galmiche@ens-paris-saclay.fr

Table of Contents

Light in Forensic Science: Issues and Applications
Edited by Giorgia Miolo, Jacqueline L. Stair and Mire Zloh

Published by the Royal Society of Chemistry, www.rsc.org

4.1. Introduction: What Is a Fingerprint?

The skin on the hand's palm and the fingers forms a complex pattern of ridges and furrows that is characteristic of each individual (Figure 4.1). This pattern is formed at an early stage of gestation, remains constant throughout the life of a person and regenerates after superficial damages to the skin. Only deep scars may affect it permanently.[1]

Underneath the skin are glands that secrete sweat or sebum depending on their nature. These liquids will be deposited and leave a trace when the skin is in contact with a surface. This is called the exchange principle in forensic science. These traces may be patent if visible with naked eyes or latent if not. Uncovering fingermarks deposited on crime scenes and evidence is one of the most important focuses in forensic science since it allows direct identification of a person if sufficiently resolved and/or DNA recovery for subsequent analysis. Since direct visualisation is rarely possible a wide range of physical and chemical techniques have been developed to highlight and fix latent traces left on crime scenes and evidence. Most physical methods rely on the use of light (ultraviolet, visible or infrared) either directly on traces or after chemical treatments. The chemical treatments used for the development of

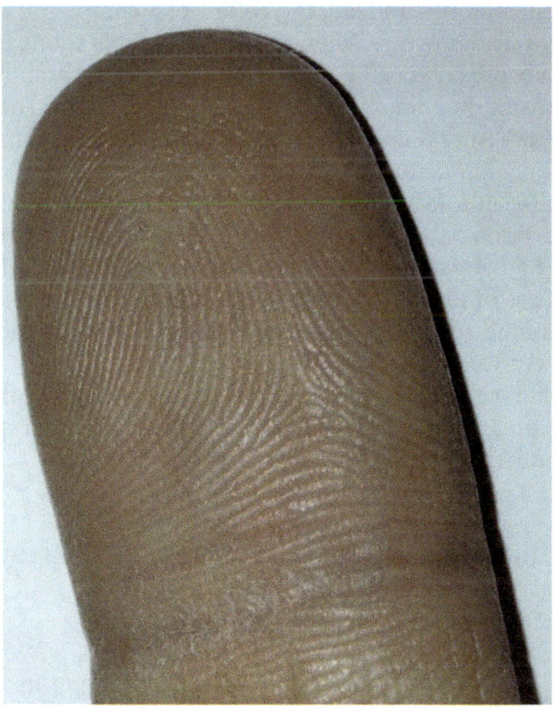

Figure 4.1. Image of a fingertip with the ridges and furrows pattern forming the fingerprint.

latent prints will mainly depend on the nature of the material being analysed. They are typically classified in three categories:[2]

- Porous materials which comprises paper, cardboard, wood, *etc*. These are absorbent and may retain traces for a very long time.
- Non-porous materials include glass, metal, plastics, rubber, *etc*. These are not absorbent and the traces remain on their surface which makes them more fragile.
- Semi-porous materials are in between the previous two since they are weakly absorbent. Such materials are glossy paper or cardboards, some treated wood, and may be treated with techniques adapted for porous or non-porous materials.

Porous materials are treated with chemicals that react with the amino acids which are one of the components of sweat to form coloured and/or fluorescent products. Non-porous materials are typically analysed with powders or cyanoacrylate fuming. When these treatments are not sufficient to visualise a print, a post-treatment may be applied.

Research into new techniques to reveal latent fingermarks is an ongoing process given their relevance in forensic science. One of the most useful approaches is to develop treatments that will make the prints fluorescent rather than just coloured. Indeed, the colour of a compound is the result of a "negative" physical process. When the visible light is shone on a particular compound it absorbs a specific portion of the spectrum and the observed colour comes from the remaining light. Thus it can only be observed under normal illumination and subject to background and surrounding interference. On the other hand, fluorescence is a "positive" process. When a fluorescent dye is excited at a particular wavelength (ultraviolet or visible) located in one of its absorption bands it emits its own light at a different wavelength of lower energy (higher wavelength). Thus fluorescence is best observed in the dark with the use of proper filters. This feature makes fluorescence a more sensitive technique since it is easier to detect than absorption because of reduced interference.

In this chapter we will present the main techniques used for latent fingermark developement as well as recent academic results with a special emphasis on treatments that give a fluorescent output.

4.2. Examination of Fingermarks Without Pretreatment

4.2.1. *Visual Examination*

The principle of visual examination is to utilise lighting in such a way as to provide the best possible contrast between the fingermark ridges and the substrate; this was the first technique proposed for the detection of fingerprints. Once it has been detected under white light, there are various techniques to illuminate the mark in a way suitable for photography, which include oblique illumination, specular illumination, dark field illumination, co-axial illumination and the use of polarised light. These techniques, as well as the procedures recommended by the British Home Office Center for Applied Science and Technology (CAST) for their operational use, are summarised in the Fingerprint Source Book – Chapter 2[3] to which the reader should refer for further details.

4.2.2. *Ultraviolet Imaging*

Latent fingermarks absorb UV and are fluorescent under UV excitation, because of their composition and potentially because of the presence of foreign contaminants present on the fingers and transferred during deposition. The UV absorption and fluorescence properties of fingerprint's constituents were first observed by Okhi in 1970.[4] The origin of these properties is not entirely identified, and depends on the excitation wavelength used. Fatty acids as well as the amino acids tyrosine and tryptophan are believed to contribute most to the absorption, especially under short-wave UV excitation (254 nm or 266 nm). When the substrate either fluoresces or reflects UV the contrast between the fingerprint ridges and the background may allow imaging.[5]

Lambrechts *et al.* studied the fluorescence of fingermarks constituents under long-wave UV excitation (365 nm) and hypothesised that tryptophan in peptides or proteins was a major contributor to the fingermark's autofluorescence.[6] The fingermark's autofluorescence was explored by Bramble *et al.*[7] who used the fourth harmonic of a Nd:YAG laser (266 nm) to illuminate fingermarks deposited on white paper. It is worth noting that in many cases the substrate is also strongly blue fluorescent under UV excitation because it contains optical brighteners: Saitoh *et al.* have shown that the combined used of a band-pass filter and of time-resolved fluorescence strongly improves the quality of the detection.[8]

Care should be taken when using UV illumination: a loss of fluorescence due to photobleaching of the fingermarks is observed upon prolonged UV illumination. In addition, exposure to UV radiation can cause damage to skin and eyes, as a consequence the operators should wear appropriate personal protective equipment when using this technique. Furthermore, exposure to UV radiation, especially short-wave UV, will cause progressive damage to the DNA potentially present on the exhibit.

4.2.3. *Infrared Imaging*

Fingerprints can also be imaged using infra-red light: Fourier transformed infrared chemical imaging has been used to directly image latent fingermarks,[9] as well as fingerprints fumed with cyanoacrylate.[10] In the case of latent fingermarks, infrared absorption is due to the presence of fatty acids esters, carboxylic acids salts, amino acids and proteins. This technique not only provides an image of the fingermark, but also gives information regarding the distribution, and total concentration of the compounds present across the fingerprint.[11] FTIR imaging has also been used to determine whether fingerprints are contaminated with explosive particulates.[12] It is worth noting that FTIR imaging has been performed on fingerprints that were collected from the crime scene using gel lifting media.[13]

4.3. Development of Latent Fingermarks on Porous Substrates

4.3.1. *Ninhydrin and Post-treatment*

Ninhydrin or 1*H*-indene-1,2,3-trione (Scheme 4.1) is an amino acid sensitive reagent that was serendipitously prepared by Ruhemann in 1910.[14] The tricarbonyl form can be isolated only in anhydrous conditions, otherwise it is stable in its hydrate form 2,2-dihydroxy-1*H*-indene-1,3(2*H*)-dione.

Ruhemann also discovered that ninhydrin reacts with amino acid to form a coloured compound that was nicknamed Ruhemann's Purple and rightfully established its structure in 1911.[15] The reaction mechanism is multistep but has not been entirely elucidated. It involves 2 ninhydrins and carbon dioxide, water and an aldehyde are the other reaction's products (Scheme 4.2).

This discovery rapidly led to the extensive use of ninhydrin in analytical chemistry to quantify amino acids in samples. It was only in 1954 that Odén and von Hofsten reported its use for the detection of prints on paper.[16]

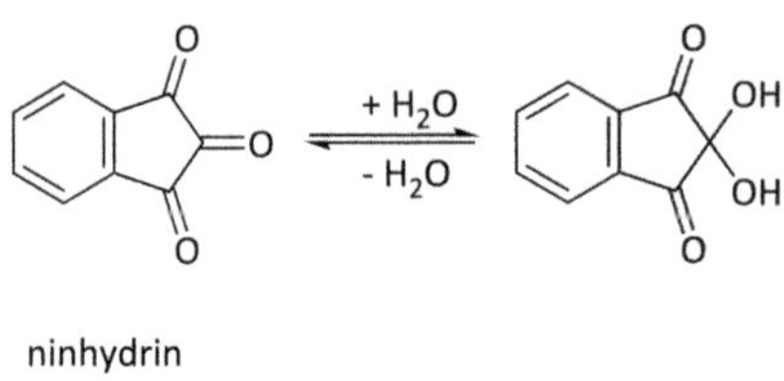

Scheme 4.1. Equilibrium between ninhydrin and its hydrate form.

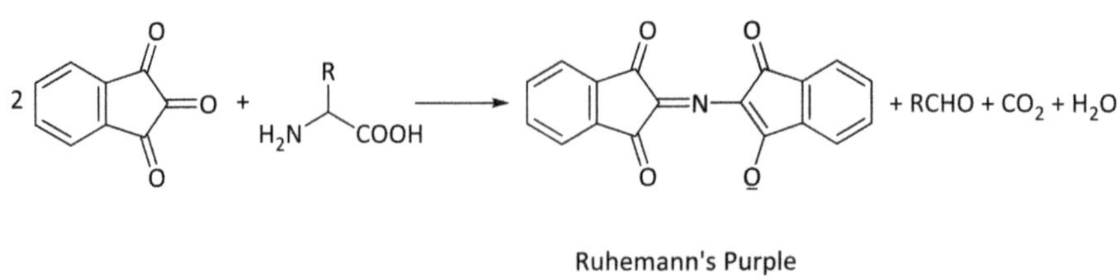

Scheme 4.2. Reaction between ninhydrin and an amino acid to form Ruhemann's Purple.

This discovery led to a rapid development of ninhydrin as a central tool to reveal latent fingerprints on porous materials. It is very efficient because the amino acids from the fingerprint have a high affinity for cellulose which is the main constituent of paper and porous materials in general. The amino acids thus do not migrate and even though their total amount decreases with time it may possible to reveal prints long after they were deposited (up to 40 years). The purple colour of the product formed can be viewed by naked eye if the support is not coloured and if the print contains enough amino acids (Figure 4.2 left).

Many formulations for treatment of evidence with ninhydrin have been developed in the past and are still in use today. They rely on the preparation of a ninhydrin solution in various types of organic solvents such as acetone, ethanol, methanol, isopropanol or petroleum ether or a mixture of them.[17] However, these formulations suffer from many drawbacks since most of these solvents are toxic and/or flammable and give low development qualities because of background staining. Furthermore, inks present on paper are often soluble in the organic solvents causing their dispersion in the material which obscures the fingerprints. Usual organic solvents were then replaced by 1,1,2,-trifluorotrichloroethane (Freon113 or CFC113) that is volatile but non-flammable, non-toxic and do not dissolve ink. But in 1992 this solvent was banned by most countries because of its impact on the ozone layer. It was then replaced by 1,1,1,2,2,3,3,4,4-nonafluoro-4-methoxybutane (commercially known as HFE-7100) which is a hydrofluoroether that is not harmful to

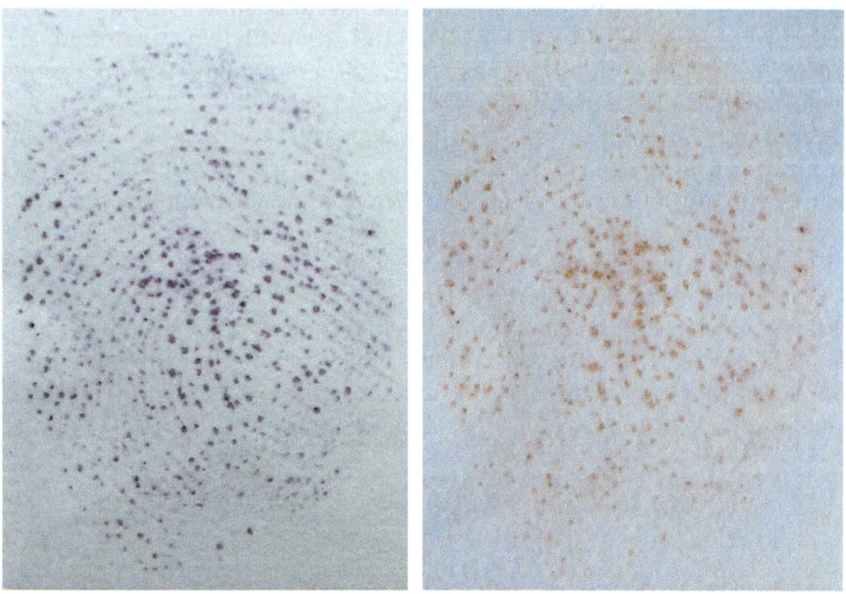

Figure 4.2. Left: white paper treated with ninhydrin; Right: the same sample post-treated with zinc chloride.

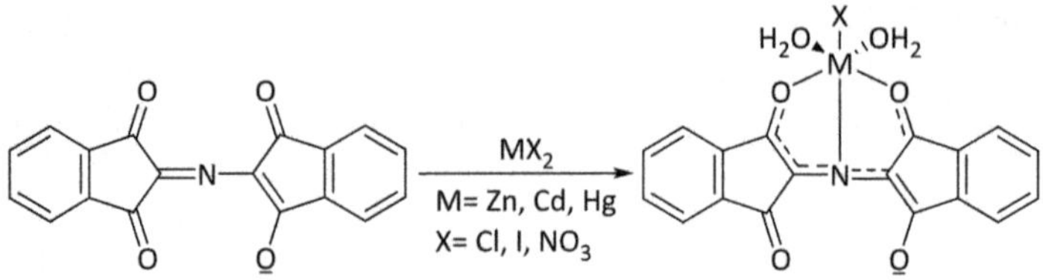

Scheme 4.3. Reaction between Ruhemann's Purple and metal salts to form a coordination complex.

the ozone layer but otherwise has similar properties to Freon113. Formulations based on HFE-7100 are the one recommended by law enforcement agencies such as the Home Office.[18] Practically, the solutions may be applied on the evidence by spraying, dipping, or painting. However, after application the fingermarks are not directly visible since the reaction is not instantaneous. Further development is needed through the application of heat and humidity after solvent evaporation. Typically, the evidence is placed into a humidity-controlled oven at 80 °C and 65% relative humidity for five to seven minutes. Alternatively, the evidence can be developed at room temperature in a humid and confined environment but the process may take weeks in that case. The marks can then be observed directly (Figure 4.2 left). In any case, after ninhydrin treatment the evidence should be kept in the dark as much as possible since Ruhemann's Purple is photosensitive and fades upon light exposure.

In some cases it may be difficult to visualise ninhydrin developed prints. Ruhemann's Purple is not fluorescent but it was long known that it can form a complex with various metal salts from the group IIb (zinc, cadmium, *etc*; Scheme 4.3). The resulting products have a different colour which depends on the metal used and may be fluorescent.[19]

Herod *et al.* showed in 1982 that it was possible to do a post-treatment on ninhydrin developed fingerprints with a zinc(II) chloride solution that resulted in the formation of an orange/red coloured complex (the main absorption band of the complex is centred at 490 nm; Figure 4.2 right) that is fluorescent in green upon irradiation with an argon laser.[20] Fluorescence can be enhanced by cooling the evidence in liquid nitrogen. However, the weak fluorescence of the complex makes it of limited interest from a practical point of view. Other metal salts such as cadmium or mercury also form complexes that are fluorescent but their toxicity has limited their operational use.

4.3.2. *Fluorescent Analogues of Ninhydrin*

4.3.2.1. 1,8-Diazafluoren-9-one (DFO). The first synthesis of 1,8-Diazafluoren-9-one (DFO) (Scheme 4.4) was described in 1950 by Druey and Schmidt[21] but its use as fingermark reagent appeared in 1990.[22,23] This ninhydrin analogue reacts with amino acids contained in the latent marks to

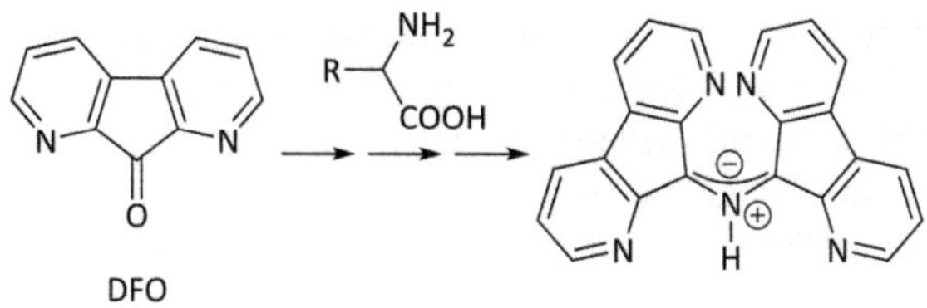

Scheme 4.4. 1,8-Diazafluoren-9-one (DFO) and fluorescent compound obtained after reaction with amino acids.

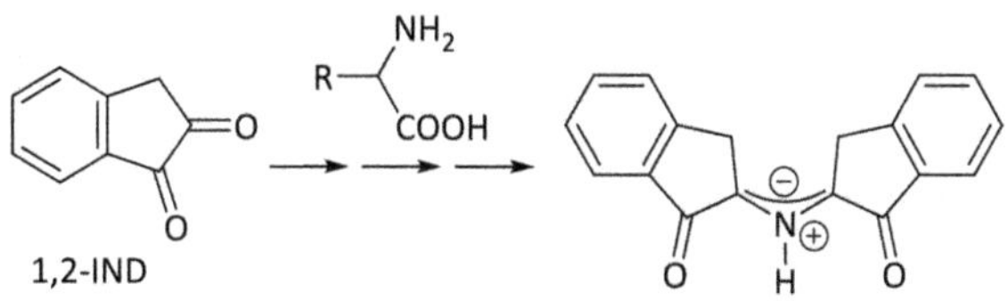

Scheme 4.5. 1,2-Indanedione (1,2-IND) and fluorescent compound obtained after reaction with amino acids.

give a red azomethine ylide derivative (Scheme 4.4) similar to the Ruhemann's purple structure which makes the print visible under normal light.

The product formed is also inherently fluorescent when excited with an intense green light (λ_{ex} = 470 nm, λ_{em} = 570 nm) without any toning with metal salts, contrary to the ninhydrin process.

DFO is dissolved in an hydrofluoroether (such as 1-Methoxynonafluorobutane) with small quantities of methanol and acetic acid, essential for the reaction to be able to take place.[24,25] A porous object with potential evidence (paper for instance) is carefully dipped in the solution and then allowed to dry before being heated in a dry oven at 100 °C for 20 minutes to reach an optimum of fluorescence.[26] In contrast to ninhydrin, metal salt post-treatment of DFO developped fingerprint do not improve its fluorescence.[27] DFO is compatible with subsequent DNA analysis.[28]

4.3.2.2. 1,2-Indanedione. Although 1,2-Indanedione (1,2-IND) (Scheme 4.5) was first described in 1912,[29] it had never been tested on latent prints before 1997.[30,31] Similar to DFO and ninhydrin, 1,2-Indanedione reacts with amino acids present in the latent marks to give a pale pink azomethine ylide derivative which is highly fluorescent under green light excitation.

Azoury *et al.*[32] have reported that 1,2-Indanedione is compatible with subsequent DNA analysis. As for DFO or ninhydrin, after dipping in an indanedione solution and air-drying, fingerprints are usually heated for a few minutes in an oven at 100 °C, even though they may develop at room temperature in 24–48 h.[33] Contrary to DFO, humidity seems important in the development process and leads to inconsistent results depending on local weather conditions.[34]

If this product is inherently fluorescent, studies[33,35] have shown that treatment of 1,2-indanedione developed fingerprints with metal salt such as zinc or cadmium chloride induces an increase of the luminescence intensity and a darker pink colour leading to a better contrast. Metal salt can be directly integrated in the DFO formulation instead of a post-treatment. In this case, fingerprints development seems less reliant on ambiant humidity.[36]

Wiesner *et al.*[37] has shown that better fingerprint development is obtained using a formulation without acetic acid and Wilkinson[25] has shown that methanol can form a stable hemiketal which inhibits the reaction with amino acids. However, disagreement surfaced over the use of methanol and acetic acid and most forensic laboratories continue to use them in their indanedione formulation. This lack of consensus and the mixed results can explain why the indanedione reagent is not recommended by the British Home Office Center for Applied Science and Technology (CAST).[34]

4.4. Development of Latent Fingermarks on Non-porous Substrates

4.4.1. Cyanoacrylate Fuming and Fluorescent Post-treatment

Cyanoacrylate esters also known as superglue are a family of liquid adhesives that have been developed in the 1950's for the aircraft industry that also found widespread applications in medicines and everyday life. The acrylate monomer reacts quickly in the presence of a nucleophile (typically water) to give a white polymer (Scheme 4.6).

In the late 1970's/early 1980's it was independently found in Japan, the United Kingdom and North America that exposure of latent marks to cyanoacrylate vapour led to their development on non-porous surfaces. The reaction mechanism was not known at the time (and is still the subject of intense research) and optimum conditions were not reached. The process was long (up to one day) and not always reproducible. Further investigations mainly by the

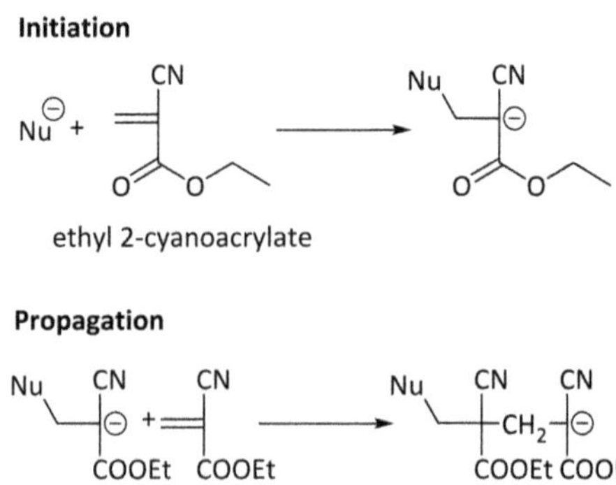

Scheme 4.6. Anionic polymerisation reaction of ethyl 2-cyanoacrylate.

Police Scientific Development Branch (PSDB) of the Home Office in England allowed it to be discovered that the process works best in a closed cabinet at a relative humidity (RH) level of approximately 80%. The process could be also accelerated by heating the cyanoacrylate monomer at 120 °C. That is why the process is now commonly known as cyanoacrylate fuming or CA fuming. This led to the development of the first commercial device in the mid 1980's by the Mason Vactron Company that was later purchased by the Foster and Freeman company.[18] Several brands that commercialise CA fuming chambers of various sizes now exist. Field devices have also been developed but the requirement for a controlled humidity for a good development limit their efficiency. However, they work in closed environments such as cars or small rooms.

Another process to deposit CA on fingerprints is to use a vacuum instead of the heat/humidity combination to generate vapours and the polymer. The vacuum deposition is far less used and gives deposits that are less visible to the naked eye.

The exact mechanism for the polymer formation is still a subject of debate. It was recently established that water is probably not the initiator but rather that the carboxylate function of lactate and alanine (two of the major components of eccrine sweat) are likely to be the initiators of the anionic polymerisation.[38] However, the process is now very reliable. Furthermore DNA can be recovered from CA fumed traces for further analysis.[39,40]

CA fuming works well on non-porous and semi-porous surfaces. Thus supports such as glass, plastics (bags, packaging, *etc*), metals (cans) ceramics or adhesives which are non-porous are well suited for the process. Semi-porous surfaces such as waxed surfaces, certain types of wall paints, papers with a glossy finish or varnished wood are also usable supports. The fingerprints appear as a white deposit that is formed of noodle-like polymer fibres, if the process was conducted in the recommended conditions of relative humidity (Figure 4.3). If the RH level is different the deposit appears more like a film by scanning electron microscopy. One of the main advantages of CA fuming is the speed of the process (approximately 20 to 30 minutes for a complete cycle). Another advantage is the high quality of the revelation. Since the reaction is initiated by the amino acids present in the deposit the polymer will only form on the ridges and not in the furrows. If the fingerprint is seen under a microscope one can also see the pores that are not developed.

The visualisation of the CA deposit is possible by visual examination through oblique lighting or UV imaging. Yet, it may be difficult to see faint marks or to differentiate them from the support (Figure 4.4 left); post-treatments are thus required to highlight them. The simplest one is to apply dark or fluorescent powders onto the prints (see Section 4.5.1). But such treatment may destroy the marks. A more popular post-treatment is dyeing with a fluorescent compound (Figure 4.4 right). The process is carried out by immersing the support in an alcoholic solution containing the dye. Alternatively spraying the dye solution is also possible. The solvent swells the polymer allowing the dye to enter the fibres. After solvent evaporation, it remains trapped and the fluorescence is visualised by external light excitation

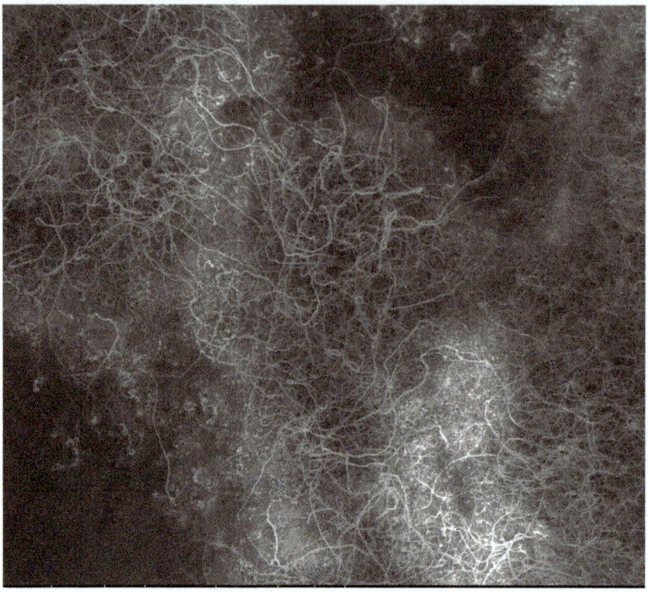

Figure 4.3. SEM picture of a latent fingerprint revealed with cyanoacrylate (magnitude 1200×).

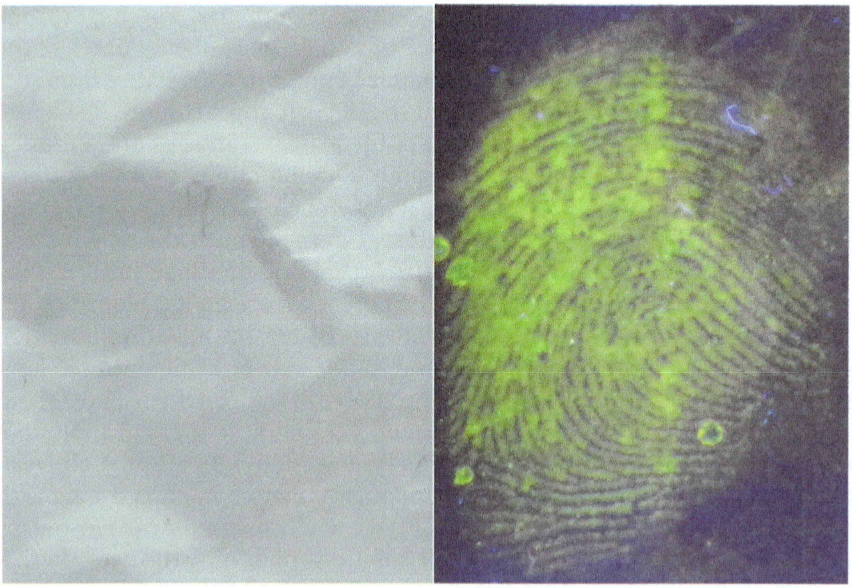

Figure 4.4. Left: white plastic bag fumed with CA under white light; Right: the same sample after post-treatment with Rhodamine 6G illuminated at 365 nm.

(Figure 4.4 right). Water may also be used in some formulations but the dye uptake is not so efficient.

Several organic dyes have been reported for the post-treatment of CA fumed fingerprints (Table 4.1). Most of them are cationic dyes. It is thought that this feature improves the dye–polymer interactions leading to better uptake and more intense luminescence. One coordination complex based on Europium is also available commercially (Thenoyl Europium complex). The choice of the dye is guided by several factors:

- contrast of the fluorescence with the background colour;
- commercial availability;
- recommendations from police forces;
- health and safety issues.

Table 4.1. Structures and properties of most common fluorescent dyes used for fingerprint staining.

Dye name	Chemical structure	Absorption band (nm)	Fluorescence emission maximum (nm)/colour
Ardrox P133D		435–480	476/Blue green
MBD	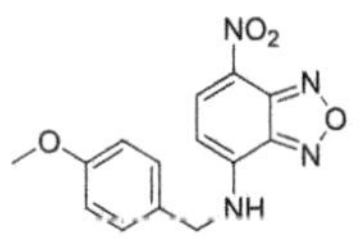	415–505	515/Green
Basic red 28	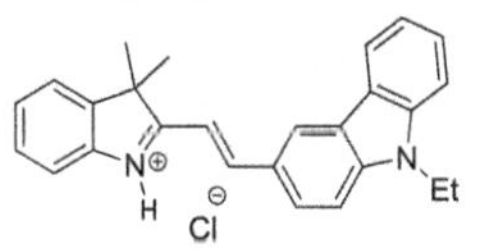	470–550	549/Yellow
Rhodamine 6G (basic red 1)	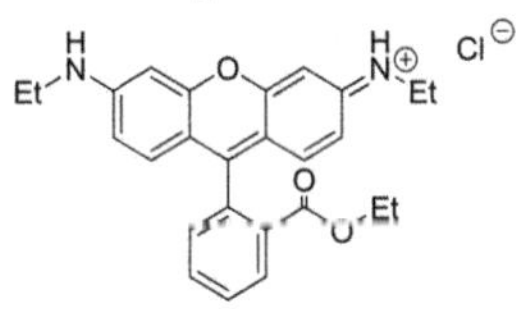	495–540	552/Yellow
Safranin O (basic red 2)	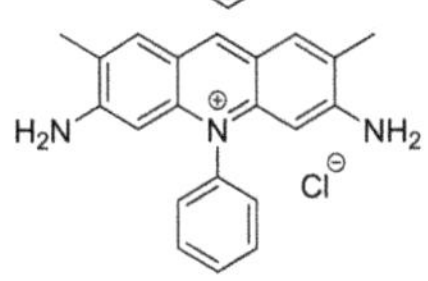	470–550 (max 512)	563/Orange-red
Nile red	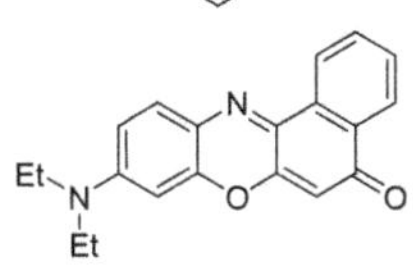	450–560	580/Red
Thenoyl Europium chelate (TEC)	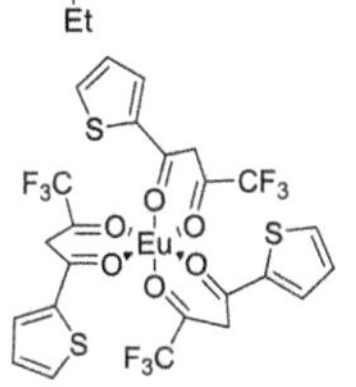	350–370	614/Red

The dyes are sold in various formulations. The main difference is the nature of the solvent that can be organic (alcohol) or water. The latter one is recommended for dyeing semi-porous surfaces since organic solvent may diffuse in such materials and lead to a colouration of the support itself and obscure the visualization of the prints. Use of water instead precludes this problem but the staining of the polymer is less efficient. Mixtures of dyes are also available that contain typically three different dyes absorbing at different wavelengths. When the dye with the absorption band in the blue (*i.e.* higher energy) is excited, it can then transfer its excited state energy to one that is excited at lower energy through a FRET process. This cascade allows the use of a blue excitation wavelength to generate an orange/red fluorescence. The large difference between the two colours makes it easier to visualize the light emitted.

4.4.2. One-step Fluorescent Cyanoacrylates

Post-treatments used to increase the contrast between a fingerprint and the substrate are efficient but they have some drawbacks. They are time consuming (between 24 and 48 h), may damage the print or the substrate during the dipping (or spraying) and the rinsing steps and they cannot be used on semi-porous surfaces due to the impregnation of the substrate with the fluorescent stain solution. Moreover, the handling of toxic chemicals enforces the use of a fume-cupboard, an expensive piece of protective equipment. Finally, these post-treatments may interfere with the subsequent DNA analysis.[41]

For all these reasons, in the early 1980's, Morton[42] came forward with the idea of adding an emissive molecule to the cyanoacrylate in order to directly obtain fluorescent prints. Even if a patent was filed, no commercial application went through the market. Over the past decades, many scientific studies have attempted to circumvent the inner limitations of the colouration step following the fumigation process. Two different strategies have been tried with widely differing results: chemical modification of the cyanoacrylate or co-evaporation of a cyanoacrylate and a fluorescent dye.

In 1995, BKA (Bundeskriminalamt, the federal criminal police office of Germany)[43] reported on the unsuccessful effort to chemically modify the cyanoacrylate monomer to make it fluorescent by a Diels–Alder reaction between anthracene and the ethyl cyanoacrylate monomer. Subjected to a thermal retro-Diels-Alder reaction, the Diels–Alder adduct was theoretically supposed to release the cyanoacrylate monomer and the fluorophore simultaneously. However, due to the high temperature of these retro-Diels-Alder reactions, adducts typically decomposed rather than producing fluorescent cyanoacrylate fumes leading to poor fingerprint development.

In 2013, Bentolila *et al.* (Israel Police)[44], have prepared modified versions of ethyl cyanoacrylate. In their study, they chemically modified the ester part of the cyanoacrylate monomers to replace it by a fluorescent moiety such as benzyl, naphthyl, anthracenyl or fluorenyl groups. The fumigation experiments were performed in a fuming chamber heated up to 150 °C. Fumes did not evolve when these new cyanoacrylates were used alone and fingermark development was not observed on the substrates. When the compounds were diluted with ethyl cyanoacrylate fingermark development was observed with white light but no fluorescence was observed under UV light, suggesting that only ethyl cyanoacrylate evaporated and developed the marks.

In 2014, Groenveld *et al.*[45] also tried to modify cyanoacrylates with fluorescent group such as methylthienyl, methylfuranyl or dansyl. The synthesis pathway was different but the result was the same, as the new cyanoacrylates did not succeed in visualizing fingermarks.

Due to a higher molecular weight than the usual ethyl cyanoacrylate, all these new cyanoacrylate esters are not volatile enough at atmospheric pressure to afford fingermark development in a regular fuming chamber and an increase of the temperature often leads to a thermal decomposition of the modified cyanoacrylate.

The first example of a one-step fuming process is attributed to Weaver and Clary[46] in 1993. They successfully co-fume a dye from the styryl family (Scheme 4.7, **I**) and a polymerised cyanoacrylate. CN-Yellow, the first commercial one-step cyanoacrylate fuming process, is an optimization of this initial study.[47] The cyanoacrylate was polymerised in presence of 4-Butylamino-*N*-butyl-1,8-naphthalimide (Scheme 4.7, **II**), a dye also known as Solvent Yellow 43. In the polymer matrix, this fluorophore absorbs in the wavelength range from about 365 nm to 470 nm with an optimum at 415 nm and fluoresces in the wavelength range from about 450 nm to 650 nm. CN Yellow is supplied in the form of a cartridge for a use with a fuming wand (a butane torch) or as a powder (CN Yellow Crystals) for use in a fuming cabinet able to heat at 230 °C. The heating induces the depolymerisation of the polymer and

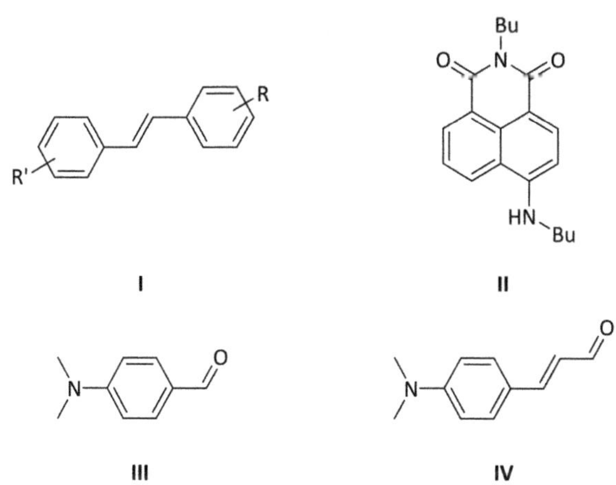

Scheme 4.7. Chemical structures of dyes used in high temperature fluorescent one-step cyanoacrylate processes.

simultaneously releases cyanoacrylate oligomers and fluorophores leading to fingerprint development.

Two other commercially available products have appeared recently on the market which only differ from CN Yellow by the fluorophore choice. PolyCyano UV[48] is a mixture of a Poly(ethyl-2-cyanoacrylate) powder with 5% of 4-Dimethylaminobenzaldehyde (DMAB, Scheme 4.7, **III**). Takatsu *et al.*[49] have already shown the ability of a vapour of this fluorophore to stain a latent mark previously developed by cyanoacrylate fuming but it was a two-step process. DMAB can be excited at 365 nm and fluoresces in the wavelength range from about 420 nm to 650 nm in the polymer matrix of the fingerprint.[48] The quality of enhanced fingermarks developed with Polycyano UV is comparable to those obtained with the two-step process.[50,51] However, the use of a UV excitation may result in background fluorescence that can limit the contrast between the latent mark and the substrate.[52]

In PECA multiband the fluorophore used is the 4-Dimethylaminocinnamaldehyde[52] (DMAC, Scheme 4.7, **IV**) which can fluoresce in the wavelength range from about 365 nm to 555 nm when excited with wavelengths up to 505 nm. The varying excitation wavelengths limit background fluorescence issues from different substrates. However, fluorescence becomes visible only after some hours of development when fingerprints are excited in the visible spectral range.

To be able to evaporate their fluorophore dye, all these commercial products require a processing temperature of 230 °C for a use in fumigation chamber. If the use of polymerised cyanoacrylate could be considered as an advantage considering its infinite shelf-life, these technologies require an expensive modification of the heating mantle of the chamber to reach the target temperature, an additional cost which limits their interest. Moreover, these products result in powdery residues on the inside of the fuming cabinet, in the circulation fans, filters and the humidifier fan which required thorough cleaning that is time consuming.[52] Finally, above 200 °C cyanoacrylate decomposes leading to highly toxic cyanide gas formation which is problematic for health and safety considerations.[53]

Lumicyano[54] has been designed as a liquid mixture of Ethyl-2-cyanoacrylate and an easily vaporizable low-molecular weight fluorophore,[55,56,57] the 3-Chloro-6-ethoxy-tetrazine (Scheme 4.8), that allows a one-step fluorescent revelation of latent marks under standard humidity and temperature fumigation conditions (80% humidity, 120 °C). Since processing temperature is maintained at 120 °C, no cyanide gas formation is possible with Lumicyano.

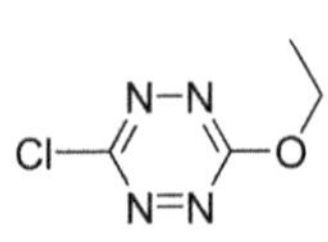

Scheme 4.8. Chemical structure of 3-chloro-6-ethoxy-tetrazine.

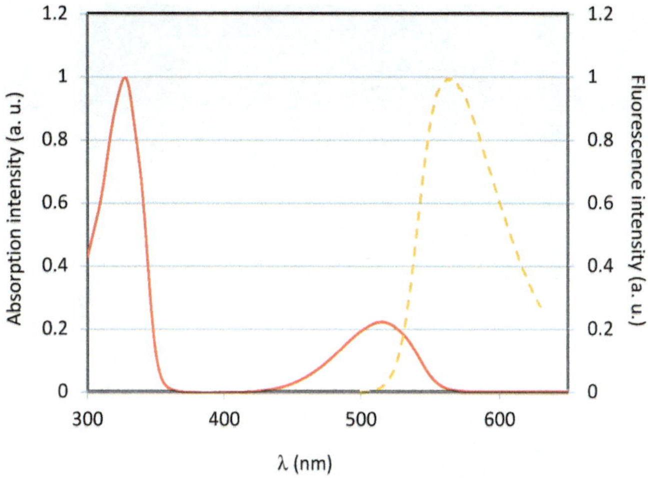

Figure 4.5. Absorption (red line) and fluorescence (yellow line) spectra of the 3-chloro-6-ethoxy-tetrazine in solution in acetonitrile. Arbitrary units.

The yellow fluorescence (520–650 nm, λ_{max} = 562 nm) of this orange-red fluorophore can be excited in the UV spectral range (315–340 nm, λ_{1max} = 326 nm) or in the visible (450–550 nm, λ_{2max} = 511 nm) as shown in Figure 4.5. This dual excitation wavelength allows one to overcome the lack of contrast between a fingerprint and its substrate sometimes observed under UV excitation due to a fluorescence background.

The first generation of the Lumicyano was supplied as a 1% (weight/weight) pink solution of tetrazine in cyanoacrylate. Fingerprints developed by this product revealed an excellent contrast due to a perfect co-deposition of the fluorescent dye and the polymer of cyanoacrylate.[54] Moreover, a pseudo-operational trial[58] using this 1% formulation reported a number of developed marks similar to the usual two-step process (cyanoacrylate/BY40) and an additional 15% detection rate when Lumicyano is followed by a subsequent BY40 staining. However, tetrazine can sublimate at room temperature due to its low molecular weight leading to a fluorescence fading of the fingerprints[54,58] which can occur in few days depending on the material of the substrate and the storage conditions. To circumvent this limitation a Lumicyano kit has been developed as a second-generation product in which the cyanoacrylate and the fluorophore are delivered in separate bottles. An extemporaneous mixture allows one to drastically increase the quantity of fluorophore up to 10% without any shelf-life consideration (Figure 4.6).

The use of a Lumicyano kit with 4% of tetrazine enhance the contrast of the fingerprint with the substrate, the level of detail of the fingerprint (definition of ridges and furrows) and the continuity of the ridges in relation to the first-generation of Lumicyano and allows the development of fingermarks until the tenth depletion.[59] A pseudo-operational study[60] using a Lumicyano

Figure 4.6. A latent fingermark on glass developed with Lumicyano with 5% of fluorophore as viewed under an excitation at 325 nm using Labino SuperXenon lamp.

kit formulation with 5% of tetrazine reported a number of developed marks similar to the usual two-step process (cyanoacrylate/BY40) and an additional 30% detection rate when a subsequent BY40 staining is processed. Moreover, the fluorescence persisted for at least four weeks when stored in appropriate conditions, away from light and in a dry place. Finally, Lumicyano does not interfere with DNA analysis thus making genetic identification possible on a single fingerprint.[61]

Thanks to these advantages, Lumicyano is now used routinely to fume an entire room or a vehicle (Figure 4.7) by French police forces (INPS and IRCGN) and LKA Dresde (Germany), a promising novelty as it facilitates the subsequent collect of samples for DNA analysis.

Recently, studies[52,62] were carried out with the aim of comparing these one-step processes. Despite the numerous advantages of the one-step processes, additional studies are necessary to fully evaluate their effectiveness and compare them to the well-established two-step process before a complete validation by the fingerprint community.

Figure 4.7. A latent fingermark developed with Lumicyano on a seat belt directly in the vehicle.

4.5. Miscellaneous Methods Based on Luminescence

4.5.1. Fluorescent Powders

Powders have been used to develop fingermarks since the end of the 19th century and probably still account nowadays for the largest number of fingerprints identified worldwide. This technique involves the application of a finely ground powder on the latent fingermark. The powder adheres to the residues composing the ridge patterns and the excess powder can be blown off the substrate, thus developing the fingerprint. Many different powder formulations have been tested, the most commonly used contain starch, kaolin, rosin or silica gel as an adhesive. Either metallic flakes or dyes are added to obtain a good contrast.[63]

Different organic dyes have been added to these powders, such as rhodamine and acridine derivatives,[63] and different formulations are commercially available. The use of near-infrared fluorescent dyes that can be excited with conventional forensic lighting sources appears as a promising way to increases the contrast between the developed fingerprint and the substrate. Chlorophyll and/or anthocyanin containing food powders, especially spirulina, have been used after micronization by ball milling, in order to develop fingerprints in the near-infrared. This improves the contrast, especially on multicoloured substrates (Figure 4.8).[64] A mixture of rhodamine 6G and styryl dye 11 has been coated onto basic aluminium oxide powder

Figure 4.8. Fingerprints developed on wrapping paper using spirulina (a) white light (b) 640 nm illumination using a 850 nm long-pass filter. Reprinted from Forensic Science International, **249**, R. S. King, P. M. Hallett and D. Foster, Seeing into the infrared: A novel IR fluorescent fingerprint powder, e21–e26, Copyright (2015), with permission from Elsevier.

and found to be a suitable powder to detect fingerprints both in the visible and near-infrared regions. One limitation for practical use expressed by the authors is that few commercial digital cameras are able to photograph NIR luminescence.[65]

Inorganic luminophores have also been incorporated into dusting powders. Lanthanide-based compounds offer several advantages, such as large Stokes shifts and long luminescence lifetimes. Green luminescent europium doped strontium aluminate, prepared by hydrothermal synthesis and ground by ball milling, has been used to develop fingermarks on various non-porous substrates.[66] Micronized Egyptian blue (calcium copper silicate $CaCuSi_4O_{10}$) has also been used for its near-infrared luminescent properties, which allows an excellent contrast with the substrate but requires a modified digital camera.[67]

4.5.2. Fluorescent NPs

Most of the powders currently used for fingerprint detection contain particles of micrometric size. In contrast, nanoparticles are particles with at least one dimension smaller than 100 nm. Nanoparticles are currently attracting significant interest for fingerprint detection, although they are not yet used in routine casework.[68,69] This interest arises from the fact that some nanoparticles, such as quantum dots or lanthanide based nanophosphors possess unique intrinsic luminescence properties. Additionally, because of their smaller size compared to traditional powders, they offer a higher surface/volume ratio and as consequence a better adhesion to fingermark ridges.

This high surface/volume ratio can also be exploited to create specific interactions with the fingermarks residues.

Different types of intrinsically luminescent nanoparticles have been evaluated for fingermark development. Since the work of Mendel *et al.* in 2000 on the use of CdS nanocrystals on fingerprints fumed with cyanoacrylate,[70] different types of quantum dots have been investigated such as CdSe,[71] CdTe,[72] and non-toxic Mn doped ZnS quantum dots.[73] These nanoparticles allow the development of fingerprints on various non-porous substrates, such as glass or foil, and on the sticky side of tape. CdTe quantum dots in aqueous solution were also claimed to be very efficient to detect blood fingermarks.[74] In each case, the substrate is immersed in an aqueous quantum dot suspension, which is also called the Small Particle Reagent (SPR) technique.

Inorganic nanophosphors doped with lanthanide cations are another class of luminescent inorganic nanoparticles displaying high luminescence quantum yields, good photostability and narrow emission spectra. A particular feature of these nanoparticles is that depending on their composition they can either display down-conversion luminescence, that is to say emission in the visible after UV excitation, or up-conversion luminescence, that is to say emission in the visible after near-IR excitation. Amongst the nanophosphors displaying down-conversion emission, red-emitting YVO_4:Eu nanocrystals and green-emitting $LaPO_4$:Ce,Tb nanowires have been evaluated on porous and non-porous substrates, using 254 nm excitation.[75] The upconverting nanoparticles, for example $NaYF_4$:Yb,Er or $Gd_2O_3:Ho^{3+}/Yb^{3+}$ (Figure 4.9), could be very efficient for substrates strongly autofluorescent[76,77] although they require near-IR lasers as excitation sources. In addition to lanthanide

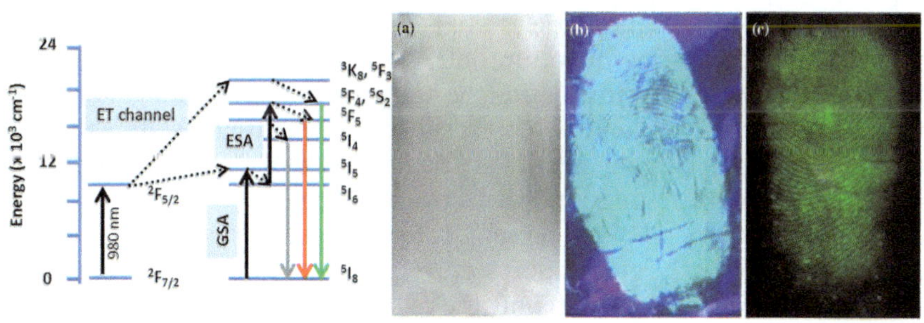

Figure 4.9. Left: the excitation and emission mechanism between sensitizer (Yb^{3+}) and activator (Ho^{3+}) dopant in up-conversion Gd_2O_3 nanoparticles. Right: development of latent fingermarks on an aluminium foil surface **a** under bright field, **b** with commercial green luminescent fingermark powder excitation with 380 nm UV lamp and **c** fingermarks developed with $Gd_2O_3:Ho^{3+}/Yb^{3+}$ nanoparticles, recorded by a short-pass filter with Nikon D 3200 camera having 20 s of shuttering speed setup. Illumination of 980 nm diode laser was used. Adapted from *Applied Physics B: Lasers and Optics,* Synthesis of $Gd_2O_3:Ho^{3+}/Yb^{3+}$ upconversion nanoparticles for latent fingermark detection on difficult surfaces, **122**, 2016, A. Kumar, S. P. Tiwari, A. K. Singh and K. Kumar. With permission of Springer.

doped inorganic particles, ZnO nanoparticles have been used to develop fingermarks on aluminium, glass and polyethylene either by the powdering technique or by the SPR technique.[78]

Fingermark development has also been performed with organic nanoparticles, such as carbon dots[79] or polymer dots[80], either by immersing the substrate in a nanoparticle suspension or by mixing a small number of nanoparticles with a conventional dusting powder.[81] Both lanthanide based nanophosphors and organic nanoparticles are less toxic than quantum dots, which is a major advantage for the use on crime scenes.

In addition to the intrinsically luminescent nanoparticles presented above, it is also possible to use silica[82] or titanium dioxide[83] nanoparticles doped with fluorescent or phosphorescent dyes. This strategy allows the combination of well known, non-toxic, and easy to synthesise nanoparticles with fluorophores displaying high brightness and tuneable emission colour. In principle, the fluorophore could also be modified to maximise the interactions between the functionalised nanoparticle and the fingermark residues.[84]

This functionalisation strategy can also be exploited to go beyond the identification of individuals: it is well known that sweat contains a variety of metabolites, which could be exploited to identify potential drug addicts. As a proof of principle, gold nanoparticles were functionalised with anti-cotinine antibody. The tagged nanoparticles were then pipetted onto fingermarks freshly deposited on glass. After rinsing, a fluorescently tagged secondary antibody fragment solution was then pipetted onto the fingermarks: only the fingermarks from smokers were fluorescently labelled.[85] This system was then extended to the detection of THC, methadone and benzoylecgonine (the major metabolite of cocaine).[86] Another example consists of the preparation of silica nanoparticles prepared by the Stöber method and first doped in their core by red emitting quantum dots. These nanoparticles are then further functionalised on their surface by green emitting quantum dots, through covalent linkage between the mercapto propionic acids coating the quantum dots and the silanol groups and by a layer of polyallylamine. This results in nanoparticles suitable for fingermark detection using the Small Particle Reagent technique. In the absence of TNT, the dominant emission comes from the green fluorescent quantum dots. In contrast, when TNT is present, it forms a complex with the polyallylamine and quenches the green fluorescence of the quantum dots present on the surface of the nanoparticles and only the red fluorescence quantum dots emit (Figure 4.10).[87] As a consequence, such as system not only allows the identification of individuals but also gives information on whether they have manipulated explosives. However, the complexity, and as a consequence the cost, of such bifunctionalised nanoparticles may limit their application in real casework.

4.5.3. Other Methods

Amongst the miscellaneous methods that have been explored to develop fingermarks, two are within the scope of this chapter since they require the use of light. The first one consists of the use of polydiacetylene derivatives.

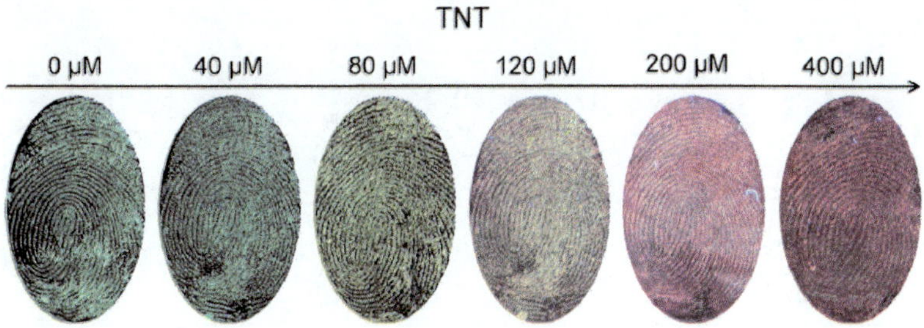

Figure 4.10. Imaging of LFPs with increasing amounts of TNT with the green/red QD nanohybrid: besides the ridge pattern of the fingerprint, the images also showed green to red colour change (traffic light-type) with increasing amounts of TNT. All fluorescence images were excited with a 365 nm UV lamp. Reproduced from ref. 87 with permission from The Royal Society of Chemistry.

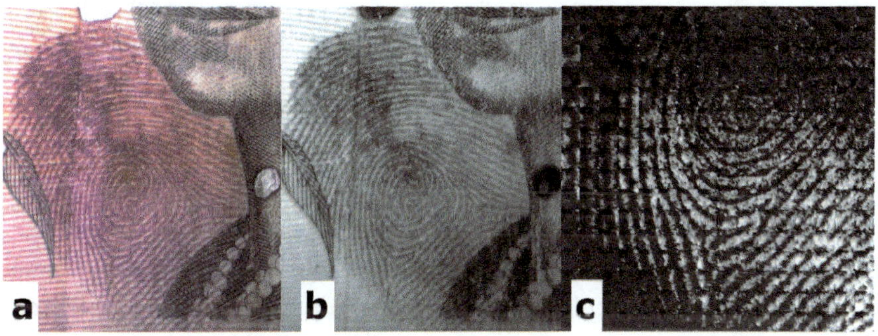

Figure 4.11. Actively developed fingermark on a polymer banknote using a mixture of diacetylene derivatives HDDPU and HDDCPCU: (a) white light photograph; (b) filtered light photograph; (c) monochrome representation of FTIR spectral image (second derivative, 1720 cm^{-1}). Reproduced from *Forensic Science International*, **216**, A. De Grazia, M. Mikhael, N. Stojanovska, B. Reedy, R. Shimmon and M. Tahtouh, Diacetylene copolymers for fingermark development, 189–197, Copyright 2012, with permission from Elsevier.

Polydiacetylene are conjugated polymers formed by topochemical polymerisation of diacetylene monomers upon UV irradiation or upon thermal annealing. To occur, the topochemical polymerisation requires the monomer to crystallise with the correct intermolecular spacing and orientation. In 1979, Miller and Patel found that that a solution of the diacetylene monomers 2,4-hexadiyne-1,6-bis(phenylurethane) (HDDPU) and 2,4-hexadiyne-1,6-bis(*p*-chlorophenylurethane) (HDDCPU) (see Scheme 4.9) could be used to develop fingermarks on non-porous surfaces.[88] In 2012, De Grazia *et al.*[89] extended the study to various surfaces: their hypothesis is that when a solution of the two monomers is sprayed on a surface with latent fingermarks, the diacetylene monomers are able to crystallise differently on the fingermark's ridges and on the substrate. They call active development

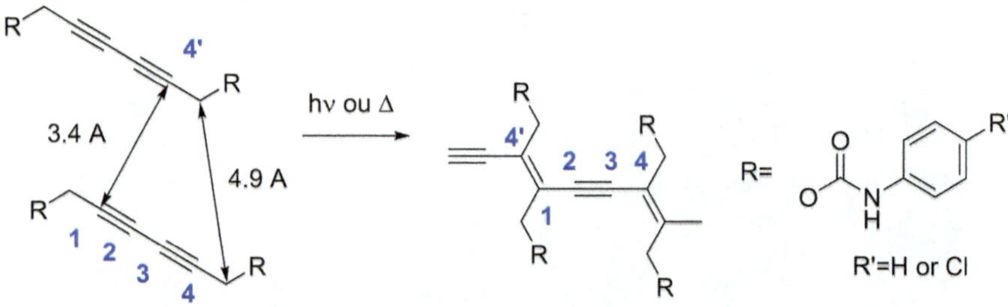

Scheme 4.9. topochemical polymerization of diacetylene derivatives HDDPU (R′=H) and HDDCPCU (R′=Cl).

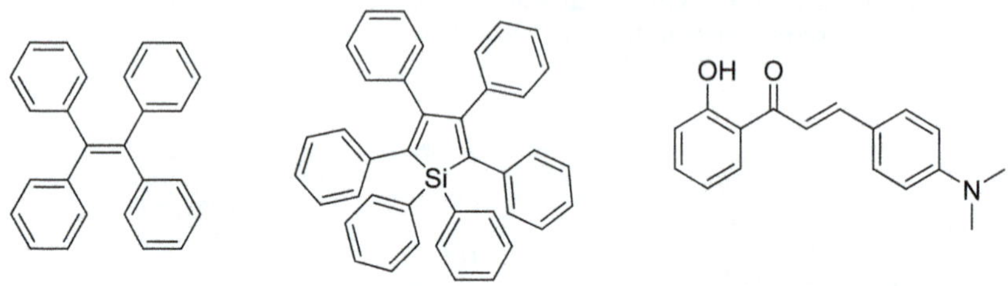

Scheme 4.10. AIE dyes used for fingermarks development: tetraphenylethene (left), hexaphenylsilole (middle) and 4-dimethylamino-2′-hydroxychalcone (right).

the case where the diacetylene crystallises on the ridges in a way suitable for polymerisation: after development the fingerprint appears purple on a white background. In contrast, the inactive development is when the fingerprint appears white on a purple background. Active or inactive development depends on the substrate and on the donor. Interestingly, fingerprints developed with this polydiacetylene reagent can be imaged with daylight but also by FTIR spectral imaging at the C=O stretch frequency of the urethane (Figure 4.11) or by Raman imaging by looking at the triple bond vibration of the polymer.[90] In another example,[91] hydrochromic polydiacetylene derivatives have been used for sweat pore mapping: when a fingertip is pressed on a film of these hydrochromic polydiacetylene (prepared by inkjet printing) the small amount of water from the sweat pores triggers the blue to red colourimetric transition (and the corresponding off to on fluorescence switching) of the polydiactetylene. This allows imaging of the sweat pores, which is known to be a reliable way to identify an individual, as sweat pores have a unique and permanent distribution on fingermark ridges.

A second method relies on the use of Aggregation Induced Emission (AIE) dyes: AIE dyes are non-fluorescent in solution and becomes fluorescent in the condensed state because of restricted molecular motions and as consequence limited non-radiative deactivation pathways. It has been shown that some AIE dyes, such as tetraphenylethene, siloles[92] or the red emitting ESIPT dye 4-Dimethylamino-2′-hydroxychalcone[93] (Scheme 4.10) can specifically label fingermarks, probably thanks to hydrophobic interactions with the residues composing the fingermark. Based on the same strategy, amphiphilic molecules containing a perylene[94] or a perylene diimide[95] core and hydrophilic side chains have also been evaluated for fingermarks development.

References

1. C. Champod, C. Lennard, P. Margot and M. Stoilovic, *Fingerprints and Other Ridge Skin Impression*, CRC Press, 2004, p. 304.
2. B. Yamashita and M. French, *The Fingerprint Sourcebook*, National Institute of Justice, Washington, 2011, pp. 1–67.
3. S. M. Bleay, V. G. Sears, H. L. Bandey, A. P. Gibson, V. J. Bowman, R. Downham, L. Fitzgerald and T. Ciuksza, Finger mark examination techniques within scope of ISO 17025, *Fingerprint Source Book*, Home Office Centre for Applied Science and Technology (CAST), 2012, ch. 2, pp. 3–38.
4. H. Ohki, *Rep. Natl. Res. Inst. Police Sci.*, 1970, **23**(1), 33.
5. S. M. Bleay, V. G. Sears, H. L. Bandey, A. P. Gibson, V. J. Bowman, R. Downham, L. Fitzgerald and T. Ciuksza, Finger mark imaging techniques, *Fingerprint Source Book*, Home Office Centre for Applied Science and Technology (CAST), 2012, ch. 4, pp. 289–321.
6. S. A. G. Lambrechts, A. van Dam, J. de Vos, A. van Weert, T. Sijen and M. C. G. Aalders, *Forensic Sci. Int.*, 2012, **222**, 89.
7. S. K. Bramble, K. E. Creer, W. G. Qiang and B. Sheard, *Forensic Sci. Int.*, 1993, **59**, 3.
8. N. Saitoh and N. Akiba, *Sci. World J.*, 2006, **6**, 691.
9. N. J. Crane, E. G. Bartick, R. S. Perlman and S. Huffman, *J. Forensic Sci.*, 2007, **52**, 48.
10. M. Tahtouh, P. Despland, R. Shimmon, J. R. Kalman and B. J. Reedy, *J. Forensic Sci.*, 2007, **52**, 1089.
11. P. Hazarika and D. A. Russell, *Angew. Chem., Int. Ed.*, 2012, **51**, 3524.
12. Y. Mou and J. W. Rabalais, *J. Forensic Sci.*, 2009, **54**, 846.
13. C. Ricci, S. Bleay and S. G. Kazarian, *Anal. Chem.*, 2007, **79**, 5771.
14. S. Ruhemann, *J. Chem. Soc., Trans.*, 1910, **97**, 1438.
15. S. Ruhemann, *J. Chem. Soc., Trans.*, 1911, **99**, 1486.
16. S. Odén and B. Von Hofsten, *Nature*, 1954, **173**, 449.
17. T. Trozzi, R. Schwartz and M. Hollars, *Processing Guide for Developing Latent Prints*, Federal Bureau of Investigation, Laboratory Division, U.S. Department of Justice, U.S. Government Printing Office, Washington, DC, 2000, pp. 33 and 56.
18. S. M. Bleay, V. G. Sears, H. L. Bandey, A. P. Gibson, V. J. Bowman, R. Downham, L. Fitzgerald and T. Ciuksza, Finger mark development techniques within scope of ISO17025, *Fingerprint Source Book*, Home Office Centre for Applied Science and Technology (CAST), 2012, ch. 3, pp. 39–289.
19. D. B. Hansen and M. M. Jouillé, *Chem. Soc. Rev.*, 2005, **34**, 408.
20. D. W. Herod and E. R. Menzel, *J. Forensic Sci.*, 1982, **27**, 200.
21. J. Druey and P. Schmidt, *Helv. Chim. Acta*, 1950, **33**, 1080.
22. R. Grigg, T. Mongkolaussavaratana, C. A. Pounds and S. Sivagnanam, *Tetrahedron Lett.*, 1990, **31**, 7215.
23. C. A. Pounds, R. Grigg and T. Mongkolaussavaratana, *J. Forensic Sci.*, 1990, **35**, 169.

24. A. A. Cantu, D. A. Leben, M. M. Jouillie, R. J. Heffner and R. R. Hark, *J. Forensic Identif.*, 1993, **1**, 43.
25. D. Wilkinson, *Forensic Sci. Int.*, 2000, **114**, 123.
26. S. Hardwick, T. Kent, V. Sears and P. Winfield, *Fingerprint Whorld*, 1993, **19**, 65.
27. C. Conn, G. Ramsay, C. Roux and C. Lennard, *Forensic Sci. Int.*, 2001, **116**, 117.
28. C. J. Frégeau, O. Germain and R. M. Fourney, *J. Forensic Sci.*, 2000, **45**, 354.
29. W. H. Perkin, W. M. Roberts and R. Robinson, *J. Chem. Soc., Trans.*, 1912, **101**, 232.
30. R. Ramotowski, A. A. Cantu, M. M. Joullie and O. Petrovskaia, *Fingerprint Whorld*, 1997, **23**, 131.
31. D. B. Hauze, O. Petrovskaia, B. Taylor, M. M. Joullie, R. Ramotowski and A. A. Cantu, *J. Forensic Sci.*, 1998, **43**, 744.
32. M. Azoury, A. Zamir, C. Oz and S. Wiesner, *J. Forensic Sci.*, 2002, **47**, 586.
33. C. Roux, N. Jones, C. Lennard and M. Stoilovic, *J. Forensic Sci.*, 2000, **45**, 761.
34. S. M. Bleay, V. G. Sears, H. L. Bandey, A. P. Gibson, V. J. Bowman, R. Downham, L. Fitzgerald and T. Ciuksza, Alternative finger mark development techniques, *Fingerprint Source Book*, Home Office Centre for Applied Science and Technology (CAST), 2012, ch. 5, pp. 322–483.
35. C. Wallace-Kunkel, C. Lennard, M. Stoilovic and C. Roux, *Forensic Sci. Int.*, 2007, **168**, 14.
36. M. Stoilovic, C. Lennard, C. Wallace-Kunkel and C. Roux, *J. Forensic Identif.*, 2007, **57**, 4.
37. S. Wiesner, E. Springer, Y. Sasson and J. Almog, *J. Forensic Sci.*, 2001, **46**, 1082.
38. S. P. Wargacki, L. A. Lewis and M. D. Dadmun, *J. Forensic Sci.*, 2007, **52**, 1057.
39. N. von Wurmb, D. Meissner and R. Wegener, *Forensic Sci. Int.*, 2001, **124**, 11.
40. P. Kumar, R. Gupta, R. Singh and O. P. Jasuja, *J. Forensic Leg. Med.*, 2015, **32**, 64.
41. B. Bhoelai, B. J. de Jong, M. de Puit and T. Sijen, *Forensic Sci. Int.: Genet. Suppl. Ser.*, 2011, **3**, e429.
42. W. P. Morton, Patent US4504408, 1985.
43. M. Spring, L. Gros and W. Deinet, *Summary of the Research Projects of the Private Fresenius Technical College*, University of Idstein Commisioned by the Bundeskriminalamt Wiesbaden, Bundeskriminalamt, Wiesbaden KI2, 1995.
44. A. Bentolila, J. Totre, I. Zozulia, M. Levin-Elad and A. J. Domb, *Macromolecules*, 2013, **46**, 4822.
45. G. Groeneveld, S. Kuijer and M. de Puit, *Sci. Justice*, 2014, **54**, 42.
46. D. E. Weaver and E. J. Clary, *J. Forensic Identif.*, 1993, **43**, 481.
47. D. E. Weaver, *Co-polymerization of Sublimation Dyes and Expanding the Micro-crystalline Vapors of Cyanoacrylate in Fingerprint Development, Research Report Submitted to the U.S. Department of Justice*, 2009.
48. Foster & Freeman, *PolyCyano UV - Cyanoacrylate Fuming Equipment - User Manual (Applications)*, 2013.
49. M. Takatsu, O. Shimoda and H. Teranishi, *J. Forensic Sci.*, 2012, **57**, 515.
50. W. Hahn and R. Ramotowski, *J. Forensic Identif.*, 2012, **62**, 279.
51. S. Chadwick, L. Xiao, P. Maynard, C. Lennard, X. Spindler and C. Roux, *Aust. J. Forensic Sci.*, 2014, 1.
52. V. Stewart, P. Deacon and K. Farrugia, *Fingerprint Whorld*, 2016, **41**, 162.
53. T. C. Fung, K. Grimwood, R. Shimmon, X. Spindler, P. Maynard, C. Lennard and C. Roux, *Forensic Sci. Int.*, 2011, **212**, 143.
54. C. Prete, L. Galmiche, F.-G. Quenum-Possy-Berry, C. Allain, N. Thiburce and T. Colard, *Forensic Sci. Int.*, 2013, **233**, 104.
55. P. Audebert, F. Miomandre, G. Clavier, M.-C. Vernières, S. Badré and R. Méallet-Renault, *Chem.–Eur. J.*, 2005, **11**, 5667.

56. Y.-H. Gong, F. Miomandre, R. Méallet-Renault, S. Badré, L. Galmiche, J. Tang, P. Audebert and G. Clavier, *Eur. J. Org. Chem.*, 2009, **35**, 6121.
57. Q. Zhou, P. Audebert, G. Clavier, F. Miomandre and J. Tang, *RSC Adv.*, 2014, **4**, 7193.
58. K. J. Farrugia, P. Deacon and J. Fraser, *Sci. Justice*, 2014, **54**, 126.
59. A. Bisotti, C. Allain, J.-L. Georges, F. Guichard, P. Audebert, I. Barbosa and L. Galmiche, *J. Forensic Identif.*, 2016, **66**, 560.
60. K. J. Farrugia, J. Fraser, N. Calder and P. Deacon, *J. Forensic Identif.*, 2014, **64**, 556.
61. K. Gicquel, *About Lumicyano and Its Compatibility with DNA Sampling*, Institut Génétique Nantes Atlantique, 2015.
62. A. Khuu, S. Chadwick, X. Spindler, R. Lam, S. Moret and C. Roux, *Forensic Sci. Int.*, 2016, **263**, 126.
63. G. S. Sodhi and J. Kaur, *Forensic Sci. Int.*, 2001, **120**, 172.
64. R. S. King, P. M. Hallett and D. Foster, *Forensic Sci. Int.*, 2015, **249**, e21.
65. S. Chadwick, P. Maynard, P. Kirkbride, C. Lennard, A. McDonagh, X. Spindler and C. Roux, *Forensic Sci. Int.*, 2012, **219**, 208.
66. L. Liu, Z. Zhang, L. Zhang and Y. Zhai, *Forensic Sci. Int.*, 2009, **183**, 45.
67. B. Errington, G. Lawson, S. W. Lewis and G. D. Smith, *Dyes Pigm.*, 2016, **132**, 310.
68. M. J. Choi, A. M. McDonagh, P. Maynard and C. Roux, *Forensic Sci. Int.*, 2008, **179**, 87.
69. J. Dilag, H. J. Kobus and A. V. Ellis, *Curr. Nanosci.*, 2011, **7**, 153.
70. E. R. Menzel, S. M. Savoy, S. J. Ulvick, K. H. Cheng, R. H. Murdock and M. R. Sudduth, *J. Forensic Sci.*, 2000, **45**, 545.
71. Y. F. Wang, R. Q. Yang, Y. J. Wang, Z. X. Shi and J. J. Liu, *Forensic Sci. Int.*, 2009, **185**, 96.
72. K. Cai, R. Yang, Y. Wang, X. Yu and J. Liu, *Forensic Sci. Int.*, 2013, **226**, 240.
73. C. Xu, R. Zhou, W. He, L. Wu, P. Wu and X. Hou, *Anal. Chem.*, 2014, **86**, 3279.
74. A. Becue, S. Moret, C. Champod and P. Margot, *Forensic Sci. Int.*, 2009, **191**, 36.
75. M. Wang, M. Li, A. Yu, J. Wu and C. Mao, *ACS Appl. Mater. Interfaces*, 2015, **7**, 28110.
76. M. Wang, Y. Zhu and C. Mao, *Langmuir*, 2015, **31**, 7084.
77. A. Kumar, S. P. Tiwari, A. K. Singh and K. Kumar, *Appl. Phys. B*, 2016, **122**, 190.
78. M. J. Choi, K. E. McBean, P. H. R. Ng, A. M. McDonagh, P. J. Maynard, C. Lennard and C. Roux, *J. Mater. Sci.*, 2008, **43**, 732.
79. H. Li, X. Guo, J. Liu and F. Li, *Opt. Mater.*, 2016, **60**, 404.
80. H. Chen, K. Chang, X. Men, K. Sun, X. Fang, C. Ma, Y. Zhao, S. Yin, W. Qin and C. Wu, *ACS Appl. Mater. Interfaces*, 2015, **7**, 14477.
81. D. Fernandes, M. J. Krysmann and A. Kelarakis, *Chem. Commun.*, 2015, **51**, 4902.
82. Y.-J. Kim, H.-S. Jung, J. Lim, S.-J. Ryu and J.-K. Lee, *Langmuir*, 2016, **32**, 8077.
83. M. J. Choi, T. Smoother, A. A. Martin, A. M. McDonagh, P. J. Maynard, C. Lennard and C. Roux, *Forensic Sci. Int.*, 2007, **173**, 154.
84. S. Moret, A. Bécue and C. Champod, *Forensic Sci. Int.*, 2016, **259**, 10.
85. R. Leggett, E. E. Lee-Smith, S. M. Jickells and D. A. Russell, *Angew. Chem., Int. Ed.*, 2007, **46**, 4100.
86. P. Hazarika, S. M. Jickells, K. Wolff and D. A. Russell, *Angew. Chem., Int. Ed.*, 2008, **47**, 10167.
87. P. Wu, C. Xu, X. Hou, J.-J. Xu and H.-Y. Chen, *Chem. Sci.*, 2015, **6**, 4445.
88. G. G. Miller and G. N. Patel, *J. Appl. Polym. Sci.*, 1979, **24**, 883.
89. A. De Grazia, M. Mikhael, N. Stojanovska, B. Reedy, R. Shimmon and M. Tahtouh, *Forensic Sci. Int.*, 2012, **216**, 189.
90. N. Stojanovska, A. De Grazia, M. Tahtouh, R. Shimmon and B. Reedy, *J. Forensic Sci.*, 2015, **60**, 619.

91. J. Lee, M. Pyo, S. H. Lee, J. Kim, M. Ra, W. Y. Kim, B. J. Park, C. W. Lee and J. M. Kim, *Nat. Commun.*, 2014, **5**, 3736.
92. L. Xu, Y. Li, S. Li, R. Hu, A. Qin, B. Z. Tang and B. Su, *Analyst*, 2014, **139**, 2332.
93. X. Jin, L. Dong, X. Di, H. Huang, J. Liu, X. Sun, X. Zhang and H. Zhu, *RSC Adv.*, 2015, **5**, 87306.
94. F. Wang, J. Chen, H. Zhou, W. Li, Q. Zhang and C. Yu, *Anal. Methods*, 2014, 654.
95. K.-R. Wang, Z.-B. Yang and X.-L. Li, *Chem.–Eur. J.*, 2015, **21**, 5680.

Chapter 5

Light as an Ionizing Agent in Mass Spectrometry. Microbial Identification by Matrix Assisted Laser Desorption Ionization–Mass Spectrometry

Donata Favretto*[a] and Pietro Traldi[b]

[a]University Hospital of Padova, Legal Medicine and Toxicology, Italy;
[b]National Research Council, Italy
*E-mail: donata.favretto@unipd.it

Table of Contents

Light in Forensic Science: Issues and Applications
Edited by Giorgia Miolo, Jacqueline L. Stair and Mire Zloh

Published by the Royal Society of Chemistry, www.rsc.org

5.1. Introduction to Mass Spectrometry

According to J. B. Fenn,

Mass spectrometry is the art of measuring atoms and molecules to determine their molecular weight. Such mass or weight information is sometimes sufficient, frequently necessary, and always useful in determining the identity of a species. To practice this art one puts charge on the molecules of interest, i.e., the analyte, then measures how the trajectories of the resulting ions respond in vacuum to various combinations of electric and magnetic fields.

Clearly, the sine qua non of such a method is the conversion of neutral analyte molecules into ions. For small and simple species the ionization is readily carried by gas-phase encounters between the neutral molecules and electrons, photons, or other ions. In recent years, the efforts of many investigators have led to new techniques for producing ions of species too large and complex to be vaporized without substantial, even catastrophic, decomposition[†]

A mass spectrometer determines the mass of a molecule by measuring the mass-to-charge ratio (m/z) of its ionic species, generated by inducing either the loss or gain of a charge from a neutral species. Once formed, ions are electrostatically directed into a mass analyzer where they are separated according to their m/z and finally detected. The result of molecular ionization, ion separation, and ion detection is a *mass spectrum* providing information on molecular mass and structure of the target analyte. There is an abundance of different instrumental configurations, but all molecules undergo the same processes regardless of instrument design.

Generally, four components are present in all mass spectrometers:

- a sample inlet,
- an ionization source,
- a mass analyzer,
- and an ion detector.

Sample molecules need to be introduced into the instrument through a *sample inlet*, converted to ions in the *ionization source*, pushed into the *mass analyzer* where ions are separated according to their m/z value, using magnetic and/or electric fields and/or radio frequency to exert forces on ions in a high vacuum region. It is also possible to have some steps occurring in the same physical space, whereas they are separated in time (ion traps). The *detector* converts the energy of ions impacting on it into electrical signals, which are then amplified and transmitted to a computer.

Therefore, a compound must be charged or ionized and the ions must be in the gas phase to be introduced into the high vacuum system of a mass spectrometer. This is easily fixed for gaseous or heat-volatile samples,

---

[†]Quote reprinted with permission from Fenn, J. B. (2003), Electrospray Wings for Molecular Elephants (Nobel Lecture). *Angewandte Chemie* International Edition, 42: 3871–3894. Copyright The Nobel Foundation 2003.

but not for analytes that decompose upon heating. For those specimens, *desorption* or *desolvation* methods have been adopted; ionization and desorption/desolvation are usually separated processes, but the term "ionization method" commonly refers to both ionization and desorption (or desolvation) steps.

The choice of ionization method depends on the nature of the sample and the type of information required from the analysis. So-called '*soft ionization*' methods such fast atom bombardment and electrospray ionization are used to generate information on the molecular weight of analytes but produce little or no fragment-ion information.

As detailed below, before discussing the role of light in the process of ionization, a brief review of the historically more important ionization agents is presented.

5.2. Electron Beams as Ionizing Agent

5.2.1. Electron Ionization

Electron ionization or *electron impact* (EI) ionization was the first of all the ionization methods to be developed and studied. When a beam of accelerated electrons passes through the gas-phase sample, an electron that collides with a neutral analyte molecule can knock off an electron from its outer electron shell, resulting in a positively charged ion. The ionization process can thus produce a *molecular ion* which will have the same molecular weight and elemental composition as the starting analyte (minus an electron), or it can break molecular bonds and produce *fragment ions* which correspond to “pieces” of the analyte molecule. The samples are usually introduced into the ion source block through a heated inlet or a gas chromatograph.

The mechanism of electron ionization is well understood, giving reproducible mass spectra of volatile compounds. Most EI sources use electrons with an energy of 70 electron volts (eV) for EI; decreasing the electron energy can reduce fragmentation, but a high degree of fragmentation is necessary to obtain structural information. Libraries of EI mass spectra are available and can be searched for mass spectral "fingerprints" of molecules. The main restrictions are that samples must be thermally volatile and stable and that in the EI mass spectrum the molecular ion may be weak or absent. Typically, molecules less than 1000 Da are amenable to EI.

5.3. Gas Phase Reactions as Ionizing Agent

5.3.1. Chemical Ionization (CI)

The chemical ionization (CI) process requires a reagent gas such as methane, isobutane, or ammonia that is first ionized by EI at a high reagent gas pressure, resulting in ion–molecule reactions between the gas ions and the gas neutrals. Some of the products of these reactions react with the analyte molecules to produce ions from the analyte in the gas phase.

Example (R = reagent, S = sample, e^- = electron, $^{\bullet}$ = radical electron, H = hydrogen):

$$R + e \rightarrow R^{+\bullet} + 2\,e^-$$

$$R^{+\bullet} + RH \rightarrow RH^+ + R^{\bullet}$$

$$RH^+ + S \rightarrow SH^+ + R$$

(of course, other reactions can occur)

CI often gives molecular weight information producing $[M + H]^+$ ions and CI mass spectra are simpler than EI, with a reduced fragmentation. However, also for CI samples must be thermally volatile and stable; obtaining less fragmentation than EI, CI mass spectra are often not informative for a library search. Typically, molecules less than 1000 Da are amenable to CI.

5.3.2. *Negative-ion Chemical Ionization (NCI)*

Many important compounds of biological interest can produce negative ions under the right conditions and negative ion mass spectrometry is, for those analytes, efficient, sensitive and selective.

Negative ions can be produced by a number of processes, among which *resonance electron capture.* In NCI, a buffer gas (usually a CI gas such as methane) is used to slow down the electrons in the electron beam until some of the electrons have just the right energy to be captured by the analyte molecules. The buffer gas also stabilizes the energetic anions and reduces their fragmentation.

NCI is generally an efficient ionization, producing less fragmentation than positive-ion EI or CI, and has a great selectivity for some environmentally or biologically important compounds. However, not all volatile compounds produce negative ions, and the spectra somehow lack reproducibility.

A better approach to NCI is the control of the energy of the ionizing electrons and the direct formation of negative ions.

Typically, molecules less than 1000 Da are amenable to NCI.

5.4. Atmospheric Pressure Ionization (API) Processes

In these methods, the analyte is dissolved in a water/organic solution, sprayed at atmospheric pressure, ionized and transferred to the vacuum of the mass spectrometer. A combination of thermal and pneumatic processes desolvate and ionize the analytes as they enter the ion source. Typical flow rates are less than 1 microliter per minute up to about a milliliter per minute. These methods are well-suited for both flow-injection analysis (FIA) and hyphenated liquid chromatography (LC-MS) techniques.

5.4.1. *Electrospray Ionization (ESI)*

The sample solution is sprayed across a high potential difference (a few kilovolts) from a needle into an orifice in the interface. Heat and gas flows are used to desolvate the ions existing or formed in the sample solution. Electrospray ionization can produce multiply charged ions from large molecules such as peptides and proteins. The number of charges on the ionic species can be determined by comparing two charge states that differ by one charge and solving algorithms, or examining the mass-to-charge ratios for resolved isotopic clusters.

ESI is suitable for charged, polar or basic compounds, permits the detection of high-mass compounds at mass-to-charge ratios that are easily determined by most mass spectrometers (*m*/*z* typically less than 2000 to 3000). It is the best method for analyzing multiply charged compounds, and can also control fragmentation by varying the interface lens potentials. It is totally compatible with LC.

Limitations: multiply charged species require interpretation and mathematical transformation; uncharged, non-basic, low-polarity compounds (*e.g.* steroids) are not ionized; ESI is very sensitive to contaminants such as alkali metals or basic compounds; the ion currents obtained are relatively low.

The molecular weight range of analytes amenable to ESI is typically up to 200 000 Da.

5.4.2. *Atmospheric Pressure Chemical Ionization (APCI)*

APCI uses a similar interface to that used for ESI. In APCI, a corona discharge is used to ionize the analyte in the atmospheric pressure region. The gas-phase ionization in APCI is more effective than ESI for less-polar species. ESI and APCI are complementary methods. It is an excellent LC-MS interface. The molecular weight range of analytes amenable to APCI is less than 2000 Da and multicharged ions are not formed.

5.5. Light as Ionizing Agent

The absorption of UV light by a neutral can result in electronically excited states which can undergo relaxation through two different phenomena: (i) emission of light or (ii) emission of an electron:

$$\mathrm{M} + h\nu \longrightarrow \mathrm{M}^{*} \longrightarrow \mathrm{M} + h\nu \quad \text{(i)}$$

$$\mathrm{M}^{*} \longrightarrow \mathrm{M}^{+\bullet} + e^{-} \quad \text{(ii)}$$

To achieve ionization, the energy absorbed must lead to a continuum state, in other words it must be at least equal or higher of the ionization energy of the neutral M.

Typical photon sources for *photoionization* (PI) are frequency–quadrupled Nd : Yag lasers delivering photons at 266 nm wavelength (4.6 eV) and ArF excimer lasers at 193 nm (6.3 eV). Both light sources clearly deliver photons well below the ionization energy (IE) of most molecules. Fortunately, the absorption of energy has not to be a one-photon process. Instead, stepwise accumulation of energy from less energetic photons is also feasible. Thus, the normal procedure to achieve PI is *multiphoton ionization* (MUPI).[1]

The next stepping-stone to photoionization is finding the electronic levels of the neutral, because nonresonant ionization has rather low cross-sections that translate into poor ionization efficiencies along with high photon flux requirements. Resonant absorption of photons is more effective by several orders of magnitude. Ideally, resonant absorption of the first photon leads to an intermediate state from where absorption of a second photon can forward the molecule into a continuum. This technique is known as 1 + 1 *resonance – enhanced multiphoton ionization* (REMPI). From a practical point of view, the second photon should be, but not necessarily has to be, of the same wavelength.

PI has attracted mass spectrometrists from the early days of the technique; it exhibits some special theoretical advantages with respect to electron ionization, but it also suffers from some severe practical limitations.[2,3] Although the ion population in electron ionization conditions exhibits a wide internal-energy distribution, the ion population in photoionization conditions (due to the usually well defined photon energy and to the kind of photon–molecule interaction) shows a very homogeneous and narrow internal-energy distribution. Then the main disadvantages of PI are related to the usually low yield of product ions, which is related to the sample vapour density and irradiation parameters (power, photon energy, irradiance). In the "classical" approach, in which the sample is vaporized inside an ion chamber at reduced pressure, the product ion yield is surely lower than that achieved by electron ionization with the same sample density. For these reasons, the use of photoionization in analytical mass spectrometry has been in the past strongly limited.

Nevertheless photoionization was widely employed in chemical physical studies related to the photochemistry of gaseous species.[2,3]

More recently, a new atmospheric pressure ionization approach has been proposed; it is based on the interaction of a "classical", not laser-produced, light beam with the vapours emanating from a nebulizer of the sample solution, and it is called atmospheric pressure photoionization (APPI).[4]

5.5.1. *Atmospheric Pressure Photoionization (APPI)*

Nowadays APPI, due to the high sample density that is present in the region close to the nebulizer, leads to a very high sensitivity, which is better than that present in atmospheric pressure chemical ionization (APCI) conditions,[5] where corona discharge represents the ionizing approach.

This point was well demonstrated by the paper from the Bruins and co-workers,[4] that shows that by APPI it is possible to obtain an analyte signal intensity eight times higher than that obtainable with a commercial APCI source. In that same paper, the operative conditions of the APPI source were simply described by considering that, in these conditions, direct photoionization can occur only for molecular species that exhibit an IE lower than the photon energy ($h\nu$). In the commercially available APPI systems a krypton discharge lamp is employed. It emits photons up to 10.2 eV (see Figure 5.1), and consequently only molecules with an IE < 10.2 eV can be directly ionized with the formation of an odd-electron molecular cation $M^{+\bullet}$. The ionization energies of some organic compounds are reported in Table 5.1. It is interesting to observe that the compounds usually employed as solvents in LC/MS experiments exhibit IE values higher than 10 eV. Consequently, it would be to expected that practically no interference should originate by them in

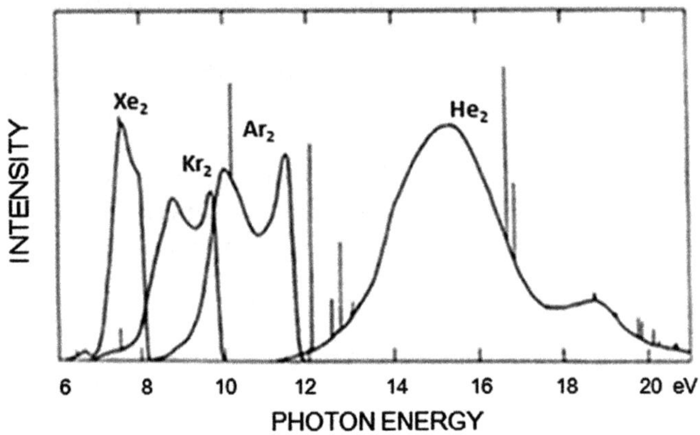

Figure 5.1. Intensity distribution as a function of photon energy in the continuum emission spectra of xenon, krypton, and argon excited by microwaves and of helium excited by a condensed discharge. The lines superposed on the continua are due to emission from hydrogen, oxygen, nitrogen, and carbon present as impurities, and provide internal calibration points. Curves are replotted from F. J. Vastola, R. O. Mumma and A. J. Pirone, Analysis of organic salts by laser ionization, *Org. Mass Spectrom.*, 1970, 3, 101–104.

Table 5.1. Energetics for Photoionization.

PhotoMate™ lamp			
Krypton 10.0 eV, 10.6 eV			
Ionization Energies (IE)		Solvent IE values	
Anthracene	7.4 eV	Toluene	8.82 eV
Fluornthene	7.8 eV	Acetone	9.70 eV
Caffeine	8.0 eV	Methanol	10.85 eV
4-Nitrotoluene	9.5 eV	Acetonitrle	12.19 eV
2,4,6-Trinitrotoluene	10.59 eV	Water	12.61 eV

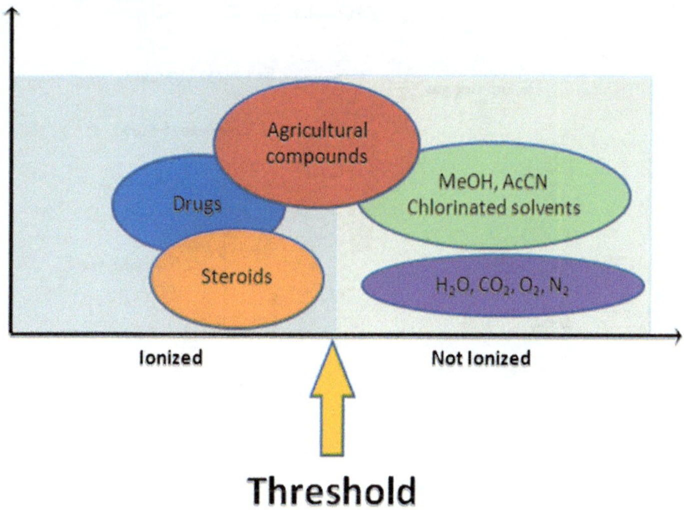

Figure 5.2. Principle of Operation: Ionizing the Analyte of Interest.

LC/APPI/MS conditions. Furthermore, APPI could be selectively employed for compounds of interest, again with minor interferences, from the complex substrate in which they are present.

The fields of applications of APPI are summarized in Figure 5.2.

The method was employed for the LC-MS analysis of a series of samples, some of which were of pharmaceutical interest, and the obtained results were particularly exciting.[4]

Those first data stimulated the interest of different research groups, as well as of the instrument producers, to make the APPI source commercially available. Many communications on APPI applications have been presented at international mass spectrometry meetings,[5–8] and some interesting papers have been published.[9–14]

Researchers agree that, at first sight, the mechanisms that operate in APPI seem to be particularly simple: for molecules with ionization energy < $h\nu$, a direct ionization is to be expected, with the formation of $M^{+\bullet}$; for molecules with an EI > $h\nu$, photoionization cannot take place and, consequently, the presence of a dopant is required. By adding to the sample solution a compound with IE < 10 eV, some ions are produced, able to activate a change-exchange or protonation reaction with the analyte. For this purpose, benzene, toluene and acetone (IE = 9.24, 8.83 and 9.50 eV, respectively[15]) have been proposed.

Before describing some applications, it is relevant to look at the experimental set up and to summarize the possible mechanisms that are operative at phenomenological level. The scheme of an APPI source is reported in Figure 5.3. The solution containing the analyte is firstly nebulized and by action of high temperature (300 °–400 °C) experimented in a well defined region, vaporized. From the vaporizer mainly molecules of solvent and analyte are

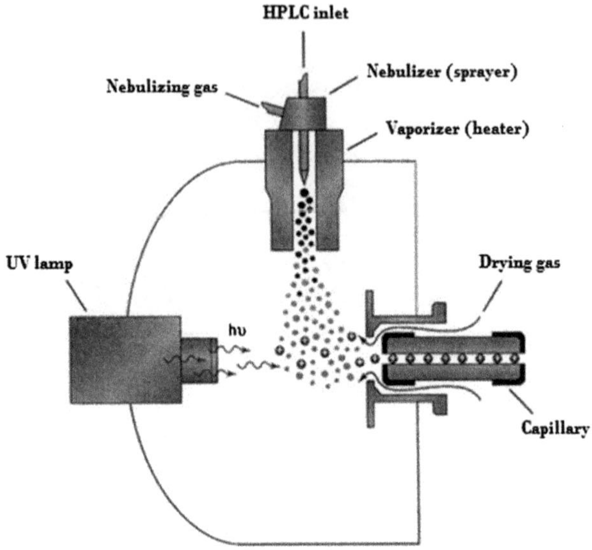

Figure 5.3. APPI source with the orthogonal geometry (data from Syagen Technologies Inc., Tustin, CA).

ejected, with only a minor contribution of clusters. Then the vapour interacts with photons generated by a Kr lamp and these interactions activate a series of phenomena, summarized as follows.

Species present inside the APPI source

Analyte: ABC

Solvent(s): S

Other gaseous species (*e.g.* N_2, O_2, water present at trace level): G

(i) First APPI step: photoexcitation:

$$\mathrm{ABC} + h\nu \rightarrow \mathrm{ABC}^*$$

$$\mathrm{S} + h\nu \rightarrow \mathrm{S}^*$$

$$\mathrm{G} + h\nu \rightarrow \mathrm{G}^*$$

(ii) Possible processes originating from excited species

$$\mathrm{ABC}^* \rightarrow \mathrm{ABC} + h\nu \text{ (radiative decay)}$$

$$\mathrm{ABC}^* \rightarrow \mathrm{AB}^{\bullet} + \mathrm{C}^{\bullet} \text{ (photodissociation)}$$

$$\mathrm{ABC}^* \rightarrow \mathrm{ACB}^* \text{ (isomerization)}$$

Collisional quenching:

$$\mathrm{ABC}^* + \mathrm{S} \rightarrow \mathrm{ABC} + \mathrm{S}^*$$

$$ABC^* + G \rightarrow ABC + G^*$$

$$S^* + ABC \rightarrow S + ABC^*$$

Only when $h\nu \geq$ IE, ionization can take place:

$$ABC^* \rightarrow ABC^{+\bullet} + e^-$$

$$S^* \rightarrow S^{+\bullet} + e^-$$

$$G^* \rightarrow G^{+\bullet} + e^-$$

As above described, for the ionization of compounds exhibiting IE values higher than 10 eV, the use of a dopant D is required, with the following proposed mechanism:

$$D + h\nu \rightarrow D^* \rightarrow D^{+\bullet}$$

$$D^{+\bullet} + ABC \rightarrow D + ABC^{+\bullet}$$

The proposed mechanism is ideologically wrong. In fact, the condition for which the reaction

$$D^{+\bullet} + ABC \rightarrow D + ABC^{+\bullet}$$

can take place is $IE_{ABC} \leq IE_D$.

Consequently, different mechanisms can be invoked.

The total ion chromatogram of a mixture of carbamazepine, acridine, naphthalene and diphenyl sulfide obtained in HPLC-APPI conditions is reported in Figure 5.4. The same mixture, analyzed in the presence

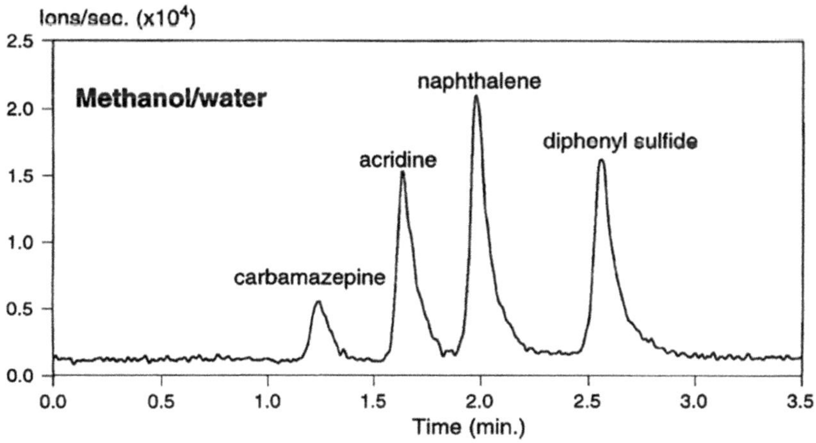

Figure 5.4. HPLC-APPI/MS chromatogram (SIM) obtained without dopant. Reprinted with permission from D. B. Robb, T. R. Covey and A. P. Bruins, Atmospheric pressure photoionisation: An ionization method for liquid chromatography-mass spectrometry, *Anal. Chem.*, 2000, 72, 3653–3659. Copyright 2000 American Chemical Society.

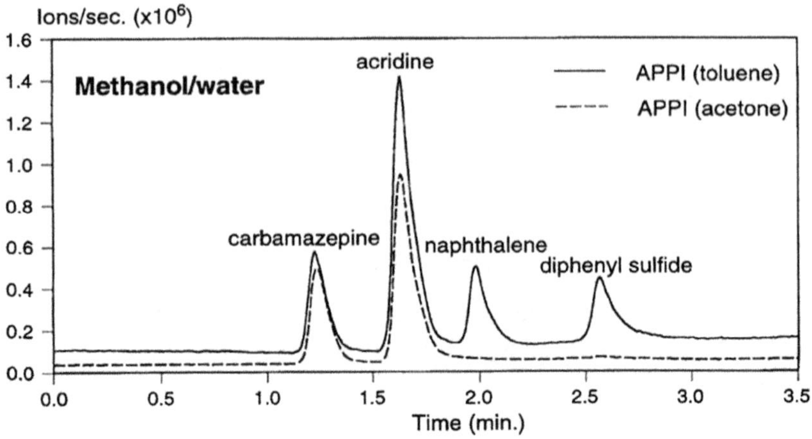

Figure 5.5. HPLC-APPI/MS chromatograms. Reprinted with permission from D. B. Robb, T. R. Covey and A. P. Bruins, Atmospheric pressure photoionisation: An ionization method for liquid chromatography-mass spectrometry, *Anal. Chem.*, 2000, **72**, 3653–3659. Copyright 2000 American Chemical Society.

of toluene or acetone as dopants, led to the chromatograms reported in Figure 5.5. It is interesting to observe that, while the presence of toluene leads mainly to a different relative abundance of the different analytes, the presence of acetone completely inhibits the ionization of naphthalene and diphenyl sulfide. These behaviors suggest that not only the odd electron molecular ion of dopant $D^{+\bullet}$ are generated, but other species are also generated, reasonably by reactions with the other molecules present in the ion source at atmospheric pressure. In order to investigate these aspects, benzene was injected in the ion source at increasing injection flows. Thus, while at low partial pressure only $M^{+\bullet}$ ions at *m*/*z* 78 are generated, by increasing partial pressure the $M^{+\bullet}$ disappears and new ionic species at *m*/*z* 94, 156, 171 and 187 appear. For the ion at *m*/*z* 94 collisional experiments confirmed the structure of a molecular ion of phenol. In principle, it could be generated by two processes: the reaction of molecular ion of benzene with O_2 or H_2O but experiments with $H_2{}^{18}O$ proved that only the former process takes place.

$$\underset{m/z\ 78}{C_6H_6^{+\bullet}} + O_2 \longrightarrow \underset{m/z\ 110}{\left[C_6H_6 - O_2\right]^{+\bullet}} \xrightarrow{-O} \underset{m/z\ 94}{C_6H_6O^{+\bullet}}$$

$$\underset{m/z\ 78}{C_6H_6^{+\bullet}} + H_2O \longrightarrow \underset{m/z\ 96}{\left[C_6H_6 - OH_2\right]^{+\bullet}} \xrightarrow{-H_2} \underset{m/z\ 94}{C_6H_6O^{+\bullet}}$$

The related mechanism is shown below

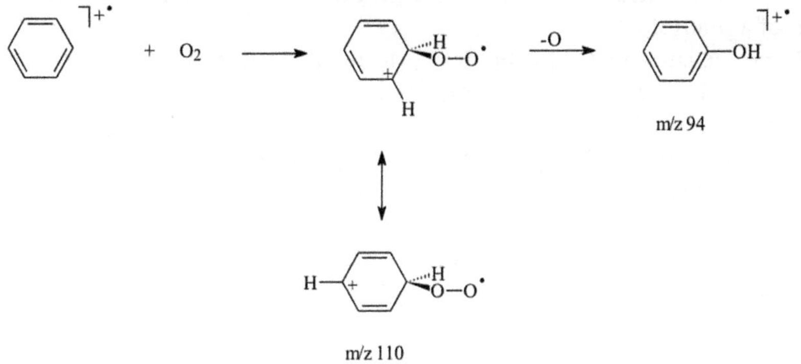

For the ions m/z 170 and 171 the following structures were proposed

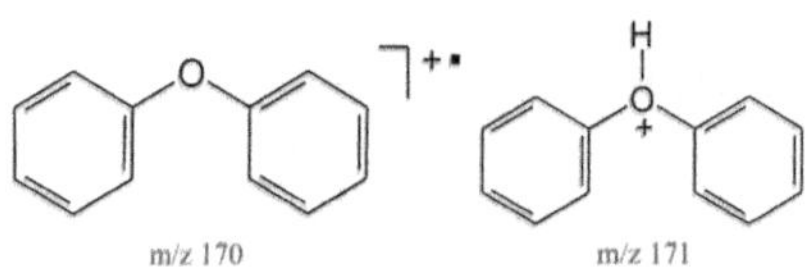

whereas the reaction of ionized phenol with neutral phenol gives rise to ions at m/z 186 and 187

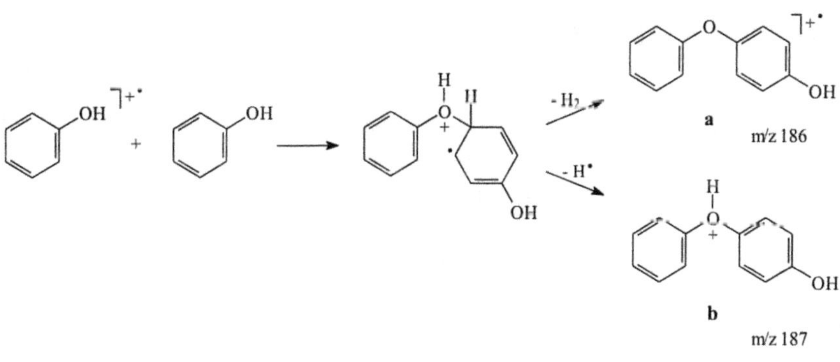

The behavior of benzene in APPI conditions stimulated the further investigation: do the above described mechanism occur only in APPI conditions or can take place also at an environmental level? Firstly, some "*in vitro*" experiments were performed, exposing a closed 4 mL glass vial containing 100 µL of benzene to sunlight for 6 h (from 10 am to 4 pm on a sunny summer day). The phenol/benzene GC/MS peak ratio increases from 1.8×10^{-4} to 0.105. In parallel another 4 mL vial containing 100 µL of benzene was covered with an aluminum foil and exposed on the same day and for the same length of time to sunlight.

The phenol/benzene peak ratio shows only a small increase (from 1.8×10^{-4} to 1.9×10^{-4}) indicating that sunlight irradiation of benzene led to the formation of a phenol.

Some measurements in the field (Messina harbour and high density traffic zone in Padova, Italy) confirmed that on a sunny day phenol is produced for irradiation of benzene ejected from cars. It should be emphasized that phenol has a markedly higher toxicity than benzene. For inhalation, the LD_{50} (mouse) is 9980 ppm for benzene and 42 ppm for phenol. These results show that the study of the light–molecule interaction in the gas phase is not only interesting from the analytical point of view, but it can be considered a valid test bench to study photochemical reactions of environmental interest.

5.5.2. *Laser Desorption/Ionization (LDI), Matrix Assisted Laser Desorption/Ionization (MALDI) and Surface Activated Laser Desorption/Ionization (SALDI)*

What happens when a photon interacts with a solid, crystalline sample? The quantized energy of the photon $\varepsilon = h\nu$ is transformed in a quantized lattice vibration or elastic wave. The quantum of energy in an elastic wave is called a phonon. Almost all of the concepts, such as the wave-particle duality, which apply to photons apply equally well to phonons. Sound waves in crystal are composed of phonons. Thermal vibrations in crystals are thermally excited phonons, analogous to the thermally excited photons of black-body electromagnetic radiation in an enclosure.

The technique of *laser desorption/ionization* (LDI) was introduced in the late 1960s,[16–18] long before the advent of field desorption (FD), *californium plasma desorption* (^{252}Cf-PD) or *fast atom bombardment* (FAB). While low-mass organic salts and light-absorbing organic molecules are easily accessible by LDI,[17,18] it takes a serious effort to obtain useful mass spectra of biomolecules,[19] in particular when the mass of the analyte exceeds 2000 u.[20,21] FAB and ^{252}Cf-PD therefore represented the standard in biomolecular and biomedical mass spectrometry until the late 1980s, while LDI was regarded as rather exotic.[22]

Unfortunately, when LDI was applied to organic compounds poor results were obtained, reasonably due to the thermal effect associated with laser irradiation, with the consequent decomposition of the analyte.

The idea, practically contemporary, of two research groups was the laser irradiation of a suitable solid matrix in which, at 10^{-4} molar level, the analyte is present. In this case most of the thermal (phonon) energy is adsorbed by the matrix crystal with the formation of reactive species in gas phase that, in the few nanoseconds in which a dense plume of matrix ions and neutral analyte is generated, ionization of analyte takes place.

Matrix-assisted laser desorption-ionization[23] is based on the interaction of a laser beam (usually generated by an ultraviolet (UV) laser, $\lambda = 337$ nm) with a crystal of suitable matrix containing, at a very low level, the analyte of interest (usually the analyte/matrix molar ratio is of the order of 10^{-4}). As depicted in Figure 5.6, the laser beam–crystal interaction leads to the vaporization of a microvolume of the solid sample, with the formation of a cloud

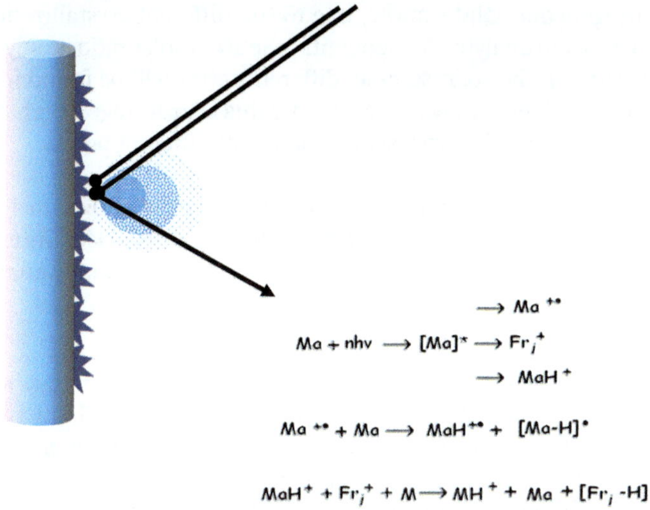

Figure 5.6. Interaction between laser beam and matrix (Ma) and analyte (M) solid samples, leading to the formation of protonated analyte molecules (MH^+).

rapidly expanding the space. The interaction gives rise to the formation of ionic species from the matrix (Ma), which exhibit an absorption band in correspondence to the laser wavelength, as $Ma^{+\bullet}$ (odd electron molecular ions), Fr_i^+ (fragment ions), MaH^+ (protonated molecules), and Ma_nH^+ (protonated matrix clusters). These species, through gas-phase, ion–molecule reactions, give rise to analyte positive ions (usually protonated molecules). Analogously, the formation of $[M–H]^-$ anions from the matrix can lead to deprotonated molecules of analyte.

A detailed description of the MALDI mechanism is highly complex, due to the presence of many different phenomena:

1. First, the choice of the matrix is relevant to obtain effective and well-reproducible data.
2. The photon-phonon transformation, obtained when a photon interacts with a crystal and gives information on the vibrational levels of the crystal lattice, cannot be applied to the laser induced vaporization observed in MALDI experiments, due to the inhomogeneity of the solid sample.
3. The laser irradiance (laser power/cm^2) is an important parameter: different irradiance values lead to a vapor cloud of different density, and consequently the ion–molecule reactions can take place with highly different yields.
4. The solid-sample preparation is usually achieved by the deposition on a metallic surface of the solution of matrix and analyte with a concentration suitable to obtain the desired analyte/matrix ratio. The solution is left to dry under different conditions (simply at atmospheric pressure, reduced pressure, or under a nitrogen stream). This method is usually called the Dried Droplet Method. In all cases, what is observed is the formation of an

inhomogeneous solid sample, due to the different crystallization rate of the matrix and analyte. Consequently, the 10^{-4} molar ratio is only a theoretical datum: in the solid sample, different ratios will be found in different positions and the only way to overcome this is to average a high number of spectra corresponding to laser irradiation of different points.

The instrumental setup for a MALDI experiment is shown in Figure 5.7. A laser beam, produced by a N_2 triplicated Na/Yag pulse laser, interacts with the solid state sample, generating analyte ions that are usually analysed by a time of flight instrument.

5.5.2.1. Time of Flight. Time of flight is surely the simplest mass analyzer.[24] In its basic form, it consists of an ion accelerator and a flight tube under vacuum. Magnetic, electrostatic, and electrodynamic fields are no longer present. In its "linear" configuration TOF is based on the acceleration, by the action of suitable acceleration voltage V, of the ions generated inside the ion source. The potential energy is transformed into kinetic energy

$$zV = \tfrac{1}{2} mv^2 \text{ from which}$$

$$v = \left(\frac{2zV}{m}\right)^{\frac{1}{2}} \tag{5.1}$$

Eqn (5.1) shows that ions of different m/z values will follow, after acceleration, linear pathways with different speeds. In other words, the m/z values are inversely related to the squared speed.

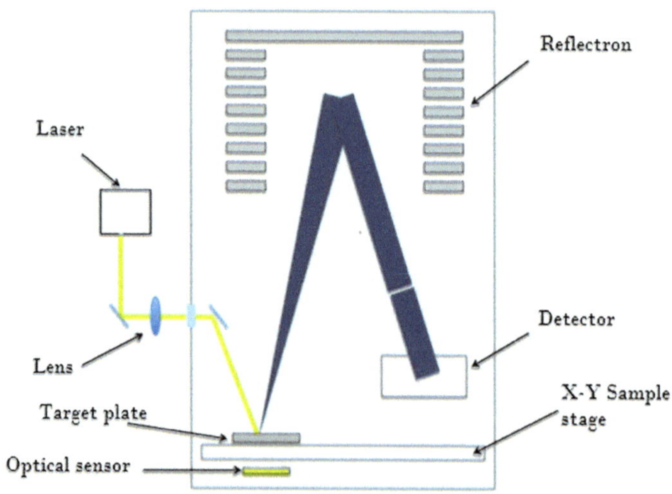

Figure 5.7. Instrumental setup for a MALDI experiment.

If the ions follow the linear pathway inside a field free region (drift tube) of length l, considering that $v = l/t \Rightarrow t = l/v$, it follows that

$$t = l\left(\frac{m}{2zV}\right)^{1/2} \tag{5.2}$$

This equation shows that ions of different m/z values reach the detector, placed at the end of the drift tube, at different times, proportional to the square root of their m/z value. By this experiment, we will obtain "arrival time" spectrum of the ions, which can be transformed into the mass spectrum by the relationship expressed by eqn (5.2). For this reason, this device is called TOF.

The simplest TOF configuration for ions to be followed is called "linear".

The ions emerging from the source are usually not homogeneous with respect to their speed (this effect mainly arises from the inhomogeneity of the acceleration field). Of course, a distribution of kinetic energy will reflect immediately on the peak shape and a wide kinetic energy distribution will lead to an enlarged peak shape, with the consequent decrease of resolution. To overcome this negative aspect, different approaches have been proposed. Those usually employed consists of a reflectron device. As shown in Figure 5.8, the reflectron consists of a series of ring electrodes and a final plate. The plate is placed at a few hundred volts over the V values employed for ion acceleration. By using a series of resistors, the different ring electrodes are placed at decreasing potentials. When an ion beam with kinetic energy $E_k \pm \Delta E_k$ interacts with this field, the ion with excess kinetic energy $(E_k + \Delta E_k)$ will penetrate the field following a pathway longer than that followed by ions with mean kinetic energy E_k. In contrast, ions with a lower kinetic energy will follow a shorter pathway. This phenomenon leads to a thickening of the ion arrival time distribution with a consequent, significant increase in mass resolution. Nowadays, TOF systems with resolutions > 10 000 are commercially available.

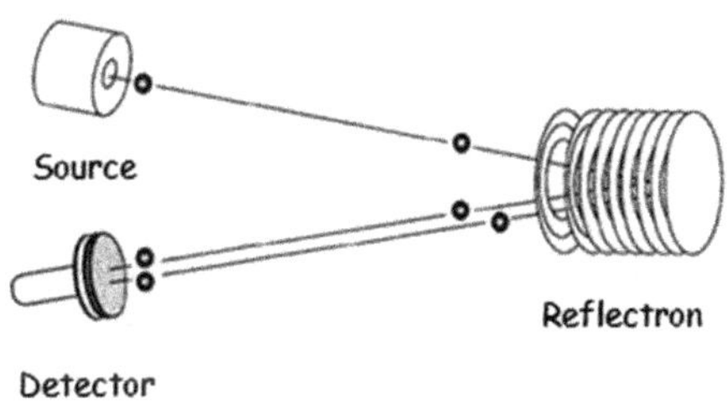

Figure 5.8. Scheme of a TOF analyzer with the reflectron device, leading to a resolution increase.

5.5.2.2. On the Primary Ionization mechanism(s) in Matrix-Assisted Laser Desorption Ionization. Different mechanisms have been proposed to explain MALDI ionization.[25] All are in agreement with the presence of two different steps: (i) primary ion formation from the matrix (Ma); (ii) secondary ion formation of the analyte (A) originating from the gas phase interaction of reactive matrix ions and neutral molecules of the analyte. Point (i) is the most debated one, due to the fact that the photon energy (for N_2 laser $h\nu = 3.68$ eV, for triplicated Nd:Yag laser $h\nu = 3.55$ eV) is much lower than that required for matrix ionization (typical ionization energies (IEs) of the most employed matrices are in the range 7–10 eV).[26]

It has been proposed[27] that multiple photon absorption, leading to multiple electronic excited matrix molecules, followed by energy pooling phenomena, would lead to the formation of the matrix odd electron molecular ion $Ma^{+\bullet}$ that, in turn, reacts with neutral matrix molecule leading to $[Ma + H]^+$. The last species must be considered responsible for the analyte (A) protonation in gas phase:

$$MaH^+ + A \rightarrow Ma + AH^+ \tag{5.3}$$

only if $PA_A > PA_{Ma}$, (PA = Proton Affinity).

The possible limitation of this mechanism lies in the low probability of multiphoton absorption and energy pooling, which makes the high ionic yield observed in MALDI experiments difficult to explain.

Alternatively, it has been considered that the electronic excitation (1 photon required) leads in general to an increase of the matrix acidity, promoting proton transfer. However, excited-state proton transfer (ESPT)[28,29] does not affect carboxylic acids, and consequently it cannot be invoked as an effective mechanism for proton transfer for the MALDI matrices; furthermore, ESPT is active for compounds that do not exhibit any activity as a MALDI matrix.

The desorption of preformed ions, already present in the solid state sample (before laser irradiation), has been proposed as a MALDI ionization mechanism.[30] The laser effect would be, in this case, just the desorption of these ions or matrix clusters containing them. In the latter case ions are made free by declustering reactions occurring in the high density desorbed plume. This mechanism can be invoked in the case of preprotonated compounds, metal complexes, and ionic compounds, and it is well supported by thermal desorption experiments, revealing the presence in gas phase of the same ions generated by laser irradiation. However, the presence of MaH^+ ions in the matrix before laser irradiation is difficult to prove, and consequently a gas phase protonation mechanism different to that reported in eqn (5.3) must be invoked.

Another mechanism is possible, related to the chemical nature of the matrices effective for MALDI experiments and their structural arrangements when crystallized. The solid–state interaction of carboxylic groups present in the matrix molecules could be considered responsible, when irradiated, for some disproportionation reactions, leading to both $[Ma + H]^+$ and $[Ma - H]^-$ ions, that is, the reactants necessary for the $[A + H]^+$ and/or $[A - H]^-$ formation.

It is interesting to note that the most of effective matrices for MALDI experiments are aromatic compounds containing a carboxyl moiety. Carboxylic acids are known as compounds which very readily form strong hydrogen bonds, and their carboxylic groups can act as both donors and acceptors.[31–33] In some cases intermolecular hydrogen bonds between such groups are formed.

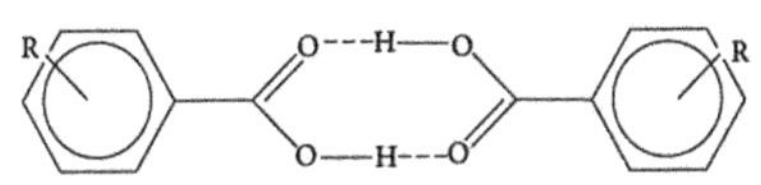

The laser irradiation of the matrix dimer with the above structure can be considered responsible for the formation of the ion couple $[Ma + H]^+$ and $[Ma - H]^-$.

In order to evaluate the role of photon irradiation, some theoretical calculations were performed. According to the first model, an absorbed UV photon by the matrix dimer induces its dissociation process into monomers accompanied by its simultancous heterolysis of one of the hydrogen bonds. This calculation model in the case of benzoic acid (as an example) is illustrated by the dissociation reaction reported in the scheme below

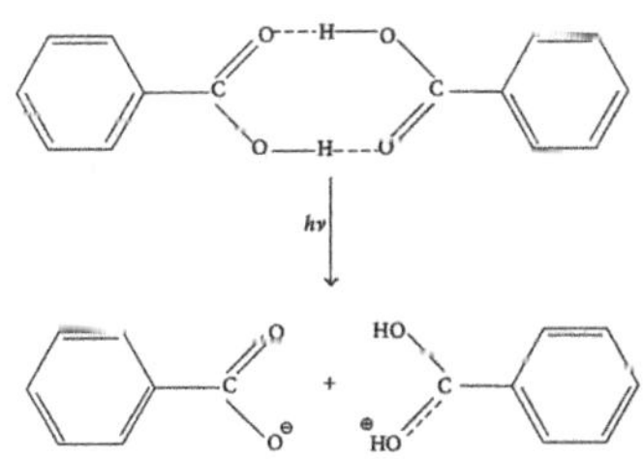

The values of energy calculated for the studied dimers are given in Table 5.2.

Table 5.2. Dissociation energy (E) of aromatic carboxylic acid dimers.

Dimers	E (eV)	E_{337} (eV)	E_{335} (eV)
Benzoic acid	7.03		
2,5-Dihydroxybenzoic acid	6.71		
4-Methoxybenzoic acid	6.79	3.68	3.50
4-Nitrobenzoic acid	6.94		
4-*N*,*N*-dimethylaminobenzoic acid	6.56		

Table 5.3. Dissociation energy for the photoionization mechanism of the matrix dimer molecules.

Dimers	*E* (eV)	E_{337} (eV)	E_{335} (eV)
Benzoic acid	3.69		
2,5-Dihydroxybenzoic acid	3.55		
4-Methoxybenzoic acid	3.43	3.68	3.50
4-Nitrobenzoic acid	3.35		
4-*N*,*N*-dimethylaminobenzoic acid	3.42		

The second proposed photoionization mechanism of the matrix dimer molecules corresponds to the reaction given below

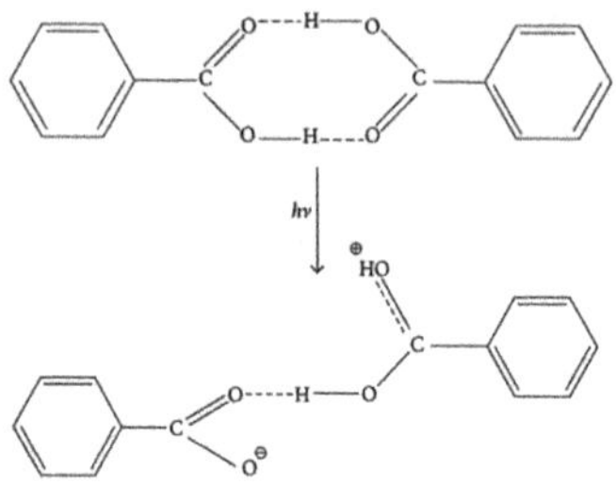

The energies of two processes, a heterodissociation of one of the both O–H bonds associated with the proton transfer to carbonyl oxygen atom and a rotation of the benzoic acid moiety about the O–H hydrogen bond of 180°, contribute to the energy requirement of this reaction. The calculated values of the energies are given in Table 5.3.

5.5.3. *Surface Activated Laser Desorption Ionization (SALDI)*

When the laser irradiation of the analyte deposited on nanostructures without the presence of classic matrices needed in MALDI experiments, interesting results are often obtained.[34–38]

The chemical and physical aspects of these methods were recently reviewed by Stolee *et al.*[39] Summarizing, it has been shown that when the nanostructure dimensions are commensurate with the laser wavelength, a series of phenomena is observed that reflects an enhanced desorption/ionization of organic and inorganic compounds. These phenomena (optical, electronic, vibrational, thermal, mechanical and chemical ones) originate by rapid energy deposition inducing highly nonlinear processes. In other words, the analogous dimensions of photon wavelength and nanostructures lead to a severe reduction in the diffusive phenomena present in "classic" MALDI experiments: the laser energy in the case of the nanosystem remains "trapped" inside the nanostructure, leading to the above-cited phenomena which are absent, or at least strongly reduced, under MALDI conditions. Furthermore, considering the nanostructure dimensions (in the range of 1–500 nm) and the applied extraction voltages (of the order of 10^4 V), this results in strong electrical fields ($E = 10^4\ \text{V}/100 \times 10^{-7}\ \text{cm} = 10^4\ \text{V}/10^{-5}\ \text{cm} = 10^9\ \text{V cm}^{-1}$) being present that are able to induce field ionization/desorption phenomena.[40]

Nanosystems of different chemical natures and with different morphologies have been tested. Thus, nanostructures based on porous silicon (desorption/ionization on silicon)[35,36,38,41,42] structured as microcolumns,[43,44] nanodots,[45] nanowires,[35,46] nanoparticles,[47] and nanofilms[48] have been proposed. Alternatively, to porous silicon, many others materials have been studied, among which are Au,[48,49] Ag,[50] Ge,[51,52] C[53,54] and Pt.[55] In particular, graphite-assisted laser desorption/ionization[53–58] found interesting applications.

Besides avoiding low molecular weight ionic species originating from "classic" MALDI matrices, in some cases with nanostructures a significant increase of sensitivity (detection limit of 800×10^{-24} mol)[59] has been observed.

A further point that has to be stressed is that in some studies effective results have been obtained by covering the nanostructures with suitable compounds, such as glycerol (G) and thioglycerol (T).[34,60] In this framework, the role of water acidity and surface morphology has been discussed by Chen *et al.*[61] We emphasize that these approaches are similar to that employed by Tanaka *et al.* in an early paper on MALDI:[62] in that case a mixture of glycerol and powdered cobalt was employed as the matrix and the cobalt crystals could be, in principle, considered as the "nano" substrate. Furthermore, it must be considered that the thin layer of glycerol and thioglycerol on the nanosystem could induce ionization mechanisms analogous to those present under fast atom bombardment (FAB) conditions.[63]

5.6. Applications of Laser Desorption Ionization in Microbiology

The accurate and rapid identification of bacteria is necessary for epidemiological surveillance, detection of biological warfare agents, of water- and food-borne pathogens, of antibiotic resistance and diagnosis of blood and urinary tract pathogens. For the classification and the identification of bacterial species, traditional approaches are routinely applied in microbiology, based on physiological, serological, biochemical, and chemotaxonomic methods; more recently genomic methods have been implemented for classification of bacteria using 16S rRNA and 18S rRNA gene sequencing. However, in the case of unknown bacterial samples PCR-based methods cannot be used for classification. Mass spectrometric proteomic approaches can be applied in microbiological applications for the identification of bacterial species, and characterization of microbial proteomes developed with both gel-based and gel-free protein separation methods. However, both SDS-PAGE protein profiling and 2D gel electrophoresis failed to become popular among microbiologists since they are laborious, time consuming and expensive. As a rapid and efficient alternative, the protein profiling of whole bacterial cells by MALDI time-of-flight mass spectrometry (MALDI-TOF MS) was proposed for detection of bacteria, regardless of their phylogeny and specific situation. All isolates, recovered either from human or animal tissues (blood, cerebrospinal fluid, biopsy, pus, wound, stool, *etc.*) or vegetal tissues or non-biological surfaces, can be analysed after incubation of specimens in appropriate media, and the number of publications related to MALDI-TOF MS applications in medical microbiology has steeply increased in the last 20 years.[64,65]

Identification of microbes by MALDI-TOF MS can be obtained by comparing the MS spectrum of unknown microbial isolates with the MS spectra of known microbial isolates contained in an extensive database. The identity of a microorganism can be established down to the genus, and in many cases to the species and strain level.[66] For species level identification, a typical mass range of *m*/*z* 2–20 kDa is used, which represents mainly ribosomal proteins, about 60–70% of the dry weight of a microbial cell.[67] Many commercial libraries of organism peptide mass fingerprints are available, making this approach simple and convenient in diagnostic laboratories.

A second option for microbial identification is matching the masses of a proteic biomarker, typical of that microbe, with the masses of proteins predicted from its genome sequence; this however requires knowledge of the complete genome sequence of an organism to create a database of its predicted proteins and their molecular masses. Microbial identification by MALDI-TOF MS is not influenced by the microbial culture conditions, although these might affect the microbial physiology and protein expression profile[68,69] nor by the culture time.[70] The sample preparation for MALDI-TOF of microorganisms depends on the source where they were isolated and the chemical constituents of their cell walls. Some microbes can be identified directly while some others require the preparation of whole cell lysates or crude cell extracts. In direct cell profiling, a single colony of microorganisms is spotted directly onto the MALDI sample plate and immediately covered with the matrix solution. Direct cell profiling was applied to identify Gram negative bacteria like *Neisseria* spp.[71] *Yersinia* spp.[72] and *Vibrio* spp.[73]

For Gram-positive species, a preparatory extraction of microbes with formic acid was useful prior to MALDI-TOF MS.[74,75] Verroken *et al.*[76] described a modified procedure for identification of aerobic actinomycetes *Nocardia* spp. due to the complex nature of their cell wall. Bacteria were lysed in boiling water, followed by ethanol precipitation of proteins. The precipitated proteins were dried, resuspended in 70% formic acid and acetonitrile and analyzed by MALDI-TOF MS. Since safety is a major concern for routine investigations of pathogens, methods have been proposed for sample preparation of mycobacteria. EI Khéchine *et al.*[77] described a procedure which combined inactivation and processing methods: mycobacterial colonies were placed in screw-cap tubes containing water and 0.5% Tween 20 and were inactivated by heating at 95 °C for 1 h. After centrifugation and disruption of the mycobacterial cell by vortexing with glass beads, the resultant pellet was re-suspended in formic acid, acetonitrile, and centrifuged again. Finally, the supernatant was deposited onto the MALDI target plate and covered with matrix. For the identification of yeasts, a preparatory extraction with formic acid was reported to be suitable.[78,79]

For fungal hyphae and spores Cassagne *et al.*[80] devised a protocol wherein fungi cultivated on Sabouraud gentamicin-chloramphenicol agar for 72 h at 27 °C were incubated in ethanol and extracted with formic acid. Acetonitrile was added, the mixture was centrifuged and supernatant was used for MALDI-TOF MS analysis. Lau *et al.*[81] reported a method based on mechanical lysis for sample preparation of fungal hyphae and spores. Analogously, analysis of intact cells of members of the genus *Penicillium* generated poor MALDI spectra, but re-suspension of the conidia and spores in trifluoroacetic acid-acetonitrile and disruption with glass beads discriminated the species with 100% accuracy.[82]

5.6.1. *Detection and Identification of Agents of Biological Warfare*

Biological warfare is formally defined as the conscious use of biological and chemical agents as weapons deliberately chosen because of their potentially harmful or lethal effects. In the current political scene, deliberate effort to engage in biological warfare is called bioterrorism. The term applies when chemical and biological weapons are used as a means of manifest or latent damage to obtain ideological, political or financial gain. To combat biological-warfare attacks, and to prevent natural outbreaks caused by pathogen organisms, rapid and accurate identification of microbes as agents of bioterrorism is required. Conventionally, organisms which pose severe threats are identified by phenotypic, genotypic, and immunological methods that are slow, burdensome and pose significant risk to the laboratory personnel. As a rapid alternative, MALDI-TOF MS succeeded in identifying highly pathogenic organisms like *Brucella* spp., *Coxiella burnetti, Bacillus anthracis, Francisella tularensis,* and *Y. pestis*. However, safe protocols are needed for inactivation of vegetative cells and spores of highly pathogenic organisms, that need to be integrated into the MALDI laboratory. Lasch *et al.*[83] proposed an inactivation protocol based on tri-fluoro acetic acid (TFA) for vegetative cells and spores, but Couderc *et al.*[84] found ethanol inactivation yielded MALDI-TOF MS spectra of significantly higher quality for Yersinia isolates. Jeong *et al.*[85] reported a direct *in situ* MALDI-TOF MS which allowed the detection and identification of aerosolized Bacillus spores, without any pre-treatment process. The spores were directly spotted on the MALDI target plate and left for air-drying; subsequently the matrix solution was applied and air-dried; the spotted samples were analyzed by MS.[85,86] *Bacillus anthracis* (genus Bacillus; family *Bacillaceae*) is a spore-forming gram- positive bacterium used in biological warfare and bio–terror activities because they are the causative agents of anthrax. For example, the mailing of *B. anthracis* spores in the US causing anthrax in 22 cases in 2001 (11 inhalational and 11 cutaneous cases; 5 of the inhalational cases were fatal). The following strains were used in the study of Jeong *et al.*:[86] *B. anthracis Sterne, B. cereus, B. globigii, B. subtilis, B. thuringiensis.* A single colony of each strain was inoculated into nutrient broth for sporulation, and incubated in a shaking incubator for 2–4 days at 32 °C. When the strains showed > 99% spore formation by optical microscopy, cultivated spores were collected by centrifugation. Purified spores were diluted to 1×10^9 colony forming unit (CFU) mL^{-1} by suspending in sterile water, and stored at 4 °C until used. The *B. anthracis* spores were inactivated by autoclave for 20 min at 212 °C. The concentration and inactivation of *B. anthracis* spores were confirmed by plate count in triplicate. For MALDI-TOF MS, about 1×10^6 CFU of purified spores of Bacillus species were directly spotted onto a single well of a target ground steel without any treatment and allowed to evaporate for 5 min at room temperature. Subsequently, 1 µL of 12 mg mL^{-1} α-cyano-4-hydroxy-cinnamic acid (HCCA) solution prepared in acetonitrile and trifluoroacetic acid (TFA) was applied to each dried spore spot on the MALDI target plate and then allowed to dry for 5 min at room temperature. The target plate was transferred to the vacuum area of the MALDI-TOF system, and mass spectra were acquired. The method was compared to other sample preparation methods: an inactivation method and a modified method combined with bead beating and TFA extraction. Each sample was spotted in triplicate. Mass spectra of spores were acquired using a Bruker autoflex speed RF MALDI-TOF. The instrument was equipped with a Nd/YAG laser operating at 355 nm with pulse rates of up to 1 kHz. The pulse ion extraction time was 200 ns. Acquisition of spectra were carried out in the linear mode in the mass to charge range between 2 and 20 kDa. *Escherichia coli* DH5α (Bruker Daltonics, Germany) was used as a reference strain for mass calibration, with a peak assignment tolerance of about 1000 ppm. At least 200 individual laser shots were summed for each spectrum. The mass spectra were processed by smoothing, baseline subtraction, and intensity normalization. Cluster analysis and principle component analysis (PCA) were performed. Mass spectrum data of each spore were obtained in 5 different runs of experiments to confirm the reproducibility of peak patterns. It was observed that to obtain biomarker peaks of *Bacillus* spores a higher laser power than generally used for microorganisms was needed to break down spores in the "direct mode" and to increase the ionization efficiency of the internal contents of the spores. The optimal conditions were found when 1 × 10 CFU/well was used to achieve the best signal-to-noise ratio in mass spectra. The spectrum for *B. anthracis* spores yielded the following main ionic species: 2503, 3089, 3376, 6684, 6698, 6753, and 6840. Ionic species at *m*/*z* 6684, 6698, 6753, and 6840 showed significantly discriminative pattern for identification of *B. anthracis*. When comparing the direct method to the inactivation and extraction methods, ionic species at *m*/*z* 3089, 6684, 6698, and 6840 were not observed when the inactivation and extraction methods were applied; the peak at *m*/*z* 2503 was not observed when the extraction method was used. Cluster analysis using mass spectra of 5 Bacillus spores and PCA analysis by mass spectra of 7 Bacillus spores and vegetative cells were carried out. The mass spectra of Bacillus spores and vegetative cells were distinguishable. However, precisely classifying the spores of *B. cereus* and *B. thuringiensis* was difficult because they showed high similarity patterns. Nonetheless, the *B. anthracis* spores could be easily distinguished from the other spores of the *B. cereus* family.

In a biological war, toxins delivered as aerosols are likely to cause maximum damage to a civilian or military population through inhalation and their early and unambiguous detection is required to initiate countermeasures. MALDI-TOF MS was applied for the detection of protein toxins, such as staphylococcal enterotoxin B, botulinum neurotoxins, *Clostridium perfringens* epsilon toxin, shiga toxin.[87,88]

Alam *et al.*[88] developed a simple method of sample processing for identification of protein toxins: a nebulizer was used to generate aerosols which were collected using a cyclone collector (see Figure 5.9). With MALDI-TOF-TOF, information from peptide sequences was used for detecting toxins from blind spiked samples. The bioinformatic analysis of seven protein toxins for elucidation of unique peptides with conservation status across all known sequences provided a high confidence for detecting toxins originating from any geographical location and source organism. ETX prototoxin could be detected from aerosol at 0.2 ppb concentration in aerosol.

Another example of a biological threat agent requiring a rapid detection and identification of the microorganism involved is the genus Brucella, that contains highly infectious species. The Brucella species primarily considered to be pathogenic for humans are *B. melitensis, B. suis* (biovars 1, 3, and 4), *B. abortus,* and occasionally *B. canis. B. suis* biovars 2 and 5 are not considered to be human pathogens because no human cases have been documented for these agents. The accurate identification of Brucella species using MALDI-TOF-MS was achieved by constructing a Brucella protein reference library based on multilocus variable-number tandem repeat analysis (MLVA) data.[89] From a frozen stock, the bacterial strains were cultured on blood agar plates for at least 48 h at 35 °C in the presence of 5% CO_2. Sample preparation included the suspension of 30 colonies for each species in 300 µl of water, inactivation with 900 µl of absolute ethanol and incubation for 90 min. The suspension samples were ultracentrifuged and the

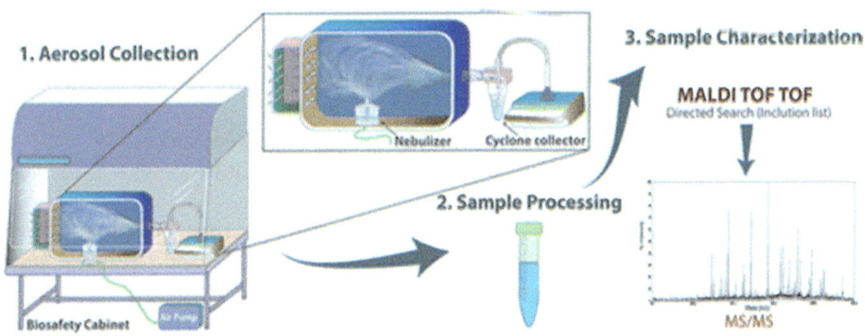

Figure 5.9. Protocol to collect toxins from aereosols for MALDI TOF–TOF analyses. Reprinted with permission from S. I. Alam, B. Kumar and D. V. Kambo, Multiplex detection of protein toxins using MALDI-TOF-TOF tandem mass spectrometry: application in unambiguous toxin detection from bioaerosol, *Anal. Chem.*, 2012, **84**, 10500–105070. Copyright 2012 American Chemical Society.

supernatants were removed. Subsequently, 50 μl of 70% formic acid was added to the pellet, and the pellet was mixed. Next, 50 μl of pure acetonitrile was added, and the suspension was mixed carefully. The particulate matter that could not be dissolved was spun down by ultracentrifugation. Finally, four spots were created, using 0.5 μl of the supernatant per spot, onto a MALDI-TOF polished steel target plate and air dried. Subsequently, the spots were overlaid with 0.5 μl of α-cyano-4-hydroxycinnamic acid and a 10 mg ml^{-1} acetonitrile/water solution (1 : 1) with 2.5% TFA and dried at room temperature. All of the mass spectra were automatically acquired on a Bruker Autoflex III instrument in linear mode using 20 kV source voltage. Composite mass spectra were generated from 10 different positions per spot using, in total, 2000 laser shots at each spot generated by a 200 Hz smartbeam laser (355 nm). The mass spectra were recorded in the *m*/*z* range 800–20 KDa. The instrument was externally calibrated with a bacterial test standard (BTS, Bruker) and *E. coli* DH5α was added during each extraction procedure. For the construction of the custom Brucella reference library, 24 MS spectra for each bacterium were generated (eight MS-spectra were generated per day on three different days).

5.7. Conclusions

MALDI-TOF-MS, exploiting light as an ionization/desorption method, can be developed into a fast and trustworthy identification method to analyze biological differences in microorganisms. The advantages of MALDI-TOF-MS over other techniques are that it is rapid, accurate and cost-effective, allowing the high-throughput identification of bacteria without preliminary identification steps. The bacteria are identified by comparing the obtained mass spectra to the mass spectra from a reference library.

References

1. U. Boesl, Laser mass spectrometry for environmental and industrial chemical trace analysis, *J. Mass Spectrom.*, 2000, **35**, 289–304.
2. R. C. Dunbar, *Gas Phase Ion Chemistry*, Academic Press, New York, 1979.
3. J. D. Morrison, *Gaseous Chemistry and Mass Spectrometry*, John Wiley & Sons, New York, 1986.
4. D. B. Robb, T. R. Covey and A. P. Bruins, Atmospheric pressure photoionisation: An ionization method for liquid chromatography-mass spectrometry, *Anal. Chem.*, 2000, **72**, 3653–3659.
5. D. B. Robb, A. P. Bruins, H. A. M. Peters and P. L. Jacobs, *Atmospheric Pressure Photoionization (APPI) for High Sensitivity LC/MS in Bioanalysis. In Proceedings of the 48th ASMS Conference in Mass Spectrometry and Allied Topics*, Long Beach, CA, 11–15 June, 2000.
6. G. Koster and A. P. Bruins, *Mechanisms for Ion Formation in LC/MS by Atmospheric Pressure Photo-ionization (APPI). In Proceedings of the 49th ASMS Conference in Mass Spectrometry and Allied Topics*, Chicago, IL, 27–31 May, 2001.

7. J. F. Alary, *Comparative Study: LC-MS/MS Analysis of Four Steroid Compounds Using a New Photoionization Source and a Conventional APCI Source. In Proceedings of the 49th ASMS Conference in Mass Spectrometry and Allied Topics*, Chicago, IL, 27–31 May, 2001.
8. G. Impey, B. Kieser and J. F. Alary. *No Title. In Proceedings of the 49th ASMS Conference in Mass Spectrometry and Allied Topics*, Chicago, IL, 27–31 May, 2001.
9. J. P. Rauha, H. Vuorela and R. Kostiainen, Effect of eluent on the ionization efficiency of flavonoids by ion spray, atmospheric pressure chemical ionization, and atmospheric pressure photoionization mass spectrometry, *J. Mass Spectrom.*, 2001, **36**, 1269–1280.
10. V. Kertesz and G. J. Van Berkel, Surface-assisted reduction of aniline oligomers, *N*-phenyl-1,4-phenylenediimine and thionin in atmospheric pressure chemical ionization and atmospheric pressure photoionization, *J. Am. Soc. Mass Spectrom.*, 2002, **13**, 109–117.
11. A. Leinonen, T. Kuuranne and R. Kostiainen, Liquid chromatography/mass spectrometry in anabolic steroid analysis–optimization and comparison of three ionization techniques: electrospray ionization, atmospheric pressure chemical ionization and atmospheric pressure photoionization, *J. Mass Spectrom.*, 2002, **37**, 693–698.
12. H. Keski-Hynnilä, *et al.*, Comparison of electrospray, atmospheric pressure chemical ionization, and atmospheric pressure photoionization in the identification of apomorphine, dobutamine, and entacapone phase II metabolites in biological samples, *Anal. Chem.*, 2002, **74**, 3449–3457.
13. C. Yang and J. Henion, Atmospheric pressure photoionization liquid chromatographic – mass spectrometric determination of idoxifene and its metabolites in human plasma, *J. Chromatogr. A*, 2002, **970**, 155–165.
14. T. J. Kauppila, *et al.*, Atmospheric pressure photoionization mass spectrometry. Ionization mechanism and the effect of solvent on the ionization of naphthalenes, *Anal. Chem.*, 2002, **74**, 5470–5479.
15. R. D. Levin and S. G. Lias, *Ionization Potential and Appearance Potential Measurements, 1971–1981*, U. S. Nat. Bur. Stand., NSRDS-NBS-71, 1982.
16. N. C. Fenner and N. R. Daly, Laser used for mass analysis, *Rev. Sci. Instrum.*, 1966, **37**, 1068–1070.
17. F. J. Vastola and A. J. Pirone, Ionization of organic solids by laser irradiation, *Adv. Mass Spectrom.*, 1968, **4**, 107–111.
18. F. J. Vastola, R. O. Mumma and A. J. Pirone, Analysis of organic salts by laser ionization, *Org. Mass Spectrom.*, 1970, **3**, 101–104.
19. M. A. Posthumus, P. G. Kistemaker, H. L. C. Meuzelaar and M. C. Ten Noever de Brauw, Laser desorption-mass spectrometry of polar nonvolatile bio-organic molecules, *Anal. Chem.*, 1978, **50**, 985–991.
20. C. L. Wilkins, D. A. Weil, C. L. C. Yang and C. F. Ijames, High mass analysis by laser desorption Fourier transform mass spectrometry, *Anal. Chem.*, 1985, **57**, 520–524.
21. M. Coates and C. Wilkins, Laser desorption Fourier transform mass spectra of malto-oligosaccharides, *Biomed. Mass Spectrom.*, 1985, **12**, 424–428.
22. R. D. Macfarlane, Mass spectrometry of biomolecules: from PDMS to MALDI, *Braz. J. Phys.*, 1999, **29**, 415–421.
23. M. Karas, U. Bahr and U. Gießmann, Matrix-assisted laser desorption ionization mass spectrometry, *Mass Spectrom. Rev.*, 1991, **10**, 335–357.
24. H. Wollnik, Time-of-flight mass analyzers, *Mass Spectrom. Rev.*, 1993, **12**, 89–114.

25. R. Knochenmuss, *Electrospray and MALDI Mass Spectrometry*, John Wiley & Sons, 2010.
26. R. Knochenmuss, Ion formation mechanisms in UV-MALDI, *Analyst*, 2006, **131**, 966–986.
27. P. D. Setz and R. Knochenmuss, Exciton mobility and trapping in a MALDI matrix, *J. Phys. Chem. A*, 2005, **109**, 4030–4037.
28. M. Karas, D. Bachmann and F. Hillenkamp, Influence of the wavelength in high-irradiance ultraviolet laser desorption mass spectrometry of organic molecules, *Anal. Chem.*, 1985, **57**, 2935–2939.
29. J. F. Ireland and P. A. H. Wyatt, Acid-base properties of electronically excited states of organic molecules, *Adv. Phys. Org. Chem.*, 1976, **12**, 131–221.
30. M. Karas, M. Glückmann and J. Schäfer, Ionization in matrix-assisted laser desorption/ionization: singly charged molecular ions are the lucky survivors, *J. Mass Spectrom.*, 2000, **35**, 1–12.
31. L. Leiserowitz, Molecular packing modes. Carboxylic acids, *Acta Crystallogr., Sect. B: Struct. Crystallogr. Cryst. Chem.*, 1976, **32**, 775–802.
32. J. K. Maurin, Oxime-carboxyl hydrogen bonds: the preferred interaction determining crystal packing of 'carboxyoximes', *Acta Crystallogr., Sect. B: Struct. Sci.*, 1998, **54**, 866–871.
33. J. D. Dunitz and P. Strickler, *Structural Chemistry and Molecular Biology*, W. H. Freeman, 1968.
34. J. Sunner, E. Dratz and Y. C. Chen, Graphite surface-assisted laser desorption/ionization time-of-flight mass spectrometry of peptides and proteins from liquid solutions, *Anal. Chem.*, 1995, **67**, 4335–4342.
35. E. P. Go, *et al.*, Desorption/ionization on silicon nanowires, *Anal. Chem.*, 2005, **77**, 1641–1646.
36. S. Alimpiev, S. Nikiforov, V. Karavanskii, T. Minton and J. Sunner, On the mechanism of laser-induced desorption-ionization of organic compounds from etched silicon and carbon surfaces, *J. Chem. Phys.*, 2001, **115**, 1891–1901.
37. M.-J. Kang, *et al.*, Nanowire-assisted laser desorption and ionization mass spectrometry for quantitative analysis of small molecules, *Rapid Commun. Mass Spectrom.*, 2005, **19**, 3166–3170.
38. N. Shenar, S. Cantel, J. Martinez and C. Enjalbal, Comparison of inert supports in laser desorption/ionization mass spectrometry of peptides: pencil lead, porous silica gel, DIOS-chip and NALDI target, *Rapid Commun. Mass Spectrom.*, 2009, **23**, 2371–2379.
39. J. A. Stolee, B. N. Walker, V. Zorba, R. E. Russo and A. Vertes, Laser–nanostructure interactions for ion production, *Phys. Chem. Chem. Phys.*, 2012, **14**, 8453–8471.
40. R. Gomer, Field emission, field ionization, and field desorption, *Surf. Sci.*, 1994, **299–300**, 129–152.
41. G. Luo, Y. Chen, G. Siuzdak and A. Vertes, Surface modification and laser pulse length effects on internal energy transfer in DIOS, *J. Phys. Chem. B*, 2005, **109**, 24450–24456.
42. J. Wei, J. M. Buriak and G. Siuzdak, Desorption-ionization mass spectrometry on porous silicon, *Nature*, 1999, **399**, 243–246.
43. Y. Chen and A. Vertes, Adjustable fragmentation in laser desorption/ionization from laser-induced silicon microcolumn arrays, *Anal. Chem.*, 2006, **78**, 5835–5844.
44. J. A. Stolee, B. N. Walker, Y. Chen and A. Vertes, Nanophotonic ion sources, *AIP Conf. Proc.*, 2010, **1278**, 98–110.

45. C.-K. Chiang, *et al.*, Nanomaterial-based surface-assisted laser desorption/ionization mass spectrometry of peptides and proteins, *J. Am. Soc. Mass Spectrom.*, 2010, **21**, 1204–1207.
46. G. Piret, H. Drobecq, Y. Coffinier, O. Melnyk and R. Boukherroub, Matrix-free laser desorption/ionization mass spectrometry on silicon nanowire arrays prepared by chemical etching of crystalline silicon, *Langmuir*, 2010, **26**, 1354–1361.
47. S. Dagan, *et al.*, Internal energy deposition with silicon nanoparticle-assisted laser desorption/ionization (SPALDI) mass spectrometry, *Int. J. Mass Spectrom.*, 2009, **283**, 200–205.
48. R. Nayak and D. R. Knapp, Matrix-free LDI mass spectrometry platform using patterned nanostructured gold thin film, *Anal. Chem.*, 2010, **82**, 7772–7778.
49. L. Colaianni, *et al.*, Laser desorption ionization-mass spectrometry detection of amino acids and peptides promoted by gold nanowires, *Sens. Lett.*, 2010, **8**, 539–544.
50. A. L. Pyayt, B. Wiley, Y. Xia, A. Chen and L. Dalton, Integration of photonic and silver nanowire plasmonic waveguides, *Nat. Nanotechnol.*, 2008, **3**, 660–665.
51. T. Seino, *et al.*, Matrix-free laser desorption/ionization-mass spectrometry using self-assembled germanium nanodots, *Anal. Chem.*, 2007, **79**, 4827–4832.
52. H. Sato, A. Nemoto, A. Yamamoto and H. Tao, Surface cleaning of germanium nanodot ionization substrate for surface-assisted laser desorption/ionization mass spectrometry, *Rapid Commun. Mass Spectrom.*, 2009, **23**, 603–610.
53. R. Ma, M. Lu, L. Ding, H. Ju and Z. Cai, Surface-assisted laser desorption/ionization mass spectrometric detection of biomolecules by using functional single-walled carbon nanohorns as the matrix, *Chem.–Eur. J.*, 2013, **19**, 102–108.
54. S. L. Candelaria, *et al.*, Nanostructured carbon for energy storage and conversion, *Nano Energy*, 2012, **1**, 195–220.
55. M. Najam-ul-Haq, *et al.*, Role of carbon nano-materials in the analysis of biological materials by laser desorption/ionization-mass spectrometry, *J. Biochem. Biophys. Methods*, 2007, **70**, 319–328.
56. S. Cha and E. S. Yeung, Colloidal graphite-assisted laser desorption/ionization mass spectrometry and MSn of small molecules. 1. Imaging of cerebrosides directly from rat brain tissue, *Anal. Chem.*, 2007, **79**, 2373–2385.
57. H. Kawasaki, T. Yonezawa, T. Watanabe and R. Arakawa, Platinum nanoflowers for surface-assisted laser desorption/ionization mass spectrometry of biomolecules, *J. Phys. Chem. C*, 2007, **111**, 16278–16283.
58. S. Cha, *et al.*, Direct profiling and imaging of plant metabolites in intact tissues by using colloidal graphite-assisted laser desorption ionization mass spectrometry, *Plant J.*, 2008, **55**, 348–360.
59. S. A. Trauger, *et al.*, High sensitivity and analyte capture with desorption/ionization mass spectrometry on silylated porous silicon, *Anal. Chem.*, 2004, **76**, 4484–4489.
60. M. J. Dale, R. Knochenmuss and R. Zenobi, Graphite/Liquid mixed matrices for laser desorption/ionization mass spectrometry, *Anal. Chem.*, 1996, **68**, 3321–3329.
61. Y. Chen, H. Chen, A. Aleksandrov and T. M. Orlando, Roles of water, acidity, and surface morphology in surface-assisted laser desorption/ionization of amino acids, *J. Phys. Chem. C*, 2008, **112**, 6953–6960.
62. K. Tanaka, *et al.*, Protein and polymer analyses up to m/z 100,000 by laser ionization time-of-flight mass spectrometry, *Rapid Commun. Mass Spectrom.*, 1988, **2**, 151–153.
63. C. Dass, The role of a liquid matrix in controlling FAB-induced fragmentation, *J. Mass Spectrom.*, 1996, **31**, 77–82.

64. N. Singhal, M. Kumar, P. K. Kanaujia and J. S. Virdi, MALDI-TOF mass spectrometry: an emerging technology for microbial identification and diagnosis, *Front. Microbiol.*, 2015, **6**, 791.
65. C. K. Fagerquist, B. R. Garbus, W. G. Miller, K. E. Williams, E. Yee and A. H. Bates, *et al.*, Rapid identification of protein biomarkers of *Escherichia coli* O157 : H7 by matrix-assisted laser desorption ionization-time-of-flight-time-of-flight mass spectrometry and top-down proteomics, *Anal. Chem.*, 2010, **82**, 2717–2725.
66. M. Welker, Proteomics for routine identification of microorganisms, *Proteomics*, 2011, **11**, 3143–3153.
67. N. Valentine, S. Wunschel, D. Wunschel, C. Petersen and K. Wahl, Effect of culture conditions on microorganism identification by matrix-assisted laser desorption ionization mass spectrometry, *Appl. Environ. Microbiol.*, 2005, **71**, 58–64.
68. E. Carbonnelle, J. L. Beretti, S. Cottyn, G. Quesne, P. Berche, X. Nassif and A. Ferroni, *J. Clin. Microbiol.*, 2007, **45**, 2156–2161.
69. E. N. Ilina, A. D. Borovskaya, M. M. Malakhova, V. A. Vereshchagin, A. A. Kubanova and A. N. Kruglov, *et al.*, Direct bacterial profiling by matrix-assisted laser desorption-ionization time-of-flight mass spectrometry for identification of pathogenic Neisseria, *J. Mol. Diagn.*, 2009, **11**, 75–86.
70. R. Stephan, N. Cernela, D. Ziegler, V. Pflüger, M. Tonolla and D. Ravasi, *et al.*, Rapid species specific identification and subtyping of Yersinia enterocolitica by MALDI-TOF mass spectrometry, *J. Microbiol. Methods*, 2011, **87**, 150–153.
71. R. Eddabra, G. Prévost and J. M. Scheftel, Rapid discrimination of environmental Vibrio by matrix-assisted laser desorption ionization time-of-flight mass spectrometry, *Microbiol. Res.*, 2012, **167**, 226–230.
72. A. A. Alatoom, S. A. Cunningham, S. M. Ihde, J. Mandrekar and R. Patel, Comparison of direct colony method *versus* extraction method for identification of Gram-Positive cocci by use of bruker biotyper matrix-assisted laser desorption ionization–time of flight mass spectrometry, *J. Clin. Microbiol.*, 2011, **49**, 2868–2873.
73. R. T. Saffert, S. A. Cunningham, S. M. Ihde, E. Kristine, J. Monson and R. Patel, *et al.*, Comparison of bruker biotyper matrix-assisted laser desorption ionization–time of flight mass spectrometer to BD phoenix automated microbiology system for identification of gram-negative bacilli, *J. Clin. Microbiol.*, 2011, **49**, 887–892.
74. A. Verroken, M. Janssens, C. Berhin, P. Bogaerts, T. D. Huang and G. Wauters, *et al.*, Evaluation of matrix-assisted laser desorption ionization–time of flight mass spectrometry for identification of *Nocardia* species, *J. Clin. Microbiol.*, 2010, **48**, 4015–4021.
75. A. E. I. Khéchine, C. Couderc, C. Flaudrops, D. Raoult and M. Drancourt, Matrix-assisted laser desorption/ionization time-of-flight mass spectrometry identification of mycobacteria in routine clinical practice, *PLoS One*, 2011, **6**, e24720.
76. L. G. Stevenson, S. K. Drake, Y. R. Shea, A. M. Zelazny and P. R. Murray, Evaluation of matrix-assisted laser desorption ionization-time of flight mass spectrometry for identification of clinically important yeast species, *J. Clin. Microbiol.*, 2010, **48**, 3482–3486.
77. E. S. Theel, B. H. Schmitt, L. Hall, S. A. Cunningham, R. C. Walchak and R. Patel, *et al.*, Formic acid-based direct, on-plate testing of yeast and Corynebacterium species by Bruker Biotyper matrix-assisted laserdesorption ionization–time of flight mass spectrometry, *J. Clin. Microbiol.*, 2012, **50**, 3093–3095.

78. C. Cassagne, F. Pratlong, F. Jeddi, R. Benikhlef, K. Aoun and A. C. Normand, *et al.*, Identification of Leishmania at the species level with matrix-assisted laser desorption ionization time-of-flight mass spectrometry, *Clin. Microbiol. Infect.*, 2014, **20**, 551–557.
79. A. F. Lau, S. K. Drake, L. B. Calhoun, C. M. Henderson and A. M. Zelazny, Development of a clinically comprehensive database and a simple procedure for identification of molds from solid media by matrix-assisted laser desorption ionization–time of flight mass spectrometry, *J. Clin. Microbiol.*, 2013, **51**, 828–834.
80. J. M. Hettick, B. J. Green, A. D. Buskirk, M. L. Kashon, J. E. Slaven and E. Janotka, *et al.*, Discrimination of Penicillium isolates by matrix-assisted laser desorption/ionization time of- flight mass spectrometry fingerprinting, *Rapid Commun. Mass Spectrom.*, 2008, **22**, 2555–2560.
81. D. B. Jernigan, P. L. Raghunathan, B. P. Bell, R. Brechner, E. A. Bresnitz, J. C. Butler, M. Cetron, M. Cohen, T. Doyle, M. Fischer, C. Greene, K. S. Griffith, J. Guarner, J. L. Hadler, J. A. Hayslett, R. Meyer, L. R. Petersen, M. Phillips, R. Pinner, T. Popovic, C. P. Quinn, J. Reefhuis, D. Reissman, N. Rosenstein, A. Schuchat, W. J. Shieh, L. Siegal, D. L. Swerdlow, F. C. Tenover, M. Traeger, J. W. Ward, I. Weisfuse, S. Wiersma, K. Yeskey, S. Zaki, D. A. Ashford, B. A. Perkins, S. Ostroff, J. Hughes, D. Fleming, J. P. Koplan, J. L. Gerberding and National Anthrax Epidemiologic Investigation Team, Investigation of bioterrorism-related anthrax, United States, 2001: epidemiologic findings, *Emerging Infect. Dis.*, 2002, **8**, 1019–1028.
82. P. Lasch, H. Nattermann, M. Erhard, M. Stämmler, R. Grunow and N. Bannert, *et al.*, MALDI-TOF mass spectrometry compatible inactivation method for highly pathogenic microbial cells and spores, *Anal. Chem.*, 2008, **80**, 2026–2034.
83. C. Couderc, C. Nappez and M. Drancourt, Comparing inactivation protocols of Yersinia organisms for identification with matrix-assisted laser desorption/ionization time-of-flight mass spectrometry, *Rapid Commun. Mass Spectrom.*, 2012, **26**, 710–714.
84. Y. S. Jeong, S. Choi, E. Chong, J. H. Kim and S. J. Kim, Rapid detection of Bacillus spore aerosol particles by direct *in situ* analysis using MALDI-TOF mass spectrometry, *Lett. Appl. Microbiol.*, 2014, **59**, 177–178.
85. Y. S. Jeong, J. Lee and S. J. Kim, Discrimination of Bacillus anthracis spores by direct *in-situ* analysis of matrix-assisted laser desorption/ionization time-of-flight mass spectrometry, *Bull. Korean Chem. Soc.*, 2013, **34**, 2635–2639.
86. S. Kull, D. Pauly, B. Störmann, S. Kirchner, M. Stämmler and M. B. Dorner, *et al.*, Multiplex detection of microbial and plant toxins by immunoaffinity enrichment and matrix-assisted laser desorption/ionization mass spectrometry, *Anal. Chem.*, 2010, **82**, 2916–2924.
87. S. I. Alam, B. Kumar and D. V. Kamboj, Multiplex detection of protein toxins using MALDI-TOF-TOF tandem mass spectrometry: application in unambiguous toxin detection from bioaerosol, *Anal. Chem.*, 2012, **84**, 10500–10507.
88. F. Lista, F. A. Reubsaet, R. De Santis, R. R. Parchen, A. L. de Jong and J. Kieboom, *et al.*, Reliable identification at the species level of *Brucella isolates* with MALDI-TOF-MS, *BMC Microbiol.*, 2011, **11**, 267.
89. F. Lista, F. A. Reubsaet, R. R. De Santis, R. R. Parchen, A. L. de Jong, J. Kieboom, A. L. van der Laaken, I. A. Voskamp-Visser, S. Fillo, H. J. Jansen, J. Van der Plas and A. Paauw, *BMC Microbiol.*, 2011, **11**, 267.

Chapter 6

The Use of Light in Forensic Glass Analysis

Claire Gwinnett*[a] and Roger Robson*[b]

[a]Staffordshire University, Forensic and Crime Science Department, The Science Centre, Leek Road, Stoke-on-Trent, ST42DF, UK;
[b]Chartered Society of Forensic Sciences, Copthall Bridge House, Station Bridge, Harrogate, HG1 1SP, UK
*E-mail: c.gwinnett@staffs.ac.uk, roger.robson@forensic-outlook.com

Table of Contents

Light in Forensic Science: Issues and Applications
Edited by Giorgia Miolo, Jacqueline L. Stair and Mire Zloh

Published by the Royal Society of Chemistry, www.rsc.org

6.1. Introduction

Glass is commonly described as a particulate or 'trace' evidence that is traditionally used to provide information regarding whether a person or object has come in contact with another person or object at a crime scene as described in Locard's fundamental principle of 'every contact leaves a trace'. The multitude of modern day objects that are made, or part-made, of glass means that there is a plethora of opportunities for glass to be broken and then subsequently transferred to the perpetrator of a crime or to a victim during the course of a criminal activity. Glass is commonly found at crime scenes where windows or doors containing glass have been broken in order to obtain entry, or exit, including both buildings and vehicles.[1,2] Glass objects such as drinking vessels, bottles, and decorative glassware are commonly broken during the commission of crimes against the person and can readily transfer to individuals or objects that come in contact. In addition to these examples, glass from light bulbs and light casings, and windscreens, particularly from vehicles, can be retrieved from vehicle collisions, including in hit and run cases and ram raids. Glass can help provide important information regarding the 'who?', 'what?', 'where?', 'when?' and 'how?' of a case. This information is gathered by determining how glass transfers and persists and by applying a range of sensitive techniques, many of which utilise light to characterise glass samples. The optical properties of glass lend itself to being scrutinised under different light sources and employing certain wavelengths of light to investigate its composition. Prior to the introduction of more sophisticated techniques, the use of light was relied upon for the identification of glass and its subsequent comparison. Today, light still holds an important place in glass analysis whilst also complementing more sensitive techniques.

At this point it is important to define 'light' in terms of the wavelength range which the techniques discussed in this chapter utilise to help characterise glass. The electromagnetic spectrum ranges from the shorter wavelengths of gamma rays to the longer wavelengths of radio waves. 'Light' typically refers to the visible range which spans from 380 nm to 700 nm (violet to red respectively). This range borders upon the near visible ranges of ultraviolet (10 nm to 380 nm) and infrared (700 nm to 1000 nm). This chapter concentrates upon the visible spectrum but as UV illumination is particularly useful for the identification of float glass, it will also be discussed. The X-ray region of the electromagnetic spectrum (0.1 to 10 nm) borders upon the ultraviolet region and is also useful for determining the trace elemental composition of glass. Techniques utilising light for the analysis of trace elements in glass will be briefly covered.

In order to truly understand how and when the use of light is significant to forensic glass analysis, it is important to know what glass is in terms of its composition and to comprehend its value as evidence, as ultimately these are determined by the techniques utilised in its characterisation.

Sections 6.1.1 to 6.1.2 outline this key information and therefore help understand the benefits and limitations of using light to analyse glass evidence.

6.1.1. *What is Glass?*

Historically, glass has been produced for thousands of years; starting in ancient Egypt until the modern day where glass now has extensive uses and applications. Developments in manufacturing methods now means that the world in which we live is a wealth of sources of glass types and articles, including but not limited to: optical glass, flat glass, container glass, tableware, industrial and laboratory glass, glass fibres and glass for electrotechnology and electronics. In forensic casework, there is no limit to the glass types that may be analysed due to the unique nature of crime scenes and the individuals involved, but it is common for glass samples to have originated from items that are highly prevalent in the environment and that commonly get broken during a crime. Due to this, forensic glass analysts may expect to see the majority of samples originating from flat glass, container glass and tableware. The most common of these is flat glass with container glass being less prevalent in casework. Some glass, due to its infrequency in the environment is a gold chalice for glass analysts, for example, historic coloured glass from church windows.

Glass is commonly understood as being a hard, brittle material which has solidified from a molten form without crystallising.[3] The vast majority of commonly found glass at crime scenes contains silica, is inorganic and non-metallic but this is not true for all glass. Although much less common, there are other glass types that are organic, that do not utilise silica as a raw ingredient (instead they use a different glass-forming oxide) or are metallic. In addition to these, glass may also not be produced by cooling from a melt, for example, glass may be formed by vapour deposition or the sol–gel process.[4,5] With these exceptions in mind, the only common characteristics of glass is that no glass has a long range periodic atomic arrangement (is an amorphous solid)[4] and exhibits a transformation range which is time and temperature dependent. This relationship between a melt, a supercooled liquid and a glass is frequently described by using a volume–temperature diagram as seen in Rawson's *Glasses and their Applications*.[6] This transformation range is important as the rate of cooling during production affects the final glass properties, for example, fast cooling through the transformation stage, creates a higher density glass at room temperature; a characteristic which is sometimes quantified in forensic glass analysis. This transformation range can also affect the homogeneity of the refractive index (RI) of the glass; without a highly controlled cooling rate, the RI of the glass can differ from one area to another due to differences in its thermal history.[6] This variation in RI is very important to consider when interpreting RI measurements of glass from casework and is further discussed in Section 6.3.2.

Ultimately, given these variations in glass composition and their potential affect upon analysis, it is adequate to define glass as a 'frozen supercooled liquid'.[5]

There are three methods of classifying glass; by chemical composition, processing behaviour or by end-use. These methods are not exclusive to each other and are all important for forensic analysis, but chemical composition and end-use are more regularly determined in casework. A broad categorisation of different chemical compositions leads to four main groups; soda-lime, borosilicate, lead and special glass. Soda-lime glass is a very common glass and frequently seen in casework. Typically, soda-lime is composed of 71–75% by weight SiO_2 (sand) with 12–16% soda and is used to make drinking glasses, bottles and window glass. Borosilicate glass typically contains 70–80% SiO_2, 7–13% B_2O_3 (boric oxide) plus NaO_2 (soda), K_2O (potash) and Al_2O_3 and has a high resistance to temperature change and chemicals resulting in this glass type being used for laboratory glassware and kitchenware. Lead crystal has 54–65% SiO_2, 18–38% PbO (lead oxide), 13–15% NaO_2 or K_2O and a variety of other oxides. This high amount of lead oxide gives a high refractive index and therefore is used for high-end tableware and drinking glasses due to its 'sparkling' appearance when cut. Lead crystal is typically more expensive to produce and therefore tableware made of such glass is expensive to the customer. Given the socio-economic influences this means that the glass analyst rarely sees lead crystal in casework. Special glasses can be incredibly varied due to the different elements included into the raw mixture and can include glasses such as aluminosilicate glass, optical glasses and glasses used in electronics.[5] The variation in composition and the large range of elements seen in these glasses can make these particularly evidentially useful if they occur in forensic casework. The chemical composition of glass may be determined in forensic casework through identifying particular characteristics such as the refractive index (RI) but trace elemental analysis is often required to confirm this due to there being an overlap in optical properties between different glass compositions. Light as a source of information, in this case, can be useful for eliminating certain glass types.

Classifying glass by its end-use is useful as it also encompasses the manufacturing process it has undergone but can be confusing when determining actual source. For example, flat glass can have different chemical compositions but also different end-uses, *e.g.* it may be used for windows or tables or picture frames. With this in mind, glass is more easily classified into broad categories for forensic purposes, such as sheet (flat glass), tableware, container, fibre glass *etc*. In glass manufacture, container glass and tableware can sometimes be described as subcategories of hollowware and glass tubing,[5] so it is important to clearly define what objects fall into each category when interpreting glass evidence. Many of these have subcategories which may be determined in a case if there are sufficient features to do so. For example, different types of processed flat glass such as safety glass and glass with different surface finishings. This classification by end-use is further discussed in Section 6.1.2.1.

6.1.2. *Evidential Value of Glass in Forensic Casework*

The evidential value of glass lies in its ability to provide the following information:

a. Source level information, *i.e.* identification of where the glass may have originated from.
b. Activity level information, *i.e.* how the glass may have transferred to the object or individual and how long it may have been retained (persistence of the glass).

Both source level and activity level information will be briefly outlined but a focus will be placed upon source level identification as it is this level of information in which light can contribute the most.

6.1.2.1. Source Level Information. Source level information is used to answer the common question of 'what?' and by associating glass found on individuals to glass items at a crime scene, also the question of 'who?'. This form of evidence will never compete with the likes of fingerprints and DNA for answering the question of 'who was at a scene?' but it holds value as corroborative evidence and increases in its importance when the latter evidence types are not present. When evaluating source level, a forensic expert may consider two propositions, such as the following:

Prosecution proposition H_p: *The recovered glass from the suspect's clothing originates from the window broken at the crime scene.*

Defence proposition H_d: *The recovered glass from the suspect's clothing does not originate from the window broken at the crime scene.*

These propositions underpin the use of the likelihood ratio approach for interpreting evidence which allows for an objective use of data to account for factors that forensic glass analysts identify as important, such as frequency of occurrence.[7,8] This chapter focuses upon the use of light in the analysis of glass, therefore the use of likelihood ratios is only addressed in terms of what information from utilising light may be incorporated into this interpretation, but many resources exist to aid the forensic glass analyst in this approach.[8–15]

As corroborative evidence, the value of glass lies in its ability to be associated with a particular source of glass whilst being differentiated from others. As glass is a mass-produced product, with many major manufacturers producing the majority of glass and different glass categories (*e.g.*

vehicle window glass and architectural window glass) being produced by the same methods,[16] there is a requirement, therefore, for the use of a combination of analysis techniques to discriminate between glass sources. To address source level propositions, a knowledge of the inter and intra variation of different characteristics (aka the between source and within source variability) is required; this is discussed in more detail in Section 6.3.2.

It is common in an investigation for glass to be analysed to determine the source by identifying its glass type (composition) and end-use.[17] This can provide intelligence information which can help inform the events at a crime scene. Sensitive techniques are required in order to characterise the glass to the extent that different sources can be identified. In addition to this, reference collections or studies of different glass objects are required to aid interpretation. Glass classification, *i.e.* the categorisation of glass evidence into a broad class that identifies to a certain extent its use, has been shown to be possible by utilising multiple analysis techniques, for example, refractive index (RI) and elemental composition. A study by Hickman in 1981 [16] indicated that with the use of a reference collection of known samples and multivariate analysis, 70% of the 349 glass samples analysed could be categorised into three main groups; sheet, container and tableware. Although these are broad categories which may be subject to overlap in terms of the items that may be included in these categories, for example, tableware may be a diverse category including drinking vessels and decorative items which may also be described as 'containers', this classification is important in narrowing down possible sources that may then be scrutinised in terms of their prevalence in the environment.

6.1.2.2. Activity Level Information for Glass Evidence. Activity level information describes the mechanisms by which glass may be transferred to a particular object or individual and the time that it may persist upon that object or person. Arguably, this level of information is more important than source level for evidence such as glass, as this level of information can be used to help reconstruct a crime scene and identify timelines of events and as such, can help answer the questions of 'what?' 'when?', 'where?' and how?' For example, the distribution of glass shards on a suspect's clothing may give an indication as to where he/she was standing, or which arm he/she used to smash a window. In the same manner as source level, when evaluating activity level, a forensic expert may consider two propositions, such as the following:

Prosecution proposition H_p: *The suspect broke the window at the crime scene.*

Defence proposition H_d: *The suspect did not break the window at the crime scene and denies any recent activity with broken glass associated with the incident.*

In order to be able to address these propositions, a deep understanding of how glass transfers and the factors that affect its persistence is required. Knowledge of what factors contribute to primary, secondary, tertiary and possibly quaternary transfer is required when someone breaks glass or is near to glass at the time of it being broken. Many studies have investigated the amount of glass transferred under different circumstances, *e.g.* different glass objects, position of person breaking glass, different clothing, number of impacts, and also analysed how much glass persists on different surfaces over time under different activities, *e.g.* walking. For example, studies have been conducted investigating the transfer and subsequent persistence of glass indirectly transferred from hands to outer jackets,[18] the transfer to upper garments from breaking bottles and drinking glasses[19] and the transfer of float glass into garments using different objects to break the glass.[20] Comprehensive studies have been conducted to understand the mechanisms of how glass breaks and the subsequent factors that affect the number and distribution of transferred glass particles.[21–29] Together these studies provide the foundation of data required to calculate the probability of the evidence given the two opposing activity level propositions, although the diverse variables involved in criminal activities mean that this is still challenging.

To identify the likelihood that glass shards may be present on a person or object by chance rather than through the activity of a crime, the prevalence of finding such glass in the population must be known.[15] For both source and activity level, it is beneficial to use robust data which can be obtained by collating the characteristics of different glass samples either analysed through research or casework, and/or manufacturer information regarding the production rates of different glass types. These days most forensic labs have their own glass reference databases; most of which are localised which may be of importance when determining geographic commonality. They may be constructed of samples collected over many years from casework and/or directly from glass manufacturers. In addition to glass databases, population studies are vital in understating how common a particular glass composition/type and collection of glass characteristics are in different environments and the likelihood that these glass samples may be there by chance alone. Studies into the frequency of glass fragments being found in the head hair and headwear of the general population *vs.* individuals who work in the glass industry,[30] on clothing,[31] clothing from individuals suspected to be involved in a crime,[32,33] high school children's clothing and footwear[33,34] all aid in the interpretation of glass evidence and indicate its potential evidential value.

By understanding the evidential value of glass, the types of questions that this form of evidence may be able to answer and the concepts of both source and activity level propositions, a glass analyst will best be in a position to develop an effective analysis strategy that employs light to its fullest potential.

6.1.3. *The Importance of Light in the Analysis of Glass*

Many reasons why we use glass in day-to-day life is due to the fact that it is one of the few solids to transmit light in the visible spectrum.[4] In its uses in light bulbs, windows and optics, the ability to transmit light is crucial. Its optical properties are fundamental to the desired characteristics of many glass objects, for example, the high RI of lead crystal for creating perceived aesthetic properties of high-end glassware. In glass manufacture, the optical properties are categorised into bulk (*e.g.* RI and optical dispersion), colour and other optical effects such as photosensitivity.[4] For forensic purposes, the focus is upon the quantitative analysis of some of these properties, namely RI and utilising light to observe any morphological features which may allow sample discrimination.

A variety of different techniques which utilise light for the analysis of glass are available. The decision as to which may actually be used in a case and the order they are employed, may be dependent upon a number of factors, these include:

- recovery methods
- laboratory protocols and equipment (and whether these techniques have been validated through the laboratory accreditation process)
- sample size available as some observations/techniques may not be possible with microscopic samples
- the glass type/characteristics
- the questions requested in the investigation, *e.g.* whether this is a comparison or whether further information about its potential source is required
- the destructive nature of any tests.

Once retrieved, the analysis of glass is typically stepwise in nature with less time consuming and therefore relatively inexpensive techniques being used first to screen samples and gain initial morphological information and then proceeding to more sensitive methods to gain quantitative data regarding the samples' optical properties, curvature and composition. If at each stage evidence samples are unable to be differentiated from the control samples, further levels of information will be sought. This process is not limitless, cost implications and resources will ultimately be included in the decision as to the extent of the analysis to be completed whilst also ensuring the needs of the investigation are sufficiently met.

It is common to employ a range of microscopy techniques, all of which utilise light to either qualitatively or quantitatively characterise the samples, these include stereomicroscopes,[35] polarizing light microscopes[36] and phase contrast microscopes with heated stage.[37–42] The latter is probably the most commonly used approach for glass analysis as it is able to quantify a key glass property using light, its refractive index. In addition to this, the use of a furnace may be employed to anneal glass samples to observe any change in refractive index.[43–47] Use of further microscopy techniques, such as interferometry and Differential Interference Contrast (DIC) microscopy use the ability to split light into two rays as a method for observing surface detail and topography.[35,48–51] The use of light in the ultraviolet (UV) region, in the form of fluorescence microscopy or a UV light source is important in identifying flat glass and any photoluminescence properties in glass.[35] It must be noted, that not all approaches used in glass analysis use light to characterise and observe, but still contribute to the knowledge of a sample. Glass density analysis, in the past, has sometimes been used in casework as an additional and relatively simple and inexpensive quantitative method to complement RI measurements but does not utilise light in its approach, rather it uses a sink/float method using calibrated density liquids.[52] In addition, a plethora of sophisticated techniques for the analysis of the trace element composition of glass are available, most of which do not use light as a source of energy or source of information, other than laser ablation ICP-MS; this and other approaches are listed in Section 6.5.

These techniques may be utilised in forensic analysis but additional observations and methods are commonly employed within the glass industry, normally during the manufacturing process and post-production analysis. This chapter does not focus upon these techniques, but is worth noting that of these, glass industries may use the technique of optical dispersion which utilises light to control chromatic aberration. Optical dispersion is the measurement of refractive index over a range of different specified wavelengths. For example, the difference in refractive index may be calculated between the RIs measured at 486.1 and 656.3 nm (the F and C emission lines of hydrogen respectively).[4]

The use of light in the analysis of glass is of paramount importance; it is a fundamental element of many of the regularly employed techniques in the forensic glass analyst's armoury. The nature of glass lends itself to the use of light to determine many of its morphological, optical and compositional properties.

6.2. Use of Light for the Initial Analysis of Glass Samples

A priority of screening evidence samples is to eliminate samples which clearly are not associated with the known samples and therefore reduce time and cost of the investigation, and also to create a strategic approach for the further analysis of potential evidence. In this stage of the examination, priorities may be made in terms of sample size and choice which will form the direction of the investigation and ensure that appropriate samples have been chosen for further analysis. An example of this may be the decision to only analyse glass shards bearing a sharp fracture edge as this is most likely to have been recently broken.[31] The use of light is fundamental to this stage of the analysis, as the use of different forms of transmitted and reflected light microscopy can reveal important features for comparison and categorisation.

An initial task in forensic glass analysis is the sorting of glass fragments out of bulk debris recovered *via* various methods of agitation from a suspect's clothing, collected debris from scenes or from vacuum sweepings, tape lifts or combings of samples from hair. These may be separated using forceps, a fine steel probe or by gently shaking collected material/brushing with a paint brush onto paper[31] or into petri dishes.[30] The use of a magnified light or low power microscopy at this stage can be invaluable for ensuring there is no loss of sample and for visualising the initial features of any glass samples for sorting. Use of a labelled grid under petri dishes or tape lifts helps create a systematic approach to searching for samples. In addition to this, different coloured backgrounds may be utilised, *e.g.* dark coloured cardboard, to improve visualisation. It is important to note at this stage that analysis of glass samples should be conducted in a clean search room with the option of work benches being covered in paper.

The majority of glass evidence in casework is very small in size and therefore can be difficult to observe with the naked eye. Due to this, the aid of stereomicroscopy techniques is needed for observing features on the original surface or within the glass. This may be accompanied by mounting the glass fragments onto stubs of Blu-tack® mixed with carbon black to give a contrasting background.[35] Low magnification is typically used first, increasing to higher magnifications for smaller samples.[30] Features observed are dependent upon the size of the glass sample, the type and whether fracture analysis is also to be carried out. The latter may involve the observations of fractures such as Wallner lines on the broken edge of the glass sample to indicate the direction of force.

The size of the glass shard may be measured if the shape of the shard is indicative of it being tempered glass (aka toughened or safety glass) as this type of glass shatters into small chunky cuboid, wedge or spiked shaped pieces rather than the very thin tapered shards sometimes described as 'beeswings' seen in non-toughened glass.[53,54] Sizes of transferred glass are typically small with ranges between 0.1 and 0.2 mm being found, although people who work with glass or prolific occupational burglars have had larger samples found upon them.[30] Studies which have surveyed different glass types have ascertained the thickness ranges of glass from different sources, for example, toughened vehicle glass was shown to typically be 4.5 or 6 mm.[2]

The use of incident light[35] can be useful in initially observing any surface features, particularly when observing samples at oblique angles as the transparent nature of glass can make these features difficult to see in transmitted light. Any samples with particular surface features may be identified initially by specular reflections from the illuminating source; this should be achieved by using co-axial illumination.[35] Surface features may be indicative of how the glass has been manufactured or may include damage caused by its use or environment. Surface features may be as a result of a surface finishing that has been applied during manufacturing to suit end-use or current trends in glass, examples of these may be a mirrored surface, crackle glass (purposely cracked surfaces for decoration) or frosted glass (which is created by applying hydrofluoric acid or sandblasting the surface[5]). These surface modifications are normally clearly visible, but small fragments and other surface features may require a stereomicroscope to view. Examples of other surface features that may be present include:

- Air bubbles: these may be formed during the melt due to small bubbles of trapped gas (these are also known as seeds)[55] or purposely included whilst the glass is still soft by pricking with a needle.
- Striae (also known as cords): these are streaks of inhomogeneous glass which have different RI values to the bulk glass.
- Discolouration: caused by impurities in the raw melt.[5]
- Stones: can be from crystalized glass or small particles of refractory material.
- Scratches: these may be due to general wear, washing or from a particular source such as windscreen wipers on a windscreen.[50]

Other techniques which may be employed to qualitatively and quantitatively characterise the surface of glass evidence that utilise light include interferometry (the use of an interference objective)[35,48,49,56] and Nomarski differential interference contrast;[35,51] these are discussed further in Section 6.4.

6.2.1. *Photoluminescence of Glass*

Many glasses when exposed to Ultraviolet (UV) light will fluoresce due to the presence of heavy metal ions and may be distinguished from each other by their fluorescent properties.[57] Fluorescence (a photoluminescence process) may be described in this instance as the emission of photons from the glass surface after excitation from a UV light source. The glass surface absorbs the UV radiation which excites orbital electrons. When the electrons return to their ground state, photons are emitted at a longer wavelength within the visible spectrum. In particular, the use of light within the UV range, particularly 254 nm, can help identify float glass. Float glass, as its name implies, is produced by floating a layer of molten glass on a bath of liquid tin. This creates a glass which has two very flat surfaces with fewer surface imperfections compared to flat glass manufacturing processes that use rollers. This lack of imperfections is due to one surface being exposed to the air (non-float side) in a clean environment and the other to the smooth surface of the liquid tin (float side). A very small amount of tin diffuses into the float side which fluoresces either a pale yellow or green when exposed to UV light.[35] This may be observed by adding a UV light attachment onto the microscope,[31] using a hand-held UV light source (for

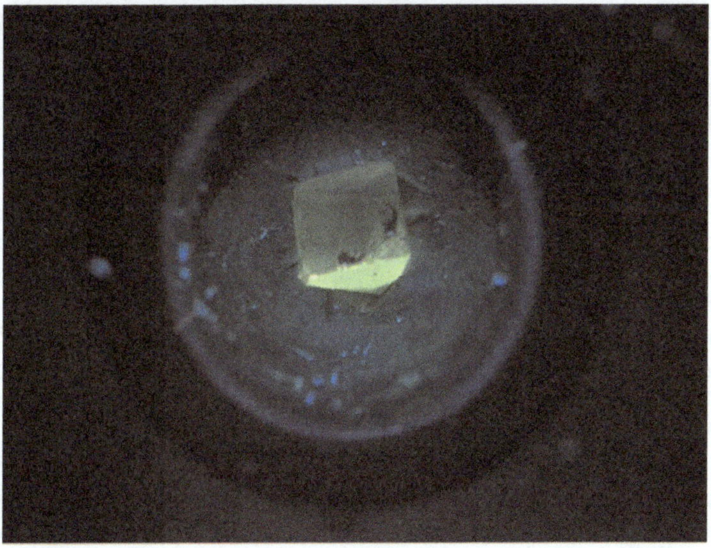

Figure 6.1. Float glass under the Flview. Image courtesy of Chris Hunter, SMCS.

larger pieces) or by utilising a specialised piece of equipment such as the Fluorescence Glass Flat Viewer (Flview) as previously made by SMCS Ltd. The use of a viewing chamber, as seen with the Flview, provides a dark environment which aids in the identification of weak fluorescence. Figure 6.1 shows a small shard of float glass exposed to UV light and demonstrates the fluorescent properties of the surface which was in contact with the tin during manufacture.

This observation is generally limited to a diagnostic test as it is difficult to assess the relative intensities of the fluorescence in small samples just using visual techniques. Fluorescence spectrometry, which quantifies fluorescence intensity at multiple wavelengths (*e.g.* 260, 280 and 340 nm), has been successfully employed to distinguish float and non-float surfaces on fragment sizes that are generally found in casework. Larger pieces of toughened glass have also demonstrated a useful level of discrimination between different glass samples which, although less likely to be found in casework, could be a useful additional characteristic for comparison.[57]

Phosphorescence of glass, also a photoluminescence process, has also been used to differentiate glass.[58] Phosphorescence is also the emission of photons upon excitation but is slower to re-emit than in fluorescence, this is because the excited electrons also incur a change in spin state. This change in spin state prevents the electrons relaxing back into their ground state straight away and therefore the glass will re-emit light for much longer.

Initial screening, observations and diagnostic tests in glass analysis would not be possible without the use of light. By using light at different angles and wavelengths, surface detail, fluorescent and phosphorescence properties are revealed providing additional layers of information for comparison and identification.

6.3. Glass Refractive Index Measurements

The analysis of refractive index (RI) is the most common characteristic to be analysed in forensic glass analysis and still maintains its position as an essential tool for screening samples, aiding glass categorisation and provenance determination and providing quantitative data for comparison.[1,37,41,59–61] It is not unusual these days, in the UK at least, to see some forensic scientists presenting findings based solely on RI results, thus demonstrating its importance.

Generally, measurement of the refractive index of the glass is the first quantitative analysis (other than measuring its size) undertaken by a forensic glass analyst and may ultimately be the only quantitative analysis conducted if sufficient in being able to discriminate between samples. It may be desirable to also observe the change in RI upon annealing (*i.e.* the heat treatment of glass in order to change its internal structure), which is described in Section 6.3.3.

Before continuing, it is important to note what refractive index is and why it is a suitable characteristic to be analysed for glass. Refractive index can be defined as the ratio of the speed of light within a vacuum to the speed of light in a transparent medium; in this case that medium is glass. It may also be described in layman's terms in court as the 'bending of light'; this effect can be seen in Figure 6.2, where the ratio between sine of the angle of incident light (i) and sine of the angle of refraction (r) equals the refractive index (n).

Refractive index is not constant and is dependent upon the wavelength of the incident light. For the quantification of RI, a filter is used which creates monochromatic light at a given wavelength, commonly 589.3 nm (sodium line) in forensic analysis.

In optical glass, where the manufacturing process is tightly controlled to create stress-free glass, the glass is a completely isotropic material in terms of its RI, meaning that it has a single refractive index regardless of orientation. In reality, many other glass types either are intentionally subjected to thermal stress to create internal stresses (*e.g.* toughened glass) or unavoidably have internal stresses created due to the manufacturing process; these glasses are therefore anisotropic in their RI, meaning that there is more than one RI present which is dependent upon the direction in which the light passes through it. This variation in RI can cause issues for the forensic scientist if it is not understood and acknowledged, this is further discussed in Section 6.3.2.

Glasses commonly occurring in casework have RI values of between 1.5125 and 1.5250.[16] but have been stated more broadly as being 'near 1.5'.[37] Glass of much lower or higher RI values, when found on a suspect or at a scene, may be

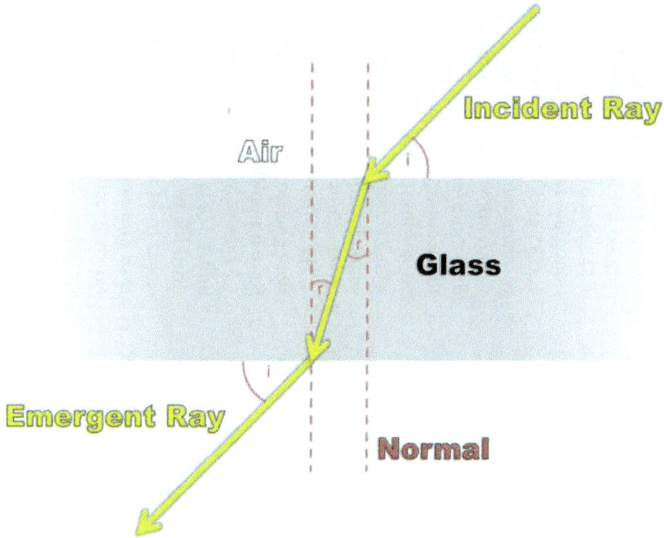

Figure 6.2. Refractive Index of Glass.

of higher evidential value. Table 6.1 provides further ranges of RI in glass samples analysed in a selection of forensic studies. For lists of common RI values for different types of glasses, there are a plethora of glass science and glass manufacture texts.

6.3.1. Methods for Determining the Refractive Index (RI) of Glass

The popular nature of using RI as a characteristic for comparing glass samples has led to many developments in improving methods to increase accuracy and precision. This has seen movement from manual observations to semi-automated systems that reduce subjectivity and human error. Along with these developments have come guidelines to optimise RI measurements, such those documented in the ASTM E1967 – 11a standard: *Standard Test Method for the Automated Determination of Refractive Index of Glass Samples Using the Oil Immersion Method and a Phase Contrast Microscope*[62] which covers accurate analysis of glasses that have RI values between 1.48–1.55, of both regular and irregular shaped fragments and as small as 300 µg in size. In forensic science, RI of glass research may be conducted for multiple reasons, such as to create databases, optimise methods, identify prevalence of different glass sources in different populations, establish the extent of variation within and between samples and to identify different methods of interpretation, *e.g.* the use of likelihood ratios. All of which are beneficial for casework. Table 6.1 provides some examples of studies investigating RI of glass since 1986. These demonstrate the different purposes, methods and RI values seen in glass analysis. Studies have been chosen to exemplify changes in

Table 6.1. A selection of glass RI analysis studies demonstrating a range of methods and results.

Authors (date)	Purpose of study	Samples analysed	RI analysis method	RI results/comments
Buckleton, Axon and Walsh (1986)[2]	To survey RI, colour and thickness of vehicle window glass in New Zealand and identify any variation between windows within the same vehicle and between the top and bottom layers of laminated windscreens	513 samples from 388 different vehicles including windscreens, side windows, sun roofs and rear windows	Oil immersion method with Nikon Labophot phase contrast microscope and Mettler FP82 hot stage. A Locke Scientific silicone oil B was used	Overall range of RI values was between 1.5117 to 1.5239. The toughened float glass ranged between 1.516 and 1.519 and toughened non-float glass ranged between 1.5135 and 1.5170. Laminated windscreens were highly varied but with two common distributions with the mean RIs around 1.5175 and 1.5184. Variations in RI and thickness were seen in glass of different years of manufacture
Zoro, Locke, Day, Badmus and Perryman (1988)[63]	To determine the presence of any variation in RI between the surface and bulk sample of float, non-float flat glass, glass containers and tableware	16 glass objects including windows (float = 5, non-float = 5), bottles ($n = 3$), and drinking vessels ($n = 3$)	GRIM 1 using silicon oil DC710. Flakes were chipped from both surfaces (inside and outside or float/anti-float) and from the centre of the glass. Glass samples were determined as having either a negative or positive deviation RI from the bulk RI	Bulk RI values ranged between 1.51773 to 1.51919 for flat float glass, 1.51553 to 1.52949 for flat non-float glass, 1.52021 to 1.52081 for bottles and 1.51844 to 1.53040 for drinking vessels. Large differences in RI were seen between float and non-float surfaces. Much smaller differences were seen between non-float glass surfaces. Container glass surfaces demonstrated differences, which included consistent higher RI values on the outer surfaces for 2 of the 3 bottles and 1 of the 3 drinking vessels

McQuillan and Edgar (1992)[31]	To ascertain the presence of and number of possible sources of glass fragments on the clothing of individuals (not suspected of a crime) utilising RI measurements, initial observations, annealing and chemical analysis (in a separate paper)	631 fragments of glass retrieved from the surfaces and pockets of 432 items of clothing, comprising of 216 trousers, 191 jackets and 25 pullovers	GRIM 1 using silicon oils DC710, DC704 and DC550. Ten measurements were made to 4 decimal places (dp) but reported to 3dp. RI measurements were grouped using a RUNG computer programme using a standard deviation of 0.00005	RI values ranged from 1.499 to a very low frequency having over 1.532. The distribution of RI values appeared to be normally distributed with the frequency maximum being between 1.517 to 1.521. The highest frequency of samples was 1.521. Grouping indicated that it was unusual to find more than 6 fragments from one source on an item of clothing although it must be noted that grouping was only completed using RI data
Cassista and Sandercock (1994)[88]	To compare the 'temperature variation' and 'double variation' methods for determining RI in glass. Also calculated the dispersion for each sample	Nine glass samples, including float, non-float and toughened glass	Temperature variation method: calculated Ri from the mean observations of the point where the Becke lines disappeared when both heating and cooling the sample in a silicone RI oil at 585.6 nm using a Mettler FP82 hotstage Double variation method: the same as above but at multiple wavelengths using a calibrated quartz wedge	RI values are dependent on the methods used. Precision for the temperature variation method was greater than for the double variation method (mean SD values = 0.00002 and 0.00004 respectively). The temperature variation method was also quicker. Dispersion data offers little advantage over RI data for the discrimination of glass

(*continued*)

Table 6.1. *(continued)*

Authors (date)	Purpose of study	Samples analysed	RI analysis method	RI results/comments
Jackson. F, Maynard. P, Cavanagh-Steer K, Dusting. T and Roux (2013)[30]	To determine the potential source of glass found in the head hair and headwear of the general population and glaziers	Six fragments from 230 head hair samples from the random population (headwear not included). 138 glass fragments were recovered from 24 of the 25 glaziers with 109 of them coming from the head hair samples and 29 from the headwear samples	GRIM 3 that was connected to a Leica DM 2500 phase contrast microscope with a video camera and a Mettler Toledo FP82HT hot stage, with a ramp rate of 4–8 °C min^{-1}. Silicone B oil was used and RI values grouped by range. A search was run for the range of the RI values ±0.0001 against a reference glass database to identify potential source	Random population: The RI range was 1.5155 to 1.5230. These were identified as most likely to be window glass Glaziers: The RI range was 1.51449 to 1.52357. These were also deemed to be most likely from window or wind-screen glass
Munger, Gates and Hamburg (2014)[37]	To analyse the variation in RI across vehicle windscreens	1800 glass samples from across different sections of 14 windscreen panes (7 interior and 7 exterior) from 7 vehicles were analysed for RI and half the samples were analysed for thickness	GRIM 3 attached to a Leica DMLB phase contrast microscope with Mettler hot stage	The range of the mean RI values of all 14 windscreen panes was 1.51864 to 1.52264. The mean SD of the windshields was 0.00004. RI values were often higher in the centre than the left and right edges and internal and exterior panes could be differentiated in 4/7 windscreens using RI, thickness or colour

equipment over time and provide a brief insight into the different purposes for analysing RI for forensic glass analysis. Levels of detail differ between studies due to the information provided in the published papers.

Whatever the purpose of analysing the RI of glass, whether for research or for casework, there are three main stages to the process regardless of whether a manual or semi-automated approach is being used (albeit these stages may differ), these are:

1. sample preparation
2. RI oil (silicon oil) choice
3. RI measurements.

Each of these stages will be briefly outlined below.

6.3.1.1. Sample Preparation. Prior to the quantification of the RI of the glass, the sample must be prepared and mounted in an appropriate oil chosen for analysis. A sufficient number of glass samples must be chosen to allow any variation in RI in the bulk sample to be determined. Further discussion of variation in RI is discussed in Section 6.3.2. If the glass shard is large or a greater number of edge counts are required for analysis (10 is normally deemed the minimum in casework), the sample can be ground to a fine powder using a pestle and mortar or broken using the edge of a blade.[31] It has been noted that by crushing the glass sample, any variation between the RI of the surface and the bulk sample, as seen in float glass, will not be captured as the resultant samples will be a mix of outer and inner glass.[63] This may be avoided by using one of three methods described by Underhill in 1980.[64]

6.3.1.2. Choosing an Appropriate RI Oil. For the RI of the glass to be determined it must be immersed in a medium of known refractive index, *i.e.* a silicone oil that has been calibrated with a minimum of three glass reference samples.[62] This is known as immersion refractometry. The RI of the silicon oil varies inversely with temperature, whilst the corresponding change in RI of the solid glass through the temperatures it is exposed to on a heated stage is negligible. Figure 6.3 demonstrates the relationship between the RI of the oil and RI of the glass.

As seen from Figure 6.3, an appropriate oil is one which is initially higher in RI than the glass sample. When using a manual system, how much higher the oil is in RI is dependent on the heating range of the heating stage; normally this should have a working range of 26 °C to 118 °C and a minimum precision of 0.1 °C,[62] but stages that have lower temperature limits may be encountered and require an oil to have an RI nearer to the RI of the glass. When using a manual system, two optical methods that utilise light, may be employed to aid the choosing of an appropriate RI oil; namely these are the 'Relief test' and the 'Becke line test'.

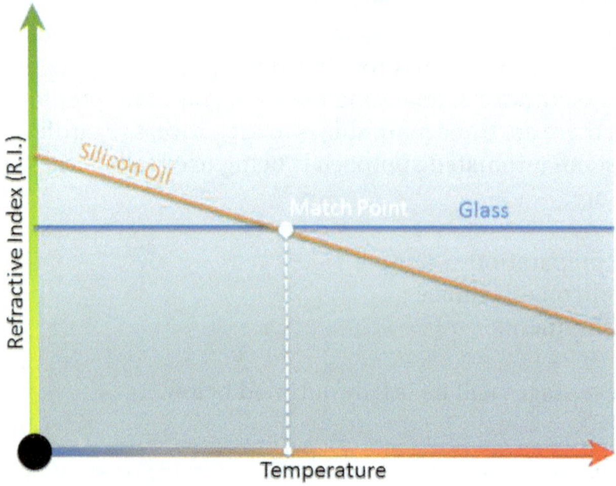

Figure 6.3. The relationship between the RI of the silicone oil and RI of the glass when temperature is increased.

Relief can be described as the contrast of the glass fragment boundaries with the oil in which it is immersed. The principle that underpins the relief test is that the greater the relief, the greater the difference in RI between the oil and the glass. For a manual approach using a hot-stage that has a low upper temperature limit, the relief of the glass should be low under plane polarized light. Whilst the relief test indicates an approximate match with an oil of known RI, it does not indicate whether the glass or the oil has the higher RI. This can be established using the Becke line test. Bright haloes, or Becke lines, can be observed around the edges of the glass edges when the condenser iris is closed down or the microscope set up in phase contrast (this involves using an illuminating annulus in the condenser to illuminate the specimen with a hollow cone of light, and a matching annulus in the objective to increase relief at the glass edge boundaries). The direction of movement of the Becke lines when moving the objective from 'low focus' to 'high focus' indicates the medium with the higher RI. In this case, the Becke lines move into the medium of higher refractive index. As depicted in Figure 6.3, for an appropriate RI oil, the Becke lines must move into the oil to work on a heated stage. This process of choosing an oil is based on trial and error and may require samples to be washed in acetone and dried to remove any adhering oil from the surface before trying an alternative oil.

Using a semi-automated system, the choosing of an oil is much less time consuming, with generally a standard set of three oils being provided that, between them, cover all glass types. Originally, Dow Corning oils were used and were named as DC710, DC704 and DC550, but more recently silicone oils obtained from Locke Scientifics named Silicon A (RI = 1.540), B (RI = 1.520) and C (RI = 1.481) oils are commonly used with the GRIM system. Other reference oils are available from other manufacturers such as Craic which range in RI from 1.48–1.54.

6.3.1.3. Refractive Index Measurement. Refractive index measurements may be analysed manually or by using a semi-automated system, the choice of which may be due to equipment availability or analyst preference. Regardless of which of these is chosen the fundamental principles are the same. Ultimately, glass is immersed in a known RI oil, is observed using a phase contrast microscope to increase the contrast of the edges, a video camera and a hot-stage under monochromatic illumination (sodium light at 589 nm) and utilises the principle that as the RI oil is heated (using a heated stage) its refractive index reduces by a constant amount per degree, also known as its temperature coefficient. As seen in Figure 6.3, when the temperature of the oil has been raised sufficiently, the RI of the oil will equal the RI of the glass and at this point the contrast between the glass edges and oil will be at its minimum.[62] This temperature match point can be translated in to the refractive index of the glass by manually calculating it using the RI details of the oil or by using a calibration curve previously created for the RI oil from analysing reference glass standards.[62] With a manual approach, the analyst observes the match point on a video screen whereas with a semi-automated approach, this is determined by a software program. Figure 6.4 shows a typical manual set-up with a Nikon polarising light microscope with phase contrast condenser and objective, heated stage and camera and video screen.

The semi-automated system for measuring the RI of glass is called the Glass Refractive Index Measurement (GRIM®) system and was initially introduced in the 1980s by Foster and Freeman. Further iterations of the system have been introduced, with the current system being in its third version, called the GRIM 3®. The evolution of this system has focussed upon increasing the speed of analysis (*i.e.* more edges analysed at a time, with quicker characterisation), accuracy and precision. The current system analyses RI to 5 decimal places with standard deviations of 0.00003RI over a 5 day period and can analyse 4 edges at a time. A contrast signal generated during the temperature cycle (which helps identify if the sample is appropriate for analysis) is converted to edge counts for each sample and is reported as a scale of 1–99; the higher the number of edge counts, the higher the contrast (relief) seen in the sample.[1] These semi-automated systems work by calculating the RI mean minimum contrast values for both the heating and cooling cycles. Figure 6.5 shows the GRIM 3® demonstrating the image processing software that analyses glass edge contrast and notes when this reaches the match point.

There are some limitations to this semi-automated system, for example it has been noted that due to the use of a semi-automated video processing equipment it may be more difficult to identify where there are edges that are

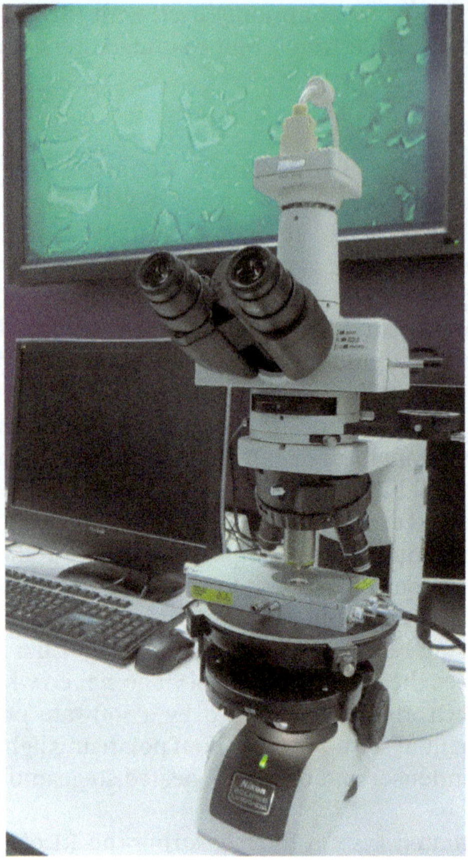

Figure 6.4. Hot-stage microscope set-up for the measurement of refractive index.

unobscured by the bulk glass when carrying out analysis of the RI of glass surfaces compared to manual visual searching.[63]

A limitation that may be encountered with both the manual and semi-automated method of RI analysis, is the potential inaccuracies and low reproducibility caused by low edge contrast,[1] described as 'low relief' earlier in the chapter; this has been associated with very thin edges of glass. This low contrast can cause issues when determining the match temperature through observing edge contrast manually. It has also been shown that when using GRIM, as the edge count values decreases due to poorer edge contrast, the variation in refractive index increases.[1]

The ability to apply strict edge count minimum requirements (cut-off points), *e.g.* 10 edge counts before the analysis is valid, and putting in place protocols to reanalyse if these cut-off points are not met, mean that high precision is still seen in semi-automated systems.

Light plays an important role in all stages of the analysis of RI in glass. By being able to harness the properties of light in the form of phase

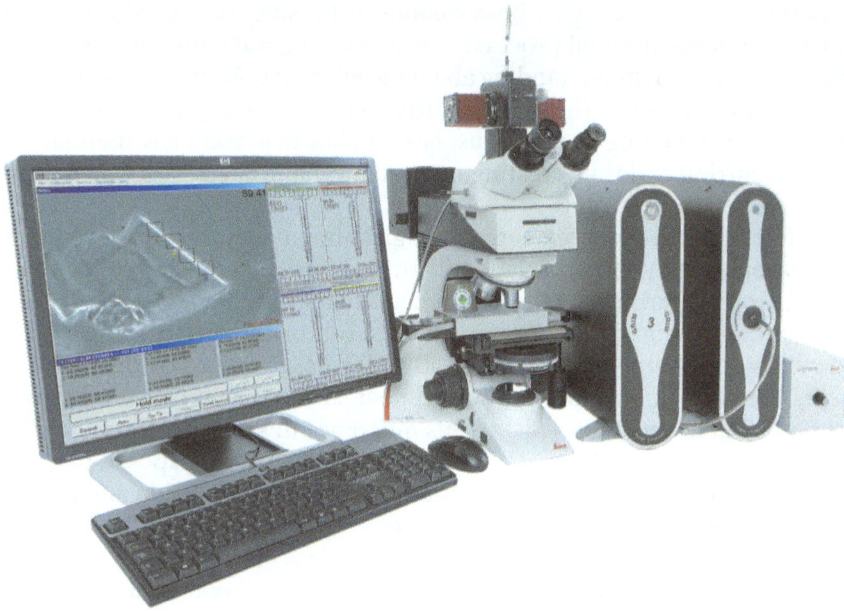

Figure 6.5. Glass Refractive Index Measurement (GRIM) system. Image courtesy of Foster and Freeman.

contrast microscopy and applying the principle of immersion refractometry, a fundamental characteristic of glass can be accurately and precisely quantified.

6.3.2. Intra-variation in Refractive Index of Glass

It is well known that many glass types, other than optical glass, exhibit variation in their refractive index within a single item (intra-variation). Variation between glass types and batches of glass (inter-variation) is beneficial for distinguishing between samples and increases evidential value, but variation within an object can cause problems for the forensic scientist. Without understanding the potential variation present across an entire object, the variation seen within a very small fragment retrieved from a suspect may underestimate the variation, possibly leading to falsely excluding the sample as coming from that object.[37] Due to this, many studies have endeavoured to quantify variation across different glass items including sealed beam headlights,[65] vehicle windows,[37,45,66] window glass,[67] toughened glass,[68] container glass,[45] and others particularly focussing upon float glass.[44,59,64,67] These studies have highlighted some of the reasons for variation within glass, including inhomogeneity in the glass melt, differences in thermal and chemical histories, weathering, the glass edge morphology and differences in surface composition due to the manufacturing process.

Certain glasses are highly homogenous in RI, such as optical glass. Consideration of the thermal processes used in the manufacture of this type of glass has led to an understanding about the effect of different manufacturing processes on the RI of the end product. To create homogenous RI across a bulk item, the cooling process must be very slow to prevent any thermal gradient occurring and thus creating internal mechanical stresses which cause the glass to be anisotropic.

Differences seen between surfaces of glass fragments can be due to multiple reasons but the main variation identified has been due to the manufacturing process, particularly the float process to make float glass. In float glass, the float surface and the opposite surface (anti-float surface) of the glass can show considerable difference in RI which has is caused by the float side being enriched in SnO, whilst the anti-float surface has reduced other constituents but is still rich in SiO.[64] This difference in surface RI for float glass has also been seen by Zoro *et al.*[63] Weathering of surfaces exposed to different environments may also cause differences in RI from inside and outside surfaces, this has been seen in a sample of shop window glass.[63] In container glass, differences in RI may be due to the surface strengthening treatment that bottles undergo when manufactured leading to a potentially higher RI in the outer surface compared to the bulk sample (centre of the glass) as seen in the study outlined in Table 6.1.

An understanding of glass, its manufacturing process and its subsequent optical properties, particularly how RI may vary within an object, is essential for interpreting this evidence in a case. Without this knowledge, the use of light for the analysis of RI may be detrimental to the outcome of an investigation.

6.3.3. *Change in Refractive Index Upon Annealing*

The variation in RI described in Section 6.3.2 may be bought about by glass samples being exposed to very high temperatures. At a crime scene, this may be caused by the glass being exposed to the high temperatures of a fire scene. More commonly, annealing is a manufacturing process to remove internal stresses in the glass by increasing its temperature above its transformation range and then cooling it down slowly,[44] this is seen in toughened glass. Glass demonstrates a transformation range with a change in temperature with the upper end of the range being described as the annealing point.[69] This transformation range of the glass can differ between glass types. The annealing point may be described as the point where the internal structural stresses present in the glass are released due to the glass being exposed to a sufficiently high temperature to lower the viscosity of the glass.

The re-annealing of glass may be used in forensic investigations (albeit, today this is rare in the UK) to gain information about the thermal history of the glass and to particularly identify toughened glass when the sample is very small and cannot be identified as this type by observing its shape alone.

Toughened glass is deliberately given greater internal stress during manufacture to ensure that once its breaking limit is reached, it completely breaks down into smaller and 'safer' shards. Upon re-annealing, the internal stress of toughened glass is dissipated resulting in the RI of the glass being increased.[44] In non-toughened glass, there is much less internal stress, therefore the change in RI is much less. Many studies have been conducted on the identification of toughened and non-toughened glass in this manner[38,43–47] and have suggested a variety of thresholds in the change in RI pre and post annealing for the identification of toughened glass.

The effect of annealing upon the intra-variability of RI of glass has been investigated with a variation in the results shown. In an early study investigating toughened and non-toughened window glass and bottles, annealing proved to reduce variability in RI across items.[45] Whereas a later study has shown that annealing non-toughened float glass and toughened float glass may have increased RI variability by a factor of 1.29–1.58 (mean = 1.43) and a factor of 0.63–0.76 (mean = 0.69) respectively.[44]

Understanding the effect of annealing upon an optical property exploited by the use of light is an important discovery for forensic glass analysis and makes a complementary analysis for the characterisation of glass.

6.4. Use of Light for Determining Curvature and Topography Detail

Observing any surface features on glass, as described in Section 6.2, can be very helpful in identifying any manufacturing characteristics or post damage detail caused through fracture or other external sources. Before glass is annealed, many processes may be applied to the glass, such as being moulded, pressed, drawn or rolled, all of which may leave features upon the surface of the glass, such as bumps and pits from rollers or moulding; ripples from spinning; craters from impact and grooves from abrasion.[35] Light microscopy, as described in Section 6.2, can be very useful at this stage but further techniques that utilise light may reveal further surface information.

Two forms of interference microscopy, Nomarski differential interference contrast (DIC) microscopy and interferometry, are useful additional tools for surface analysis. The use of DIC microscopy and interferometry to analyse these surface details and quantify curvature is an effective use of light which, although it not utilised in casework regularly, can provide valuable intelligence information in certain cases. For example, curvature quantification can help identify the original source of the glass, and surface scratches may be linked to a particular cause, *e.g.* windscreen wipers.[50]

Differential interference contrast (DIC) microscopy is a refined version of phase contrast microscopy and works by utilising a polarizer which creates plane polarized light and a beam splitter which splits the light beam from the lamp into two, which in this case is a prism. These two beams of light are displaced when split into two, travelling perpendicularly to each other, and when passing through the prism reach the glass surface at different points. The light is reflected back through the prism and an analyser which recombines the rays and makes them vibrate in the same direction. If the surface of glass is textured or curved, the two rays will travel different distances (*vs.* the same distance if the surface is flat), called optical path difference, and cause a phase displacement.[51] This optical path difference exhibits itself as interference colours (polarization colours) which can be used to clearly observe the fine surface detail.

Interferometry allows the measurement of surface features through observing interference fringes and allows the analysis of curvature by observing the overall interference fringe pattern. Whereas DIC microscopy is good for fine surface detail, interferometry observes a greater area and creates a contour map of the surface.[35] Surface contour (whether concave or convex) and estimates of fringe curvature, general shape of the fragment, and any surface scratches may be determined by using an interference objective under a two microscope system.[31,53] Fringe curvature can be estimated by analysing the presence of any Newton's rings (a concentric ringed interference pattern created by curved surfaces) resulting in an estimated radius of curvature. A recommendation of using Newton rings at a spacing of 15 fringes per mm has been suggested by Locke in 1984.[49]

By using interference microscopy to observe the presence of any curvature upon broken vehicle windows, it has been identified that there may be sufficient redistribution of the internal stress of toughened glass when broken to make the surface convex when originally it was concave.[35,53] This, along with the knowledge that very thin sheet glass may curve when being mounted for analysis, causing curved fringe patterns, mean that care must be taken when interpreting these interference fringes so as not to identify the glass as from a curved object when in fact it is flat.[35]

These two uses of interference microscopy, although rarely used in casework these days, demonstrate how light may be manipulated to provide additional data that can help identify the source of a glass sample.

6.5. The Use of Light to Quantify Trace Elemental Composition

Previously, RI characterisation was capable of showing great discrimination between glass items due to three main reasons outlined by Brown in 1985;[70] these include different manufacturing procedures for different types of glass, changes in manufacturing procedures with time, and different manufacturing procedures in different countries. It is now widely recognised that analysing the RI of glass alone does not allow differentiation of glass between recently produced batches of glass[41] and, as quality control in glass increases, the variation in optical features such as RI will decrease thus reducing its discriminating power.[71] It has been noted that certain glass types cannot be discriminated by RI alone, *e.g.* float glass, and by doing so, vastly increases the chance of type II errors (where two samples are incorrectly associated with a common source).[72]

Trace elemental analysis can provide the discriminating power that RI fails to provide. A range of spectroscopic techniques are available for the forensic glass analyst including inductively coupled plasma mass spectrometry (ICP-MS)[71] and multi-collector inductively coupled plasma mass spectrometry (MC-ICP-MS),[73] inductively coupled plasma atomic emission spectrometry (ICP-AES), spark source mass spectrometry,[74–76] atomic absorption, neutron activation,[77,78] scanning electron microscopy/energy dispersive X-ray spectrometry (SEM/EDX),[79,80] micro X-ray fluorescence, dc arc emission spectrometry and laser ablation (LA-ICP-MS).[81] The latter of these techniques is relatively new compared to others and utilises light within the visible range, due to this, laser ablation ICP-MS will be discussed further.

Many techniques employed for the determination of the elemental composition of glass do not use light as the source of energy in which to breakdown the sample into its relative components for analysis. Light amplification by stimulated emission of radiation (LASER) was developed in 1960 and provides the opportunity to directly analyse the elemental content of a solid glass sample.[82] Laser ablation ICP-MS utilises a monochromatic beam of high intensity monochromatic light (213 nm or 266 nm UV light source[72]) to bombard the glass sample and produce ions (ionisation) which then are detected and characterised using a mass spectrometer.

LA-ICP-MS has been proven to be an effective alternative to solution ICP-MS also known as inductively coupled plasma mass spectrometry (ICP-MS), which requires the sample to be digested and analysed as a solution, due to the reduced sample preparation time, reduced sample consumption[81] and its ability to analyse irregular shaped fragments and contaminated surfaces without issues.[83] LA-ICP-MS has acknowledged as having increased sensitivity compared to other techniques,[41,72,81,84] *e.g.* SEM/EDX, and allows the quantification of trace and ultra-trace elements as well as the major and minor elements quantified by other approaches,[83] for example 30 major and trace elements were used to discriminate between different float glasses effectively[72] and 39 used in a method evaluation study.[85] Validation studies and guidelines are in existence to provide optimal sampling procedures, instrumental set-up,[85] data processing and to aid the forensic glass analyst in their interpretation.[81,86] Organisations, such as the European Natural Isotopes and Trace Elements in Criminalistics and Environmental Forensics (NITECRIME) network, have worked to develop standardised methods.[87] There are a variety of different LA-ICP-MS instruments available to the glass analyst, including a relatively new laser induced breakdown spectroscopy (LIBS) system by Foster and Freeman.

In addition to its ability to discriminate samples more effectively, LA-ICP-MS offers many advantages over other trace elemental techniques, *e.g.* including the small sample size, reduced issues with contamination, less spectral interferences and provides greater information, *e.g.* lateral distribution of the elements in a sample.[72] One potential limitation is that upon using this technique, there is a possibility of the RI value of the analyses glass being modified, which would need to be taken into account in the analysis strategy.[86]

Ultimately, LA-ICP-MS shows how the use of light combined with a robust analysis method allows glass to be quantified and profiled in terms of its composition. This powerful technique has already been introduced into some laboratories and ensures that glass evidence may still be important in a criminal case even when it is from a common source.

6.6. Conclusion

Glass evidence can provide important corroborative evidence in a criminal case, its benefits being its relative stability in terms of its characteristics compared to other evidence and its ability to provide information towards both source and activity level propositions. Light plays an essential part in the forensic analysis of glass both historically and in current practice. A range of wavelengths may be employed from the early stages of sample screening to the analysis of sample trace elemental composition. Glass is one of the few transparent forms of evidence that exhibits optical features that can be quantified by using light in order to reliably identify its type and provide robust data for comparison to known samples. In addition to this, extensive research and effort has been made to understand these optical properties for the benefit of the courts. Refractive index is probably the most well-known characteristic to be examined in forensic glass analysis but other techniques exploiting the use of light, when used in combination, provide a comprehensive collection of both qualitative and quantitative data that allow for more confident conclusions. Although many forms of evidence utilise light to characterise and discriminate, none other than glass employ it so readily.

Acknowledgements

Thank you to Chris Hunter and SMCS for supplying Figure 6.1, William Bailey for the production of Figures 6.2 and 6.3 and Foster and Freeman for Figure 6.5.

References

1. A. W. N. Newton and J. S. Buckleton, *Forensic Sci. Int.*, 2008, **177**, 24–31.
2. J. S. Buckleton, B. W. Axon and K. A. J. Walsh, *Forensic Sci. Int.*, 1986, **32**, 161–170.
3. Oxford English Dictionary, *Glass*, Oxford University Press, 2016.
4. J. Shelby, *Introduction to Glass Science and Technology*, The Royal Society of Chemistry, 2nd edn, 2005.
5. H. Pfaender, *Schott Guide to Glass*, Chapman and Hall, London, 2nd edn, 1996.
6. R. Rawson, *Glasses and Their Applications*, The Institute of Metals, London, 1991.

7. I. W. Evett and J. S. Buckleton, *J. Forensic Sci. Soc.*, 1989, **29**, 317–324.
8. R. Cook, I. Evett, G. Jackson, P. Jones and J. Lambert, *Sci. Justice*, 1998, **38**, 231–239.
9. I. W. Evett, *J. Forensic Sci. Soc.*, 1986, **26**, 3–18.
10. R. Cook, I. W. Evett, G. Jackson, P. J. Jones and J. A. Lambert, *Sci. Justice*, 1998, **38**, 151–156.
11. I. W. Evett, G. Jackson, J. A. Lambert and S. McCrossan, *Sci. Justice*, 2000, **40**, 233–239.
12. R. Cook, I. W. Evett, G. Jackson and M. Rogers, *J. Forensic Sci. Soc.*, 1993, **33**, 149–152.
13. J. M. Curran, J. Buckleton and C. M. Triggs, *Forensic Sci. Int.*, 1999, **104**, 91–103.
14. J. M. Curran, C. M. Triggs, J. S. Buckleton, K. Walsh and T. Hicks, *Sci. Justice*, 1998, **38**, 15–21.
15. F. Taroni, A. Biedermann, S. Bozza, J. Comte and P. Garbolino, *Forensic Sci. Int.*, 2012, **220**, 173–179.
16. D. A. Hickman, *Forensic Sci. Int.*, 1981, **17**, 265–281.
17. G. Zadora, Z. Brozek-Mucha and A. Parczewski, *Probl. Forensic Sci.*, 2001, **47**, 144–146.
18. G. Cooper, *Sci. Justice*, 2013, **53**, 166–170.
19. M. Irwin, *Sci. Justice*, 2011, **51**, 16–18.
20. T. Hicks, R. Vanina and P. Margot, *Sci. Justice*, 1996, **36**, 101–107.
21. C. A. Pounds and K. W. Smalldon, *J. Forensic Sci. Soc.*, 1978, **18**, 197–203.
22. T. J. Allen and J. K. Scranage, *Forensic Sci. Int.*, 1998, **93**, 167–174.
23. T. J. Allen, K. Hoefler and S. J. Rose, *Forensic Sci. Int.*, 1998, **93**, 175–193.
24. T. J. Allen, K. Hoefler and S. J. Rose, *Forensic Sci. Int.*, 1998, **93**, 195–200.
25. T. J. Allen, A. R. Cox, S. Barton, P. Messam and J. A. Lambert, *Forensic Sci. Int.*, 1998, **93**, 201–208.
26. T. J. Allen, J. Locke and J. K. Scranage, *Forensic Sci. Int.*, 1998, **93**, 209–218.
27. E. Springer, B. Sc, A. Zeichner, D. Ph, R. Springer, T. Breaking and G. Vehicle, *J. Forensic Sci.*, 1986, 691–694.
28. J. Locke and J. A. Unikowski, *Forensic Sci. Int.*, 1992, **56**, 95–106.
29. J. Locke and J. K. Scranage, *Forensic Sci. Int.*, 1992, **57**, 73–80.
30. F. Jackson, P. Maynard, K. Cavanagh-Steer, T. Dusting and C. Roux, *Forensic Sci. Int.*, 2013, **226**, 125–131.
31. J. McQuillan and K. Edgar, *J. Forensic Sci. Soc.*, 1992, **32**, 333–348.
32. P. H. Harrison, J. A. Lambert and J. A. Zoro, *Forensic Sci. Int.*, 1985, **27**, 171–187.
33. S. A. Coulson, J. S. Buckleton, A. B. Gummer and C. M. Triggs, *Sci. Justice*, 2001, **41**, 39–48.
34. R. J. Davis and J. D. DeHaan, *J. Forensic Sci. Soc.*, 1977, **17**, 271–285.
35. B. R. Elliott, D. G. Goodwin, P. S. Hamer, P. M. Hayes, M. Underhill and J. Locke, *J. Forensic Sci. Soc.*, 1985, **25**, 459–471.
36. D. G. Sanger, *J. Forensic Sci. Soc.*, 1973, **13**, 29–32.
37. C. Munger, K. M. Gates and C. Hamburg, *J. Forensic Sci.*, 2014, **59**, 1351–1357.
38. G. Zadora and D. Wilk, *Ocena wartościdowodowej współczynnika załamania światła zmierzonego przed i po Wykon. procesu odprężania okruchów szkła opakowaniowego, okiennego i z szyb samochodowych*, 2009, vol. 80, pp. 365–385.
39. J. Locke and M. Underhill, *Forensic Sci. Int.*, 1985, **27**, 247–260.
40. A. Zeichner, D. Ph, G. Feingold, M. S. R. Zeichner and I. Sample, *J. Forensic Sci.*, 1989, 1003–1006.
41. C. D. May and R. J. Watling, *Forensic Sci., Med., Pathol.*, 2009, **5**, 66–76.
42. S. M. Ojena and P. R. De Forest, *J. Forensic Sci. Soc.*, 1972, **12**, 315–329.

43. J. M. Marcouiller, *J. Forensic Sci.*, 1990, **35**, 554–559.
44. K. P. Rushton, S. A. Coulson, A. W. N. Newton and J. M. Curran, *Forensic Sci. Int.*, 2011, **209**, 102–107.
45. J. Locke and C. A. Hayes, *Forensic Sci. Int.*, 1984, **26**, 147–157.
46. J. Locke, D. G. Sanger and G. Roopnarine, *Forensic Sci. Int.*, 1982, **20**, 295–301.
47. J. Locke and L. A. Rockett, *Forensic Sci. Int.*, 1985, **29**, 237–245.
48. H. O. Forensic and P. Forensic, *Forensic Sci. Int.*, 1984, **2**, 53–66.
49. J. Locke, *Forensic Sci. Int.*, 1984, **26**, 53–66.
50. J. Locke and L. A. Rockett, *Forensic Sci. Int.*, 1986, **31**, 253–260.
51. B. Aj and C. Gv, *J. Microsc. (Oxford, U. K.)*, 1979, **24**, 13–15.
52. K. W. Smalldon and C. Brown, *J. Forensic Sci. Soc.*, 1973, **13**, 307–309.
53. J. Locke, H. O. Badmus and A. C. Perryman, *Forensic Sci. Int.*, 1986, **31**, 79–85.
54. J. A. Zoro, *Forensic Sci. Int.*, 1983, **22**, 213–219.
55. C. Bray, *Ceramic and Glass: A Basic Technology*, Society of Glass Technology, 2000.
56. P. House, *Forensic Sci.*, 1983, **22**, 221–230.
57. J. B. F. Lloyd, *Forensic Sci.*, 1981, 325–342.
58. C. Ar and J. Pf, *J. Forensic Sci.*, 1978, **23**, 263–273.
59. R. L. Bennett, N. D. Kim, J. M. Curran, S. A. Coulson and A. W. Newton, *Sci. Justice*, 2003, **43**, 71–76.
60. E. J. Garvin and R. D. Koons, *J. Forensic Sci.*, 2011, **56**, 491–500.
61. M. Pawluk-Kołc, J. Zieba-Palus and A. Parczewski, *Forensic Sci. Int.*, 2006, **160**, 53–58.
62. ASTM International, *ASTM E1967-11a: Standard Test Method for the Automated Determination of Refractive Index of Glass Samples Using the Oil Immersion Method and a Phase Contrast Microscope*, 2016.
63. J. A. Zoro, J. Locke, R. S. Day and A. C. Perryman, *Forensic Sci. Int.*, 1988, **39**, 127–141.
64. M. Underhill, *J. Forensic Sci. Soc.*, 1980, **20**, 169–176.
65. P. R. De Forest, *J. Forensic Sci.*, 1972, **17**, 409–425.
66. G. D. Renshaw and P. D. B. Clarke, *J. Forensic Sci. Soc.*, 1974, **14**, 311–317.
67. M. M. Davies, R. J. Dudley and K. W. Smalldon, *Forensic Sci. Int.*, 1980, **16**, 125–137.
68. J. S. Crockett and M. E. Taylor, *J. Forensic Sci. Soc.*, 1969, **9**, 119–122.
69. A. W. N. Newton, L. Kitto and J. S. Buckleton, *Forensic Sci. Int.*, 2005, **155**, 119–125.
70. G. A. Brown, *J. Forensic Sci.*, 1985, **30**, 806–813.
71. D. C. Duckworth, C. K. Bayne, S. J. Morton and J. Almirall, *J. Anal. At. Spectrom.*, 2000, **15**, 821–828.
72. P. Weis and S. Becker, *Laser Ablation ICP-MS in Forensic Glass Analysis: A Decade of Experience*, available at http://projects.nfstc.org/trace/docs/final/Becker.pdf.
73. A. Martyna, K. E. Sjastad, G. Zadora and D. Ramos, *Talanta*, 2013, **105**, 158–166.
74. M. D. G. Dabbs, B. German, E. F. Pearson and A. W. Scaplehorn, *J. Forensic Sci. Soc.*, 1972, **4**, 281–286.
75. J. Locke, D. Boase and K. W. Smalldon, *J. Forensic Sci. Soc.*, 1978, **18**, 123–131.
76. A. Haney, *J. Forensic Sci.*, 1977, **22**, 534–544.
77. M. Olejnik, T. Szprengier-Juszkiewicz, P. Jedziniak and J. Zmudzki, *Accredit. Qual. Assur.*, 2007, **12**, 637–641.
78. S. J. Pitts and B. Kratochvil, *J. Forensic Sci.*, 1991, **36**, 122–137.
79. G. Zadora and Z. Brozek-Mucha, *Mater. Chem. Phys.*, 2003, **81**, 345–348.
80. G. Zadora, *J. Forensic Sci.*, 2009, **54**, 49–59.
81. K. Smith, T. Trejos, R. J. Watling and J. Almirall, *At. Spectrosc.*, 2006, **27**, 69–75.

82. F. A. Orellana, C. G. Gálvez, F. A. Orellana, C. G. Gálvez, M. T. Roldán, C. García-Ruiz, M. T. Roldán and C. García-Ruiz, *TrAC, Trends Anal. Chem.*, 2013, **42**, 1–34.
83. P. Weis, The transparent witness: forensic examination of glass evidence at the Bundeskriminalamt, *Spectrosc. Eur.*, 2016, available at https://www.spectroscopyeurope.com/article/transparent-witness-forensic-examination-glass-evidence-bundeskriminalamt.
84. S. Umpierrez, T. Trejos, K. Neubauer and J. Almirall, *At. Spectrosc.*, 2006, **27**, 76–79.
85. C. Larkoczy, M. Diicking, J. A. Hoogewerff, J. R. Almirall, J. Buscaglia, A. Dobney, R. D. Koons, S. Monrero, G. J. Q van der Peij, W. R. S Sroecklein, T. Trejos, J. R. Warling and V. S. Zclanowicz, *J. Forensic Sci.*, 2005, **50**, 1–15.
86. S. Berends-Montero, W. Wiarda, P. de Joode and G. van der Peijl, *J. Anal. At. Spectrom.*, 2006, **21**, 1185.
87. *European Natural Isotopes and Trace Elements in Criminalistics and Environmental Forensics Network: Natural isotopes and trace elements in criminalistics and environmental forensics*, European Commissions: Community Research and Development Information Service, 2005, available at: http://cordis.europa.eu/project/rcn/56995_en.html.
88. A. R. Cassista and P. M. L. Sandercock, *J.-Can. Soc. Forensic Sci.*, 1994, **27**, 203–208.

Chapter 7

A Review on Analytical Techniques Used for Forensic Fiber Analysis

Nirvani Mujumdar and Andres D. Campiglia*

Department of Chemistry, University of Central Florida, 4111 Libra Drive, P. O. Box 25000, Orlando, Florida 32816-2366, United States

*E-mail: Andres.Campiglia@ucf.edu

Table of Contents

Light in Forensic Science: Issues and Applications
Edited by Giorgia Miolo, Jacqueline L. Stair and Mire Zloh

Published by the Royal Society of Chemistry, www.rsc.org

7.1. Background

For decades forensic science has been an integral part of criminal investigations around the world. At present, nearly 400 laboratories dedicated to forensic science have been accredited under the American Society of Crime Laboratory Directors/Laboratory Accreditation Board (ASCLD/LAB) in the United States.[1] Crimes such as robbery, burglary or any type of assault, which could probably leave trace levels of physical or biological evidence behind, may now provide evidence linking the offender to the crime.[2] Trace evidence can comprise a variety of materials such as hair, fibers (from clothing, car seat, carpet, *etc.*), paint chips, soil, glass pieces, pollen, dust, *etc.*[3,4] Their forensic significance arises from the fact that the transfer of trace material can provide a valuable insight into the relationship between an offender (suspect) and a victim or a crime scene. The likelihood of establishing a partnership based on prior contact is based on Locard's exchange principle which suggests that whenever two objects come into contact, each of them transfers a part of its own material onto the other object.[5] Hence, while conducting forensic examination of textile fibers, examiners are interested in the comparison of individual fibers recovered from the crime scene and from an unknown source, with other fibers obtained from clothes obtained from a suspect who might possibly be related to the crime scene.[5–7]

Out of all the forensic laboratories in the USA, 85% reported of having between 1 to 3 fiber analysts; out of which, 75% spend 30% or less time on fiber analysis.[8] This is reflected in the average number of cases solved or worked upon each year, which is 27 cases per analyst per year.[8] Wiggins further states that it is likely that with so many cases being crimes against an individual within the USA, very little (if any) time is available for an examiner to deal with less serious cases where fiber contact may have occurred. However, textile fibers are considered as one of the most important forms of trace evidence because they are ubiquitous and can be transferred easily.[2,9] Moreover in a criminal investigation, trace textile fiber evidence has a high evidential value because, although produced in large numbers, the fibers are not homogenous or indistinguishable products,[9,10] meaning that if fibers are recovered from a crime scene, it is possible to analyze them and distinguish them based on their characteristics. Fibers can be a part of direct (primary) or indirect (secondary) transfer, and forensic laboratories can examine such transferred fibers by comparing them with a standard or a known source to discover the possible common origins.[3,4,11] In other words, when a fiber is transferred from a fabric directly onto a person's clothing, it is known as a direct or primary type of transfer; whereas a secondary or indirect transfer occurs when previously transferred fibers from one person's clothing is transferred to another person's clothes. Although, the value of fiber transfer evidence has been well accepted in the forensic community, research and further examination is mostly conducted on primary transfer.[12]

Fibers have several characteristics such as material type, texture, dyes of a variety of colors and shades used to dye the fibers, *etc.* that could make them distinguishable from one another, and also play a crucial role in fiber analysis. Fibers are the smallest units or components used to manufacture textiles for clothing, household furniture, upholstery, floor coverings, rugs and carpets, car seats, tents, sails, ropes, cordages, *etc.*[4,9] Even though several other classifications and subtypes of fibers exist, they can essentially be classified as either natural or man-made.[9,13] Natural fibers such as wool, cotton, silk, asbestos, *etc.* occur or are produced in nature, whereas man-made fibers are manufactured by humans from either naturally occurring fiber-forming polymers (*e.g.* viscose) or synthetic fiber-forming polymers (*e.g.* polyester).[9] Further, natural fibers are divided into animal (protein), vegetable (cellulose) or mineral (asbestos) origin; whereas man-made fibers are classified into either synthetic polymers (polyvinyl, polyamide, polyester, *etc.*), natural polymers (viscose, acetate, *etc.*) or other types of fibers (glass, metal, carbon, *etc.*), depending on their base material.[4,9,13,14]

Another crucial aspect of forensic fiber analysis involves the analysis of dyestuffs or pigments used for imparting color onto the textile fibers.[15] Since different types of dyes have different types of affinities for the substrate onto which they are being applied, dyeing is the process employed by industries to imparting a certain color to a textile material by the interaction between the fabric substrate and a dye.[3] In short, a dye is a colored chemical compound used to evoke the visual sensation of a specific color by permanently adhering to a substrate (textile fiber), so that it is able to absorb and reflect certain wavelengths of light in the visible (vis) region of the spectrum ranging from 400–700 nm.[3,16] The ability of a dye to absorb a particular wavelength of light is based on its molecular structure, which depends on the chromophores present within the dye and their degree of unsaturation. As the degree of conjugation or the number of conjugated double bonds increases, the maximum wavelength of absorption also increases.[17] Chromophores include double bonds (C=C) and are the components within dye molecules that absorb UV-vis radiation and are responsible for color emission from the dye.[16–18] When the molecule is exposed to light, the structure of chromophore oscillates, light is absorbed and color becomes visible.[18] In contrast to dyes, a pigment is a compound having no affinity for the textile fiber substrate, hence it is incorporated into the fiber while manufacturing it, or bonded to the surface.[3,19] Pigments are seen as colorants in a variety of fiber types (*e.g.* polypropylene, viscose, acrylic, polyamide, polyester, *etc.*). Since pigments do not have any affinity for fibers, they are generally added in the melt or during production, and are therefore not subsequently extractable.[19]

There are various ways of classifying fiber dyes, including their method of application, chemical class, or the type of fiber substrate onto which they are applied. The important or major classes of dyes are: acid, basic, direct, disperse, azoic, metallized, reactive, sulfur and vat dyes.[3,4,18] Acid dyes are applied under acidic conditions so that the basic functional groups (*e.g.* amino) on the fiber substrate get protonated and positively charged. These groups form ionic bonds with the deprotonated functional groups (*e.g.* sulfonate) of the dye molecule.[3,18,19] Basic dyes are also applied under acidic conditions where the protonated functional groups on the dye form ionic bonds with the negatively charged functional groups on the fiber substrate. In the presence of heat and an electrolyte, direct dyes are directly incorporated into the cellulose during fiber production; whereas disperse dyes are also incorporated directly into synthetic fibers by linking with the substrate through weak van der Waal's force and some hydrogen bonding.[3,18,19] In brief, different dyes are bound to the fibers either through covalent and ionic bonding, Van der Waals forces or impregnation of colloidal dye particles into the fiber substrates.[18,19] Some of the important fiber dye classes and the fiber types to which they are applied are summarized in Table 7.1, with a detailed discussion associated with different fiber dye classes and their applications in the literature.[19]

In 2012, the global fiber production was reported to be around 85.9 million metric tons (MT), out of which man-made fibers consisted of 58.6 MT (68%), of that 26.3 MT (31%) comprised of cotton and 1.07 MT (1%) was wool.[20] Synthetic or man-made fibers are the largest subgroups of fibers, and polyester fibers are the most popular among synthetic fibers.[4] In the studies related to international fiber population, cotton was reported as one of the most consumed textiles in the world, with blue and black/grey colored cotton being the most commonly encountered combination.[21] The world-wide production of man-made fibers during 2012 is presented in Table 7.2.[4]

Table 7.1. Different types of dye classes and the fiber substrates to which they are applied.

Dye class	Fiber substrate
Acid	Wool, silk, polyamide, protein, polyacrylonitrile, polypropylene
Basic	Polyacrylonitrile, acrylic, polyester, polyamide
Direct	Cotton, viscose
Disperse	Polyester, polyacrylonitrile, polyamide, polypropylene, acetate
Reactive	Cotton, wool, polyamide
Sulfur	Cotton
Vat	Cotton
Azoic	Cotton, viscose
Metallized	Wool, polypropylene

Table 7.2. Global man-made (synthetic) fiber production in 2012.

Synthetic fiber material	Million metric tons	Percentage of production
Polyester	43.30	75%
Cellulosic	4.95	9%
Polyamide	4.01	7%
Polypropylene	2.43	4%
Polyacrylic	1.92	3%
Others	0.91	2%

As stated earlier, fibers are mass produced and are used in large quantities. Forensic examiners must be able to examine transferred fibers and compare them with a standard or a known fiber source to discover the possible common origins. Previous studies on fiber persistence have concluded that a very high proportion (about 80%) of fibers may be lost during the first 4 hours after the initial transfer; hence there should not be a delay in collecting fiber evidence.[3,4,22] Some factors – such as the type of fibers transferred, the type of receiving material and the extent of the usage of receiving material after transfer – influence how the transferred fibers adhere to the receiving material (clothes).[11] Therefore, it is very important for examiners to analyze or characterize the questioned fibers in as much details as possible to prove whether the fibers belong to the same source or not.[23] The 'analytical aspects' of a crime-related case can be optimized by adopting some strategies such as: (i) choosing the best target fibers, (ii) using the most efficient method for recovering fibers from a crime scene, (iii) examining the most relevant exhibits first, and (iv) using the most discriminating analytical techniques appropriate for the type of fibers recovered from a crime scene.[24] The upcoming sections will provide a detailed review on the various fiber recovery methods as well as different analytical techniques used to date for the forensic identification, comparison and examination of trace textile fiber evidence.

7.2. Analytical Methods Used for Forensic Fiber Examination

The determination of evidential value, *i.e.* the extent to which trace fibers of different origins can be differentiated, is directly related to the analytical methods used for their analysis. These methods could either be destructive or non-destructive in nature; hence, if a sample size is limited, non-destructive methods must be exhausted before subjecting the sample to any destructive tests (*e.g.* pyrolysis, HPLC, *etc.*).[25] On the other hand, non-destructive methods of fiber comparison and analysis (*e.g.* microscopy) provide information about the fiber polymer (or the substrate), refractive indices, luster, birefringence, elemental composition of fibers,[4] and also aid in discriminating between different classes and subclasses of fibers (single fiber comparison) by comparing specific spectral features.[10,26] Some other techniques (SEM) can be used primarily to examine the elemental content of the fibers. A standard forensic examination of trace textile fiber evidence, according to their types (*e.g.* synthetic polymer) and subtypes (*e.g.* nylon), are based on the principles of microscopy, spectroscopy, chromatography and mass spectrometry (MS).[3,4] Several different analytical techniques that have been reported earlier and applied to the forensic examination of fibers will be discussed in this review and are presented in Figure 7.1. The primary focus of this chapter is to review the analytical techniques that use any form(s) of light sources for illuminating and/or detecting the sample of interest.

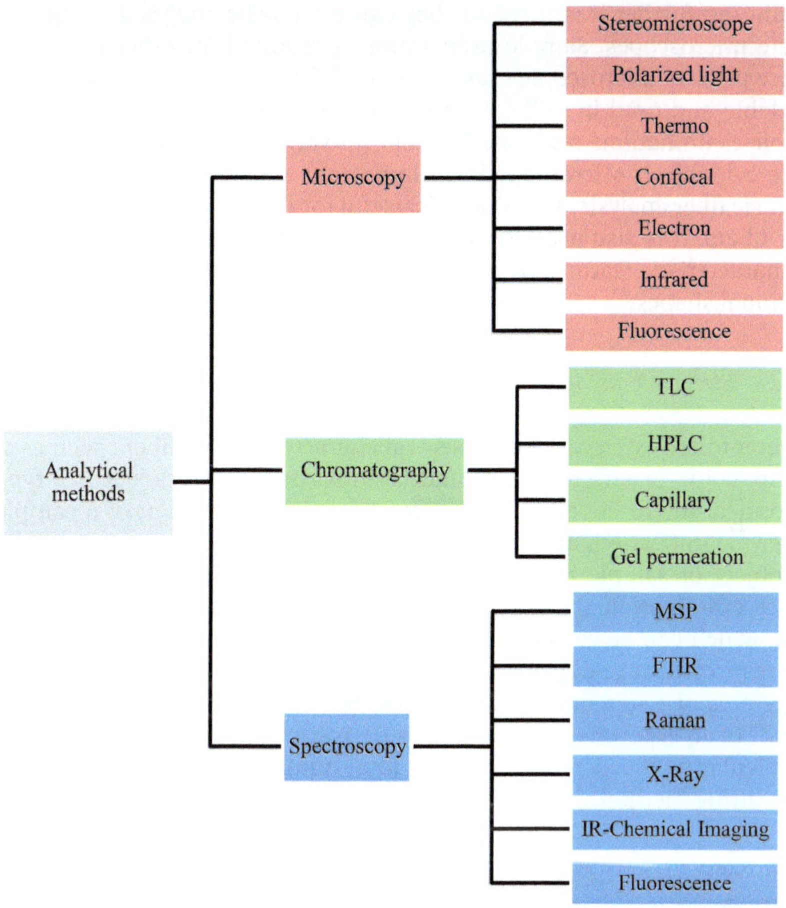

Figure 7.1. Analytical techniques applied towards the analysis of forensic fiber evidence.

7.3. Microscopy

Several decades ago, the identification and comparison of fibers in forensic science laboratories were at a relatively simple level but relied heavily on microscopy.[22] The application of microscopy in the field of forensic fiber analysis is unlimited and is of great importance. This is mainly due to the ability of microscopes to identify, image and resolve a single tiny piece of fiber without any structural modification. For example, a trace analyst uses microscopes for comparison of trace evidence such as fibers, hairs, soil/dust, paint, glass *etc.*; a firearms specialist compares marks or striations on bullets, tools, weapons, *etc.*, a document examiner studies ink line patterns and crossings; whereas a serologist studies bloodstains and their patterns using different types of microscopes. Evidence such as ink, bloodstains, or bullets require no further treatment as they can directly be analyzed using appropriate microscopes. Sample preparation is required for other types of evidence prior to microscopic examination. Different kinds of microscopes are available and used in well-equipped forensic laboratories. In short, microscopic examinations remain a key tool for various types of comparisons of trace evidence.[10] Microscopy is an important analytical tool in the field of forensic fiber analysis, and is mainly useful for the analysis of unrelated polymer fibers. It is also used to examine fraying and cuts in the identification of characteristic features of a polymer such as striations, crimps, and cross-sectional shapes.[27]

7.3.1. Stereomicroscopy

A stereomicroscope can record several characteristics of fibers such as size, crimp, color, luster,[3] striations and cross-sectional shapes.[27] It is a type of an optical microscope that is designed for the observation of a sample at lower magnification, and typically uses a light reflected from the surface of an object or sample rather than transmitting through it. Preliminary evidence examination should include an overall viewing of all items in question, which can be accomplished through stereomicroscopy. In contrast to the restricted stage-objective distance in compound microscopes, the open stage area of a stereomicroscope allows for viewing a large number of samples.[28] In an experiment conducted by De Wael *et al.*, stereomicroscopy aided in calculating the % recovery, or the total number of fibers retrieved by the tape-lifting method.[29]

7.3.2. Polarized Light Microscopy

Polarized light microscopy (PLM) is used to compare manufactured and synthetic dyed fibers, because it reveals the polymer class or the substrate based on the rotation of incident polarized light by the fiber.[3,4,30] PLM involves the use of a polarized light to investigate the optical properties of a sample. Polarized filters are used to obtain polarized light, and to configure the movement of light waves and forcing their vibration in a single direction. Being omnidirectional, light waves will vibrate out at an angle perpendicular from the direction in which it is transmitted. A polarizer is used to allow a beam of light to move in one direction only, causing all light passing through to be blocked except for the light waves that vibrate in parallel with their privileged (light passing) directions.[31,32]

This method can differentiate between common synthetic and regenerated fibers based on the characteristics of a fiber's surface and diameter, their optical properties by means of relative refractive index, birefringence, determination of the sign of elongation.[26] Goodpaster[3] stated that PLM can be used to detect dichroism in cases where dye molecules are linear and oriented along the axis of a fiber, and also to distinguish two fibers based on characteristics such as fluorescence of the dyes, or optical brighteners added to the fibers. A target fiber study using cinema and car seat fibers determined that fluorescence microscopy was more accurate compared to white light microscopy while comparing the number of apparent matches with green cotton target fibers.[33]

7.3.3. *Confocal Microscopy*

Confocal microscopy can capture fiber images of the cross section at any point in the field of view along the length of a fiber. It is an optical imaging technique used for increasing the optical resolution and contrast of a micrograph by adding a spatial pinhole placed at the confocal plane of the lens to eliminate any light that is out of focus. It creates a three-dimensional structure from the obtained images by collecting sets of images at different depths (optical selection) from a sample.[34] An excitation light or laser source is used where the blue light reflects off a dichoric mirror. From there the laser hits two mirrors which are mounted on motors; these mirrors scan the beam of light across the sample. The dye from the sample fluoresces and the emitted light (green) gets de-scanned by the same mirrors that are used to scan the excitation light (blue) from the light source. The emitted light then passes through the dichroic mirror and gets directed to the pinhole, later on measured by a detector such as a photomultiplier tube (PMT).[35]

Since it is a fluorescence-based technique it is destructive in nature, and the images obtained are auto-fluorescence of the specimen, meaning that fluorescence staining is not required.[36] Lepot *et al.* previously coupled a confocal microscope to Raman spectrometer for forensic fiber analysis caseworks.[37] The study of dye-diffusion or fluorescein dye uptake (as a function of time) into nylon 66 fibers was studied using laser scanning confocal microscopy.[38] Using a 408 nm excitation source, it takes under 5 minutes to cross-section a range of man-made fibers (with a thickness of 130 mm) such as nylon, acrylic, polyester, and acetate; as well as natural fibers such as linen, cotton and silk using confocal microscopy. No specific mounting technique is needed; hence this method can be performed on fibers that have been prepared for other forensic microscopic examinations.[36]

7.3.4. *Thermo-microscopy*

Thermo-microscopy has previously been reported, to have aided in the differentiation of synthetic fibers (*e.g.* nylon and olefin carpet fibers) by measuring their melting points. A thermo microscope or hot stage microscope is a microscope coupled with a hot-stage accessory (either open or closed), which provides valuable information regarding the physical characteristics of a sample. It is an analytical technique that combines the properties of microscopy and thermal analysis to enable the solid state characterization of materials as a function of temperature. A sample under investigation is placed onto a microscopic stage which consists of a large area of temperature control element with excellent heating and cooling systems with temperature varying from −200 °C to 500 °C. A color camera is attached to the microscope to observe visual changes, and the hot stage controller (which is attached to the system) monitors the temperature program and transmits the thermal results to a computer for further data analysis.[39,40]

The melting points of nylon 6 and nylon 6,6 are around 213 °C and 250 °C, respectively. Polythene olefin melts at around 135 °C whereas polypropylene melts at 170 °C.[41] After thermo-microscopic analysis, if all the characteristics of the two fibers being compared are identical, the next step is to conduct an examination using a comparison microscope.[25] A comparison microscope is used to identify and compare the questioned fibers to the ones obtained from a crime scene. This technique has a comparison bridge that rests on two microscopes, has two sets of optics built in one instrument, and is used to compare features such as color, crimp, pigmentation, thickness, luster and cross-sectional shapes.[8,25,41] Fiber identification, sub-classification of the fiber type, and identification and differentiation of fiber subclass based on differences in their melting points – are the three most common reasons for observing the thermal behavior of fibers *via* thermal microscopy.[26]

7.3.5. Electron Microscopy

Electron microscopes have an advantage over other conventional microscopes, as they have a wide range of magnification, resolving capacity, and the ability to perform elemental analysis when equipped with an energy dispersive X-ray spectrometer.[4] Scanning electron microscopy (SEM) coupled with energy dispersive spectrometry (SEM-EDS) is used primarily to examine the elemental contents of the fibers, by analyzing the inorganic materials arising from the residues of the manufacturing process, additives, or environmental contaminants.[3,42] SEM is useful in revealing the morphological features of the surface, cross-section or the tips of a fiber.[42] Moreover, surface imaging using SEM was previously reported to have aided in the identification of animal hair structure,[43] and can also help in the detection of trace debris on a fiber surface. A study was conducted for the recognition and identification of modified acrylic fibers from unmodified acrylics, using SEM in conjugation with energy dispersive X-ray analysis (SEM-EDX). 'Teklan' acrylic fibers showed a strong chlorine peak, 'Acribel' acrylics of various compositions tested showed a complete absence of chlorine, and 'Elura' modac fibers showed the presence of bromine.[44] In short, modified acrylics could be distinguished from one another, based on the manufacturing companies.

7.3.6. *Infrared Microscopy*

An infrared (IR) microscopy or microspectroscopy is widely used in forensic laboratories for the identification, comparison and analysis of single fibers.[30] IR microscopy can easily be performed on fibers less than 100 μm long. The infrared spectra of different types of acrylic fiber samples have previously been examined by using a scanning IR microscopy, where 20 different varieties of undyed acrylic fiber types could be spectrally differentiated.[45] It is a minimally destructive technique, where the physical morphology of a fiber is altered when a sample is flattened prior to analysis.[30,46] Identification of generic classes (*e.g.* nylon, acrylic) and sub-generic classes (*e.g.* polyacrylonitrile, nylon 6,6) of fibers is possible using this kind of microscopy.[30] Given the polymeric nature of synthetic and natural fibers, if the sample size is small, then pyrolysis coupled to gas chromatography (GC) or mass spectrometry (MS) is an informative and a minimally destructive technique.[47]

7.3.7. *Fluorescence Microscopy*

Another practical and convenient method of fiber comparison is the use of fluorescence microscopy. Several dyed fibers exhibit a surprising effect of fluorescence – when illuminated with light with shorter wavelengths (near UV region) they emit light of a longer wavelength than that with which they were excited. This phenomenon of the emission stopping when the exciting radiation is cut off, is known as fluorescence.[26] Macrae *et al.* reported the importance of adding fluorescence examination in the UV region to increase the discriminating power of fiber comparisons from about 50% for white light bright field comparison microscopy to as much as 80% with the addition of UV fluorescence.[10,48] Dye identification from two pairs of single indistinguishable fibers was reported by Appalaneni *et al.* using fluorescence microscopy where an epifluorescence microscope was fiber-optically coupled to a commercially available spectrofluorimeter (FluoroMax).[49] More recently, the same methodology was implemented towards the discrimination of four different pairs of visually indistinguishable single fibers dyed with dyes having similar molecular structures[50] (additional details are provided in Section 7.6.6). Dyes and pigments frequently fluoresce, and since a fiber's color is usually achieved by adding more than one colorant, we can expect that some of these components will fluoresce. Normal dyes fluoresce in a variety of colors with different excitation wavelengths, hence a complete range of excitation wavelengths should be employed and all the emission spectra must be recorded.[26]

7.4. Chromatography

Chromatography is the science of separating and identifying mixtures of substances into their individual components. Figure 7.2 represents a schematic diagram of a typical chromatographic system.[51] Color or dye comparison is

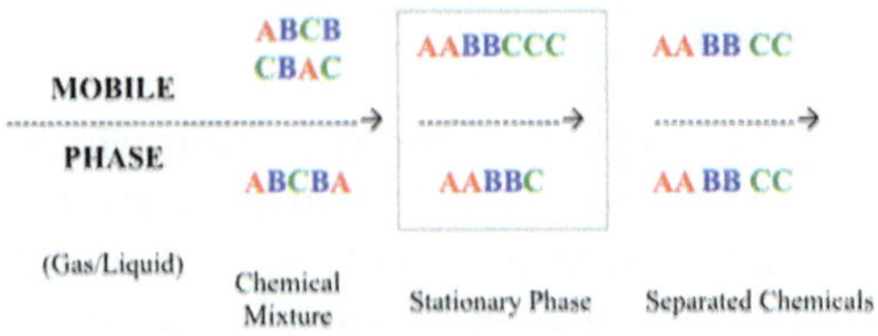

Figure 7.2. Schematic diagram of a basic chromatographic system.

a critical factor to connect a fiber collected from a crime scene and a fiber obtained from a suspect or a known source. The color of a fiber, present naturally or imparted through dyes or pigments, can be extremely discriminating for forensic examiners. Even though they are partially destructive in nature, chromatographic methods of analysis are regularly used by forensic scientists to compare and examine fibers because they can provide a much greater degree of discrimination than physical or optical methods alone. This section will provide an overview of various chromatographic methods implemented towards the analysis of dyes in a fiber such as thin layer chromatography (TLC), high performance liquid chromatography (HPLC) and reverse-phase HPLC, capillary electrophoresis (CE), ultra-performance liquid chromatography (UPLC) and gel permeation chromatography (GPC).

7.4.1. Thin Layered Chromatography

For over 25 years, thin layered chromatography or TLC has been used to compare control and recovered fibers from crime scenes, and is a commonly used tool for the comparison of fiber dyes and the separation of dye classes in several forensic laboratories.[19] It can be coupled with other instrumentation techniques such as microscopy and UV-vis MSP for more specific comparison of fibers.[52–54] A typical TLC separation involves the application of a solution of the sample to be analyzed as a discrete spot or smear upon the stationary (chromatographic) plate, typically coated with silica gel or alumina. Once the solvent in which the sample was applied has evaporated, the plate is developed by allowing a mobile phase to move by capillary action upwards through the stationary phase thus leading to separation.[51] The characteristics of the eluted compounds can be reported as R_f values, which are a measure of the relative distance travelled by each compound from the original spot with respect to the solvent front.[55] Briefly, if a mixture (in a solution) is applied to a TLC plate and a solvent is allowed to pass across it, different components from the mixture will travel with the solvent at different rates depending on their physical and chemical properties. Visually similar looking colors can be made of different dye components and hence can be easily and quickly distinguished using TLC. Owing to the spectral properties of dye molecules and their high absorptivities, the detection levels required can generally be achieved by the human eye.[19] TLC eluent systems have been previously used

Table 7.3. Extraction solutions commonly used for fiber-dye analysis.

Dye class	Fiber type	Extraction solution
Acid	Wool, silk, polyamide, protein, polypropylene, polyacrylonitrile	Pyridine/water (4 : 3 v/v)
Azoic	Cotton, viscose	Pyridine/water (4 : 3 v/v)
Basic	Polyacrylonitrile, modified acrylic polyamide, polyester	Formic acid : water (1 : 1 v/v) Pyridine/water (4 : 3 v/v)
Direct	Cotton, viscose	Pyridine/water (4 : 3 v/v)
Disperse	Polyacrylonitrile, polyester, acetate, polyamide, polypropylene	Pyridine/water (4 : 3 v/v)
Metallized	Triacetate	Pyridine/water (4 : 3 v/v)
	Wool, polypropylene	Aqueous oxalic acid (2%) then pyridine/water (4 : 3 v/v)

for separating dyes used on fibers such as polyester, nylon and acrylic,[56] cotton,[57,58] cellulosic,[59] polypropylene,[60] and wool.[61]

As mentioned earlier, fibers are most likely to be dyed with a combination of several different types of dye; hence the first step in the analytical process is the extraction of dye(s) from the fiber. Several different extraction methods utilizing a variety of solvents have been reported previously.[56,59–65] Multiple solvent system was used to separate several classes of dyes such as acid, basic, disperse and reactive. These enabled researchers to select the most appropriate pair of solvent or extraction systems for a particular dye class.[57,66] A summary of the extraction solutions used for different types of fibers and varied classes of dyes is listed in Table 7.3.[19,55] Wiggins *et al.* employed the TLC method to examine the dye batch variations in textile fibers, and reported that in many instances, TLC highlighted the variations between dye batches that were not detected with microscopy or visible spectroscopy.[53] Additional literature specified using TLC for the analysis of reactive colored (blue, black and red) dyes in wool and cotton fibers, where TLC provided additional information (compared to comparison microscopy and UV-vis MSP) and gave greater individuality to the fibers.[54,67] In brief, comparison microscopy and UV-vis MSP are used for primary analysis and comparisons of dyes that are encountered in textile fibers in the field of forensic science. If relatively large quantities of dyes are extracted from fibers, TLC could extensively be used as an additional comparative technique for forensic fiber analysis because it is cheap, and with practice, relatively easy to perform.[4,19]

7.4.2. High Performance Liquid Chromatography (HPLC)

HPLC has been in use since the late 1960s and is arguably the most widely used of all the analytical separation or chromatographic techniques. Its popularity lies in its applicability in several fields and to a wide range of analytes such as amino acids, proteins, nucleic acids, carbohydrates, drugs, metabolites, pesticides, steroids, explosives, dyes, inks, *etc.*[55,68] A typical HPLC system consists of a solvent reservoir containing the mobile phase,

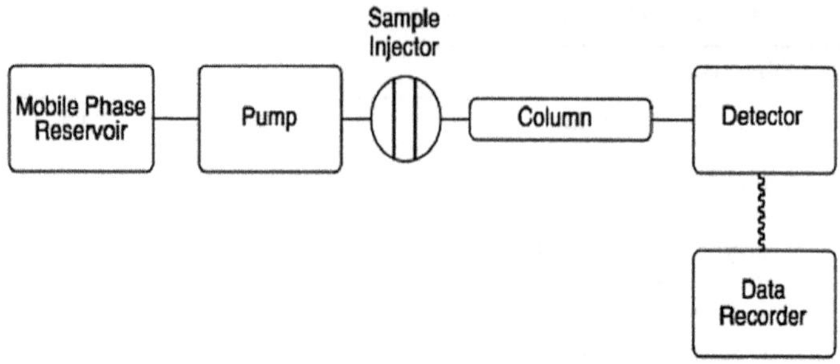

Figure 7.3. Schematic diagram of the components of a typical HPLC system. Adapted from *Identification of Textile Fibres*, 1st edn, M. M. Houck, Analysis of dyes using chromatography, p. 16, Copyright © 2009 Woodhead Publishing Ltd., with permission from Elsevier.

pump, injection system, separation column (3–25 cm long; 3–5 mm internal diameter), detector and a data recording system. The mobile phase is pumped through the stationary phase at high pressure (up to 6000 psi) with flow rates of 0.1–10 ml min^{-1}. A sample containing a mixture of compounds is introduced into the separation column *via* the injection system. Components of a mixture are resolved in the column depending on their selectivity towards either the stationary or the mobile phase contained within the column, known as retention time. A detector records the separation in form of a graph or a chromatogram, which is further analyzed by the data recording system.[55,68] A schematic of a typical HPLC system is presented in Figure 7.3.[55]

HPLC has advantages over MSP and TLC for dye analysis, because MSP is limited when attempting to analyze highly absorbing dark colored fibers and TLC requires relatively large quantities of dyes and different eluent systems for various classes of dyes. On the other hand, HPLC exhibits better chromatographic resolution, greater sensitivity and can be used for qualitative and quantitative analysis.[3,4] There are different forms of HPLC depending on the retention mechanism involved. Partition chromatography is the most common type, where the stationary phase is chemically bonded to silica particles, and can further be divided into normal phase and reverse phase chromatography. In normal phase chromatography, the stationary phase is relatively polar, allowing the non-polar compounds to elute first; whereas in reverse phase chromatography, the stationary phase is non-polar and allows polar compounds to elute first.[55]

There are several considerations to make while developing an HPLC system for fiber dye analysis, such as the general chemical nature of the dye (acid, basic or neutral), probable extraction solvents to use, and the possibility of sample degradation. For example, organic solvents used for dye extraction can interfere with a UV-vis detector.[68] Some of the extracted dye classes might degrade at high temperatures, hence lower temperatures, or the use of antioxidants is necessary.[68] A summary of the HPLC separation systems developed for the analysis of acid, basic and neutral and other types of dyes is tabulated for this purpose.[69] For example, organic solvents used to extract dyes can interfere with the UV-vis detector. Some dyes can degrade at higher temperatures, hence low extraction temperatures and/or antioxidants must be used.

Reverse phase HPLC was used earlier for the separation of dyes belonging to the charged-dye class, and anionic dyes were separated more successfully compared to cationic dyes. Basic dyes can be separated efficiently by employing a separation system based on the ion-exchange property of silica. HPLC by itself can only characterize a dye based on its retention time; hence a sensitive and specific detection method must be incorporated along with chromatography. Using a multiwavelength detector helps in obtaining a complete UV-vis spectrum from dyes, determining peak purity, and assisting in generating a spectral database.[3,4] The powerful combination of HPLC coupled with mass spectrometry has previously been reported and applied to forensic fiber dye analysis. Petrick *et al.*[70] combined HPLC with electrospray ionization mass spectrometry (discussed in the next section) for separating a mixture of 15 basic and 13 disperse dyes extracted from acrylic and polyester fibers of 0.5 cm length. Combining HPLC with different detection methods enables the determination of molecular structure information of the eluted bands, hence providing an extra dimension of information.[4]

In a recent study, a highly sensitive chromatographic instrumentation method (HPLC) coupled with diode array detector (DAD) and a mass spectrometer, or HPLC-DAD-MS, was implemented for the separation and detection of nine different types of dyes, their varieties ranging from neutral, positively and negatively charged dyes; dyes with single and/or multiple charges; hydrophilic and hydrophobic dyes; and dyes with a wide range of sizes and molecular masses; all in a single analytical run.[71] A single set of chromatographic conditions was applied to separate dyes of nearly all relevant fiber dye classes (from fibers as small as 1 mm in length) such as acidic, basic, direct, disperse and reactive dyes. The authors reported this method to have a very high specificity due to analysis by high-resolution MS and a DAD, and were able to analyze single textile fibers of a length of few millimeters or less. Previous studies reporting fiber-dye analysis using HPLC was preferred on synthetic fibers due to easy extraction of dyes from these fibers, however, besides synthetic fibers (such as polyamide, polyester, acrylic and regenerated cellulose) Carey *et al.* used natural fibers for their study, such as cotton and wool. Three different digestion procedures were implemented to isolate the dye molecules from fibers.[71] Kretschmer *et al.* reported a combination of HPLC methods such as size exclusion liquid chromatography (SEC) and reversed-phase chromatography (RPC) for the analysis and differentiation of six different types of PET polyester fibers obtained from different manufacturers. Chromatograms obtained from these fibers differed insignificantly in terms of their peak structure and area distribution. Data obtained was subjected to principle component cluster analysis, which was successful in differentiating the classes of fibers based on the manufacturing company.[72]

van Bommel *et al.* reported using chromatographic and spectroscopic methods for the analysis of 65 synthetic dyestuffs that were developed and used in the 1800s.[73] For the HPLC analysis, two different solvent systems (the mobile phase of the first system consisted of a gradient of water, methanol and 5% phosphoric acid in water; whereas the second system used a water and methanol gradient with a 5 mM tetra butyl ammonium hydroxide) were evaluated. Detection was performed using a 996 photodiode array detection (DAD) system. The combination of UV-Vis spectra (ranging from 200–700 nm) along with chromatographic analysis of the colorants is a very strong tool for identification. Limits of quantitation (LOQs) were calculated at a signal-to-noise (S/N) of 10 at the maximum adsorption wavelengths of the dyes in the visible region. Dyes were not only distinguished from one another by their retention times, but also were identified by their unique UV-vis absorption spectra.[73]

Another technique for the sensitive comparison of dyes extracted from fibers is ultra-performance liquid chromatography (UPLC). According to the van Deemter equation, as the sample size decreases to less than 2.5 μm, there a significant gain in efficiency and it does not diminish at increased flow rates or linear velocities. By using smaller particles, the elution speed and peak capacity (number of peaks resolved per unit time in gradient separations) can be extended to new limits, which is UPLC.[74] Rapid analysis can be performed by UPLC by using high pressure pumps (<15 000 psi) capable of moving samples and mobile phases through columns packed with increasingly smaller diameter (1.7 μm) stationary phase particles.[75] Hence, this technique is able to achieve rapid separation at a very high resolution in a very short analysis time (<5 min). Decreased band-broadening, sharper peaks and higher signal-to-noise ratio measurements are possible to achieve using UPLC.[75]

7.4.3. *Capillary Electrophoresis*

Separation of components in a chemical mixture occurs in a narrow bore capillary under the influence of an electric field, and is based upon the differential migration of charged species through a fused silica capillary filled with electrolyte.[55] Detection is typically carried out by using UV-vis absorbance spectrophotometry, resulting in a graph or an electropherogram which looks similar to a chromatographic separation. Details pertaining to the composition of the capillary columns and the mechanism are explained elsewhere.[76] A schematic diagram of a typical capillary electrophoresis (CE) instrumental setup is described in Figure 7.4.[55,76]

CE analysis is very popular in the field of forensics due to its exceptional separation power, rapid analysis, low sample requirement (<50 nl) and minimal sample preparation.[4] As reported earlier in various literature articles,[77–79] CE was employed for the examination of acidic, basic and reactive dyes, and was successful in separating them with a high resolution of separation.

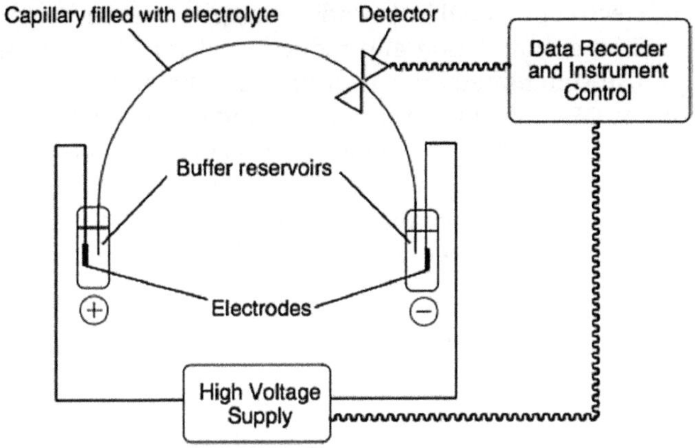

Figure 7.4. Schematic diagram of a typical CE system. Adapted from *Identification of Textile Fibres*, 1st edn, M. M. Houck, Analysis of dyes using chromatography, p. 17, Copyright © 2009 Woodhead Publishing Ltd., with permission from Elsevier.

However, it was also reported that earlier attempts to develop rigorous CE methods for fiber-dye analysis were unsuccessful due to lack of reproducible migration times, poor sensitivity and the inability to separate non-ionizable dyes.[68] Hence, the introduction of micelles into the buffer now allows for separation of water-insoluble dyes; this modification is known as micellar electrokinetic chromatography (MEKC), and is reported in detail elsewhere.[76] Initially reported by Terabe *et al.*,[80] MEKC provides a method where neutral molecules can be resolved. Li *et al.* reported the ability of MEKC to separate different ionic species based on their M/Z ratio, hydrophobicity, and charge interactions at the surface of micelles.[81] Natural dyes such as flavonoids and anthraquinones were extracted from wool samples and analyzed using MEKC.[82] On the other hand, disperse dye (on polyester) and vat dye (on cotton) were difficult to ionize and separate using CE.[83,84]

Oxspring *et al.* compared the selectivity of two techniques namely – reversed-polarity capillary electrophoresis and adsorptive stripping voltammetry (AdSV) – for the separation and identification of various reactive textile dyes.[85] Anionic reactive dyes have an electrophoretic mobility towards the anode, which is opposite to the electroosmotic flow (EOF); hence they have long migration times compared to cations and neutral species. Hence, reversing the polarities of the electrodes by applying a negative voltage to the capillary reverses the direction of EOF and changes the order in which species migrate through the capillary to anions eluting first, followed by neutral and cationic species. The dyes now migrate against the EOF, towards the anode and are detected using a UV/vis diode array detector. Using this technique, it was possible to separate the dyes according to differences in their electrophoretic mobilities.[85] Stripping voltammetry using a hanging mercury drop electrode (HMDE) was also employed for the separation and determination of reactive dyes by Oxspring and group where cell current was measured as a function of electric potential. Adsorptive stripping measurement is based on the formation, accumulation and reduction of the absorbed species by differential pulse voltammetry.[85] To summarize, the authors reported that reversed-polarity CE could effectively analyze and separate all six textile dyes under investigation, and provided a fast method of analysis with all six dyes eluting from the capillary within 7.5 minutes. On the other hand, AdSV was shown to be a very sensitive technique to monitor individual dyes, but lacked the selectivity to deal with a complex mixture of samples.[85]

7.4.4. Gel Permeation Chromatography

Gel permeation chromatography (GPC) or size exclusion chromatography is a type of liquid chromatography (using a solid stationary phase and a liquid mobile phase), which separates analytes based on their size.[86] In the late 1970s, GPC was first used as an analytical tool for the forensic identification and discrimination of polymer products, and to determine the molecular weight distribution of polymers.[86,87] Briefly, GPC is used to either characterize polymers and/or separate mixtures into discrete fractions *e.g.* polymers, oligomers, monomers, *etc.* Evaluation of the molecular weight distribution amongst PET fibers was reported in previous literature,[88,89] where various solvent mixtures such as m-cresol/chloroform, o-chlorophenol/chloroform and 1,1,2,2-tetrachlorothane/nitrobenzene were used as a GPC solvent for PET fibers. Also, 1,1,2,2,3,3-hexafluoroisopropanol was described as an effective solvent for determining molecular weights of PET and nylon-6,6 fibers.[90] Farah *et al.* employed GPC for the analysis of 14 PET fibers and used hexafluoroisopropanol to dissolve the fibers. Results showed that distinct differences in the molecular weight of different fiber samples were found which may have potential use in forensic fiber comparison.[91] PET fibers with average molecular weights between 20 000–70 000 g mol^{-1} were determined, and the detection limits for this method were tested by using samples with concentrations as low as 1 μg ml^{-1}. Molecular analysis of PET fibers by GPC allowed for fiber comparison that could not be otherwise distinguished with high confidence, and this method could be extended to forensic comparison of other synthetic fibers such as acrylics and polyamides.[91]

7.5. Spectroscopy

Microscopic screening is essential to discriminate between fibers obtained from the clothes of a victim or a suspect at a crime scene. However, beyond microscopic analysis, sometimes fibers could still be indistinguishable; and in such cases spectroscopic techniques are required for further examination. Spectroscopy is the study of interactions between a matter or a sample and electromagnetic radiation and the spectral measurement devices are commonly known as spectrometers or spectrophotometers.[17] Such techniques are used to identify the type of polymer or substrate, along with their composition. Due to the wide range of samples available at a crime scene, a variety of spectroscopic techniques have previously been employed for their identification in a forensic investigation. During recent years, the application of spectroscopy to the identification and discrimination of textile fibers has increased gradually. Some of the important spectroscopic techniques such as Fourier transformed infrared spectroscopy (FTIR), Raman spectroscopy (RS) and micro-spectrophotometry in the UV-vis region (UV-vis MSP) are used in the study and analysis of textile fibers, mainly man-made polymers such as polyester, polyamides and polyacrylics.[4] Also, there are other less used spectroscopic techniques such as X-ray fluorescence spectroscopy (XRFS) and infrared chemical imaging (IRCI), which are emerging in the field of forensic fiber analysis.[5]

7.5.1. Microspectrophotometry

Microspectrophotometry or MSP is a special method within the UV-vis spectroscopy category, where a microscope collects light from a sample and transmits it to a UV-vis spectrometer.[3,92] This technique is more suitable for the discrimination of dyed fibers and is more suitable for dyed fibers that appear similar to the naked eye.[4] Since dyes are substances with conjugated systems of excitable electrons, UV-vis spectroscopy in general, but MSP in particular, is the most suitable method for the analysis of different dyes on fibers. Hence, in forensic science, MSP has been established as one of the most important methods for analysis because it is a rapid, repeatable and a non-destructive method (with little or no sample preparation needed) for examining trace evidence which consists of colored microscopic samples such as fibers, paints and inks.[92] MSP can distinguish between indistinguishable colored fibers by identifying the spectral characteristics two fibers in comparison, based on difference in the molecular structures of their chromophores and also on the environment in which the chromophore is found.[3,4,93]

Color can be a highly discriminating feature in forensic fiber comparison, and the advent of visible MSP increases an examiner's ability to objectively discriminate color over and above microscopic comparisons. Even though further fiber discrimination was attained in the UV spectral range and reported in the literature previously,[94–100] there has been little published data on the validation of analytical parameters and protocols for UV-vis MSP in forensic fiber analysis. Fibers of fabrics such as wool, silk, polyester and acrylics with methylvinylpyridine (MVP) are not suitable for analysis in the UV range below 300 nm due to the polymer absorbance interference.[101,102] However, colored fibers with no polymer interference such as cotton, viscose, olefin and acrylics without MVP are suitable for analysis and colored fibers could be differentiated.[103] Morgan *et al.* employed fluorescence MSP in addition to UV-vis MSP on 4 types of yellow and 3 types of red fibers (indistinguishable under light microscopy), since fluorescence was found to add considerable discrimination even without common fiber class/color combinations.[104] Multivariate statistical methods of spectral analysis such as unsupervised principal component analysis (PCA) and supervised linear discriminant analysis (LDA) were used to evaluate the discrimination between similar fibers of various textile fiber types (nylon, acrylic, polyester and cotton).[104] Was-Gubala *et al.* used this method for discriminating single cotton fibers dyed with reactive dyes coming from the same manufacturer, and confirmed that all of the analyzed samples were distinguishable between each other with the use of MSP, mostly in the visible, and also in ultraviolet range.[105]

7.5.2. *Fourier Transform Infrared (FTIR) Spectroscopy*

This method is used classically to compare fibers in forensic and industrial analysis due to its sensitivity to minor variations in the polymer structure. FTIR is commonly used for the determination of a fiber's composition (natural *versus* synthetic, organic *versus* inorganic); class (such as polyester, polyamides and acrylics); or subclass (nylon 6, nylon 6,6, *etc.*),[106–109] but not for identifying the types of dyes used to color a fiber.[3,30] This is due to the fact that inherently there are low concentrations of dyes found in most textiles (less than 5% of the total weight of the fiber). However, a modified version of this technique known as diffuse reflectance infrared Fourier transform spectroscopy (DRIFTS) has been reported as being more successful in characterizing fiber dyes. DRIFTS has the advantage of not following Beer's law, hence reflectance measurements are more sensitive than transmission at lower dye concentrations.[110] Previously, this method was shown to have successfully discriminated between both dye color and reactive dye state on cotton fibers.[111,112] Some advantages of the FTIR method are that it is rapid, easy, selective, and non-destructive in nature,[4] meaning that other analytical methods can be implemented further after FTIR analysis. The only disadvantage is that FTIR analysis cannot be accomplished if a fiber has a diameter thicker than 30 μm, since the IR light is poorly transmitted.[109] Hence, fibers are flattened before analysis, changing their morphology. IR is a very powerful tool for the discrimination of fiber types, and to enhance the evidential value of a sample obtained at a crime scene.

Polyethylene terephthalate (PET) is the most common type of polyester fiber, and these fibers can be differentiated based on the evaluation of two features in the IR spectra: (i) the *trans*-gauche conformation, and (ii) the O–H end-group content of the molecule.[113] Initially, *trans*-conformers of the PET molecules produced Raman peaks at 846, 973, and 1340 cm^{-1}; and subsequently the ratios of the last two peaks were calculated because this is where the gauche conformation originates. The end-group content was estimated by calculating the ratios of peaks at 3340 and 874 cm^{-1}. By calculating the relative standard deviation of the polyester fibers, a way for sub-classifying PET fibers was reported.[6,113] Causin *et al.* utilized IR spectroscopy for the quantitative differentiation of undyed acrylic fibers by obtaining the absorbance peaks from functional groups such as nitrile (2240 cm^{-1}), carbonyl (1730 cm^{-1}) and

C–H (1370 cm^{-1}). Ratios of A_{1730}/A_{2240} and A_{1730}/A_{1370}, a relative measure of the co-monomer content in the fiber, were used to differentiate the samples.[114]

7.5.3. Raman Spectroscopy

Also known as vibrational spectroscopy, this is a very good analytical technique for the efficient discrimination between different dyes used for coloring textile fibers, as the molecular structures of many dyes are characterized by non-polar bonds.[5] Most of the Raman bands from dyed fibers come from the dyes themselves, meanwhile there are only a few weaker signals obtained from textile polymer substrates. Jochem *et al.* reported that Raman spectra in the visible region provided detailed data on the pigments but gave little or no information about the sub-class or types within the polymer; and in contrast, the opposite holds true for IR spectroscopy which was well suited for characterizing the polymer but failed to provide useful data on the pigments.[115]

By comparing the Raman spectra obtained from cotton (natural) *versus* rayon (synthetic) fibers, major difference between these fibers were reported at Raman peak at 650 cm^{-1}, obtained from rayon spectrum (the C–S–C stretch).[6] This could possibly result from incomplete regeneration of xanthate derivative back into the form of cellulose during manufacturing rayon.[116] Rayon fibers are man-made and are generated from wood cellulose and have an identical chemical structure to cotton (which is made up of cellulose); yet Raman spectroscopy was able to distinguish between these two fiber types.[5,6] Moreover, differentiation between two animal (natural) fibers such as wool and silk was reported,[6] indicating that the main difference between them is the disulphide bond (S–S) which is present in silk, exhibiting a Raman peak at 523 cm^{-1}. Additional differentiation of polymer subtypes was successful after the principal spectral differences were reviewed between the three most common subtypes of nylon (nylon 6, nylon 6.6 and nylon 6.12) and also acrylic subtypes (acrylic Creslan 61 and modacrylic).[6] These findings agree with the results obtained by Miller *et al.*, where subgeneric fiber identification and differentiation by Raman analysis was possible for undyed fibers such as nylon 6 and nylon 6,6.[109] It was also indicated that the differences between the nylon subtypes were more visible in Raman than in IR spectra. Several such studies have been conducted previously, reporting detailed differences between Raman bands obtained from nylon, acrylic, polyester and other subtypes of fibers.[6,7,109,115,117,118]

A study conducted to discriminate between linen and jute[119] reported that a specific Raman band was obtained from jute samples at 1736 cm^{-1} and another band from linen at 1578 cm^{-1}. Real casework has been reported by Lepot *et al.* where Raman spectroscopy efficiently categorized questioned fibers to the set of known fibers, while also characterizing a detailed dye spectrum from all the questioned fibers. These are some of the examples of the highly specific identification capabilities of Raman spectroscopy.

This method is of special interest for analyzing natural fibers, where cotton fibers are often encountered at crime scenes. Since dye composition is the only characteristic feature of these fibers, Raman measurements could be useful in improving the discriminating power of these fiber types.[37]

Raman spectroscopy has also been successful in differentiating between several types of synthetic/man-made fibers. Polyester is a generic term that comprises different man-made textile materials, where synthetic fibers are usually produced from polyethylene terephthalate (PET).[5] By studying the relative Raman band intensities from 350–1100 cm^{-1}, it is possible to observe variances between different subtypes of PET fibers that have similar spectral features.[5,6] During a real case, molten end colorless polyester fibers were searched for in a suspect's blue denim textiles. These fibers had forensic interest because they are rarely present in blue denim textiles. Raman analysis of the colorless part of these fibers confirmed these fibers belonged to the PET polymer class.[120] In another murder case, the suspect's sweatshirt with flock printing was the garment of interest. A particular fiber type known as mauve polyamide caught the examiner's attention, because several of these fibers were present on the victim. These mauve polyamide fibers produced reproducible and resolved Raman bands at 1400 cm^{-1}. Lepot *et al.* were able to differentiate between three undyed fibers such as cotton, polyester (PET) and polyamide (PA6.6).[37]

The implications of Raman spectroscopy mainly for the study of dyes on fibers, and for the examination of natural and manufactured fibers has been extensively performed in the past. In the research conducted by Jochem *et al.*, Raman bands were obtained from dyed and undyed polyacrylonitrile (PAN) fibers. Undyed fibers provided only a few bands at the excitation wavelengths used, whereas spectra from strongly pigmented fibers were solely attributed to the pigment or dye on the fiber (without bands from the matrix or undyed fibers).[115] Raman bands of dyes exhibited strong/intense peaks at low uptake, and increased in intensity at medium uptake, however intensity of bands decreased at high dye uptake – indicating an optimum concentration for observing the spectrum.[5] A collaborative study of the evaluation of Raman spectroscopy was conducted by Massonnet *et al.* on two red colored acrylic fibers (with different polymer types) using 9 different laser wavelengths ranging from blue (λ = 458 nm) to near IR (λ = 1064 nm). Results stated that red lasers (λ = 633 and 685 nm) gave the poorest spectral quality due to fluorescence interference, near IR (785, 830 and 1064 nm) lasers provided average results, whereas blue (488 nm) and green (532 nm) lasers provided spectra with the best quality.[118]

Thomas *et al.* used a laser with excitation wavelength of 632.8 nm provided the best results with little acquisition time and no spectral degradation.[21] Overall, a Raman instrument equipped with multiple lasers is recommended to analyze forensic fiber samples, since certain excitation wavelengths can cause fluorescence to interfere with the spectra. The quality of spectra obtained greatly depends on the laser wavelength that is selected, which must be an important parameter to consider when choosing a Raman instrument. To summarize, Raman spectroscopy has been established as a method for fiber analysis, and identified as a priority research area by the European Fibers Group (EFG).[121] Even though techniques such as MSP, IR and Raman are used as non-destructive methods to measure color, very little information is gained about the specific dye. Raman is well suited for analyzing colored fibers since it requires no sample preparation, is a no-contact technique, needs only a small sample size, reproducible results can be obtained in less than a minute, and multiple tests can be performed on the same fiber.[3–6] Raman measurements are insensitive to water or moisture, hence it can be used to examine even a wet sample through a plastic pack or even in a glass container without interference.[4] Lastly, fluorescence interference can be reduced, or removed, by mathematically post-processing of the data, by altering the laser wavelengths, or using other techniques such as surface enhanced resonance Raman spectroscopy (SERRS), which increases the signal by several orders of magnitude while reducing the background noise or fluorescence.[3,118]

7.5.4. X-Ray Fluorescence (XRF) Spectroscopy

XRF is a powerful quantitative and qualitative analytical tool used for the elemental characterization and composition determination of fibers. This method is fast, accurate, non-destructive, and usually requires minimum sample preparation.[122–124] When elements in a certain sample are exposed to a high intensity X-ray source, the fluorescent X-rays will be emitted from the sample at energy levels unique to those elements.[125] During the study of a hit-and-run case, Sano *et al.* employed XRF spectroscopy to analyze the inorganic contents from smooth surface artificial leather fibers. Results showed important differences in the detected elements and characteristic X-ray intensities, even for samples of similar color.[124]

Another study focused on analyzing fibers of different textile materials (polyester, viscose and wool) using total-XRF spectroscopy, yielding a 'fingerprint-type' trace elemental pattern.[123] Prange *et al.* stated that there are trace-element patterns present in textile fibers, allowing examiners to clearly distinguish not only between different fiber materials, but also between sub-groups of fibers made from the same material type. Total XRF is able to analyze samples as small as 3 mm in length and mass of less than 1 µg.[123] Koons used energy dispersive (ED-XRF) for the elemental characterization of individual carpet fibers from various automobiles. It is common to find several metallic elements in automotive carpet fibers, which are less frequently obtained from residential carpet fibers; allowing for source classification of the questioned fibers.[122] Results indicated that automotive and residential fibers could be differentiated based on the presence of certain transition metals (such as Co, Cr and Zn) which were obtained from automotive fibers. This method could provide promising classification results for nylon fibers, which typically do not contain metallic elements when intended for residential use.[122] Quantitative identification of many elements introduced into cotton textiles by addition or modification can be accomplished using X-ray fluorescent spectra, and almost all elements (metal or nonmetal) in the periodic chart can be detected, identified or quantitatively analyzed using this method.[126] Moreover, elemental analysis was conducted on chemically modified cotton textiles using X-ray fluorescence for the quantitative determination of several elements such as P, S, Cl, Ca, K, Al, Mg, *etc.*[127]

7.5.5. *IR-Chemical Imaging (IRCI) Spectroscopy*

Certain tools for sample mapping and generation of chemical images present significant advantage in the field of forensic science, by providing alternate visualization methods that enable access to scientific information. In order to elaborate the chemical images, multiple spectra are recorded using a variety of excitation wavelengths. Such imaging techniques provide information about the distribution of the chemical species along sample surfaces, aiding in the identification of its components.[5] Flynn *et al.* utilized IRCI spectroscopy for the analysis of bi-component fibers, which essentially are a special class of man-made fibers that are comprised of two polymers of different chemical and/or physical properties and exist within the same filament. In short, bi-component fibers are a combination of two or more polymeric materials and present great value as forensic evidence, due to their low frequency in the population and the specific spatial configuration of their components.[128] The authors stated that IRCI has been established as method useful to recognize and provide spatially resolved chemical information for those types of bi-component fibers where it is likely to detect spectral differences between the two components present.

When one component is flattened over the other, a laser passes through both components at the same time, providing a characteristic IR spectra of each component, along with providing images that illustrate the side-by-side configuration of two components of a fiber. The two side-by-side components were clearly observed for Monvelle fibers, whose IR chemical images were produced from the integrated spectral intensity under bands near 1641 cm^{-1} (amide I in polyamide) and 1735 cm^{-1} (C=O in polyurethane). Fibers such as Belson F040 and Cashmilon G4K also had a side-by-side configuration, since both were formed by PAN-MA-AA and PAN-MA subtypes of polyacrylic material.[128]

7.5.6. *Fluorescence Spectroscopy*

Techniques used for the non-destructive analysis of fibers (such as diffuse reflectance infrared Fourier transform spectroscopy (DRIFTS), FTIR or Raman spectroscopy) base fiber comparisons on spectral characteristics of textile dyes imbedded in the fibers. When coupled to chemometric algorithms for spectral interpretation such as principal component analysis (PCA) and soft independent modeling of class analogies (SIMCA), DRIFTS was reported to be able to discriminate both dye color and reactive dye state on cotton fabrics.[110–112] Attempts to distinguish between single fibers were, however, not reported. Raman microprobe spectroscopy was able to characterize dyes in both natural and synthetic fibers *via* a combination of Fourier transform-Raman spectra and PCA analysis.[115,129,130] Due to the inherently weak nature of the infrared absorption signal and the Raman signal, the identification of minor fiber dyes *via* FTIR and Raman spectroscopy has not been possible. The inability to detect smaller dye concentrations that could add valuable information to the signature of fibers certainly reduces the discrimination power of these techniques. The weak nature of their signals also poses serious limitations in the analysis of lightly dyed fibers.[6,42,48,92,118,131,132]

Details of the principles of fluorescence spectroscopy are reported elsewhere.[133] To the extent of our literature search, investigation of the full potential of fluorescence microscopy for forensic fiber examination has not been reported. Most fluorescence microscopy articles report measurements made with excitation and emission band-pass filters,[33,60] *i.e.* an approach that takes no advantage of the information content that exists in the spectral signatures of textile fibers. A highly discriminating approach based on fluorescence microscopy which focuses on the total fluorescence emission obtained from different dyed fibers has been reported earlier.[50] In addition to the contribution of the textile dye(s) to the fluorescence spectrum of the fiber, de la Pena *et al.* examined the contribution of intrinsic fluorescence impurities – *i.e.* impurities imbedded into the fibers during the fabrication of garments – as a reproducible source of fiber comparison. Fiber comparison was conducted *via* acquiring data formats known as room-temperature fluorescence excitation-emission matrices (RTF-EEM). Collection of 2D fluorescence emission spectra and 3D EEMs directly from the fiber is carried out using an epi-fluorescence microscope coupled with a commercial spectrofluorimeter using fiber optic probes. Statistical figures of merit for correct identification of fiber dyes *via* principal component cluster analysis have been reported so that matching single evidential fibers to other single fibers, threads, or bulk materials may be accomplished with 99% confidence.[49]

Single textile fiber dye identification has been reported using excitation-emission fluorescence spectroscopy.[49] This is a non-destructive technique where no dye extraction from the fibers is required for analysis. Two pairs of fibers with closely matched dye pairs namely acid blue 25 and 41, and direct blue 1 and 53 were analyzed, compared and differentiated with high accuracy using principle component analysis, and no false-positive identifications were observed.[49] More recently, successful discrimination between four pairs of visually and microscopically indistinguishable fibers was achieved using non-destructive total excitation-emission fluorescence spectroscopy paired with multi-way chemometric analysis.[50] The four pairs of visually indistinguishable fibers consisted of nylon 361 dyed with acid yellow 17 and acid yellow 23, acetate satin 105B dyed with disperse blue 3 and disperse blue 14, polyester 777 dyed with disperse red 1 and disperse red 19, and acrylic 864 dyed with basic green 1 and basic green 4. Different chemometric algorithms were employed to process the fluorescence data obtained from the selected dyed fibers. In brief, the ability of these algorithms to distinguish between all the four pairs of fibers excludes the possibility that the visually indistinguishable fibers originated from a common source.[50]

The same methodology of fluorescence microscopy was utilized for enhancing textile fiber identification using the fluorescence spectral data obtained from different fibers that were sequentially laundered with commercially available detergents.[134] This technique aided in the examination of the alterations in the fluorescence spectral fingerprints of single textile fibers resulting from exposure to commonly used laundry detergents that contain fluorescent whitening agents. Lastly, PCA was used to determine that the spectra of laundered fibers are distinct from the spectra of dyed and unwashed fibers such as cotton, nylon and acrylic. More recently, discriminant unfolded partial least squares (DU-PLS) method – a chemometric algorithm for data analysis – was used to analyze the 3D EEMs obtained from different dyed fibers, and to investigate the effects of weathering on the spectral features of these fibers. The fibers investigated were cotton 400 pre-dyed with DB1, nylon 361 pre-dyed with AY17 and acrylic 864 pre-dyed with BG4. The investigated fibers were exposed to humid (Florida) and dry (Arizona) weathering conditions for a time period of zero, three, six, nine and twelve months. In all cases, this algorithm was unable to differentiate non-exposed acrylic fibers from exposed acrylic fibers. DU-PLS was able to differentiate non-exposed cotton and nylon fibers from exposed fibers to Florida and Arizona weathering conditions. It was possible to determine the period of exposure to either Florida or Arizona conditions. It was also possible to discriminate between fibers exposed to Florida or Arizona weathering conditions for the same period of time. In summary, these results provide the foundation for future studies towards a non-destructive approach capable of providing information on the history of the fiber.

7.6. Other Methods for Forensic Fiber Analysis

Thermal techniques, such as differential thermal analysis (DTA) and thermogravimetric analysis (TGA) were employed by Gray *et al.* for the characterization of polymers and fibers.[135] A particular thermal characteristic from a fiber specimen may provide fingerprinting information for forensic characterization purposes. Features of a fiber sample that can be identified include the glass transition temperature; % crystallinity; melting temperature; cold-crystallinity temperature; thermal process associated with the release of water, additives or other solvent residues; phase transformations; and thermal degradation.[135] The authors stated that TGA is useful for observing the thermal events that occur with loss of fiber mass (*e.g.* associated with release of adsorbed water or thermal degradation); whereas other thermal events such as glass transition, melting or solid-phase transformations can be observed using DTA. Differences in spectral features between natural (linen, wool, rayon, silk) and synthetic (nylon, vinyon, olefin, spandex, modacrylic) fibers were observed from the thermograms that were obtained from thermal analysis of the selected fibers. These thermal techniques helped Gray *et al.* observe the thermal behavior of synthetic fiber properties, which was in sharp contrast to that of natural fiber materials.[135]

7.7. Conclusion

The importance of the analysis of trace textile fiber evidence in forensic crime scene investigations has stimulated a substantial amount of research by examiners using several microscopic and other analytical instrumentation techniques. Since the identification and discrimination of fiber evidence is crucial, different methods have become available over the past few decades for fiber analysis ranging from simple visualization or microscopic techniques to the use of advanced analytical instrumentation techniques as described throughout this chapter. Chromatographic (*e.g.* HPLC) and mass-spectrometric (*e.g.* GC-MS) techniques are selective yet destructive in nature; however, they are employed when fibers cannot be discriminated *via* non-destructive methods. In such cases, a reasonable procedure would be to perform analysis by comparing questioned and known fibers *via* dye extraction. Chromatographic techniques primarily focus on the dyes used to color the fibers and do not investigate other potential discriminating components present on the fiber. Once a dye is extracted from a fiber, the original integrity of the sample is lost. Differentiating among commercial dyes with very similar chromatographic behaviors and almost identical absorption spectra and/or fragmentation patterns could be a challenging task. Most of the microscopic techniques are non-destructive and are used by investigators to examine the physical properties of fibers. IR and thermo-microscopy are destructive because, for the former method, the physical morphology of a fiber is altered when a sample is flattened prior to analysis, whereas in the latter, a hot stage is used to hold the sample and to compare melting points of different fibers; both these methods destroy the fibers.

Being non-destructive in nature, spectroscopic techniques preserve the integrity of a sample, and fibers can be used again for further court examinations. For example, Raman, FTIR, MSP and fluorescence spectroscopy have been reported to be non-destructive, and have the advantage of requiring only a small sample for analysis, and these methods need no time for sample preparation. Our research group was successful in accomplishing the non-destructive analysis of several fibers with an instrumental set-up combining a commercially available spectrofluorimeter and an epi-fluorescence microscope, a technique known as fluorescence. With the 40X visible objective lens, the open pinhole diameter corresponds to a 6000 μm (6 mm) fiber length, which is considerably shorter than the ~2 mm fiber length typically encountered at crime scenes. A smaller pinhole diameter

(400 μm or 0.4 mm) was utilized for analyzing several dyed and undyed fibers, and the results are reported in various articles elsewhere.[49,50,134] Investigation of spectral changes that could occur in textile fibers as a result of exposure to different environmental conditions such as repetitive launderings and altering weather conditions over various time intervals was conducted using fluorescence microscopy. Moreover, four different pairs of visually and microscopically indistinguishable fibers with similar dye colors and molecular structures were successfully differentiated from one another using this method paired with different chemometric algorithms.[50] Such examinations conducted on forensic trace textile fibers can provide us with a better understanding of the physical, chemical and spectral changes that might affect textile fibers after they are analyzed, compared and differentiated using the non-destructive approach of fluorescence microscopy paired various chemometric algorithms for data analysis.

References

1. http://www.ascld-lab.org/accredited-laboratory-index/, last accessed 2016.
2. J. M. Taupin and C. Cwiklik, *Scientific Protocols for Forensic Examination of Clothing*, CRC Press, Boca Raton, FL, 2011.
3. J. V. Goodpaster and E. A. Liszewski, *Anal. Bioanal. Chem.*, 2009, **394**(8), 2009.
4. S. Farah, K. R. Kunduru, T. Tsach, A. Bentolila and A. J. Domb, *Polym. Adv. Technol.*, 2015, **26**(7), 785.
5. P. P. Meleiro and C. Garcia-Ruiz, *Appl. Spectrosc. Rev.*, 2016, **51**(4), 258.
6. L.-L. Cho, *Forensic Sci. J.*, 2007, **6**(1), 55.
7. L. C. Abbott, S. N. Batchelor, J. R. L. Smith and J. N. Moore, *Forensic Sci. Int.*, 2010, **202**(1–3), 54.
8. K. G. Wiggins, *J. Forensic Sci.*, 2001, **46**(6), 1303.
9. J. Robertson and M. Grieve, *Forensic Examination of Fibres*, ed. F. P. Adolf, Taylor & Francis, Ltd., London, UK, 1999.
10. J. Roberston, *Forensic Examination of Fibres*, Ellis Horwood Ltd., West Sussex, England, 1992.
11. J. Robertson and M. Grieve, *Forensic Examination of Fibres*, ed. J. Robertson and C. Roux, CRC Press, New York, 2nd edn, 1999.
12. J. M. Taupin, *J. Forensic Sci.*, 1996, **41**(4), 697.
13. S. K. David and M. T. Pailthorpe, *Classification of Textile Fibres: Production, Structure, and Properties*, ed. J. Robertson and M. Grieve, CRC Press, New York, 2nd edn, 1999.
14. D. F. Rendle, *Chem. Soc. Rev.*, 2005, **34**(12), 1021.
15. R. D. Blackledge, *Forensic Analysis on the Cutting Edge: New Methods for Trace Evidence Analysis*, ed. L. A. Lewis and M. E. Sigman, John Wiley & Sons, Inc., Hoboken, New Jersey, 2007.
16. H. Zollinger, *Color Chemistry: Synthesis, Properties and Applications of Organic Dyes and Pigments*, Wiley-VCH, Zurich, Switzerland, 3rd edn, 2003.
17. D. A. Skoog, F. J. Holler and S. R. Crouch, *Principles of Instrumental Analysis*, Thomson Brookes/Cole, Belmont, CA, 6th edn, 2007.
18. A. Akbari, J. C. Remigy and P. Aptel, *Chem. Eng. Process.*, 2002, **41**(7), 601.
19. J. Robertson and M. Grieve, *Forensic Examination of Fibres*, ed. K. G. Wiggins, CRC Press, New York, NY, 1999.

20. F. J. Horn, Fiber Economics Bureau, Arlington, VA, USA, 2016.
21. J. Thomas, P. Buzzini, G. Massonnet, B. Reedy and C. Roux, *Forensic Sci. Int.*, 2005, **152**(2–3), 189.
22. M. C. Grieve, *J. Forensic Sci.*, 1983, **28**(4), 877.
23. S. Bell, *Encyclopedia of Forensic Science: Facts on File*, New York, 2008.
24. M. C. Grieve and K. G. Wiggins, *J. Forensic Sci.*, 2001, **46**(4), 835.
25. https://www.fbi.gov/about-us/lab/forensic-science-communications/fsc/april1999/houcktoc.htm/houckch1.htm, 1999.
26. J. Robertson and M. Grieve, *Forensic Examination of Fibres*, ed. S. Palenik, CRC Press, New York, 1999.
27. K. H. Cochran, J. A. Barry, D. C. Muddiman and D. Hinks, *Anal. Chem.*, 2013, **85**(2), 831.
28. https://www.fbi.gov/about-us/lab/forensic-science-communications/fsc/april2005/standards/SWGMAT_fiber_training_program.pdf, 2004.
29. K. De Wael, F. G. C. S. J. Gason and C. A. V. Baes, *J. Forensic Sci.*, 2008, **53**(1), 168.
30. J. Robertson and M. Grieve, *Forensic Examination of Fibres*, ed. K. Kirkbride and M. Tungol, CRC Press, New York, 1999.
31. P. Torok, P. D. Higdon and T. Wilson, *Polarized Light Microscopy*, 1998.
32. L. Finzi and D. D. Dunlap, *Polarized Light Microscopy*, eLS, John Wiley & Sons, Ltd, 2001.
33. R. Palmer and V. Chinherende, *J. Forensic Sci.*, 1996, **41**(5), 802.
34. J. Pawley, *Handbook of Biological Confocal Microscopy*, Springer, US, 2010.
35. V. Prasad, D. Semwogerere and E. R. Weeks, *J. Phys.: Condens. Matter*, 2007, **19**(11), 7.
36. K. P. Kirkbride and S. R. Tridico, *Forensic Sci. Int.*, 2010, **195**(1–3), 28.
37. L. Lepot, K. De Wael, F. Gason and B. Gilbert, *Sci. Justice*, 2008, **48**(3), 109.
38. Y. Song, M. Srinivasarao, A. Tonelli, C. M. Balik and R. McGregor, *Macromolecules*, 2000, **33**(12), 4478.
39. A. Méndez Vilas, *Current Microscopy Contributions to Advances in Science and Technology*, ed. R. Chadha, P. Arora, S. Bhandari and M. Bala, Formatex Research Center, 2012.
40. I. M. Vitez, A. W. Newman, M. Davidovich and C. Kiesnowski, *Thermochim. Acta*, 1998, **324**(1–2), 187.
41. R. E. Bisbing, *The Forensic Laboratory Hand Book: Procedures and Practice*, ed. A. Mozayani and C. Noziglia, 2007.
42. J. Robertson and M. Grieve, *Forensic Examination of Fibres*, ed. C. Roux, CRC Press, New York, 2nd edn, 1999.
43. D. Robson, *J. Forensic Sci. Soc.*, 1994, **34**(3), 187.
44. M. C. Grieve and L. R. Cabiness, *Forensic Sci. Int.*, 1985, **29**(1–2), 129.
45. M. C. Grieve, *Sci. Justice*, 1995, **35**(3), 179.
46. M. W. Tungol, E. G. Bartick and A. Montaser, *Forensic Examination of Synthetic Textile Fibers*, ed. H. Humecki and M. Decker, New York, 1995.
47. J. Robertson and M. Grieve, *Forensic Examination of Fibres*, ed. J. M. Challinor, CRC Press, New York, 1999.
48. R. Macrae, R. J. Dudley and K. W. Smalldon, *J. Forensic Sci.*, 1979, **24**(1), 117.
49. K. Appalaneni, E. C. Heider, A. F. T. Moore and A. D. Campiglia, *Anal. Chem.*, 2014, **86**(14), 6774.
50. A. M. de la Pena, N. Mujumdar, E. C. Heider, H. C. Goicoechea, D. M. de la Pena and A. D. Campiglia, *Anal. Chem.*, 2016, **88**(5), 2967.

51. J. W. Weaver, *Analytical Methods for a Textile Laboratory*, ed. C. M. Player and J. A. Dunn, American Association of Textile Chemists and Colorists, Research Triangle Park, N.C., 1984.
52. M. C. Grieve, J. Dunlop and P. Haddock, *J. Forensic Sci.*, 1990, **35**(2), 301.
53. K. G. Wiggins, R. Cook and Y. J. Turner, *J. Forensic Sci.*, 1988, **33**(4), 998–1007.
54. K. G. Wiggins, S. R. Crabtree and B. M. March, *J. Forensic Sci.*, 1996, **41**(6), 1042.
55. M. M. Houck, *Identification of Textile Fibres*, ed. S. W. Lewis, Woodhead Publishing Ltd, Cambridge, UK, 1st edn, 2009.
56. I. B. Beattie, H. L. Roberts and R. J. Dudley, *Forensic Sci. Int.*, 1981, **17**(1), 57.
57. G. M. Golding and S. Kokot, *J. Forensic Sci.*, 1989, **34**(5), 1156.
58. D. F. Rendle, S. R. Crabtree, K. G. Wiggins and M. T. Salter, *J. Soc. Dyers Colour.*, 1994, **110**(11), 338.
59. J. M. Home and R. J. Dudley, *Forensic Sci. Int.*, 1981, **17**(1), 71.
60. A. W. Hartshorne and D. K. Laing, *Forensic Sci. Int.*, 1984, **25**(2), 133.
61. R. Macrae and K. W. Smalldon, *J. Forensic Sci.*, 1979, **24**(1), 109.
62. I. B. Beattie, R. J. Dudley and K. W. Smalldon, *J. Soc. Dyers Colour.*, 1979, **95**(8), 295.
63. J. Cheng, S. O. Wanogho, N. D. Watson and B. Caddy, *J. Forensic Sci. Soc.*, 1991, **31**(1), 31.
64. D. K. Laing, R. J. Dudley, A. W. Hartshorne, J. M. Home, R. A. Rickard and D. C. Bennett, *Forensic Sci. Int.*, 1991, **50**(1), 23.
65. R. Resua, *J. Forensic Sci.*, 1980, **25**(1), 168.
66. R. Resua, P. R. Deforest and H. Harris, *J. Forensic Sci.*, 1981, **26**(3), 515.
67. K. G. Wiggins, J. A. Holness and B. M. March, *J. Forensic Sci.*, 2005, **50**(2), 364.
68. J. Robertson and M. Grieve, *Forensic Examination of Fibres*, ed. R. S. Griffin, CRC Press, Boca Raton, FL, 2nd edn, 1999.
69. J. Robertson, *Forensic Examination of Fibres*, ed. P. C. White and I. R. Tebbett, Taylor & Francis Ltd., West Sussex, England, 1st edn, 1992.
70. L. M. Petrick, T. A. Wilson and W. R. Fawcett, *J. Forensic Sci.*, 2006, **51**(4), 771.
71. A. Carey, N. Rodewijk, X. M. Xu and J. van der Weerd, *Anal. Chem.*, 2013, **85**(23), 11335.
72. K. Kretschmer and W. Helbig, *J. Forensic Sci.*, 1992, **37**(3), 727.
73. M. R. van Bommel, I. Vanden Berghe, A. M. Wallert, R. Boitelle and J. Wouters, *J. Chromatogr. A.*, 2007, **1157**(1–2), 260.
74. http://www.waters.com/webassets/cms/library/docs/720001136en.pdf, 2004.
75. S. J. Hoy, *Development and Figures of Merit of Microextraction and Ultra-performance Liquid Chromatography for Forensic Characterization of Dye Profiles on Trace Acrylic, Nylon, Polyester, and Cotton Textile Fibers*, University of South Carolina, 2013.
76. J. Robertson and M. Grieve, *Forensic Examination of Fibres*, ed. J. Robertson, CRC Press, New York, 2nd edn, 1999.
77. S. N. Croft and D. Hinks, *J. Soc. Dyers Colour.*, 1992, **108**(12), 546.
78. H. Siren and R. Sulkava, *J. Chromatogr. A.*, 1995, **717**(1–2), 149.
79. X. Xu, H. Leijenhorst, P. Van den Hoven, J. A. De Koeijer and H. Logtenberg, *Sci. Justice*, 2001, **41**(2), 93.
80. S. Terabe, K. Otsuka, K. Ichikawa, A. Tsuchiya and T. Ando, *Anal. Chem.*, 1984, **56**(1), 111.
81. S. F. Li, *Capillary Electrophoresis - Principles, Practice and Applications*, Elseiver, Amsterdam, 1994.

82. M. Trojanowicz, L. Wojcik and K. Urbaniak-Walczak, *Chem. Anal. (Warsaw, Pol.)*, 2003, **48**(3), 607.
83. C. R. Dockery, A. R. Stefan, A. A. Nieuwland, S. N. Roberson, B. M. Baguley and J. E. Hendrix, *Anal. Bioanal. Chem.*, 2009, **394**(8), 2095.
84. A. R. Stefan, C. R. Dockery, B. M. Baguley, B. C. Vann, A. A. Nieuwland and J. E. Hendrix, *Anal. Bioanal. Chem.*, 2009, **394**(8), 2087.
85. D. A. Oxspring, W. F. Smyth and R. Marchant, *Analyst*, 1995, **120**(7), 1995.
86. www.agilent.com/cs/library/primers/Public/5990-6969EN%20GPC%20SEC%20Chrom%20Guide.pdf.
87. W. W. Mcgee, K. Coraine and J. Strimaitis, *J. Liq. Chromatogr.*, 1979, **2**(2), 287.
88. K. Weisskopf, *J. Polym. Sci., Part A: Polym. Chem.*, 1988, **26**(7), 1919.
89. G. Montaudo, M. S. Montaudo, C. Puglisi and F. Samperi, *Anal. Chem.*, 1994, **66**(23), 4366.
90. E. E. Drott and R. A. Mendelso, *Sep. Sci.*, 1971, **6**(1), 137.
91. S. Farah, T. Tsach, A. Bentolila and A. J. Domb, *Talanta*, 2014, **123**, 54.
92. J. Robertson and M. Grieve, *Forensic Examination of Fibres*, ed. F.-P. Adolf and J. Dunlop, Taylor & Francis, Ltd., London, UK, 1999.
93. M. B. Eyring, *Visile Microscopical Spectrophotometry in the Forensic Sciences*, ed. R. Saferstein, Regents/Prentice Hall, Upper Saddle River, NJ, 2nd edn, 1993.
94. T. W. Biermann, *Sci. Justice*, 2007, **47**(2), 68–87.
95. M. C. Grieve, T. Biermann and M. Davignon, *Sci. Justice*, 2003, **43**(1), 5.
96. M. C. Grieve, T. Biermann and M. Davignon, *Sci. Justice*, 2001, **41**(4), 245.
97. M. C. Grieve, T. W. Biermann and K. Schaub, *Sci. Justice*, 2006, **46**(1), 15.
98. R. Palmer, W. Hutchinson and V. Fryer, *Sci. Justice*, 2009, **49**(1), 12.
99. S. Suzuki, Y. Suzuki, H. Ohta, R. Sugita and Y. Marumo, *Sci. Justice*, 2001, **41**(2), 107.
100. K. Wiggins and A. J. Holness, *Sci. Justice*, 2005, **45**(2), 93.
101. M. C. Grieve, T. W. Biermann and K. Schaub, *Sci. Justice*, 2005, **45**(1), 13.
102. K. Wiggins and P. Drummond, *Sci. Justice*, 2005, **45**(3), 157.
103. J. Almer, E. McAnsh and B. Doupe, *Can. Soc. Forensic Sci. J.*, 2010, **43**(1), 16.
104. S. L. Morgan, A. A. Nieuwland, C. R. Mubarek, J. E. Hendrix, E. M. Enlow and B. J. Vasser, *Forensic Discrimination of Dyed Textile Fibers Using UV-VIS and Fluorescence Microspectrophotometry, Proceedings of the European Fibres Group Annual Meeting*, Prague, Czechoslovakia, 2004.
105. J. Was-Gubala and R. Starczak, *Spectrochim. Acta, Part A*, 2015, **142**, 118.
106. M. W. Tungol, E. G. Bartick and A. Montaser, *J. Forensic Sci.*, 1991, **36**(4), 1027.
107. M. W. Tungol, E. G. Bartick and A. Montaser, *Spectrochim. Acta, Part B*, 1991, **46**(11), E1535.
108. M. W. Tungol, E. G. Bartick and A. Montaser, *Appl. Spectrosc.*, 1993, **47**(10), 1655.
109. J. V. Miller and E. G. Bartick, *Appl. Spectrosc.*, 2001, **55**(12), 1729.
110. E. Suzuki, *Forensic Applications of Infrared Spectroscopy*, ed. R. Saferstein, Regents/Prentice Hall, Upper Saddle River, NJ, 1993.
111. C. Gilbert and S. Kokot, *Vib. Spectrosc.*, 1995, **9**(2), 161.
112. S. Kokot, K. Crawford, L. Rintoul and U. Meyer, *Vib. Spectrosc.*, 1997, **15**(1), 103.
113. V. Causin, C. Marega, G. Guzzini and A. Marigo, *Appl. Spectrosc.*, 2004, **58**(11), 1272.
114. V. Causin, C. Marega, S. Schiavone and A. Marigo, *Forensic Sci. Int.*, 2005, **151** (2–3), 125.
115. G. Jochem and R. J. Lehnert, *Sci. Justice*, 2002, **42**(4), 215.

116. M. Osterberg, U. Schmidt and A. S. Jaaskelainen, *Colloids Surf., A*, 2006, **291**(1–3), 197.
117. P. Buzzini and G. Massonnet, *J. Forensic Sci.*, 2015, **60**(3), 712.
118. G. Massonnet, P. Buzzini, G. Jochem, M. Stauber, T. Coyle, C. Roux, J. Thomas, H. Leijenhorst, Z. Van Zanten, K. Wiggins, C. Russell, S. Chabli and A. Rosengarten, *J. Forensic Sci.*, 2005, **50**(5), 1028.
119. H. G. M. Edwards, D. W. Farwell and D. Webster, *Spectrochim. Acta, Part A*, 1997, **53**(13), 2383.
120. K. De Wael, C. Baes, L. Lepot and F. Gason, *Sci. Justice*, 2011, **51**(4), 154.
121. K. G. Wiggins, *Anal. Bioanal. Chem.*, 2003, **376**(8), 1172.
122. R. D. Koons, *J. Forensic Sci.*, 1996, **41**(2), 199.
123. A. Prange, U. Reus, H. Boddeker, R. Fischer and F. P. Adolf, *Anal. Sci.*, 1995, **11**(3), 483.
124. T. Sano and S. Suzuki, *Forensic Sci. Int.*, 2009, **192**(1–3), E27.
125. M. S. Shackley, An Introduction to X-Ray Fluorescence (XRF) Analysis in Archaeology, in *X-Ray Fluorescence Spectrometry (XRF) in Geoarchaeology*, ed. S. M. Shackley, Springer, New York, 2011, p. 7.
126. J. W. Weaver, *Analytical Methods for a Textile Laboratory*, ed. R. J. Berni, V. W. Tripp and J. J. Herbert, American Association of Textile Chemists and Colorists, Research Triangle Park, N. C., 3rd edn, 1984.
127. B. Piccolo, D. Mitcham, V. W. Tripp and R. T. Oconnor, *Appl. Spectrosc.*, 1966, **20**(5), 326.
128. K. Flynn, R. O'Leary, C. Roux and B. J. Reedy, *J. Forensic Sci.*, 2006, **51**(3), 586.
129. I. P. Keen, G. W. White and P. M. Fredericks, *J. Forensic Sci.*, 1998, **43**(1), 82.
130. S. Kokot, N. A. Tuan and L. Rintoul, *Appl. Spectrosc.*, 1997, **51**(3), 387.
131. L. E. Jurdana, K. P. Ghiggino, K. W. Nugent and I. H. Leaver, *Text. Res. J.*, 1995, **65**(10), 593.
132. P. C. White, C. H. Munro and W. E. Smith, *Analyst*, 1996, **121**(6), 835.
133. J. R. Lakowicz, *Principles of Fluorescence Spectroscopy*, Kluwer Academic/Plenum, New York, 2nd edn, 1999.
134. N. Mujumdar, E. C. Heider and A. D. Campiglia, *Appl. Spectrosc.*, 2015, **69**(12), 1390.
135. F. M. Gray, M. J. Smith and M. B. Silva, *J. Chem. Educ.*, 2011, **88**(4), 476.

Chapter 8

Novel Psychoactive Substances (NPS) and Recent Scenarios: Epidemiological, Anthropological and Clinical Pharmacological Issues

J. M. Corkery*, L. Orsolini, D. Papanti and F. Schifano

Psychopharmacology, Drug Misuse & Novel Psychoactive Substances Research Unit, University of Hertfordshire, Department of Pharmacy, Pharmacology & Post-Graduate Medicine, College Lane, Hatfield, AL10 9AB, United Kingdom

*E-mail: j.corkery@herts.ac.uk

Table of Contents

Light in Forensic Science: Issues and Applications
Edited by Giorgia Miolo, Jacqueline L. Stair and Mire Zloh

Published by the Royal Society of Chemistry, www.rsc.org

8.1. Introduction

Over the last decade, the 'traditional' drug scene has been supplemented – but not replaced – by the emergence of a series of psychoactive substances which are either newly created or existing drugs (including medications) which are now being used in novel ways. These are referred to as 'new/novel psychoactive substances' (NPS). Commonly, the term 'legal high' has been used to describe such substances. However, this is misleading since often these molecules are already subject to regulation and, moreover, because they are 'legal' they are incorrectly assumed to be safe.

This chapter looks at the current and emerging issues and challenges faced by those dealing with the issue of NPS. A whole range of topics could be considered important aspects for investigation, *e.g.* clinical, pharmacological, analytical chemistry, toxicological, legislative, and criminological issues. All of these are interlinked because of the key issue of a continual need for information in a timely manner so as to understand the ever-changing phenomenon of NPS. However, here we are going to focus on epidemiological, anthropological and clinical pharmacological issues. Following chapters look at the way in which light can be used to assist in screening for different molecules, distinguishing between them, and identifying both known and unknown NPS.

Whilst the issues discussed here are more relevant to understanding the NPS phenomenon, the practical challenges associated with trying to achieve this also need to be appreciated. It is necessary for all those generating and/or using information on NPS to comprehend both what data are available and what their limitations are, so as to draw valid conclusions that can feed into decision making in relevant fields.

8.1.1. Definitions of NPS

The key starting point is to define what we mean by NPS. There are a number of definitions currently in use both at international, national and sub-national levels. Whilst there are some commonalities across them, there are also differences. These differences reflect the varying needs and agendas of the organisations and researchers employing them, and these can, in turn, be influenced by the nature of these different actors. Thus, the needs of law enforcement agencies are likely to be different to clinicians, who in turn would be interested in aspects different to those of analytical chemists or sociologists. Alternative terms can also be used instead of NPS, *e.g.* 'legal highs' or 'designer drugs'.

At the international level the focus is on new psychoactive substances which may pose a threat to public health. The United Nations (UN) uses the following approach: "New psychoactive substances are substances of abuse, either in a pure form or a preparation, that are not controlled by the 1961 Single Convention on Narcotic Drugs or the 1971 Convention on Psychotropic

Substances, but which may pose a public health threat".[1] The European Council uses a similar definition but extends the principle on the basis of comparable threats: a new psychoactive substance is defined as "a new narcotic or psychotropic drug, in pure form or in preparation, that is not controlled by the United Nations drug conventions, but which may pose a public health threat comparable to that posed by substances listed in these conventions".[2] In the United Kingdom (UK), the Advisory Council on the Misuse of Drugs (ACMD) proposed a more pragmatic approach, referring both to the international conventions and to the main UK drugs legislation embodying them as well as slightly different purpose for consumption: "Psychoactive drugs which are not prohibited by the United Nations Single Convention on Narcotic Drugs or by the Misuse of Drugs Act 1971, and which people in the UK are seeking for intoxicant use".[3]

Some also make distinctions between 'new' and 'novel' psychoactive substances. For example, the UN notes in its definition of an NPS: "In this context, the term 'new' does not necessarily refer to new inventions but to substances that have been recently become available".[1] Here one can see an implied differentiation between 'new' and 'novel', in the sense that 'new' can mean both something newly invented and something that is new to the market, such as a failed pharmaceutical or an old patent which has been 'rediscovered' and marketed for its potential use as a 'recreational' substance. The United Nations Office on Drugs and Crime (UNODC) even regards khat as "a particular NPS [New Psychoactive Substance] of concern".[4] 'Novel' also expresses a plurality of meanings: something newly created, or something that has come back into fashion after a period of absence from the recreational drug scene, or indeed a known NPS molecule being used in an innovative or unusual way, *i.e.* has a 'novelty' appeal.

The key point here is to recognise that the research community and others that write about NPS should be as clear as possible about what they are referring when using terms such as NPS, 'legal highs', 'designer drugs', *etc.* In that way, there should be no ambiguity over what is being described or investigated.

8.1.2. Sources, Numbers and Categories of NPS

The ideas for sources of NPS are many and varied. The starting point might be a natural product (typically plant or fungus), a 'rediscovered' discontinued pharmaceutical or patent, alterations to an existing molecular structure to avoid regulatory or other controls, abuse of medicines, experimenting with new ways of using existing drugs, *etc.*

Whatever the original inspiration for NPS, the speed with which they emerge onto drug market(s) has been accelerating over the past decade or so. The UNODC suggests that between 2009 and 2014 the number of NPS reported increased from 126 to 450. By the end of 2016, 739 substances had been reported by 106 countries through the global Synthetics

Monitoring: Analysis, Reporting and Trends (SMART) programme.[5] The EMCDDA's latest report shows that of the 98 NPS notified during 2015, 26 were cathinones and 24 synthetic cannabinoids; these together with phenethylamines account for the largest number of NPS.[6] By the end of 2016 the number of NPS notified to the agency since 2005 was 629. Of these, 171 were synthetic cannabinoids, 118 were synthetic cathinones; 25 were synthetic opioids and 20 were benzodiazepines. In 2016, 64 new benzodiazepines were reported, together with 9 opioids, 10 cannabinoids and 14 cathinones.

NPS can be categorised or classified based on what particular aspect or characteristic is of concern to legislators, forensic analysts, health professionals, (potential) users, and other stake-holders. Thus, a number of domains (and combinations thereof) could be utilised to investigate them, including:

- Legal status (controlled internationally/nationally), un-regulated, licensed medication, 'over the counter' medications, natural products (from plants, fungi or animals);
- Physiological, psychological or psychiatric effects (desired and/or adverse);
- Natural or synthetic;
- Chemical structure and/or characteristics;
- Source of supply or availability;
- User characteristics;
- Addictive potential; *etc.*

Both the UNODC and European Centre for Drugs and Drug Addiction (EMCDDA) use categories primarily based on chemical properties (Table 8.1). The UNODC also looks at their effects, but these overlap with their chemical characteristics: classic hallucinogens; dissociatives; opioids; sedatives/hypnotics; stimulants; synthetic cannabinoid receptor agonists; and those not yet assigned.

8.2. Epidemiology of NPS

As already indicated, the main groups in terms of numbers are synthetic cannabinoids, synthetic cathinones and phenethylamines. These are primarily stimulants, as are a number of molecules in other groups. They have accounted for most NPS over the past decade.

However, recent years have seen the appearance of medications withdrawn from the market because of adverse effects or lack of therapeutic benefit being 'rediscovered' for recreational use. Unlicensed medications from other countries are being made available in Europe. Alongside these, analogues of existing medications are being developed to avoid drug control or regulation. These are primarily benzodiazepines and opioids, as well

Table 8.1. Main NPS categories used by the UNODC and EMCDDA.

UNODC	EMCDDA	Examples
Aminoindanes	Aminoindanes	5,6-Methylenedioxy-2-aminoindane (MDAI)
	Arylalkylamines	5-APB (5-(2-aminopropyl)benzofuran); MPA (methylthienylpropamine)
Ketamine and phencyclidine-type substances	Arylcyclohexylamines	Methoxetamine (MXE)
Synthetic cannabinoids	Cannabinoids	APINACA, JWH-018
Synthetic cathinones	Cathinones	4-Methylethcathinone (4-MEC); α-pyrrolidinopentiophenone (α –PVP)
Tryptamines	Indolalkylamines (inc. Tryptamines)	α-methyltryptamine (AMT)
Phenethylamines	Phenethylamines	2C-E; 25H-NBOMe
Piperazines	Piperazine derivates	Benzylpiperazine (BZP); 1-(3-chlorophenyl) piperazine (mCPP)
	Piperidines & pyrrolidines	2-DPMP (Desoxypipradrol); Ethylphenidate
Plant-based substances	Plants & extracts	Kratom (mitragyna speciosa Korth); salvia divinorum; khat (Catha edulis);
Others: hallucinogens/psychedelics; opioids; sedatives/hypnotics; stimulants	Others	1,3-Dimethylamylamine (DMAA); 4,4′-DMAR (para-Methyl-4-methylaminorex); Diphenidine; Gabapentin; GHB; Modafinil; 'poppers'; Quetiapine; Zopiclone
	Benzodiazepines	Diclazepam, Etizolam; Phenazepam
	Opioids	Carfentanyl; U47,700; MT-45; AH-7921

as ADHD therapy analogues. Examples include: benzodiazepines - Phenazepam, Etizolam, Diclazepam; opioids - AH-7921, MT-45; U47,700; fentanyl analogues - carfentanyl; ADHD drugs analogues - Ethylphenidate. It has also become apparent in recent years that anti-depressants, anti-epileptics, and anti-psychotics are being used recreationally.[7] Examples include: anti-depressants – Venlafaxine; anti-epileptics – gabapentinoids; anti-psychotics: Olanzapine, Quetiapine.

Over the last decade there has been a growing trend in the use of poly-substance consumption more generally, and this is now also evident in terms of NPS use, as evidenced by deaths reported to the EU-MADNESS research project (see below), where these are being used in conjunction with 'traditional' recreational drugs. There are other emerging concerns in respect of poly-substance use related to the unwitting/unintended simultaneous consumption of several molecules. This arises in a number of ways, including the emergence of products containing several molecules of the same chemical class. More worryingly, some products contain molecules from the same category which affect different receptors in the brain.

Perhaps, of even more concern, are products containing molecules from different categories with unknown interactions both with each other, and other drugs that the user may be consuming. The use of adulterants and 'cutting' agents can also present unknown risks,[8] especially if users are taking prescribed or 'over the counter' medications. Furthermore, some NPS have been used as adulterants in controlled drugs.[9,10]

8.2.1. *Prevalence of NPS*

At the international level there is very limited information on the prevalence of NPS, especially over time, and the variety of definitions and ways of measuring use means that it is difficult to make comparisons over time and between countries.[11] Some key issues in this respect are:

- The length of time it takes to get a new question added to international and national survey instruments, means that if you ask about a specific substance it may well have disappeared from use before the survey question is in the field. One exception is mephedrone which often appears as a substance consistently asked about.
- The definitions of NPS may well vary and include drugs which some researchers would count as 'traditional' drugs of misuse (*e.g.* GHB/GBL, ketamine), whilst others may ask about 'legal highs', which some would regard as ceasing to be 'legal' once they have become controlled and therefore no longer included in the definition of NPS, *etc.*
- Use over different time periods may be measured in different ways: lifetime, last year/last 12 months, last month, last week, *etc.* and across varying age-groups. In particular, the age-ranges of 'young people' may vary.
- The populations surveyed may be different and not comparable, as usage rates may be higher in those outside of household units as they are away from the scrutiny of their parents or carers, *e.g.* those at university, living alone, on the streets, in prison, holidaying in Ibiza or Ayia Naipa, *etc.*
- Even global surveys such as the Global Drugs Survey[12] have methodological issues, notably the self-selected nature of respondents.

However, some country level data are available. At the European level these indicate that lifetime use of synthetic cannabinoids ranges from 0.5% of 15–64 year olds in Spain to 13.2% of 15–16 year-olds in Latvia.[11] The most recent Eurobarometer repor[13] looking at the use of 'new substances' in Europe reported that on average 3% (range 0–5%) of 15–24 year-olds had used such substances in the previous 12 months. Most got them from a friend (68%), drug dealer (27%), 'head'/'smart' shop (10%) and only 3% from the Internet. The majority of these 'legal highs' were taken at a party or event (65%), with friends (60%), whilst 15% used them when they were alone. The ESPAD survey for 2015,[14] covering 35 European countries, found that average lifetime experience with NPS by 15–16 year-olds was 4%, with the highest rates in Estonia and Poland (10% each), and the lowest in Belgium (Flanders), Denmark, Finland, Norway and Portugal, with rates of 1%. The average prevalence of lifetime use was 5% among boys and 4% among girls.

In the UK context, the Crime Survey for England and Wales 2016/7[15] indicates that 147 000 adults (0.4% of the population aged 16–59) reported the use of NPS in the previous year. Among young adults (ages 16–24), the prevalence of NPS use was much higher (1.2%), the majority of the users being young men.

Herbal smoking mixtures were the most commonly used form of NPS, with 40% of those aged 16–59 reporting their use the last time they had taken an NPS. NPS (and nitrous oxide) were most commonly obtained from a friend, neighbour or colleague (39%). Other common sources were shops (13%), and a stranger, acquaintance, family member or a dealer not known personally (18%). The Drug Prevalence Survey for Northern Ireland[16] found that the level of last month NPS use remained level at 0.2% of the general population aged 15–64 between 2010/11 and 2014/5, whereas the level for mephedrone rose from 0.1% to 0.3%. However, last year levels for NPS fell from 1.0% to 0.3%, and that for mephedrone fell from 1.1% to 0.5%. By contrast last year NPS use in the Republic of Ireland was 0.7%. In Scotland lifetime use of NPS in 2015 by school students aged 15 was 5%, last month use being 2%. 13% had been offered NPS. 27% of those who had used drugs in the last month thought it was alright to take legal highs, compared to 10% who hadn't used drugs.[17] In Wales, 1% of 11–16 year-olds admitted to taking mephedrone at least once in 2013/4.[18] Lifetime use of this drug amongst the same age-group in late 2015 was 1.3%, but was double this rate (2.6%) for those aged 16.[19]

8.2.2. *Reasons for Using NPS*

The reasons for using NPS and in what contexts are many and varied, and are likely to reflect those for illegal drugs: recreational use; self-medication; and experimentation (*e.g.* psychonauts). However, additional reasons for using NPS, especially 'legal highs', can include: no stigma attached to their use; their easy availability; can be purchased anonymously online; and there is no need to have a prescription for medications.

Some contexts of use are not going to be revealed by household surveys (as outlined in the previous section). So, for some at-risk groups, other types of survey have to be used *e.g.* online surveys advertised on user fora. The latter can also be used to find out about what people are using, how and when, why and where, effects (desired and unwanted) using a netnographic approach (see below).[20] User fora on the open Web and the Deep Web can also be used to find out about prices, purity, availability, suppliers, *etc.*[21]

Other methods of finding out about drug use and related issues in other settings include: work-place drug testing; drug testing in penal establishments, police stations and 'traffic stops', in-patient psychiatric units; testing confiscations and contents of 'amnesty bins' at clubs, raves, festivals, *etc.*; as well as confiscations at ports and mail/parcel sorting centres.

8.2.3. *What NPS are Being Seized?*

Identifying what NPS are being seized provides another means of knowing what is available to dealers and (potential) users. Theoretically, it can also assist in quantifying the 'market' for NPS. However, there are issues when it comes to knowing how many and what amounts of NPS have been confiscated.

Investigators may be initially faced with white or brown 'powders'. White powders could be anything from heroin, cocaine, synthetic cathinones to crushed paracetamol tablets; whilst brown powders could be amphetamines, heroin, kratom, *etc*. Whilst field-testing kits are available for the 'traditional' drugs, there are only a few developed so far for identifying NPS. Some field tests may only be able to detect classes of NPS rather than specific molecules and seized materials will thus need further investigation.

Forensic providers/services need access to NPS reference samples against which they can compare the 'seizures'; such samples can take a long time to get hold of and are initially very expensive. There are difficulties in collating and sharing information on NPS between investigating agencies. Data are held in various formats across different databases as are spectral data. The existence of several protocols to apply systematic chemical names to compounds can lead to the same molecule having multiple names, rather than a unique nomenclature. There is even less information on translating NPS quantities into doses, frequency of dosing and thus what is 'personal supply' – a critical issue in terms of law enforcement.

There is limited up to date information in the public domain on NPS confiscations. In part, this is due to the length of time it takes to collate and publish what information is available – as with 'traditional' drugs. On top of this, some may argue that it is not a good idea to put out information which is too recent as it may act as a price guide for would-be vendors and buyers. This argument certainly prevailed for 'traditional drugs' in the 1990s, when the Home Office published price data which lagged a year behind the seizure and purity data.[22] Perhaps there was some justification for this approach, but DrugScope, IDMS, Mixmag magazine and the Global Drug Survey, and more recently DrugWise, have released such information.[12,23–26] Now, in the late 2010s, information can be easily found on the Surface Net for NPS and for NPS and 'traditional' drugs on the Deep Web.

NPS confiscations in Europe continue to rise to record levels. The most recent information published by the EMCDDA on NPS confiscations are for 2015.[6] This shows there were nearly 80 000 seizures reported: about; some 25 000 (33%) were of synthetic cathinones, and 22 000 (29%) were of synthetic cannabinoids. Non-controlled benzodiazepines accounted for 11% and phenethylamines and piperazines accounted for 6% each. Other notable groups represented were: arylalkylamines (4%), arylcyclohexylamines (2%), tryptamines (1%), piperidines and pyrrolidines (1%) and potent narcotic (opioid) analgesics (1%), especially fentanyls. In terms of weight, synthetic cannabinoids were the largest NPS group represented, accounting for 2.5 tonnes. Most seizures (64%) in this group were herbal mixtures (plant material with the psychoactive substance sprayed on it), powders accounting for 13%. The top five cannabinoids seized in powder form were: 5F-AMB (61 kg), 5F-AKB48 (61 kg) and ADB- FUBINACA (57 kg). Synthetic cathinones weighing 1.8 tonnes, typically as powders, were confiscated. The top five cathinones seized included: alpha-PVP, 3-MMC, ethylone, 4-CMC and pentedrone. Two-fifths (42%) of the synthetic cathinones seized were 2-MMC (156 kg) and 3-MMC (616 kg), which are chemically related to mephedrone (4-MMC). Figure 8.1 shows

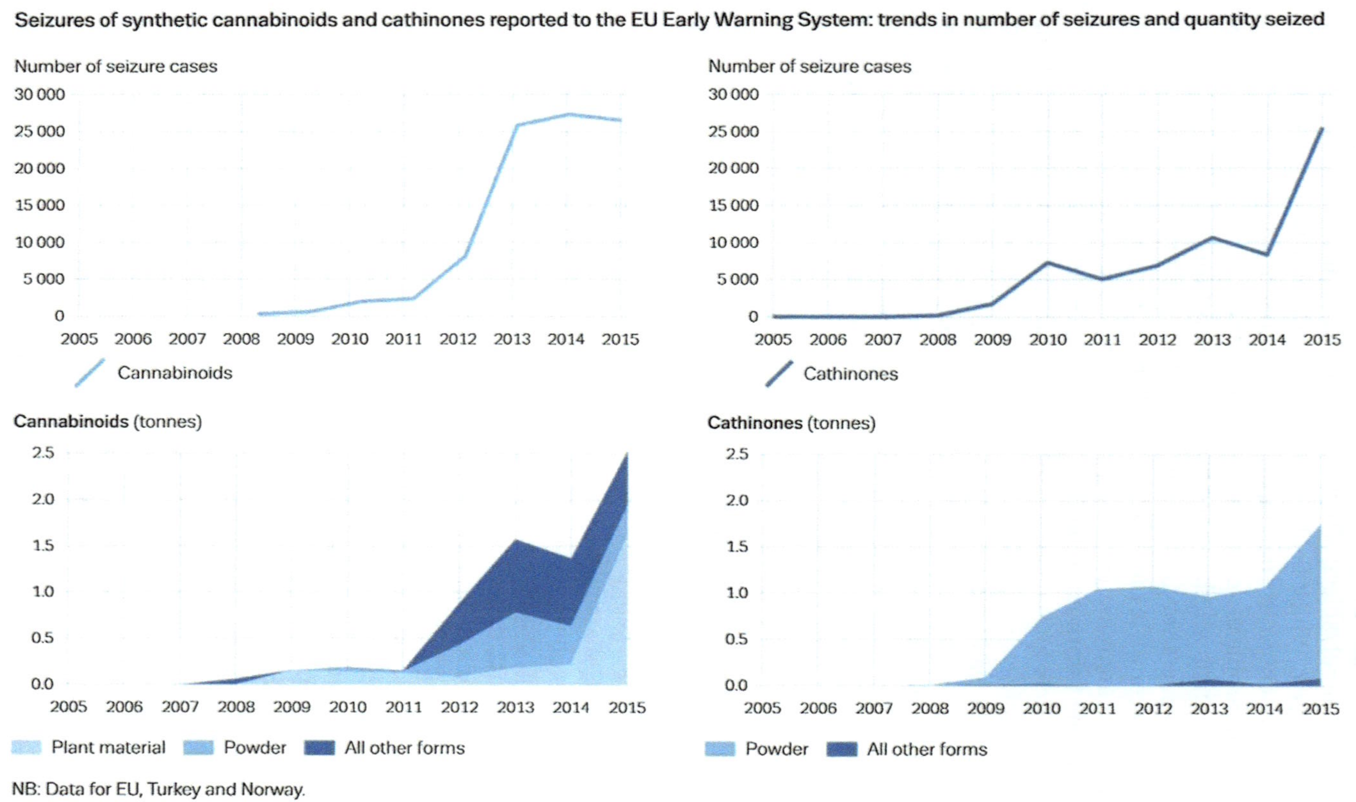

Figure 8.1. Seizures of synthetic cannabinoids and cathinones reported to the EU. Data from EMCDDA (2017),[6] reproduced with permission. © European Monitoring Centre for Drugs and Drug Addiction, 2017.

that whilst the numbers of synthetic cannabinoids increased most rapidly between 2009 and 2013 before levelling off, quantities continued to rise. For synthetic cathinones the numbers and quantities rose rapidly over the period 2009–11, before levelling off until 2014, followed by a steep increase in 2015.

The latest drug seizures data for England and Wales released by the Home Office on 3 November 2016 only contains information on mephedrone seizures for the last two years. It is only through making Freedom of Information Act requests that data for earlier years have been made available or for other NPS groups. Data for Northern Ireland have also been published, but again to get some information for specific NPS/classes of NPS requires such approaches. Although some information has been published by the Scottish Government in respect of police drug seizures in 2014/5 and 2015/6, these are presented broken down by estimates of amounts seized associated with possession and supply offences and little detail is given in terms of NPS, apart from GHB, Ketamine and mephedrone; these are only given in terms of supply offences.[27,28] Scottish drug seizure data are not available for the period 2007/8 to 2009/10. A key issue here is that the statistical requirement placed on UK police forces is to identify only specific molecules and it may take time for such processes to be implemented, and it is unlikely that retrospective information will be obtained from forensic providers.

Allowing for these limitations it is possible to present some data on NPS seizures in England and Wales (Figure 8.2). Only drugs controlled under the Misuse of Drugs Act 1971 appear in Home Office seizure statistics.[29] The earliest data available for NPS relate to GHB/GBL and ketamine. GHB/GBL seizures have remained at a low and fairly steady level over the period 2008/09 to 2015/16,

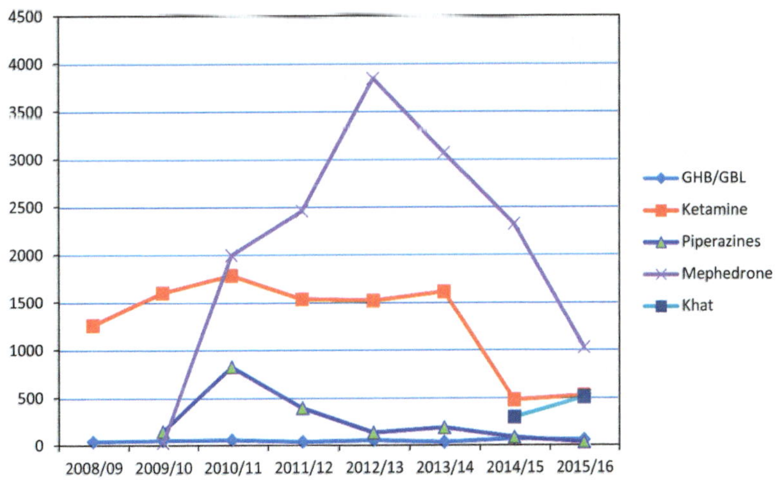

Figure 8.2. Number of NPS seizures, England & Wales, 2008/9 to 2015/6. Data from Hargreaves & Smith (2016)[18] published under the Open Government Licence 3.0, © Crown Copyright, and Freedom of Information requests to Home Office by lead author.

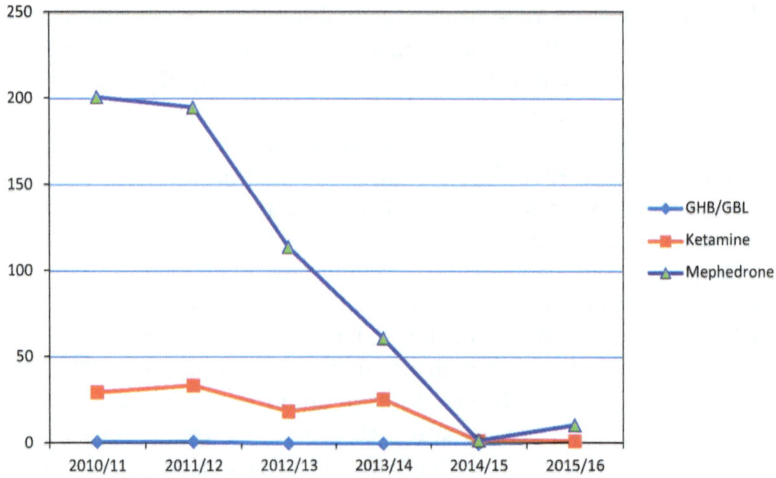

Figure 8.3. Number of NPS seizures, Scotland, 2010/11 to 2015/6. Data from the Scottish Government (2016 and 2017),[27,28] © Crown Copyright.

varying between 41 and 75 per year. By contrast, ketamine seizures peaked in 2010/11 at 1793 and remained stable until 2013/14, after which there was a sharp drop to 485 followed by a levelling off. Although piperazines started to appear in the UK market in 2005 they were not controlled until several years later, following which they fell from 830 in 2010/11 to 42 in 2015/16. Mephedrone was controlled in early 2010, yet the number of seizures did not peak until 2012/3, at 3850, since when they have fallen to 1031 in 2015/16. This decline may be due to the availability of other synthetic cathinones. Khat was only controlled from June 2014, yet the number of seizures in 2015/16 stood at 510.

In respect of Scotland, it is evident that the number of GHB/GBL seizures were at very low levels between 2010/1 and 2015/6. Seizures of ketamine remained stable between 2010/1 and 2013/4 but have then shown a fall. A more rapid decline is evident for mephedrone seizures between 2011/2 and 2014/5, although there was a small rise in 2015/6 (see Figure 8.3).

Some limited information available for Northern Ireland is presented in Figure 8.4. This shows that seizures of mephedrone have declined since they were first controlled, but they are still being made. The peak number of seizures for piperazines were in 2010/11–2011/12, and those for ketamine in 2011/12. Seizures of GHB/GBL are relatively rare. The sole seizure of khat occurred in 2015/16.

8.2.4. *Price, Purity and Potency of NPS*

There are no European Union wide data on these aspects available in the public domain on a regular basis, and such data are not published by the EMCDDA. However, it is safe to assume that data on price, purity and potency

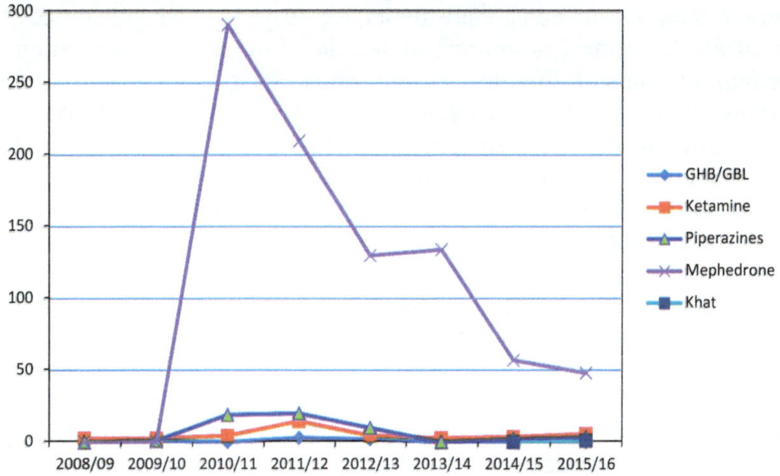

Figure 8.4. Number of NPS seizures, Northern Ireland, 2008/9 to 2015/6. Data from Freedom of Information requests to the Police Service of Northern Ireland by the lead author.

are collected by individual law enforcement agencies, forensic providers, and some academic institutions for intelligence and research purposes. These are only likely to release on an *ad hoc* basis, *e.g.* as academic papers or conference presentations, and typically are based on netnographic studies of user discussion fora, websites offering NPS for sale, by conducting online surveys or visits to 'head' or 'smart' shops. As such, these will mostly be snap-shots in time rather than regular epidemiological data. This is also likely to be the case for formal risk assessments.

The quantity of a pure drug required to produce 10 000 doses has fallen over time as more potent molecules have been produced. For example, in terms of 'traditional' stimulants, 750 g of MDMA, 200 g of cocaine or 100 g of amphetamine are needed to produce that number of doses. However, in terms of NPS the upper limit of 100 g can be exemplified by PB-22 (synthetic cannabinoid), whereas only 5 g of 25I-NBOMe (phenethylamine derivative) will give the required number of doses. At the very extreme of this spectrum is the opioid carfentanyl, which requires only 0.1 g to produce 10 000 doses.[30] These developments have implications in terms of the difficulty of dealers/retailers and (potential) consumers in accurately measuring doses, as well as for law enforcement agencies trying to interdict supplies, and clinicians dealing with emergency hospital admissions.

8.2.5. *Control and Regulation of NPS*

As we have already seen, UN international conventions on drugs provide a key as to whether or not a substance is regarded as an NPS. However, at regional or national levels other legislative systems also affect the defining of substances as NPS or quasi-equivalents, *e.g.* 'psychoactive substances'. The EMCDDA's Scientific Committe[31] undertakes formal Risk Assessments on specific molecules or chemical groups which are giving cause for concern in terms of social and severe health adverse effects, toxicity or even death. These reports are then considered by the European Council (EC) who may then decide to issue a Decision on controlling that molecule or chemical group. It is then up to individual Member States to take action in the ways they deem most appropriate to their particular situation in terms of control measures and criminal penalties.

National governments do not have to wait until a substance is controlled or regulated at the international/EU level before taking action. At a general level, countries have either adapted their primary illegal drug control laws to ban NPS by introducing a degree of flexibility in the way individual substances are listed, or through the introduction of additional regulatory mechanisms that provide quick and flexible responses but typically narrowly focused on the importation or sale of NPS.[32] Different options may be tried at different times by an individual country.

In the UK Statutory Instruments (secondary legislation) are used to add substances to the appropriate schedules of the Misuse of Drugs Act 1971. Rapid and temporary bans can be introduced. The latter typically last 12 months, during which time a thorough risk assessment is undertaken and/or legislation prepared in the eventuality that a decision is made to control the substance. In the UK the ACMD can advise the Government on the need for temporary regulation and undertakes the risk assessment and makes appropriate recommendations. This has led to Temporary Class Drug Orders (TCDOs) being introduced in November 2011 under provisions of the 1971 Act. Measures made under rapid procedures are permanent controls and do not lapse after a specified period; new substances are added to national legislation.

In order to pre-empt circumvention/evasion of controls, national legislation may extend controls to other substances using analogue and/or generic approaches, which are not explicitly mentioned in the legislation but which have similarities with controlled substances in respect of chemical structure and/or effects. In the UK these approaches are used together with listing of individual substances (including isomers, esters, ethers, and salts).

Some countries have adopted an all embracing approach to control NPS by trying to ban 'psychoactive substances'. In Ireland the law was changed to deal with 'head' shops and indirectly importation/exportation *via* online channels. However, the Criminal Justice (Psychoactive Substances) Act 2010 did not fully realise its aims; the number of 'head' shops did fall, but the contents of products offered for sale changed so that products remained legal.[33] The Irish approach formed a major part of the UK's Psychoactive Substances Act which came into force on 26 May 2016.[34] This aims to bring an end to lawful sale of NPS by 'head' shops and UK-based websites.

These differences in approach mean that the legal status of an NPS may vary between countries in Europe, even within EU Member States, resulting in producers, suppliers/dealers and users of NPS responding in a variety of ways. This means that it makes the work of epidemiologists and others monitoring the NPS phenomenon much harder in understanding what is going on and to react in appropriate ways.

8.2.6. Sanctions for NPS

A range of criminal and civil sanctions can be brought to bear on the cultivation, production, manufacture, as well as the advertising, supply and sale of NPS - and in some countries their use. From an epidemiological perspective, cautions/convictions are not a good measure of when the incidents occurred as published statistics relate to when they were finally dealt with by the police and/or courts.

No figures on drug law offences are published by the EMCDDA. This is probably due to the difficulty of collating data when many different definitions are used of such offences and also in the types of sanctions available. Data are however collected and submitted by National Focal Points. Even within the UK it has been difficult to assemble such data for the last decade or more because of the Home Office ceasing to collate drug seizure and offence data at a UK-level.[35] The UK Focal Point's most recent UK figures for the main Controlled Drugs, including mephedrone, are for 2013. Subsequent information is not broken down by drug type.

Such issues are magnified when one attempts to do this for offences/incidents related to NPS. There are differences in the level of detail of data collected, *e.g.* specific molecule or class of NPS – or indeed if such data are collected in the first instance. The requirements for UK agencies on reporting offences related to specific substances are similar to those outlined above for drug seizures. In the UK context, the most recent data on NPS currently available to the lead author are: Northern Ireland - not published; Scotland - 2015/16; and England & Wales - 2012/13. The way in which some of these datasets are compiled changed in 2014, restricting the level of detail available *via* Freedom of Information requests. Therefore, any attempt at analysis of recent data is not possible at a UK level. However, data for Scotland (see Table 8.2) show that there have been few convictions for GHB/GBL and only a couple each year for ketamine. To date, there has only been one conviction for khat. Disposals for piperazines peaked in 2011/2, whereas those for methcathinones have stayed fairly constant since that year. Controls on these two drug classes were introduced in April 2010, but the length of the prosecution process means that cases were probably dealt with in the following financial year.

8.2.7. Knowing the Size of the NPS Market

Estimating the size of the drug market takes into account a range of data sources and denominators, including prevalence, quantities available, seizures and offences. There are several difficulties in knowing the size of the

Table 8.2. Convictions for NPS drug offences, Scotland, 2006/7 – 2015/6. *Source*: Scottish Government Criminal Proceedings Database, Freedom of Information response, 14 February 2017.

Drug	Offence group	2006/7	2007/8	2008/9	2009/10	2010/1	2011/2	2012/3	2013/4	2014/5	2015/6
GHB/GBL	Possession				2		1		2	1	
	Supply, *etc.*							1			
Ketamine	Possession		2	2		4	2	2	6	1	2
	Supply, *etc.*	2	1				1	1	2		
Khat	Possession										1
Methcathinones	Other drug offence							1			
	Illegal importation								2	1	1
	Possession					14	12	13	16	19	23
	Production, *etc.*									1	
	Supply, *etc.*					6	22	22	21	16	25
Piperazines	Possession				1	5	7	4	1	6	2
	Supply, *etc.*					5	13	6	1	2	

market for 'traditional' substances of abuse. Although the amount of cannabis, heroin, cocaine and other natural products cultivated can be reasonably estimated at a general level, the quantities of manufactured products such as amphetamine-type substances, or diverted medications are far more difficult to get a handle on. This is largely due to the clandestine nature of these illicit industries and producers, distributors, *etc*. The number of users and potential users are difficult to both capture and estimate (see above). The amount consumed by individuals varies and frequency of use is largely unknown and thus the amount needed to meet demand cannot be accurately gauged.

Although the amounts of illicit substances interdicted, confiscated, seized by law enforcement and other agencies can be quantified, at best it is only possible to make a crude estimation as to what proportion of the total supply that represents. This approach has an obvious flaw in that seizures reflect the priorities of such agencies, the resources available to them, and their expertise in identifying targets, *etc*. More structured approaches include using this wider range of inputs to model the market.[36,37]

There are further complications when it comes to trying to estimate the size of the market for NPS. Users do not usually know what they have been supplied in terms of actual active ingredients, especially with 'branded products' - although 'research chemicals' are often of a very high purity. As we have seen, there is a lack of information on what people actually use and why, and even less data on doses and frequency of use.

8.2.8. NPS-related Health Issues

There are a number of theoretical sources of information on the health harms related to the use of NPS. However, only a few have so far been explored or exploited in terms of providing evidence on emerging patterns of use and associated potential health harms: calls to ambulances/paramedics/emergency services; hospital admissions - Emergency Rooms/A&E visits, in-patient psychiatric care units, Intensive Care Units, *etc*.; out-patient/community care treatment units; rehabilitation/detoxification units - residential/non-residential; General Practitioner consultations; calls to help-lines/poison centres; and fatalities.

There are a number of common practical issues across these potential sources. The International Classification of Diseases (ICD) codes used for compiling statistics on hospital admissions and discharges along with fatalities do not cater for all 'traditional' substances of misuse or all misused medications, especially in terms of specific ones, let alone NPS. To capture such cases a supplementary coding list or additional information fields have to be introduced. Individual clients/patients do not necessarily know what they have consumed and/or its level of purity/potency. Branded products may well contain other psychoactive and/or toxic ingredients. The main priority of health professionals in emergency situations is to deal with the acute symptoms, and they may not have the time and/or resources to take samples and get them analysed in a timely manner. Even if the name of a particular molecule is known, there may not be any clinical information on it readily available, or indeed such information may not yet exist. Similarly, any other products used may not be known, and potential interactions with NPS consumed may not have been established.

Emergency presentations to 16 sentinel hospitals in 10 European countries provide some insights in the extent which NPS are implicated in such events, and the nature of those presentations. The Euro-DEN Project looked at presentations over two years between October 2013 and September 2015.[38] NPS accounted for 7.0% of drugs involved, increasing from 5.6% in year 1 to 8.5% in year 2. Cathinones, especially mephedrone, were the commonest NPS encountered. However, GHB/GBL were the 4th most often substances seen, compared to mephedrone in 8th position and ketamine in 19th position in the top 20. NPS were involved in 9 of the 49 deaths reported, 4 involved mephedrone.

Recent figures on new presentations by adults to drug treatment centres in England[39] show an increasing trend in the period up to 2015/16 for GHB/GBL and NPS other than mephedrone. Presentations for mephedrone peaked in 2014/15. The majority of presentations are for synthetic cannabinoids, followed by stimulants (see Table 8.3). For young people less than 18 years old presentations for mephedrone peaked in 2012/13 (Table 8.4). Most presentations in this age-group are for stimulants followed by synthetic cannabinoids. The proportion of young people presenting for drug treatment citing club drugs and NPS rose from 7% in 2009/10 to 19% in 2015/16.[40]

Much more limited treatment data are available for Scotland. In 2015/6, the latest year for which data are currently available,[41] 206 individuals were notified to the Scottish Drug Misuse Database as currently using mephedrone, up from 191 in 2014/5. In 2015/6 61 of these individuals reported it as their

Table 8.3. Trends in numbers of new presentations citing club drugs or new psychoactive substances, adults, England, 2010/1–2015/6.[a] *Source:* Derived from PHE, 2016, Table 7.3.1,[39] published under the Open Government Licence v3.0. © Crown Copyright.

NPS mentions	2010/1	2011/2	2012/3	2013/4	2014/5	2015/6
GHB/GBL	145	201	246	255	321	389
Ketamine	927	844	969	1043	426	550
Mephedrone	953	1044	1836	1895	2024	1647
NPS (any)				320	1154	2042
Further breakdown of NPS						
Predominantly stimulant				141	346	414
Predominantly cannabinoid				67	449	1024
Predominantly hallucinogenic				19	56	83
Predominantly sedative/opioid				14	65	94
Predominantly dissociative				12	18	27
Other				77	262	451

[a]Notes: A code for mephedrone was added to the core data set in 2010/1; codes for NPS were added in 2013/4. An individual may report more than one NPS drug and thus the sum of individual drugs should not be added together to derive a total.

Table 8.4. Trends in numbers of new presentations citing club drugs or new psychoactive substances, young people (<18 years), England, 2010/1–2015/6.[a] *Source:* Derived from PHE, 2017, Table 5.3.1,[40] published under the Open Government Licence v3.0. © Crown Copyright.

NPS mentions	2010/1	2011/2	2012/3	2013/4	2014/5	2015/6
Ketamine	405	387	345	419	228	125
Mephedrone	972	1065	1788	1519	1360	823
NPS				324	887	1056
Further breakdown of NPS						
Predominantly stimulant				60	154	121
Predominantly cannabinoid				203	557	695
Predominantly hallucinogenic				29	46	37
Predominantly sedative/opioid				<5	9	13
Predominantly dissociative				<5	5	32
Other				39	139	175

[a]Notes: A code for mephedrone was added to the core data set in 2010/1; codes for NPS were added in 2013/4. An individual may report more than one NPS drug and thus the sum of individual drugs may exceed the total reported for NPS (any).

main drug, compared to 50 in 2014/5. Figures available for previous years in respect of mephedrone as a main drug are: 2006/7–0; 2007/8–0; 2009/10–34; 2010/1–71; and 2011/2–35.[42]

8.2.9. NPS-related Deaths

Arguably, perhaps the ultimate indicator of health harms of psychoactive substances are deaths associated with their consumption. The EMCDDA focuses on deaths directly related to drug use with the following case definition: "a death happening shortly after consumption of one or more illicit psychoactive drugs, and directly related to this consumption". Often these deaths are referred to as 'overdoses', or 'poisonings' or 'drug-induced deaths'. From 2008 they started using the latter term.[43] Here the focus is on cases where "the underlying cause of death was mental and behavioural disorders due to psychoactive substance use (harmful use, dependence, and other mental and behavioural disorders - F-codes) due to a number of drugs of abuse, or the underlying cause of death was poisoning (accidental, intentional or of undetermined intent - X- and Y-codes) due to a number of drugs of abuse".[43] As a consequence, other fatalities associated with drug use are not typically taken into consideration in headline statistics.[44]

In terms of applying this approach to NPS, it means that indirect NPS deaths are going to be missed from the head-line indicators not only in terms of the EMCDDA Drug-Related Deaths (DRD) Standard but also national DRD definitions (drug misuse) as well as its wider 'drug poisoning' one, and thus will not contribute to the total burden of mortality at a population level. Several types of NPS deaths are unlikely to be caught by the 'drug-induced' approach: (a) infections caused by injecting cathinones and other classes of drugs *e.g.* HIV/AIDS, hepatitis, clostridium novyi, anthrax, septicaemia,

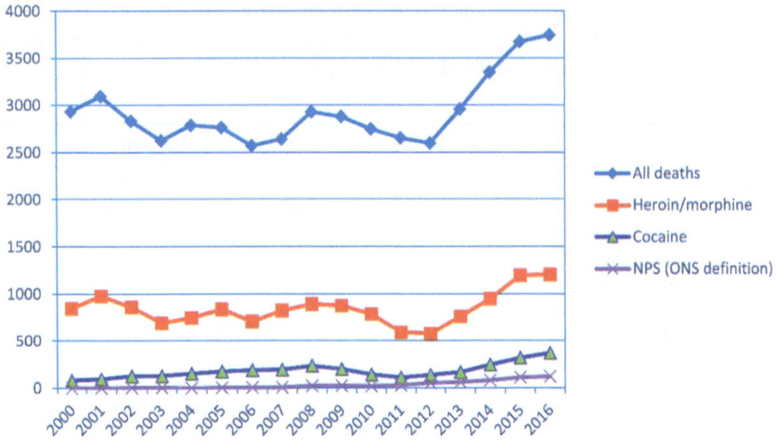

Figure 8.5. NPS death registrations in England and Wales. Data from the Office for National Statistics (ONS, 2017),[45] published under the Open Government Licence v3.0. © Crown Copyright.

necrotising fasciitis, *etc.*; (b) road traffic and other accidents, fall from heights, drownings, hypothermia, *etc.* where the perception or assessment of risk/speed is impaired and/or reaction times slowed; and (c) some deaths indirectly attributed to drugs through violence, *e.g.* acquisitive crime, organised crime, homicide.

Due to the differences in defining NPS, there are no commonly agreed definitions of NPS-related fatalities. This means it is impossible to make internal UK comparisons let alone international ones, unless one has access to the details of each potential NPS case and are able to collate data according to a study protocol. However, it is possible to illustrate the contribution made by NPS to all drug poisoning deaths (*i.e.* where specific substances were mentioned in the medical cause of death) for each of the three General Mortality Registers covering the UK.[45–47] Figures 8.5–8.7 demonstrate that there are different patterns evident within the UK. When data for individual substances are investigated, this feature becomes even more apparent. For example, there are geographical concentrations of some phenomena, *e.g.* deaths resulting from injecting of cathinones in Budapest, Scotland and south Wales, use of Ethylphenidate and Benzodiazepine analogues in Scotland, and use of 4,4′-DMAR in Northern Ireland and Hungary.

8.3. Anthropological Issues – Characteristics of Users

Having described some of the issues around obtaining an understanding of the epidemiology of NPS, we now turn to examine one of the key drivers in the NPS phenomenon – the Internet and its users.

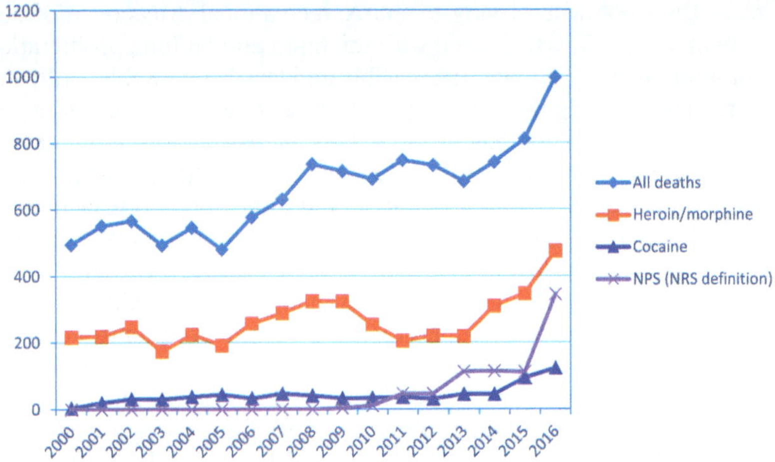

Figure 8.6. NPS death registrations in Scotland. Data from National Records of Scotland (NRS, 2017),[46] published under the Open Government Licence v3.0. © Crown Copyright.

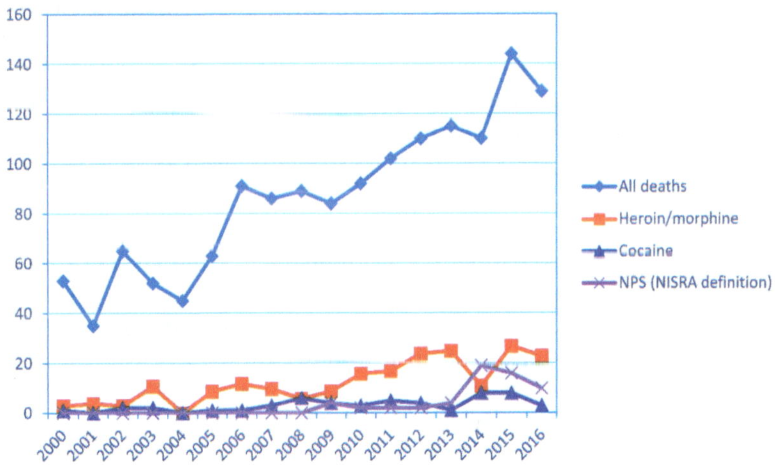

Figure 8.7. Death registrations in Northern Ireland. Data from Northern Ireland Statistics & Research Agency (NISRA, 2016),[47] published under the Open Government Licence v.3.0, © Crown Copyright, and personal communication to lead author for provisional 2016 data.

8.3.1. *Explanations About the Recent Emergence of NPS: The 'Drugs 2.0.' Internet Revolution*

It has been well-documented that Internet usage *per se* has rapidly increased over the last few years, by contributing, indeed, in shaping current drug scenarios.[6,21,48,49] Moreover, for a growing number of people the Internet is now the first place they look when trying to source recreational drugs or information about them, especially when faced with the rapid and baffling proliferation of new compounds.[6,7,48,50] In fact, the feasible and 'easily renewable' online market is gradually shifting the drug market from a street perspective to a virtual one. The virtual landscape has facilitated an increasing and 'more easy/safe' availability of recreational/internationally controlled (illicit) drugs,[6,7] through the development of electronic currencies and anonymous transaction infrastructures.[48,49] Furthermore, a new online 'quasi-legal' drug market spread due to the dissemination, sale and distribution of NPS through virtual platforms.[6] Completely untested compounds are part of an international online market that has become too fast and too complex for any government to control properly.[51]

Indeed, the advent of the so-called 'web 2.0.' has rapidly increased the availability of interactive and communicative tools able to share and disseminate information regarding NPS, including pro-drug fora, blogs and drug/legal highs/NPS selling websites.[48,49] This nexus could be termed a 'SWIM centre': a virtual space where (potential) users pool and discuss their experiences of NPS, 'dip their toe in' or immerse themselves in new types of discourse. (The allusion here is to those anonymous experimenters and commentators who go by the alias of 'SWIM' – Someone Who Isn't Me.)

Platforms such as Twitter®, Facebook®, Google+®, LinkedIn®, Tumblr®, YouTube® and so on, facilitate the exchange of short messages, *via* desktop, laptop, tablet or smartphone between members of inter-related groups and, therefore, offer opportunities for people to share opinions, information, links and experiences related to their substance abuse. There is already evidence that social networks have been employed by drug abusers for discussing and obtaining drugs.[52,53]

Some authors stated that the current growing virtual drug market generation, might be also called 'Drugs 2.0', an anarchic free-market world in which drug legislation is being outpaced by chemistry and technology.[51] In parallel, whilst the NPS market was firstly spread on the so-called 'surface web', recently it has been proposed that the Deep Web (also known as Deepnet, invisible web, hidden web, *etc.*) plays an important role in the NPS supply chain.[21,49] The Deep Web is the part of World Wide Web content that is not indexed by standard search engines (*i.e.* Google, Yahoo and so on), and is several orders of magnitude larger than the surface web.[54] Most online drug marketplaces, particularly those selling illegal substances, as well as illegal/criminal activities, are located on the 'dark net of the deep web'. The further shifting of the NPS market to the Deep Web is a further cause of concern as the magnitude/availability/dissemination of NPS may became greater and more uncontrollable.[49]

8.3.2. A New Generation of Drug/NPS Users (e-Psychonauts): Anthropological, Evolutionary and Psychological Issues

The history of the Internet is bound up with counterculture, and counterculture finds some of its richest expression in the use of psychoactive chemicals. The drugs, the technological protocols and cultural pipedreams informed and inspired each other, mainly being aimed at young people, particularly those belonging to the Nerd subculture, inspired/attracted by virtual reality and IT technologies. Within this new drug scenario, there are some 'educated and informed' users who possess a range of pharmacological/chemical notions about previously unknown compounds (NPS)/combinations of previously (un)known compounds.[51,55] The advent of the 'Drugs 2.0.' revolution facilitated the growth and diffusion of a 'new generation' of drug users/consumers (also called 'e-*Psychonauts*') who mainly use virtual platforms to provide information, sell and buy a range of new compounds/NPS from the cybermarket.[48,51,55] Psychonauts are able to provide quasi reliable information about previously unknown compounds/combinations because they possess a very large range of pharmacological/chemical notions relating to the most recent NPS available from the cyber market.[51,55]

The term 'Psychonaut', originally from the German *psychonauten,* is a neologism coined by the German writer and expert on psychoactive chemicals Ernst Jünger (1895–1998) who used the term in his 1970 essay *Annäherungen: Drogen und Rausch* (*Approaches*: *Drugs and Inebriation*), in describing Arthur Heffter's self-experimentation with the four pure alkaloids (isolated by himself from peyotl (peyote) in the late 1890s). Heffter's experience marked the world's first 'trip' with a purified chemical compound. Consequently, the term was then applied to all human self-experimentation with psychoactives/NPS, typically carried out to explore the 'inner universe'/'Psychocosmos'.[56] In a broader sense, *Psychonautics* could refer to all human usage of psychotropic ('turning towards the psyche') drugs.[57]

Nowadays, the term 'Psychonaut' refers to pharmacological 'trippers'. In fact, Psychonaut (a modern portmanteau of the Greek *ψυχή* (soul) and *ναύτης* (sailor), namely 'a sailor of the mind/soul') is a subject who experiences intentionally drug-induced altered states of consciousness,[57,58] in an attempt to investigate his/her mind, and possibly address spiritual questions, through a direct experience.[51,58] Moreover, the term was also associated with neo-shamanic practices, even though some NPS users clearly distinguish between the 'mental exploration' of the Psychonaut and 'healing-oriented shamanic practice'.[51] Some psychonauts even define themselves as the 'new shamans' of the Western world, an evolution of 'the Original Psychonauts' during an Age of Information.

The cultural reputation and attraction of NPS mainly derives from their use among e-Psychonauts.[59] From an anthropological and evolutionary perspective, psychonauts' drug intake is somewhat similar to ancient shamanic ritual plant consumption, as psychonauts appear mainly to prefer NPS with entheogen/hallucinogenic properties.[49,60] 'On-drug hallucinatory states' are seen as 'revelatory' states of mind, hence useful in the quest for the 'inner soul'/self-consciousness, in order to allow psychonauts to better understand themselves through a process of self-exploration.[49]

Like shamans, e-Psychonauts may be conceived as a 'sub-cultural virtual group' which has its own specific 'initiation rites', *e.g.* the practices required by the virtual communities to the 'novel psychonaut applicant'. These requirements include the need to formally subscribe to the index forum, so that both the applicant's contribution/posts/entries (which are anyway limited by the forum itself), but also the threads proactively s/he follows over time, can be properly scrutinised. Following this initial phase, the proper 'trial phase' can start. During this phase, the applicant will need to post a given number of 'trip reports' relating to a different range of psychoactive molecules, but mainly plant-derived substances and entheogens. These posts are then assessed in terms of their quality and accuracy. Applicant psychonauts will need to demonstrate acceptable standards of practice overtime, including the need to continue aiming at achieving psychedelic states and pursue mind exploration activities ("...you cannot be a psychonaut without tripping on drugs that helps in exploring the mind").[61] The applicant will possess excellent language/descriptive skills, so that his/her on-drug/trip reports can be made available in an accurate and precise way ("[...] a tool perhaps a psychonaut must hold close is language...only you can really describe your trip...trip reports our essential [...]").[62] Furthermore, applicants will need to demonstrate appropriate care of the trip 'preparation phase', including the accurate choice of: the compound to be ingested; the company/mentor; the setting; the goal of the experience; and the so-called 'trip-toys' (*e.g.* crayons, paper, *etc.* to express the consequent creativity). In planning the 'on-drug experience', the psychonaut will need to demonstrate appropriate knowledge of the drug's interactions and of the possible untoward medical/psychopathological effects.[63] Once recognised as an 'approved psychonaut', the applicant will be able to access a range of areas within the forum whose access was previously denied. In association with this, the successful applicant will be provided with a range of codes which will allow him/her access to a range of websites located on the Darknet/Deep Web, which provide a vast range of NPS purchase opportunities. Otherwise, s/he will be banned from the forum itself.[49,51]

Similar to shamans, e-psychonauts' main beliefs and goals often have more to do with the achievement of altered states of consciousness rather than with recreational purposes.[61,64] Furthermore, the extensive level of web-sharing activities facilitates the spread and selection of the 'psychonaut cultural phenotype'.[61] Cooperation, sharing and coordination with others are associated with evolutionary benefits.[65] Within the psychonauts' communities, the intake of psychedelics is considered the *trait d'union* of e-psychonauts, a ritual form of bonding which also facilitates the flow of psychedelic information.[66] Consistent with these observations, there is some evidence that psychedelic drug users report greater concern for others compared to both other drugs' users and drug-naïve subjects.[67] According to the 'handicap principle' theory,[65] increased levels of reciprocal altruism may give some evolutionary advantages. The capability to orientate towards altruism and compassion requires, however, an ability to attend to it both in the verbal and non-verbal presentations of others.[68] In social interactions, approval and disapproval are most typically signalled by facial expressions[69] which represent the main ways we use to communicate emotions and intentions to others.[70] Indeed, the exclusive usage of both online channels and 'avatars'/pseudonyms/hidden identities for sharing a range of human, including psychedelic, experiences inhibits the exchange of facial expressions. From this point of view, e-psychonauts mainly rely on their written linguistic skills for sharing their experiences.[63]

8.4. Clinical Pharmacological Concerns

These online exchanges provide a qualitative insight into the emerging and evolving world of NPS. Looking at such information can assist in understanding motives for using particular NPS molecules, the expected effects, dosages and frequency of use, unwanted side-effects, as well as providing a 'heads-up' of what may become more widely available. This section examines briefly what is known from netnographic studies as well as other resources. It provides an overview for the main classes of NPS in terms of pharmacological and clinical aspects.

8.4.1. *Synthetic Tryptamines*

Synthetic tryptamines comprise several substances, derivatives of the legal/controlled tryptamines, specifically designed to induce predominantly hallucinogenic effects.[71] Recently, data coming from a self-reported NPS use survey carried out by the National Survey of Drug Use and Health (2009–2013) in the USA, identified that the most common class of drugs reported was represented by tryptamines (86.1%), followed by phenethylamines (17.5%) and synthetic cannabinoids (3.6%).[72]

Psilocybin (O-phosphoryl-4-hydroxy-*N*,*N*-dimethyltryptamine), a natural hallucinogen found in certain species of mushrooms, is the precursor of this class of synthetic compounds.[73] Some tryptamines are compounds naturally produced by humans as the neurotransmitters serotonin (5-hydroxytryptamine or 5-HT), melatonin (*N*-acetyl-5-methoxy-tryptamine), bufotenin (*N*,*N*-dimethylserotonin or 5-OH-DMT or 5-hydroxy-*N*,*N*-dimethyltryptamine), DMT (*N*,*N*-dimethyltryptamine) and 5-MeO-DMT.[74–76] DMT and 5-MeO-DMT have been naturally found in some *Delosperma* species; whilst psilocin (4-OH-DMT, 4-hydroxy-*N*,*N*-dimethyltryptamine) and psilocybin have been identified in some hallucinogenic fungi; and, bufotenin and 5-Hydroxy-indolethylamines are common constituents of the venoms of the genre *Hyla*, *Leptodactylus*, Rana and *Bufo Alvarius.*[77–82] Furthermore, tryptamines have been as well synthesised for pharmaceutical/medical purposes (*e.g.* sumatriptan and zolmitriptan for migraine) and as research chemicals.[71,83] DMT, psilocin, bufotenin and DET (*N*,*N*-diethyltryptamine) were originally listed as Class A drugs in Part 1(a) of Schedule 2 of the Misuse of Drugs Act 1971. As esters and ethers of Class A drugs are also controlled, 5-MeO DMT and psilocybin (the phosphate ester of psilocin) are also under control.[84] In 1998, etryptamine (alpha-ethyl tryptamine, AET), developed as an antidepressant 'Monase' by Upjohn in the 1960s, but later withdrawn after it was found to be linked to agranulocytosis, was also added to Class A.

Synthetic/derivatives of tryptamines appeared on illicit drug markets throughout the 1990s, and dominated the online drug market until 2007, when they were then listed as 'narcotics' or 'designated substances' and quickly replaced by cathinones, phenethylamines and piperazines.[6,85] Nevertheless, novel tryptamines continue to appear on the online drug market, sold as 'legal highs' such as 5-Meo-DALT (*N,N-diallyl-5-methoxytryptamine*), AMT (*alpha-methyltryptamine, aka IT-90*), 5-MeO-aMT (*5-methoxy-α-methyltryptamine*), *5-MeO-DMT* (*5-methoxy-N,N-dimethyltryptamine*), *5-MeO-DPT* (*5-methoxy-N,N-dipropyltryptamine*), 5-IT [*5-(2-aminopropyl) indole*] *etc.*[6,7,86] Three of the NPS formally notified for the first time in 2015 to the EMCDDA were tryptamines.[6]

From a pharmacological point of view, tryptamines are extensively metabolised by monoamine oxidase inhibitors (MAOIs)[87,88] and mainly act as agonist at $5HT_{1A}$, $5HT_{2A}$ and $5HT_{2C}$ receptors,[89–91] even though they exhibit less selectivity and affinity for $5HT_{2A}$ receptors compared to phenethylamines.[92] Other receptors implicated in tryptamine interaction include vesicular monoamine transporter 2 (VMAT2), σ-1, serotonin transporter (SERT) and trace amine-associated (TAR) receptors.[88,90,93–95]

Amongst the clinical symptomatology, visual hallucinations have been predominantly reported following the use of most tryptamines; whilst prevalent auditory hallucinations have been described following the intake of DiPT (*N,N-diisopropyltryptamine*).[63,71] Altered sensory perception, colour intensification, distorted body image, depersonalisation, marked mood lability, euphoria, relaxation, entactogenic properties, and anxiety ranging from mild apprehension to panic disorder have been also described.[63,71,96] Untoward effects include agitation, tachyarrhythmia, hyperpyrexia, serotoninergic neurotoxicity and death.[88] Some fatalities have been reported and confirmed after post-mortem toxicological analysis following the intake of tryptamines.[97]

Natural tryptamines are commonly available in preparation of dried or brewed mushrooms, while tryptamine derivatives are usually sold in capsule, tablet, powder or liquid form. Synthetic tryptamines are generally swallowed, sniffed, smoked or injected. Common street names reported include 'Foxy-Methoxy' (5-Meo-DIPT), 'alpha-O', 'alpha' and 'O-DMS' (5-Meo-AMT), '5-MEO' (5-MeO-DMT), 'spice' or 'changa' (DMT), 'T-9' (DET).[63,98]

8.4.2. Synthetic Phenethylamines

Phenethylamines (also known as amphetamine/α-methyl-phenethylamine designer drugs) comprise a wide range of natural or synthetic substances which possess psychostimulant, entactogenic and hallucinogenic effects.[99] Mescaline, a naturally occurred alkaloid isolated from the peyote cactus (*Lophophora williamsii*) and traditionally chewed by Indians in Mexico as a medicine, amulet and/or hallucinogenic religious sacrament, represents the prototype psychoactive phenethylamine compound.[100] The endogenous catecholamines and neurotransmitters have the chemical structure of phenethylamines.[101]

In the 1960s and 1970s, the synthetic chemist Alexander Shulgin developed a variety of methoxylated amphetamine compounds derived from mescaline, including TMA (3,4,5-trimethoxyamphetamine), DOM (3-methoxy-4,5-methylenedioxyamphetamine), DOET (2,4,4-trimethoxyamphetamine), DOI (4-iodo-2,5-dimethoxyamphetamine), DOC (4-chloro-2,5-dimethoxyamphetamine), *etc.*; having consciousness-altering and distorted sensory perceptional properties.[99] Substituted phenethylamines were then explicitly described in the 1990s.[99]

Alongside these substances, the group of phenethylamines also include some 179 'classical' phenethylamines/MDMA (3,4-methylenedioxymethamphetamine)-like drugs (*e.g.* MDA [3,4-methylenedioxy-amphetamine]; MDEA [3,4-methylenedioxyethamphetamine]; MBDB [*N*-methyl-1-(1,3-benzodioxol-5-yl)-2-butanamine]); and the latest generation phenethylamine derivatives: *e.g.* 'Bromodragonfly' (1-(8-bromobenzo[1,2-*b*; 4,5-*b*′]difuran-4-yl)-2-aminopropane); NBOMe derivatives; indanes; benzofurans; 2C molecules such as 2,5-dimethoxy-4-bromophenethylamine (2-CB), 2,5-dimethoxy-4-iodophenethylamine (2C-I), and 2,5-dimethoxy-4-ethylphenethylamine (2C-E).[7] Nine of the 98 NPS notified for the first time in 2015 to the EMCCDA were phenethylamines.[6] Five per cent of NPS seizures of NPS reported to the EMCDDA in 2014 were from this class of substances.[6]

Bromo-Dragonfly (aka 'DOB-Dragonfly'/'3C-Bromo-Dragonfly'), firstly synthesised by Parker, in 1998,[102,103] displays a high affinity for 5-HT_{2A} agonist (Ki = 0.04 nM), 5-HT_{2B} (Ki = 0.19 nM) and 5-HT_{2C} (Ki = 0.02 nM).[103] Long-standing hallucinations, mood elevation, paranoid ideation, confusion, anxiety and flashbacks have been described after its intake.[102] Several cases of acute intoxications and fatalities have been described.[102]

Most of the 2C series phenethylamines were first synthesised by Shulgin in the 1970s and 1980s. The '2C' refers to the 2-carbon atoms separating the primary amine functionality from the phenyl ring.[99] These compounds are structurally related to mescaline, so they exert stimulating/hallucinogenic and psychedelic effects.[104] 2C series compounds are frequently ingested as MDMA substitutes and with other club drugs (*e.g.* MDMA, MBDB, ketamine, *etc.*). These compounds act primarily at the 5-HT_{2A} over 5-HT_{1A} receptors with binding ratios of 14-94.[105] Moreover, most of the 2C drugs also display low-affinity binding to D_2 receptors; a low inhibitory activity to monoamine transporters (Ki > 10 μM).[7,86,105,106] The 5-HT_{2A} receptor mediates hallucinogenic effects of these compounds.[107,108] Headache, fever, dysphoria, sometimes hallucinations, mydriasis, seizures, severe agitation, mydriasis and apnoea have been reported as untoward effects. Several deaths have been reported as well. The first and the most common compound of this group is the 2-CB (aka 'Nexus'/'Venus'/'Bees'/'Bromo Mescaline'/'Toonies'/-'Bromo'/'Erox'/'Synergy'/'XTC'). After its scheduling by the Drug Enforcement Agency

(DEA) in 1995, a large range of new 2C compounds were synthesised and marketed, such as 2C-D, 2C-E, 2C-B, 2C-N, 2C-H, *N*-ethyl-2C-B, *etc.*[6]

Recently, several highly active *N*-benzyl substituted phenethylamines (NBOMes) have become available, *e.g.* 25B-NBOMe ((2-(4-bromo-1,5-dimethoxyphenyl)-*N*-(2-methoxybenzyl) ethanamine)), 25C-NBOMe ((2-(4-chloro-2,5-dimethoxyphenyl)-*N*-[(2-methoxyphenyl) methyl] ethanamine); aka '*N*-Bomb'/'Pandora'), 25I-NBOMe ((4-iodo-2,5-dimethoxy-*N*-(2-methoxybenzyl)phenethylamine); aka '*N*-bomb'/'Smiles'/'CIMBI-5').[31,109–111] *In vitro* receptors studies have demonstrated that NBOMe compounds act as potent full agonists at the 5-HT_{2A} and 5-HT_{2C} receptors.[107] In addition, 25I-NBOMe and 25C-NBOMe are pharmacologically active at very low sub-milligram doses.[31] They also display high binding affinity to adrenergic α_{1A} (Ki < 1 μM) and α_{2A} (Ki < 1 μM) receptors as well as histamine H_1 receptors.[105] They show low-affinity binding to D_2 and D_3 receptors.[105] Their higher 5-HT_{2A} receptor affinity, compared to their 2C derivatives, explain their higher incidents of hallucinations and delusions.[112] NBOMe containing products are usually sold as tablets, capsules, powder, liquid, spray, and preloaded paper doses (blotters). They are usually orally taken (sublingually or bucally) or *via* nasal insufflations.[6] Some of them are sold as LSD replacements,[6,109] even though they are only slightly more potent than LSD. Depending on the route of administration, their effects may last from 3–8 h and 4–6 h by insufflations of 25C-NBOMe and 25I-NBOMe, respectively, and 4–10 h (25C-NBOMe) and 6–10 h (25I-NBOMe) by sublingual or buccal intake.[108,109] Psychotropic effects commonly reported include: euphoria, increased sociability, visual/auditory/olfactory/tactile sensations, hallucinations, life-changing spiritual experiences, empathy, psychedelic experiences, depersonalisation, dissociation and derealisation. Amongst the untoward effects are included: nausea, vomiting, headache, temporary dysuria, sweating, panic, severe agitation, aggressiveness, seizures, insomnia, and, sometimes, rhabdomyolysis with subsequent acidosis and renal failure, muscle rigidity, tremulousness and sudden and unexpected cardiopulmonary arrest.[6,109,111] Several fatalities have been reported following the intake of NBOMe compounds.[113–115]

8.4.3. *Piperazines*

Piperazines comprise many compounds able to mimic ecstasy-like effects. In fact, they were originally designed to represent the 'legal' alternative to ecstasy and they may be identified as cutting agents in some ecstasy pills.[6] Piperazines are commonly marketed as pills in various colours (*i.e.*, red, blue, pink, white, purple, orange, *etc.*), forms and impressions (*i.e.*, smiley face, bull's head, bird flying, Mickey Mouse, *etc.*).[6,7,86] However, piperazines have been as well sold as an off-white powder, in capsules and as a liquid.[6] Three new piperazine molecules were first notified to the EMCDDA in 2015.[6] Amongst the main clinical desired effects are included: feelings of arousal, euphoria, wakefulness, a sense of wellbeing and empathy. Whilst agitation, vomiting, stomach pain, fits, irregular heart rhythms, diarrhoea, fever and allergic reactions have been reported as untoward effects of this class of drugs,[7,86] many deaths associated with their consumption have been recorded.[97]

8.4.4. Benzofurans

Benzofurans are phenethylamines, structurally related to MDMA and MDA (3,4-methylenedioxyamphetamine), which comprise 6-APB (6-(2-aminopropyl) benzofuran; aka 'BenzoFury'); 5-APB (5-(2-aminopropyl)benzofuran); 6-APDB (6-(2-Aminopropyl)-2,3-dihydrobenzofuran; aka '4-Desoxy-MDA'/'EMA-3'); 5-APDB (5-(2-Aminopropyl)-2,3-dihydrobenzofuran; aka '3-Desoxy-MDA'/'EMA-4'); *etc.*[6] Amongst the benzofurans, the most investigated compounds are mainly represented by 6-APB and 5-APB.[116] Their main route of administration is oral since their nasal insufflation is reported as extremely painful.[117] They may exhibit effects like MDA and MDMA, *i.e.* stimulation, euphoria, entactogenic and visual/auditory hallucinations.[116] Untoward effects commonly reported include dry mouth, nausea, jaw and teeth clenching, dry eyes, insomnia, diarrhoea, light hypersensitivity, hot flushes, headache, drowsiness, panic attacks, anxiety, depression, severe paranoia and psychosis. Furthermore, some reported an unpleasant 'come-down' that may last up to several days after oral intake.[117] Several deaths have been identified as mainly related to benzofurans 5- or 6-APB.[97,118,119]

8.4.5. Amphetamine-type Substances (ATS) and Novel Psychostimulants

The 'amphetamine-type substances' (ATS), *e.g.* PMA (4-methoxyamphetamine/'*Dr Death*'), PMMA (4-methoxymeth-amphetamine), 4-MTA (4-methylthioamphetamine/'*flatliners*'), DMA (2,5-dimethoxyamphetamine), aminorex derivatives; diclofensine, methiopropamine (MPA), *etc.* have been marketed as para-substituted methoxy drugs pharmacologically and clinically comparable to amphetamines.[6,7,86] ATS are commonly supplied as tablets/crystals/powder, sometimes mixed with other substances.[120] They are usually swallowed, smoked or snorted and their common 'street names' include 'crystal met' (aka 'ice'); 'speed', 'crystal' *etc.*[120] Obsessive delusions, paranoia, hallucinations, suicidal ideation, anxiety, insomnia, depression and psychotic symptoms have been commonly reported following the intake of ATS.[120,121]

PMA and PMMA (aka 'red Mitsubishi'/'pink McDonalds'/'pink ecstasy'/'Mitsubishi turbo'/'killer'/'Dr Death'/'double stacked'/'chicken yellow'/'chicken fever') are similar to MDMA, even though they are extremely active at lower doses. They may cause a fatal rise in body temperature, euphoria, a sense of full energy, an amplification of sounds/colours and feelings of great love, sociability *etc.* An increasing number of fatalities have been described following their consumption.[122,123]

Diclofensine is a stimulant that acts as a dopamine, norepinephrine and serotonin reuptake inhibitor.[124–127] It was originally developed as an antidepressant, abandoned possibly due to concerns about its potential for abuse.[128–131]

Methiopropamine (MPA), chemically related to methamphetamine, was firstly synthesised in 1942 and advertised on the Internet as a 'research chemical' in late 2010, also branded as 'Blow'.[63,132–134] It acts as a selective norepinephrine-dopamine reuptake inhibitor.[135] At low doses, it is a functional stimulant creating mild euphoria, alertness, sexual arousal, loss of appetite, tachycardia, anxiety, *etc.* It has been reported to have a potential psychological dependence.[98] Adverse effects comprise a significant hangover after extended use with nausea, headache, dizziness, lack of energy, skin irritation and difficulty urinating.[63,98] It is taken by oral, intranasal and smoking routes.[98]

EUROPOL and the EMCDDA have issued a joint report and risk assessment about serious harms, particularly 27 deaths, occurring in Europe between 2013 and 2014, associated with 4,4-DMAR.[136,137] This is the para-methyl derivative of 4-methylaminorex, also known as '4-methul-euphoria', '4-methyl-U4Euh', '4-M-4-MAR', '4,4-dimethylaminorex' or 'Serotoni'. It is a derivative of 4-methylaminorex (4-MAR, also known as 'ice', 'euphoria' or 'U4Euh') and its desmethyl derivative *aminorex*, both controlled under the Misuse of Drugs Act 1971.[138,139] It is a potent releaser at DAT and NET similar to that of d-amphetamine and aminorex, also showing a more potent action at SERT.[140] Nasal and oral administration have been reported.[141–143] It has been sold as a powder and in tablet form. Clinical effects arise within 10-60 minutes and may last several hours.[142] Symptoms are mainly related to stimulant-type drugs (euphoria, anorectic effect, alertness, *etc.*). Untoward effects include agitation, hyperthermia, foaming at the mouth, breathing problems and cardiac arrest.[138,144]

8.4.6. *Synthetic Cannabinoids*

Legal availability is considered an important motivation for consumption. Many users do not seem to be aware of the serious adverse effects related to use of synthetic cannabinoids, since these compounds are being perceived to be somehow equivalent to marijuana and hence 'safe' and 'natural'.[145] Users are often young males, choosing 'Spice' and similar products instead of cannabis because of their low cost, easy availability, and undetectability in routine urine screening tests.

Although there are similarities, in terms of psychoactive effects, between low-dosage synthetic cannabinoids and cannabis/THC intake, both visual/auditory hallucinations and intense paranoid feelings typically occur only with higher levels of synthetic cannabinoids intake. In comparison to cannabis, synthetic cannabinoid intake may also be associated with: quicker 'kick off' effects, shorter duration of action, and larger levels of hangover effects.[145]

Synthetic cannabinoid intake has been associated with a range of psychotomimetic disturbances (*e.g.* paranoia, delusions, hallucinations), the occurrence of florid/acute transient psychosis, relapse/worsening of a pre-existing psychosis, persistent psychotic disorder/'Spiceophrenia'; and relapse of a pre-existing bipolar disorder.[86]

Synthetic cannabinoid withdrawal syndrome may be characterised by drug craving, tachycardia, tremor, profuse sweating, nightmares or insomnia, headache, anxiety/irritability, feelings of emptiness or depressive symptoms, and somatic complaints. Adverse synthetic cannabinoids related side-effects may be severe; indeed, their intake is associated with a 30-fold higher risk of seeking emergency room care as compared with traditional cannabis.[146] A short-lasting, potentially life-threatening, serotonin-like syndrome may be observed, with reported signs/symptoms of: elevated heart rate and blood pressure, mydriasis, agitation/anxiety, hyperglycaemia, dyspnoea or tachypnoea, nausea or vomiting, and seizures.[147] Other acute adverse effects associated with synthetic cannabinoids include: self-injurious or aggressive behaviour, hypermemesis, nystagmus, stroke, myocardial infarction, and acute kidneys injuries.[86] A number of deaths have been related to synthetic cannabinoid ingestion, either on their own or in combination.[97]

8.4.7. *Synthetic Cathinones*

Synthetic cathinones are stimulant NPS compounds often used by consumers for various reasons, including: hallucinogenic experiences, euphoria, mood enhancement, openness, empathy, mental clarity, stimulation, increased energy, and increased libido.[86] Whilst some cathinones have behavioural effects, respectively, similar to methamphetamine and cocaine, other ones are closer to those of MDMA. Many synthetic cathinones users report tolerance, dependence and withdrawal symptoms. The most common cathinones' adverse reactions are restlessness and anxiety, ranging from mild agitation to severe psychosis. Furthermore, tachycardia, hypertension, abdominal pain, chills, flushing, sweating, hyperthermia, renal failure, rhabdomyolysis, and seizures can also be observed.[145] Paranoid ideation and mood disturbances have been observed in chronic users of both natural and synthetic cathinones.

An increasing area of concern in Europe is the injecting of NPS, especially synthetic cathinones, at a time when injection of heroin is falling in some countries.[148] Within the Chemsex context, synthetic cathinones/mephedrone misuse, either on their own or together with methamphetamine and/or GHB/GBL, especially in the men who have sex with men (MSM) sub-population, has been associated with unprotected sex practices, sharing syringes, and highly elevated risk of the spread of blood-borne and sexually transmitted diseases.[149]

Deaths have been associated with a range of synthetic cathinones, including: mephedrone, methylone and butylone, ethylone, α-PVP, MDPV, and methedrone.[145]

8.4.8. Opioid NPS

Opioid NPS elicit their morphine-like effects through binding to opioid receptors. Little is known about in-depth pharmacology of all the different NPS opioids. However, online users' reports suggest that some compounds can have much longer durations of action.[150]

Novel agents such as AH-7921, MT-45, U-47700, U-50488, and the novel fentanyls seem to have similar mechanisms of action, with the fentanyls being especially known for their high potency, and risk for overdose and mortality.[151–155] Synthetic opioids, such as illicitly manufactured fentanyls, have in fact recently been associated with a sharp increase in the US opioid-involved overdose deaths epidemic;[156] there is also concern about such developments in Europe. Naloxone can reverse the toxicity seen with novel opioids, although the doses of naloxone required may be higher than for traditional opioids, particularly in cases of novel fentanyl toxicity.[150] Enhancing naloxone distribution and other harm reduction approaches, increasing opioid use disorder treatment capacity, improving linkage into treatment are needed in order to counter NPS opioid misuse and addiction.[156]

8.5. Clinical Approaches to NPS

We have noted earlier that emerging NPS pose a challenge in terms of their typically unknown effects. This challenge is heightened because of problems in identifying exactly what has been taken or administered, dosage, *etc*. This can be of grave significance if clinicians are presented with intoxications that may have been caused by NPS, especially in a context of poly-substance use. It is often the case that fatalities and 'near-misses', as well as intoxications, are occurring because of the consumption/administration of multiple psychoactive substances. This can be in the form of different molecules from the same chemical class being taken together, *e.g.* synthetic cathinones or synthetic cannabinoids or 'designer' benzodiazepines. Such events could occur deliberately through individuals knowingly/deliberately taking or consuming more than one molecule from the same class, or unwittingly through using branded products whose contents might not necessarily match the ingredients stated on the packaging. Indeed, branded products may well include mixtures of both NPS and controlled drugs, along with adulterants and cutting agents[157–159] – all of which may not be declared or misleadingly described. Some products may contain molecules from completely different chemical classes, *e.g.* Diphenidine is known as an *N*-methyl-D-aspartate receptor channel blocker together with the synthetic cannabinoid 5-fluoro-AB-PINACA.[160]

Even molecules from the same chemical class may affect the brain in different ways. For example, synthetic cathinones can be categorised as: (a) substrates for dopamine, serotonin and norepinephrine transporters; (b) monoamine transporter substrates with dopamine-selective profiles similar to amphetamines and methamphetamine; and (c) non-substrate transporter inhibitors.[145] Advancing knowledge of the chemical structures of cannabinoids has prompted the need to revise the traditional classification of such molecules, so as to avoid misunderstandings between forensic providers and the legal profession.[161]

An individual molecule may have its properties manipulated to act on different receptors *via* dual ligands. For example, PhAR (phenylacetylrinvanyl) has been manipulated to produce the first CB_2/TRPV1 dual ligand, and a fresh category of inverse agonists at the CB_2 receptor.[162] This demonstrates that 'hybrid' ligands can be developed and can theoretically lead to the development of further variations in NPS molecules.

Clinicians should consider direct inquiry about NPS use, particularly among young adults presenting for acute medical care with signs or symptoms that could indicate substance-related toxicity; the simultaneous misuse of other drugs should also be actively investigated.[147,163] Most NPS are usually not detected by conventional drug screening tests. Advanced tests like gas chromatography or liquid chromatography-tandem mass spectrometry can be performed on a blood or urine specimen at reference laboratories, but they are not generally available in the acute setting and are used mainly for forensic investigations. Even with the most advanced tests, not all NPS compounds can be detected.[147,164]

For patients presenting for acute medical complications of NPS use, laboratory testing should include standardised urine drug testing (which can be helpful in identifying concomitant intake of classic illicit substances), a complete blood cell count, complete metabolic panel, ECG baseline, and cardiac enzymes. Creatine phosphokinase is helpful if rhabdomyolysis is suspected on the basis of severe muscle spasms, swelling and pain in the extremities, or severe seizures.[163,164]

In the acute setting, clinicians are unlikely to have the luxury of being able to identify known, let alone unknown, NPS in a timely manner. Hopefully, some of the developments described later in the book will help alleviate this situation, through the development of 'in-field' recognition technologies and associated softwares. Difficulties in ascribing effects to a specific drug/class of drug in the context of polysubstance use, especially where NPS use is known or suspected, mean that the clinician is generally advised to treat the symptoms with which patients present.

Placing the patient in a quiet, dark environment can be helpful. In case of agitation, the initial management of agitated patients should focus on decreasing levels of both self- and outward-directed aggression and agitation. Verbal de-escalation can always be considered with more cooperative patients.[86] In case of NPS intoxication it is advisable to contact the local poison control centre, which may provide assistance in the acute setting.[147]

No specific antidotes are available for most NPS, apart from benzodiazepine and opioid compounds.[163] Given the complex, and at times unknown, pharmacology of the substances possibly ingested by the client, benzodiazepines may be considered as agents of choice.[7] Benzodiazepine use, either intramuscular (IM) or intranasal midazolam (5–10 mg IM/intranasal or

2.5–5 mg IV), because of its fast absorption/onset of action; or lorazepam (4–8 mg IM; 2–4 mg IV) may be considered.[7,165] Benzodiazepines, however, need frequent re-dosing/high dosages to achieve adequate sedative effects, and this may be a problem if clients have co-ingested alcohol.[86]

Where patients cannot be controlled with benzodiazepines alone, propofol, dexmedetomidine, barbiturates, and/or antipsychotics may be considered, although this may further contribute to the acute toxicity effects of the abused substances.[7,164] In case of cathinones intoxication with sympathomimetic complications and agitation, the use of benzodiazepines seems appropriate, in case of hyperthermia and/or the serotonin syndrome cooling and 5-HT antagonists (cyproeptadine, chlorpromazine) can be administered.[166,167]

In the case of GHB/GBL overdose with respiratory impairment, supportive airway management may be required. Renal function and creatine kinase should be checked, given reports of rhabdomyolysis.[166] Acute GHB/GBL withdrawal is a serious issue, with complications such as seizures and death. High-dose benzodiazepines should be used (titrated to avoid over-sedation, *e.g.* 40–100 mg per day diazepam in divided doses) with baclofen 10 mg three times daily.[166] In case of ketamine/methoxetamine intoxication with agitation, observation and symptomatic treatment with benzodiazepines may be required.

In psychosis induced by synthetic cannabinoids associated with agitation, the use of oral or parenteral benzodiazepines seems to be an appropriate option, providing the extra advantage of preventing the possible occurrence of seizures, which often are associated with synthetic cannabinoids intoxication.[168] According to existing reviews for the treatment of synthetic cannabinoid induced psychosis there is not a specific preference for the use of one antipsychotic over another.[169,170] In the case of antipsychotic administration, an electrocardiogram may need to be performed, and QTc interval monitored, also because synthetic cannabinoid users may present with vomiting and hypokalaemia, further complicating the clinical picture. Finally, antipsychotics may also lower the threshold for seizures.[147,171]

From a more general point of view, in treating substance-induced psychotic disorders second generation antipsychotics (SGAs) may present with some advantages over first-generation ones[172] being less frequently associated with interference at the drug reward system level, hence not increasing levels of craving.[173] Additionally, SGAs act as antagonists of the serotonin $5HT_{2A}$ receptor, the main target of most hallucinogenic drugs.[174]

8.6. What Might the Future Hold?

There are numerous possible challenges that may face the different constituencies dealing with NPS, and which may further compound the problems of getting a comprehensive understanding of the phenomenon (or indeed phenomena that may develop).

More synthetic cannabinoids (theoretically there are more than 700) and synthetic cathinones are likely to appear, leading to not only intoxications but also psychoses, paranoia and other psychiatric issues. 'Ecstasy-type' substances containing molecules associated with neurotoxicity (*e.g.* 4-CMA) or fatalities (*e.g.* PMMA) continue to be manufactured and circulated across Europe, including the UK.

There is continuing growth in the creation and use of opioid (especially fentanyls such as acetyl fentanyl, acroyloyl fentanyl, carfentanyl, furanyl fentanyl) and 'designer' benzodiazepine analogues (*e.g.*, Etizolam, Diclazepam, Flubromazepam and Pyrazolam). Both these groups have caused significant health issues in Europe, including the UK in the case of benzodiazepines. As a result, many in this latter class became controlled as Class C drugs on 31 May 2017: Etizolam, Diclazepam, Flubromazepam, Pyrazolam, Deschloroetizolam, Flubromazolam, Nitrazolam, Nifoxipam, Clonazolam, 4′-Chlorodiazepam, Bromazolam, Meclonazepam, Adinazolam, Metizolam, 3-Hydroxyphenazepam, Fonazepam.[175] Some of the substances in these classes are 'failed' medications (*e.g.* U47,700, WT-45, AH-7921). Increasingly, substance users are taking a wider range of medications recreationally. In addition, e-psychonauts are showing more interest in and experimentation with 'natural' products (plants, fungi and animals). There are already reports of 'vaping' equipment being used to administer NPS/'legal highs' *e.g.* synthetic cannabinoids.[176,177] All of this is against a background of continued poly-substance use.

Following the introduction of 'blanket ban' legislation, *e.g.* in the UK and Ireland, and disappearance of 'head' or 'smart' shops, it is likely that increased use of the Internet (Surface and Deep) will occur. There are already indications of NPS now being traded alongside traditional 'street' drugs such heroin, cocaine and cannabis in illicit street drug markets.[26]

8.7. Conclusions

This chapter has looked at the current and emerging issues and challenges of NPS from epidemiological, anthropological and clinical pharmacological perspectives. With so much remaining unknown about a wide range of issues concerned with NPS, the need for inter-disciplinary approaches to the ever-changing phenomenon which is NPS has been clearly demonstrated. Other disciplines have a role in the quest for timely, up to date information and developing methods to understand the potential characteristics and properties of new molecules. Various forensic approaches are required in order to achieve a deeper understanding of the NPS phenomenon and its implications. These are examined in the following chapters.

However, arguably, the key dimensions that need to be documented are the human ones of why people are looking for new psychoactive substances, what effects they are seeking, what they are actually taking (how, where and when), and what the consequences of their actions may be.

In addition, external factors or influences which may play a determinant role in how these play out need to be understood. The use of historic or traditional methods of epidemiology play a part, but new and innovative approaches need to be developed in a world of increasing reliance on rapid electronic exchange and dissemination of information. As we will see, this challenge is being taken up by the traditional 'hard' fields in physical science in conjunction with 'soft' science approaches such as mathematical modelling.

Whilst more light is being focused on specific aspects of identifying and understanding of NPS, we need to develop the scientific and epidemiological knowledge-bases in a timely manner. Moreover, as researchers we need to move away from the relative safety of only seeing what we are looking for – the 'lamp-post problem'[178] - into the darker, less well lit, areas of the NPS phenomenon. The potential dangers, and possible benefits, of NPS need to be illuminated. Researchers need to be encouraged, and supported financially, to explore new avenues employing innovative techniques and equipment. Hopefully, along the way we will experience some 'Eureka' moments, when the 'light-bulb' in our collective brain gets switched on.

Acknowledgements

This chapter was supported in part by grants of the European Commission (Drug Prevention and Information Programme 2014–16; contract no. JUST/2013/DPIP/AG/4823; *EU-MADNESS project*). Further financial support was provided by the EU Commission-targeted call on cross border law enforcement cooperation in the field of drug trafficking - DG Justice/DG Migrations and Home Affairs (JUST/2013/ISEC/DRUGS/AG/6429) *Project EPS/NPS* (Enhancing Police Skills concerning Novel Psychoactive Substances; NPS).

Some of this material has been presented as oral presentations by the lead author as follows:

Corkery, J. M. (2016). Oral presentation, The world of NPS and overdose. European Cities Network for Drugfree Societies - Baltic Spice Alert Project Conference #3: Synthetic opioids and cannabinoids – A challenge for the region? Stockholm City Hall, Sweden. 2–3 November 2016.

Corkery, J. M. (2016). Oral presentation, NPS issues: epidemiological data. European Conference on: Novel psychoactive substances/NPS; clinical, pharmacological, analytic chemistry, toxicity, legislative, and criminological issues. Fielder Centre, University of Hertfordshire, Hatfield, UK. 16 November 2016.

Corkery, J. M. (2016). Oral presentation, Definitions, Categorisation and Characteristics of NPS. Joint International Meeting on Drug Trafficking and NPS Online Market INTERPOL General Secretariat, Lyon, France. 6–7 December 2016.

References

1. UNODC, *The Challenge of New Psychoactive Substances: A Report from the Global SMART Programme*, United Nations Office on Drugs and Crime, Vienna, 2013, available at: http://www.unodc.org/documents/scientific/NPS_2013_SMART.pdf, accessed 27 December 2016.
2. EMCDDA-Europol, *EMCDDA-Europol 2015 Annual Report on the Implementation of Council Decision 2005/387/JHA*, Publications Office of the European Union, Luxembourg, July, 2016, DOI: 10.2810/932574, available at: http://emcdda.europa.eu/system/files/publications/2880/TDAS16001ENN.pdf, accessed 27 December 2016.
3. ACMD, *Consideration of the Novel Psychoactive Substances ('Legal Highs')*, Advisory Council on the Misuse of Drugs, London, October, 2011, available at: https://www.gov.uk/government/uploads/system/uploads/attachment_data/file/119139/acmdnps2011.pdf, accessed 27 December 2016.
4. UNODC, *Global Synthetic Drugs Assessment 2014: Amphetamine-type Stimulants and New Psychoactive Substances*, United Nations Office on Drugs and Crime, Vienna, May, 2014, available at: https://www.unodc.org/documents/scientific/2014_Global_Synthetic_Drugs_Assessment_web.pdf, accessed 27 December 2016.
5. UNODC, *World Drug Report 2017 Volume 4-Market Analysis of Synthetic Drugs: Amphetamine-type Stimulants, New Psychoactive Substances*, United Nations Office on Drugs and Crime, Vienna, 2017, available at: https://www.unodc.org/wdr2017/field/Booklet_4_ATSNPS.pdf, accessed 12 September 2017.
6. EMCDDA, *European Drug Report – Trends and Developments 2017*, Publications Office of the European Union, Luxembourg, 6 June, 2017, DOI: 10.2810/610791, available at: http://www.emcdda.europa.eu/system/files/publications/4541/TDAT17001ENN.pdf, accessed 6 June 2017.
7. F. Schifano, L. Orsolini, G. Duccio Papanti and J. M. Corkery, Novel psychoactive substances of interest for psychiatry, *World Psychiatry*, 2015, **14**(1), 15–26.
8. A. Guirguis, J. M. Corkery, J. Stair, S. Kirton, M. Zloh and F. Schifano, Intended and Unintended Use of Cathinone Mixtures, *Hum. Psychopharmacol.*, 2017, **32**(3), e2598.
9. C. Cole, L. Jones, J. McVeigh, A. Kicman, Q. Syed and M. Bellis, Adulterants in illicit drugs: a review of empirical evidence, *Drug Test. Anal.*, 2011, **3**(2), 89–96.
10. C. V. Giné, I. F. Espinosa and M. V. Vilamala, New psychoactive substances as adulterants of controlled drugs. A worrying phenomenon? *Drug Test. Anal.*, 2014, **6**(7–8), 819–824.
11. UNODC, *World Drug Report 2016*, United Nations Office on Drugs and Crime, Vienna, 23 June 2016, available at: https://www.unodc.org/doc/wdr2016/WORLD_DRUG_REPORT_2016_web.pdf, accessed 28 December 2016.
12. A. Winstock, M. Barratt, J. Ferris and L. Maier, *Global Drug Survey 2017: Global Overview and Highlights*, London, globaldrugsurvey.com, 2017, available at: https://www.globaldrugsurvey.com/wp-content/themes/globaldrugsurvey/results/GDS2017_key-findings-report_final.pdf, accessed 31 May 2017.
13. European Commission, *Report: Flash Eurobarometer 401: Young People and Drugs*, TNS Political & Social, Luxembourg, 2014, available at: http://ec.europa.eu/public_opinion/flash/fl_401_present_en.pdf, accessed 28 December 2016.

14. ESPAD Group, *ESPAD Report 2015-Results from the European School Survey Project on Alcohol and Other Drugs*, Publications Office of the European Union, Luxembourg, 2016, available at: http://www.espad.org/sites/espad.org/files/ESPAD_report_2015.pdf, accessed 28 December 2016.
15. *Drug Misuse: Findings from the 2016/17 Crime Survey for England and Wales*, ed. D. Broadfield, Statistical Bulletin 11/17, Home Office, London, 27 July 2016, available at: https://www.gov.uk/government/uploads/system/uploads/attachment_data/file/642738/drug-misuse-2017-hosb1117.pdf, accessed 12 September 2017.
16. NACDA & DHNI, *Prevalence of Drug Use and Gambling in Ireland and Drug Use in Northern Ireland*, National Advisory Committee on Drugs and Alcohol (NACDA) Dublin & Department of Health, Northern Ireland, 2016, available at: https://www.health-ni.gov.uk/sites/default/files/publications/health/all-ireland-dps-2014-15-bulletin-1.pdf, accessed 28 December 2016.
17. C. Black, L. Setterfield and L. Murray, *Scottish Schools Adolescent Lifestyle and Substance Use Survey (SALSUS) – Drug Use Report (2015)*, Scottish Government, Edinburgh, 2016, available at: http://www.gov.scot/Resource/0050/00508357.pdf, accessed 28 December 2016.
18. Ipsos MORI, *2013/14 Health Behaviour in School-aged Children (HBSC) Wales: Key Findings*, Welsh Government Social Research, Cardiff, 2015, available at: http://gov.wales/docs/caecd/research/2015/151022-health-behaviour-school-children-2013-14-key-findings-en.pdf, accessed 28 December 2016.
19. E. de Lacy, A. Fletcher, G. Hewitt, S. Murphy and G. Morre, Cross-sectional study examining the prevalence, correlates and sequencing of electronic cigarette and tobacco use among 11-16-year olds in schools in Wales, *BMJ Open*, 2017, **6**, e012784.
20. L. Orsolini, G. D. Papanti, G. Francesconi and F. Schifano, Mind navigators of chemicals' experimenters? A web-based description of e-psychonauts, *Cyberpsychol. Behav. Soc. Netw.*, 2015, **18**(5), 296–300.
21. J. M. Corkery, L. Orsolini, G. D. Papanti and F. Schifano, From concept(ion) to life after death/the grave: the 'natural' history and life-cycle(s) of Novel Psychoactive Substances (NPS), *Hum. Psychopharmacol.*, 2017, **32**(3), e2566.
22. J. M. Corkery, *Drug Seizure and Offender Statistics, United Kingdom, 1999*, Home Office Statistical Bulletin 5/01: Area Tables & Supplementary Tables, Home Office Research, Development and Statistics Directorate, London, 30 March 2001, https://web.archive.org/web/20030212224302/http://www.homeoffice.gov.uk/rds/pdfs/hosb501.pdf, https://web.archive.org/web/20030212224302/http://www.homeoffice.gov.uk/rds/pdfs/hosb501area.pdf, https://web.archive.org/web/20030212224302/http://www.homeoffice.gov.uk/rds/pdfs/hosb501supp.pdf, accessed 11 June 2017.
23. DrugScope, *Druglink Street Drug Prices 2006*, DrugScope, London, 2006, available at: http://www.drugwise.org.uk/wp-content/uploads/Factsheet-Street-drug-prices-2006.pdf, accessed 9 February 2017.
24. IDMU, *Drug Prices*, Independent Drug Monitoring Unit, Liverpool, 2017, available at: http://www.idmu.co.uk/prices.htm, accessed 9 February 2017.
25. Mixmag, The 2011 Mixmag drugs survey, *Mixmag*, 2011, **238**, 49–59.
26. H. Shapiro and M. Daly, *Highways and Buyways: A Snapshot of UK Drug Scenes 2016*, DrugWise, London, 2017, available at: http://www.drugwise.org.uk/wp-content/uploads/Highwaysandbyways.pdf, accessed 9 February 2017.

27. Scottish Government, *Drug Seizures and Offender Characteristics, 2014–15 and 2015–16*, 28 June 2017, available at: http://www.gov.scot/Resource/0052/00521660.pdf, the tables and charts, http://www.gov.scot/Resource/0052/00521661.xls, accessed 12 September 2017.
28. Scottish Government, *Drug Seizures by Police Scotland, 2013–14*, 26 October 2015, available at: http://www.gov.scot/Resource/0048/00487894.pdf, Tables, http://www.gov.scot/Resource/0048/00487971.xlsx, accessed 12 September 2017.
29. J. Hargreaves and K. Smith, *Seizures of Drugs in England and Wales, Year Ending 31 March 2016*, Home Office, London, November, 2016, available at: https://www.gov.uk/government/uploads/system/uploads/attachment_data/file/564412/seizures-drugs-hosb1316.pdf, https://www.gov.uk/government/uploads/system/uploads/attachment_data/file/564380/seizures-drugs-hosb1316-tables.ods, accessed 28 December 2016.
30. EMCDDA, *European Drug Report: Trends and Developments 2014*, Publications Office of the European Union, Luxembourg, 2014, DOI: 10.2810/32306, available at: http://www.emcdda.europa.eu/attachements.cfm/att_228272_EN_TDAT14001ENN.pdf, accessed 28 December 2016.
31. EMCDDA, *Action on New Drugs*, European Monitoring Centre for Drugs and Drug Addiction, Lisbon, 2015, available at: http://www.emcdda.europa.eu/activities/action-on-new-drugs, accessed 28 December 2016.
32. UNODC, *UNODC Early Warning Advisory on New Psychoactive Substances: Legal Responses*, United Nations Office on Drugs and Crime, Vienna, 2015, available at: https://www.unodc.org/LSS/Page/NPS/LegalResponses, accessed 28 December 2016.
33. P. V. Kavanagh and J. D. Power, New psychoactive substances legislation in Ireland - Perspectives from academia, *Drug Test. Anal.*, 2015, (7–8), 884–891.
34. UK Parliament, *Psychoactive Substances Act, C2 2016*, 2016, available at: http://www.legislation.gov.uk/ukpga/2016/2/pdfs/ukpga_20160002_en.pdf, accessed 28 December 2016.
35. R. M. Taylor, *Review of Drug Seizure and Offender Statistics*, National Statistics Quality Review Series, Report No 29, Home Office, London, 2004, available at: https://www.ons.gov.uk/ons/guide-method/method-quality/quality/nsqr/theme/crime-and-justice/nsqr-series-report-no-29--review-of-drug-seizure-and-offender-statistics.doc, accessed 9 February 2017.
36. S. Pudney, C. Badillo, M. Bryan, J. Burton, G. Conti and M. Iacovou, Estimating the size of the UK illicit drug market, in *Measuring Different Aspects of Problem Drug Use: Methodological Developments*, ed. N. Singleton, R. Murray and L. Tinsley, Home Office Online Report 16/06, Home Office, London, 2nd edn, 2006, available at: http://webarchive.nationalarchives.gov.uk/20110218135832/rds.homeoffice.gov.uk/rds/pdfs06/rdsolr1606.pdf, accessed 29 December 2016.
37. Matrix Knowledge Group, *The Illicit Drug Trade in the United Kingdom*, Home Office Online Report 2007, Home Office, London, 2nd edn, 2007, available at: http://webarchive.nationalarchives.gov.uk/20110220105210/rds.homeoffice.gov.uk/rds/pdfs07/rdsolr2007.pdf, accessed 28 December 2016.
38. EMCDDA, *Hospital Emergency Presentations and Acute Drug Toxicity in Europe: Update from the Euro-DEN Plus Research Group and the EMCDDA*, Publications Office of the European Union, Luxembourg, August, 2016, DOI: 10.2810/894142, available at: http://www.emcdda.europa.eu/system/files/publications/2973/TD0216713ENN-1_Final%20pdf.pdf, accessed 28 December 2016.

39. PHE, *Adult Substance Misuse Statistics from the National Drug Treatment Monitoring System (NDTMS): 1st April 2015 to 31st March 2016*, Public Health England, London, 2016, available at: http://www.nta.nhs.uk/uploads/adult-statistics-from-the-national-drug-treatment-monitoring-system-2015-2016.xls, accessed 28 December 2016.
40. PHE, *Young People's Statistics from the National Drug Treatment Monitoring System (NDTMS) 1 April 2015 to 31 March 2016*, 17 January 2017, available at: https://www.ndtms.net/Publications/downloads/Young%20People/young-people-statistics-from-the-national-drug-treatment-montitoring-system-2015-16.pdf, accessed 9 February 2017.
41. ISD, *Scottish Drug Misuse Database Dashboard*, ISD Scotland, Edinburgh, 2017, available at: http://www.isdscotland.org/Health-Topics/Drugs-and-Alcohol-Misuse/Publications/2017-04-04/SDMD_dashboard.swf?10:15:05, data extracted 12 September 2017.
42. ISD, *Scottish Drug Misuse Database Dashboard*, ISD Scotland, Edinburgh, 2016, available at: http://www.isdscotland.org/Health-Topics/Drugs-and-Alcohol-Misuse/Publications/2016-05-17/SDMD_dashboard.swf?13:13:01, data extracted 11 November 2016.
43. EMCDDA, *An Overview of the Drug-related Deaths and Mortality Among Drug Users (DRD) Key Indicator. January*, European Monitoring Centre for Drugs and Drug Addiction, Lisbon, January 2009, available at: http://www.emcdda.europa.eu/system/files/publications/565/EMCDDA-DRD-overview_127972.pdf, accessed 29 December 2016.
44. J. Corkery, UK drug-related mortality – issues in definition and classification, *Drugs Alcohol Today*, 2008, **8**(2), 17–25.
45. ONS, *Deaths Related to Drug Poisoning in England and Wales: 2016 Registrations*, Office for National Statistics, Newport Gwent, 2 August 2017, available at: https://www.ons.gov.uk/peoplepopulationandcommunity/birthsdeathsandmarriages/deaths/bulletins/deathsrelatedtodrugpoisoninginenglandandwales/2016registrations/pdf, https://www.ons.gov.uk/file?uri=/peoplepopulationandcommunity/birthsdeathsandmarriages/deaths/datasets/deathsrelatedtodrugpoisoningenglandandwalesreferencetable/current/deathsrelatedtodrugpoisoningenglandandwales.xls, accessed 6 August 2017.
46. NRS, *Drug-related Deaths in Scotland in 2016*, National Records of Scotland, Edinburgh, 2017, available at: https://www.nrscotland.gov.uk/statistics-and-data/statistics/statistics-by-theme/vital-events/deaths/drug-related-deaths-in-scotland/2016, https://www.nrscotland.gov.uk/statistics-and-data/statistics/statistics-by-theme/vital-events/deaths/drug-related-deaths-in-scotland/2016/list-of-tables-and-figures, accessed 16 August 2017.
47. NISRA, *Drug-related Deaths and Deaths Due to Drug Misuse Registered in Northern Ireland (2005–2015)*, Northern Ireland Statistics & Research Agency, Belfast, 2016, available at: https://www.nisra.gov.uk/sites/nisra.gov.uk/files/publications/Drug_Tables_15.xls, accessed 19 May 2017.
48. L. Orsolini, F. Francesconi, G. D. Papanti, A. Giorgetti and F. Schifano, Profiling the online recreational/prescription drugs' customers and overview of the drug vending virtual marketplaces, *Hum. Psychopharmacol.*, 2015, **30**(4), 302–318.
49. L. Orsolini, D. Papanti, J. M. Corkery and F. Schifano, An insight into the deep web; why it matters for addiction psychiatry? *Hum. Psychopharmacol.*, 2017, **32**(3), e2573.

50. L. Orsolini, D. Papanti, G. Francesconi and F. Schifano, Mind navigators of chemicals' experimenters? A web-based profiling of the e-psychonauts, *Cyberpsychol. Behav. Soc. Netw.*, 2015, **18**(5), 296–300.
51. M. Power, *Drugs 2.0. The Web Revolution That's Changing How the World Gets High*, Portobello Books Ltd, London, 2013, ISBN: 978-1846274596.
52. E. W. Boyer and J. D. Wines Jr, Impact of internet pharmacy regulation on opioid analgesic availability, *J. Stud. Alcohol Drugs*, 2008, **69**(5), 703–708.
53. M. J. Barratt, Silk Road: eBay for drugs, *Addiction*, 2012, **107**(3), 683.
54. M. K. Bergman, The deep web: surfacing hidden value, *J. Electron. Publishing*, 2001, **7**(1), DOI: 10.3998/3336451.0007.104.
55. F. Schifano, P. DeLuca, A. Baldacchino, T. Peltoniemi, N. Scherbaum, M. Torrens, M. Farre, I. Flores, M. Rossi, D. Eastwood, C. Guionnet, S. Rawaf, L. Agosti, L. Di Furia, R. Brigada, A. Majava, H. Siemann, M. Leoni, A. Tomasin, F. Rovetto and A. H. Ghodse, Drugs on the web; The Psychonaut 2002 EU project, *Prog. Neuropsychopharmacol. Biol. Psychiatry*, 2006, **30**(4), 640–646.
56. E. Jünger, *Annäherungen: Drogen und Rausch [Approaches: Drugs and intoxication]*, Verlag IL Kunst, Köln, NRW, Germany, 1970.
57. *The Internationalization of Ayahuasca*, ed. B. C. Labate and H. Jungaberle, Lit Verlag, Zurich, Switzerland, 2011, ISBN: 9783643901484.
58. P. J. Carroll, *Liber Null & Psychonaut: An Introduction to Chaos Magic*, Red Wheel/Weiser, LLC, York Beach, ME, 1987, ISBN: 978-0877286394.
59. P. Deluca, Z. Davey, O. Corazza, L. Di Furia, M. Farre, L. H. Flesland, M. Mannonen, A. Majava, T. Peltoniemi, M. Pasinetti, C. Pezzolesi, N. Scherbaum, H. Siemann, A. Skutle, M. Torrens, P. van der Jreeft, E. Iversen and F. Schifano, Identifying emerging trends in recreational drug use: outcomes from the Psychonaut Web Mapping Project, *Prog. Neuropsychopharmacol. Biol. Psychiatry*, 2012, **39**(2), 221–226.
60. L. Orsolini, P. St John-Smith, D. McQueen, D. Papanti, J. Corkery and F. Schifano, Evolutionary considerations on the emerging subculture of the e-psychonauts and the novel psychoactive substances: a comeback to the shamanism? *Curr. Neuropharmacol.*, 2017, **15**(5), 731–737.
61. Psychonauts.com, available at: http://www.psychonauts.com, accessed 27 December 2016.
62. Shroomery.org, available at: http://www.shroomery.org, accessed 27 December 2016.
63. Erowid.org, available at: http://www.erowid.org, accessed 27 December 2016.
64. Reddit.com, available at: http://www.reddit.com/r/onions/comments/22z3qe/grams_beta_version/, accessed 27 December 2016.
65. P. Gilbert, The origins and nature of compassion focused therapy, *Br. J. Clin. Psychol.*, 2014, **53**(1), 6–41.
66. J. L. Kent, *Psychedelic Information Theory. Shamanism in the Age of Reason*, PIT press/Supermassive, LLC, Seattle, WA, 2010, ISBN: 1453760172, available at: http://psychedelic-information-theory.com/pdf/PIT-Print-Web.pdf, accessed 11 June 2017.
67. M. Lerner and M. Lyvers, Values and beliefs of psychedelic drug users: a cross-cultural study, *J. Psychoact. Drugs*, 2006, **38**(2), 143–147.
68. K. McEwan, P. Gilbert, S. Dandeneau, S. Lipka, F. Maratos, K. B. Paterson and M. Baldwin, Facial expressions depicting compassionate and critical emotions: the development and validation of a new emotional face stimulus set, *PLoS One*, 2014, **9**(2), e88783.

69. I.-T. Kolassa, S. Kolassa, S. Bergmann, R. Lauche, S. Dilger, W. H. R. Miltner and F. Misual, Interpretive bias in social phobia: An ERP study with morphed emotional schematic faces, *Cognit. Emotion*, 2009, **23**(1), 69–95.
70. P. Ekman and W. V. Friesen, Constants across cultures in the face and emotion, *J. Pers. Soc. Psychol.*, 1971, **17**(2), 124–129.
71. A. Shulgin and A. Shulgin, *TiHKAL [Tryptamines I Have Known and Loved]: The Continuation*, Transform Press, Berkeley, CA, 1997, ISBN: 978-0963009692, available at: https://erowid.org/library/books_online/tihkal/tihkal.shtml, accessed 11 June 2017.
72. J. J. Palamar, S. S. Martins, M. K. Su and D. C. Ompad, Self-reported use of novel psychoactive substances in a US nationally representative survey: Prevalence, correlates, and a call for new survey methods to prevent underreporting, *Drug Alcohol Depend.*, 2015, **156**, 112–119.
73. A. Hoffmann, R. Heim, A. Brack, H. Kobel, A. Frey, H. Ott, T. Petrzilka and F. Troxler, Psilocybin und Psilocin, zwei psychotrope Wirkstoffe aus mexikanischen Rauschpilzen [Psilocybin and psilocin, two psychotropic drugs from Mexican molluscs], *Helv. Chim. Acta*, 1959, **42**(5), 1557–1572.
74. R. B. Guichhait, Biogenesis of 5-methoxy-*N*,*N*-dimethyltryptamine in human pineal gland, *J. Neurochem.*, 1976, **26**(1), 187–190.
75. S. A. Barker, J. A. Monti and S. T. Christian, *N*,*N*-dimethyltryptamine: an endogenous hallucinogen, *Int. Rev. Neurobiol.*, 1981, **22**, 83–110.
76. J. Kärkkäinen, M. Räisäinen, H. Naukarinen, J. Spoov and R. Rimon, Urinary excretion of free bufotenin by psychiatric patients, *Biol Psychiatry.*, 1988, **24**(4), 441–446.
77. G. Cimino and S. De Stefano, Chemistry of Mediterranean gorgonians. Simple indole derivatives from Paramuricea chamaeleon, *Comp. Biochem. Physiol. Ser. C. Comp. Pharmacol*, 1978, **61**(2), 361–362.
78. R. E. Schultes, Hallucinogenic plants: their earliest botanical descriptions, *J. Psychedelic Drugs*, 1979, **11**(1–2), 13–24.
79. Y. Koike, K. Wada, G. Kusano, S. Nozoe and K. Yokoyama, Isolation of Psilocybin from Psilocybe argentipes and its Determination in Specimens of some Mushrooms, *J. Nat. Prod.*, 1981, **44**(3), 362–365.
80. D. J. McKenna and G. H. N. Towers, Biochemistry and pharmacology of tryptamines and beta-carbolines. A minireview, *J. Psychoact. Drugs*, 1984, **16**(4), 347–358.
81. *Ayahuasca Analogues and Plant-based Tryptamines. The Best of the Entheogen Review 1992–1999*, ed. J. DeKorne, D. Aardvark and K. Trout, Sacramento, CA, 2nd edn, 2000, available at: https://www.erowid.org/library/books_online/ayahuasca_analogues_dekorne_complete.pdf, accessed 11 June 2017.
82. M. Collins, Some new psychoactive substances: precursor chemicals and synthesis-driven end-products, *Drug Test. Anal.*, 2011, **3**(7–8), 404–416.
83. J. L. Brandes, D. Kudrow, S. R. Stark, C. P. O'Carroll, J. U. Adelman, F. J. O'Donnell, W. J. Alexander, S. E. Spruill, P. S. Barrett and S. E. Lener, Sumatriptan-naproxen for acute treatment of migraine: a randomized trial, *JAMA*, 2007, **297**(13), 1443–1454.
84. UK Parliament, *Drugs Act 2005*, 2005, ch. 17, available at: http://www.legislation.gov.uk/ukpga/2005/17/pdfs/ukpga_20050017_en.pdf, accessed 11 June 2017.
85. B. Sanders, S. E. Lankenau, J. J. Bloom and D. Hathazi, "Research chemicals": tryptamine and phenethylamine use among high-risk youth, *Subst. Use Misuse*, 2008, **43**(3–4), 389–402.

86. F. Schifano, G. D. Papanti, L. Orsolini and J. M. Corkery, Novel psychoactive substances: the pharmacology of stimulants and hallucinogens, *Expert Rev. Clin. Pharmacol.*, 2016, **9**(7), 943–954.
87. A. W. Lessin, R. F. Long and M. W. Parkes, Central stimulant actions of alpha-alkyl substituted tryptamine in mice, *Br. J. Pharmacol. Chemother.*, 1965, **24**, 49–67.
88. *Novel Psychoactive Substances: Classification, Pharmacology and Toxicology*, ed. P. I. Dargan and D. M. Wood, Academic Press, London, 2013, ISBN: 978-0124158160.
89. W. E. Fantegrossi, A. C. Murnane and C. J. Reissig, The behavioural pharmacology of hallucinogens, *Biochem. Pharmacol.*, 2008, **75**(1), 17–33.
90. P. A. Pierce and S. J. Peroutka, Hallucinogenic drug interactions with neurotransmitter receptor binding sites in human cortex, *Psychopharmacology*, 1989, **97**(1), 118–122.
91. T. S. Ray, Psychedelics and the human receptorome, *PLoS One*, 2010, **5**(2), e9019, Erratum in: *PLoS One*, **5**(3).
92. D. E. Nichols, Hallucinogens, *Pharmacol. Ther.*, 2004, **101**(2), 131–181.
93. N. V. Cozzi, A. Gopalakrishnan, L. L. Anderson, J. T. Feih, A. T. Shulgin, P. F. Daley and A. E. Ruoho, Dimethyltryptamine and other hallucinogenic tryptamines exhibit substrate behavior at the serotonin uptake transporter and the vesicle monoamine transporter, *J. Neural. Transm. (Vienna)*, 2009, **16**(12), 1591–1599.
94. D. Fontanilla, M. Johannessen, A. R. Hajipour, N. V. Cozzi, M. B. Jackson and A. E. Ruoho, The hallucinogen *N,N*-dimethyltryptamine (DMT) is an endogenous sigma-1 receptor regulator, *Science*, 2009, **323**(5916), 934–937.
95. T. P. Su, T. Hayashi and D. B. Vaupel, When the endogenous hallucinogenic trace amine *N,N*-dimethyltryptamine meets the sigma-1 receptor, *Sci. Signal.*, 2009, **2**(61), pe12.
96. C. Sogawa, N. Sogawa, J. Tagawa, A. Fujino, K. Ohyama, M. Asanuma, M. Funada and S. Kitayama, 5-Methoxy-*N,N*-diisopropyltryptamine (Foxy), a selective and high affinity inhibitor of serotonin transporter, *Toxicol. Lett.*, 2007, **170**(1), 75–82.
97. J. Corkery, H. Claridge, B. Loi, C. Goodair and F. Schifano, *Drug-related Deaths in the UK: Annual Report 2013, Drug-related deaths reported by Coroners in England, Wales, Northern Ireland, Guernsey, Jersey and the Isle of Man*, Police forces in Scotland; & the Northern Ireland Statistics and Research Agency – Annual Report January-December 2012, 12 February, London, International Centre for Drug Policy, St George's University of London, 2014, ISBN: 978-1-897778-9-2, available at: http://www.sgul.ac.uk/images/docs/idcp%20pdfs/National%20programme%20on%20substance%20abuse%20deaths/National_Programme_on_Substance_Abuse_Deaths-Annual_Report_2013_on_Drug-related_Deaths_in_the_UK_January-December_2012_PDF.pdf, accessed 11 June 2017.
98. Bluelight.com, available at: www.bluelight.com, accessed 5 May 2017.
99. A. Shulgin and A. Shulgin, *PiHKAL [Phenethylamines I Have Known and Loved]: A Chemical Love Story*, Transform Press, Berkeley, CA, 1995, ISBN: 978-0963009609, available at: https://erowid.org/library/books_online/pihkal/pihkal.shtml, accessed 11 June 2017.
100. J. L. McLaughlin, Peyote: an introduction, *Lloydia*, 1973, **36**(1), 1–8.

101. F. Nagai, R. Nonaka and K. Satoh Hisashi Kamimura, The effects of non-medically used psychoactive drugs on monoamine neurotransmission in rat brain, *Eur. J. Pharmacol.*, 2007, **559**(2–3), 132–137.
102. O. Corazza, F. Schifano, M. Farré, P. Deluca, Z. Davey, M. Torrens, Z. Demetrovics, L. Di Furia, L. Flesland, H. Siemann, A. Skutle, P. Van Der Kreeft and N. Scherbaum, Designer drugs on the internet: a phenomenon out-of-control? the emergence of hallucinogenic drug Bromo-Dragonfly, *Curr. Clin. Pharmacol.*, 2011, **6**(2), 125–129.
103. R. E. O'Connor and J. J. Keating, Characterization of synthetic routes to 'Bromo-DragonFLY' and benzodifuranyl isopropylamine homologues utilizing ketone intermediates. Part 1: synthesis of ketone precursors, *Drug Test. Anal.*, 2014, **6**(7–8), 658–667.
104. B. V. Dean, S. J. Stellpflug, A. M. Burnett and K. M. Engebretsen, 2C or not 2C: phenethylamine designer drug review, *J. Med. Toxicol.*, 2013, **9**(2), 172–178.
105. A. Rickli, D. Luethi, J. Reinisch, D. Buhcy, M. C. Hoener and M. E. Liechti, Receptor interaction profiles of novel *N*-2-methoxybenzyl (NBOMe) derivatives of 2,5-dimethoxy-substituted phenethylamines (2C drugs), *Neuropharmacology*, 2015, **99**, 546–553.
106. M. E. Nelson, S. M. Bryant and S. E. Aks, Emerging drugs of abuse, *Emerg. Med. Clin. North Am.*, 2014, **32**(1), 1–28.
107. M. R. Braden, J. C. Parrish, J. C. Naylor and D. E. Nichols, Molecular interaction of serotonin 5-HT2A receptor residues Phe339(6.51) and Phe340(6.52) with superpotent *N*-benzyl phenethylamine agonists, *Mol. Pharmacol.*, 2006, **70**(6), 1956–1964.
108. A. L. Halberstadt and M. A. Geyer, Multiple receptors contribute to the behavioral effects of indoleamine hallucinogens, *Neuropharmacology*, 2011, **61**(3), 364–381.
109. F. S. Bersani, O. Corazza, G. Albano, G. Valeriani, R. Santacroce, F. Bolzan Mariotti Posocco, E. Cinosi, P. Simonato, G. Martinotti, G. Bersani and F. Schifano, 25C-NBOMe: preliminary data on pharmacology, psychoactive effects, and toxicity of a new potent and dangerous hallucinogenic drug, *Biomed. Res. Int.*, 2014, **2014**, 734749.
110. W. Lawn, M. Barratt, M. Williams, A. Horne and A. Winstock, The NBOMe hallucinogenic drug series: Patterns of use, characteristics of users and self-reported effects in a large international sample, *J. Psychopharmacol.*, 2014, **28**(8), 780–788.
111. D. M. Wood, R. Sedefov, A. Cunningham and P. I. Dargan, Prevalence of use and acute toxicity associated with the use of NBOMe drugs, *Clin. Toxicol. (Phila.)*, 2015, **53**(2), 85–92.
112. S. Srisuma, A. C. Bronstein and C. O. Hoyte, NBOMe and 2C substitute phenylethylamine exposures reported to the National Poison Data System, *Clin. Toxicol. (Phila.)*, 2015, **53**(7), 624–628.
113. K. G. Shanks, T. Sozio and G. S. Behonick, Fatal Intoxications with 25B-NBOMe and 25I-NBOMe in Indiana during 2014, *J. Anal. Toxicol.*, 2015, **39**(8), 602–606.
114. J. Suzuki, M. A. Dekker, E. S. Valenti, F. A. Arbelo Cruz, A. M. Correa, J. L. Poklis and A. Poklis, Toxicities associated with NBOMe ingestion-a novel class of potent hallucinogens: a review of the literature, *Psychosomatics*, 2015, **56**(2), 129–139.
115. V. B. Kueppers and C. T. Cooke, 25I-NBOMe related death in Australia: a case report, *Forensic Sci. Int.*, 2015, **249**, e15–e18.
116. J. J. Nugteren-van Lonkhuyzen, A. J. van Riel, T. M. Brunt and L. Hondebrink, Pharmacokinetics, pharmacodynamics and toxicology of new psychoactive

substances (NPS): 2C-B, 4-fluoroamphetamine and benzofurans, *Drug Alcohol Depend*, 2015, **157**, 18–27, Erratum in: *Drug Alcohol. Depend,* 2016 Feb 1, **159**, 287–288.

117. J. Jebadurai, F. Schifano and P. Deluca, Recreational use of 1-(2-naphthyl)-2-(1-pyrrolidinyl)-1-pentanone hydrochloride (NRG-1), 6-(2-aminopropyl) benzofuran (benzofury/6-APB) and NRG-2 with review of available evidence-based literature, *Hum. Psychopharmacol.*, 2013, **28**(4), 356–364.
118. C. G. Clemente, S. Chiappini, H. Claridge, C. Goodair and B. Loi, *Deaths Involving Benzo Fury, United Kingdom 2011–2012*, International Centre for Drug Policy, University of London, St George's, 2013, Poster presentation, available at: http://mededlive.co.uk/nps_social/2013/09/deaths-involving-benzo-fury-united-kingdom-2011-2012/, accessed 11 June 2017.
119. S. Elliott and J. Evans, A 3-year review of new psychoactive substances in casework, *Forensic Sci. Int.*, 2014, **243**, 55–60.
120. G. W. Fulde and S. L. Forster, The impact of amphetamine-type stimulants on emergency services, *Curr. Opin. Psychiatry.*, 2015, **28**(4), 275–279.
121. N. Stojanovska, T. Kelly, M. Tahtouh, A. Beavis and S. Fu, Analysis of amphetamine-type substances and piperazine analogues using desorption electrospray ionisation mass spectrometry, *Rapid Commun. Mass Spectrom.*, 2014, **28**(7), 731–740.
122. Y. Lurie, A. Gopher, O. Lavon, S. Almog, L. Sulimani and Y. Bentur, Severe paramethoxymethamphetamine (PMMA) and paramethoxyamphetamine (PMA) outbreak in Israel, *Clin. Toxicol. (Phila.)*, 2012, **50**(1), 39–43.
123. M. J. Barratt, M. Allen and S. Lenton, "PMA sounds fun": negotiating drug discourses online, *Subst. Use Misuse*, 2014, **49**(8), 987–998.
124. H. H. Keller, R. Schaffner, M. O. Carruba, W. P. Burkard, M. Pieri, E. P. Bonetti, R. Scherschlicht, M. Da Prada and W. E. Haefely, Diclofensine (Ro 8-4650) - a potent inhibitor of monoamine uptake: biochemical and behavioural effects in comparison with nomifensine, *Adv. Biochem. Psychopharmacol.*, 1982, **31**, 249–263.
125. L. M. Omer, Pilot trials with diclofensine, a new psychoactive drug in depressed patients, *Int. J. Clin. Pharmacol. Ther. Toxicol.*, 1982, **20**(7), 320–326.
126. S. Gasić, A. Korn and H. G. Eichler, Effect of diclofensine, a novel antidepressant, on peripheral adrenergic function, *Clin. Pharmacol. Ther.*, 1986, **39**(5), 582–585.
127. G. Di Renzo, S. Amoroso, M. Taglialatela, L. M. Canzoniero, P. Maida, G. Lombardi and L. Annunziato, Pure uptake blockers of dopamine can reduce prolactin secretion: studies with diclofensine, *Life Sci.*, 1988, **42**(21), 2161–2169.
128. J. Culig, R. S. Ehsanullah, C. Hallett, A. Iliopoulou, I. Matheson and P. Turner, A clinical pharmacological comparison of diclofensine (Ro 8-4650) with nomifensine and amitriptyline in normal human volunteers, *Br. J. Clin. Pharmacol.*, 1983, **15**(5), 537–543.
129. S. H. Jackson, P. Turner and R. S. Ehsanullah, A comparison of diclofensine and desmethylimipramine using tyramine pressor tests in normal subjects, *Br. J. Clin. Pharmacol.*, 1983, **16**(4), 427–429.
130. R. J. Lamb and R. R. Griffiths, Self-administration in baboons and the discriminative stimulus effects in rats of bupropion, nomifensine, diclofensine and imipramine, *Psychopharmacology (Berl.)*, 1990, **102**(2), 183–190.
131. N. Nakachi, Y. Kiuchi, M. Inagaki, M. Inazu, Y. Yamazaki and K. Oguchi, Effects of various dopamine uptake inhibitors on striatal extracellular dopamine levels and behaviours in rats, *Eur. J. Pharmacol.*, 1995, **281**(2), 195–203.

132. F. F. Blicke and J. H. Burckhalter, α-Thienylaminoalkanes, *J. Am. Chem. Soc.*, 1942, **64**(3), 477.
133. D. Angelov, J. O'Brien and P. Kavanagh, The syntheses of 1-(2-thienyl)-2-(methylamino) propane (methiopropamine) and its 3-thienyl isomer for use as reference standards, *Drug Test. Anal.*, 2013, **5**(3), 145–149.
134. E. D. Bouso, E. A. Gardner, J. E. O'Brien, B. Talbot and P. V. Kavanagh, Characterization of the pyrolysis products of methiopropamine, *Drug Test. Anal.*, 2014, **6**(7–8), 676–683.
135. L. Iversen, S. Gibbons, R. Treble, V. Setola, X. P. Huang and B. L. Roth, Neurochemical profiles of some novel psychoactive substances, *Eur. J. Pharmacol.*, 2013, **700**(1–3), 147–151.
136. EMCDDA-Europol, *EMCDDA–Europol Joint Report on a New Psychoactive Substance: 4,4′-DMAR (4-methyl-5-(4-methylphenyl)-4,5-dihydrooxazol-2-amine)*, Publications Office of the European Union, Luxembourg, July 2014, DOI: 10.2810/51897, ISBN 978-92-9168-743-5, available at: http://www.emcdda.europa.eu/system/files/publications/812/TDAS14006ENN_476050.pdf, accessed 8 June 2017.
137. EMCDDA, *Report on the Risk Assessment of 4,4′-DMAR in the Framework of the Council Decision on New Psychoactive Substances*, Publications Office of the European Union, Luxembourg, October 2015, DOI: 10.2810/58185, ISBN 978-92-9168-751-0, available at: http://www.emcdda.europa.eu/system/files/publications/1864/TDAK14005ENN.pdf, accessed 8 June 2017.
138. F. T. Davis and M. E. Brewster, A fatality involving U4Euh, a cyclic derivative of phenylpropanolamine, *J. Forensic Sci.*, 1988, **33**(2), 549–553.
139. M. E. Brewster and F. T. Davis, Appearance of Aminorex as a designer analog of 4-methylaminorex, *J. Forensic Sci.*, 1991, **36**(2), 587–592.
140. S. D. Brandt, M. H. Baumann, J. S. Partilla, P. V. Kavanagh, J. D. Power, B. Talbot, B. Twamley, O. Mahony, J. O'Brien, S. P. Elliott, R. P. Archer, J. Patrick, K. Singh, N. M. Dempster and S. H. Cosbey, Characterization of a novel and potentially lethal designer drug (±)-cis-para-methyl-4-methylaminorex (4,4′-DMAR, or 'Serotoni'), *Drug Test. Anal.*, 2014, **6**(7–8), 684–695.
141. Chemrus.com, available at: www.chemrus.com, accessed 5 May 2017.
142. Drugs-forum.com, available at: www.drugs-forum.com, accessed 5 May 2017.
143. Serotinin.info, available at: www.serotonin.info, accessed 28 September 2014.
144. B. Loi, M. Zloh, M. A. De Luca, N. Pintori, J. M. Corkery and F. Schifano, 4,4′-dimethylaminorex ('4,4'-DMAR'; 'Serotoni') misuse; a web-based study, *Hum. Psychopharmacol*, 2017, **32**(30), e2575.
145. F. Schifano, L. Orsolini, D. Papanti and J. Corkery, NPS: Medical Consequences Associated with Their Intake, *Curr. Top. Behav. Neurosci.*, 2017, **32**, 351–380.
146. A. Winstock, M. Lynskey, R. Borschmann and J. Waldron, Risk of emergency medical treatment following consumption of cannabis or synthetic cannabinoids in a large global sample, *J. Psychopharmacol.*, 2015, **29**(6), 698–703.
147. S. Naviglio, D. Papanti, V. Moressa and A. Ventura, An adolescent with an altered state of mind, *BMJ*, 2015, **350**, h299.
148. EMCDDA, *Injection of Synthetic Cathinones. Perspectives on Drugs*, European Monitoring for Drugs and Drug Addiction, Lisbon, 28 May 2015, available at: http://www.emcdda.europa.eu/system/files/publications/2754/Synthetic%20 cathinones_updated2015.pdf, accessed 11 June 2017.
149. A. Bourne, D. Reid, F. Hickson, S. Torres Rueda and P. Weatherburn, *The Chemsex Study: Drug Use in Sexual Settings Among Gay and Bisexual Men in*

Lambeth, Southwark & Lewisham, Sigma Research, London School of Hygiene & Tropical Medicine, London, 2014, available at: http://researchonline.lshtm.ac.uk/2197245/1/report2014a.pdf, accessed 26 June 2016.

150. D. K. Tracy, D. M. Wood and D. Baumeister, Novel psychoactive substances: types, mechanisms of action, and effects, *BMJ*, 2017, **356**, i6848.
151. A. Schneir, I. G. Metushi, C. Sloane, D. J. Benaron and R. L. Fitzgerald, Near death from a novel synthetic opioid labeled U-47700: emergence of a new opioid class, *Clin. Toxicol (Phila.)*, 2017, **55**(1), 51–54.
152. EMCDDA, *Report on the Risk Assessment of 3,4-dichloro-n-{[1-(dimethylamino) cyclohexyl]methyl}benzamide (AH-7921) in the Framework of the Council Decision on New Psychoactive Substances*, Luxembourg: Publications Office of the European Union, May 2014, DOI: 10.2810/5691, ISBN 978-92-9168-746-6, available at: http://www.emcdda.europa.eu/system/files/publications/774/TDAK-14002ENN_480892.pdf, accessed 11 June 2017.
153. EMCDDA, *Report on the Risk Assessment of MT-45 in the Framework of the Council Decision on New Psychoactive Substances*, Publications Office of the European Union, Luxembourg, October 2015, DOI: 10.2810/59253, ISBN: 978-92-9168-752-7, available at: http://emcdda.europa.eu/system/files/publications/1865/TDAK14006ENN.pdf, accessed 11 June 2017.
154. EMCDDA-Europol, *EMCDDA–Europol Joint Report on a New Psychoactive Substance: N-(1-phenethylpiperidin-4-yl)-N-phenylacrylamide (Acryloylfentanyl)*, Publications Office of the European Union, Luxembourg, February 2017, DOI: 10.2810/87713, ISBN 978-92-9497-044-2, available at: http://emcdda.europa.eu/system/files/publications/3873/TI_PUBPDF_TDAS17001ENN_PDF-WEB_20170221105322.pdf, accessed 11 June 2017.
155. J. Mounteney, I. Giraudon, G. Denissov and P. Griffiths, Fentanyls: Are we missing the signs? Highly potent and on the rise in Europe, *Int. J. Drug Policy*, 2015, **26**(7), 626–631.
156. R. A. Rudd, P. Seth, F. David and L. Scholl, Increases in Drug and Opioid-Involved Overdose Deaths - United States, 2010-2015, *MMWR. Morb. Mortal. Wkly Rep.*, 2016, **65**(5051), 1445–1452.
157. S. Davies, D. M. Wood, G. Smith, J. Button, J. Ramsey, R. Archer, D. W. Holt and P. I. Dargan, Purchasing 'legal highs' on the Internet–is there consistency in what you get? *QJM*, 2010, **103**(7), 489–493.
158. J. Ramsey, P. I. Dargan, M. Smyllie, S. Davies, J. Button, D. W. Holt and D. M. Wood, Buying 'legal' recreational drugs does not mean that you are not breaking the law, *QJM*, 2010, **103**(10), 777–783.
159. Home Office, *Annual Report on the Home Office Forensic Early Warning System (FEWS) - a System to Identify New Psychoactive Substances (NPS) in the UK*, September 2015, available at: https://www.gov.uk/government/uploads/system/uploads/attachment_data/file/461333/1280_EL_FEWS_Annual_Report_2015_WEB.pdf, accessed 12 February 2017.
160. A. Wurita, K. Hasegawa, K. Minakata, K. Watanabe and O. Suzuki, A large amount of new designer drug diphenidine coexisting with a synthetic cannabinoid 5-fluoro-AB-PINACA found in a dubious herbal product, *Forensic Toxicol.*, 2014, **32**(2), 331–337.
161. V. Shevyrin, V. Melkpzerov, G. W. Endres, Y. Shafran and Y. Morzherin, On a new cannabinoid classification system: a sight on the illegal market of novel psychoactive substances, *Cannabis Cannabinoid Res.*, 2016, **1**(1), 186–194.

162. G. Appendino, M. G. Cascio, S. Bacchiega, A. S. Moriello, A. Minassi, A. Thomas, R. Ross, R. Pertwee, L. De Petrocellis and V. Di Marzo, First "hybrid" ligands of vanilloid TRPV1 and cannabinoid CB2 receptors and non-polyunsaturated fatty acid-derived CB2-selective ligands, *FEBS Lett.*, 2006, **580**(2), 568–574.
163. M. F. Weaver, J. A. Hopper and E. W. Gunderson, Designer drugs 2015: assessment and management, *Addict. Sci. Clin. Pract.*, 2015, **10**, 8.
164. B. P. Kersten and M. E. McLaughlin, Toxicology and management of novel psychoactive drugs, *J. Pharm. Pract.*, 2015, **28**(1), 50–65.
165. L. Velez and F. Benitez, *New Troublesome Drugs: The Synthetics*, 2015, available at: http://www.emdocs.net/new-troublesome-drugs-the-synthetics/, accessed 11 June 2017.
166. C. D. Smith and S. Robert, 'Designer drugs': update on the management of novel psychoactive substance misuse in the acute care setting, *Clin. Med. (Lond.)*, 2014, **14**(4), 409–415.
167. M. E. Musselman and J. P. Hampton, "Not for human consumption": a review of emerging designer drugs, *Pharmacotherapy*, 2014, **34**(7), 745–757.
168. M. W. Manseau, in *Synthetic Cannabinoids Emergence, Epidemiology, Clinical Effects, and Management*, ed. M. T. Compton, Marijuana and Mental Health, American Psychiatric Association Publishing, Arlington, VA, 2016, pp. 149–169, ISBN: 978-1-61537-008-5.
169. D. Papanti, F. Schifano, G. Botteon, F. Bertossi, J. Mannix, D. Vidoni, M. Impagnatiello, E. Pascolo-Fabrici and T. Bonavigo, "Spiceophrenia": a systematic overview of "spice"-related psychopathological issues and a case report, *Hum. Psychopharmacol.*, 2013, **28**(4), 379–389.
170. R. J. Tait, D. Caldicott, D. Mountain, S. L. Hill and S. Lenton, A systematic review of adverse events arising from the use of synthetic cannabinoids and their associated treatment, *Clin. Toxicol. (Phila.)*, 2016, **54**(1), 1–13.
171. A. A. Monte, A. C. Bronstein, D. J. Cao, K. J. Heard, J. A. Hoppe, C. O. Hoyte, J. L. Iwanicki and E. J. Lavonas, An outbreak of exposure to a novel synthetic cannabinoid, *N. Engl. J. Med.*, 2014, **370**(4), 389–390.
172. L. San, B. Arranz and J. Martinez-Raga, Antipsychotic drug treatment of schizophrenic patients with substance abuse disorders, *Eur. Addict. Res.*, 2007, **13**(4), 230–243.
173. G. Juckel, F. Schlagenhauf, M. Koslowski, D. Filonov, T. Wüstenberg, A. Villringer, B. Knutson, T. Kienast, J. Gallinat, J. Wrase and A. Heinz, Dysfunction of ventral striatal reward prediction in schizophrenic patients treated with typical, not atypical, neuroleptics, *Psychopharmacology (Berl.)*, 2006, **187**(2), 222–228.
174. G. Valeriani, O. Corazza, F. S. Bersani, C. Melcore, A. Metastasio, G. Bersani and F. Schifano, Olanzapine as the ideal "trip terminator"? Analysis of online reports relating to antipsychotics' use and misuse following occurrence of novel psychoactive substance-related psychotic symptoms, *Hum. Psychopharmacol.*, 2015, **30**(4), 249–254.
175. U. K. Parliament, *The Misuse of Drugs Act 1971 (Amendment) Order 2017, Statutory Instruments 2017 No 634*, 3 May 2017, available at: http://www.legislation.gov.uk/uksi/2017/634/introduction/made?view=plain, accessed 11 June 2017.
176. N. Thurtle, R. Abouchedid, J. R. Archer, J. Ho, T. Yamamoto, P. I. Dargan and D. M. Wood, Prevalence of Use of Electronic Nicotine Delivery Systems (ENDS) to Vape Recreational Drugs by Club Patrons in South London, *J. Med. Toxicol.*, 2017, **13**(1), 61–65.

177. I. Jones, High on vapour, *New Sci.*, 2016, **232**(3104–3106), 21.
178. J. Mounteney, M. Ruiz, A. Bo, T. Groshkova, J. Matias, L. Vandam, F. Cardoso, R. Christie, L. Montanari, I. Giraudon and P. Griffiths, *High-risk Drug Use and New Psychoactive Substances, EMCDDA Rapid Communication*, Luxembourg, Publications Office of the European Union, June 2017, DOI: 10.2810/807363, ISBN 978-92-9497-118-0, available at: http://www.emcdda.europa.eu/system/files/publications/4540/TD0217575ENN.pdf, accessed 8 June 2017.

Chapter 9

Raman Spectroscopy for the Analysis of Novel Psychoactive Substances (NPS)

Jesus Calvo-Castro, Amira Guirguis, Mire Zloh and Jacqueline L. Stair*

University of Hertfordshire, Department of Pharmacy, Pharmacology and Postgraduate Medicine, College Lane, Hatfield, AL10 9AB, UK

*E-mail: j.stair@herts.ac.uk

Table of Contents

Light in Forensic Science: Issues and Applications
Edited by Giorgia Miolo, Jacqueline L. Stair and Mire Zloh

Published by the Royal Society of Chemistry, www.rsc.org

9.1. Introduction

In the last decade, the use of Raman spectroscopy has seen a surge of interest regarding its application in the detection/identification of novel psychoactive substances (NPS), particularly in relation to its forensic applications.[1–9] Raman spectroscopy exploits the 'Raman effect', which is the inelastic scattering of *light* (*i.e.*, radiation) from matter. The light irradiated on a sample and scattered with a resultant change in energy can be used to decipher vibrational transitions in molecules that may be used advantageously to identify chemical functional groups. Thus, monitoring NPS *via* vibrational spectroscopy with potential to decipher slight chemical modifications that are associated with NPS proliferation is very desirable. Raman spectroscopy affords the ability to analyse samples through packaging and with reduced interferences from cutting agents, which will be discussed in later sections of this chapter. In addition, approaches to enhance the Raman signal from NPS will be discussed with reference to crime scene investigations. This section intends to give an overview of the principles governing Raman spectroscopy, as well as initially a comparison with other vibrational spectroscopic techniques used for NPS detection in forensic applications, the less experienced reader and accordingly, those seeking a more in-depth treatment are referred to appropriate textbooks.[10–14]

In general, vibrational spectroscopic techniques such as Raman, Infrared (IR), and Near Infrared (NIR) spectroscopy detect changes in molecules which are associated to nuclear motions following interaction with *light*. This interaction between radiation (*light*) and matter can result in the former either being absorbed by the material, transmitted or scattered. If light is absorbed, the molecules of the material are promoted to higher energy states, so-called excited states. Therefore, absorption is a process that can only occur when the energy of the incident radiation equates to that of the energy gap between the energy levels. In the case of promotion to a higher vibrational energy state *via* absorption, the light used is infrared radiation. In addition, the vibration must occur with a change in dipole moment. Specifically, these vibrational transitions associated with NIR and IR spectroscopy pertain to radiation in the ranges of 0.78–2.5 mm (*i.e.*, 12 800–4000 cm^{-1}) and 2.5–50 mm (*i.e.*, 4000–200 cm^{-1}) respectively. Although these techniques have been used for the identification and characterisation of NPS,[8] they have limitations. In particular, polar functional groups often associated with cutting agents such as C–O and O–H are strong infrared absorbers which can mask a response from the NPS of interest. Additionally, it is often of interest to analyse a sample in the packaging such as glass or plastic, which would not be transparent to IR spectroscopy. In the case of NIR spectroscopy, the resulting spectrum is dominated by overtone and combination vibrational bands, which makes interpretation of the functional groups present difficult, as well as the spectrum being influenced by the physical properties of the powder analysed. In the context of screening and identification of NPS, Raman spectroscopy offers a number of advantages over IR and NIR for in-field testing such as non-contact; non-destructive analysis through packaging;[15,16] and low sensitivity to cutting agents,[8,17] moisture[14,15] and physical properties.[18] A direct comparison for the identification of NPS related substances using handheld IR, NIR, and Raman spectroscopy also highlighted these differences.[8]

Raman spectroscopy, as opposed to IR and NIR spectroscopy which are absorption based techniques, utilises the scattered *light* produced following the distortion of the electron cloud (following nuclear motion) upon interaction between the incident radiation and the material. Accordingly, in Raman the incident radiation, which is subsequently scattered, does not have to match that of the energy gap between the vibrational energy states. This results in the molecules, following irradiation, being promoted to higher energy virtual states as opposed to the case outlined for IR spectroscopy. The energy of these virtual states is therefore determined by the frequency of the incident radiation (*e.g.*, 785 or 1064 nm). The phenomenon of Raman scattering can be considered an inelastic process, where the energy of the incident photon and that of the scattered one differ. It is of note that in the ample majority of Raman work carried out nowadays the energy of the scattered photon is lower than for the incident one, a process known as Stokes scattering. As the thermal energy increases, the population of upper vibrational energy levels of the ground state increases, the energy of the scattered photon can be greater than that of the incident one and the anti-Stokes scattering phenomenon can be observed. Figure 9.1 shows the energy transitions to the virtual state that occur with Stokes and anti-Stokes Raman scattering as well as with Rayleigh scattering (*i.e.*, the elastic process).

Due to the intrinsically low probability of the Raman scattering phenomenon (only one in *ca.* 10^6–10^8 photons are Raman scattered), high energy monochromatic light sources such as lasers are employed in the irradiation of the materials subject to analysis, given that the probability of scattered photons increases with the fourth power of the excitation energy.[10] However, higher energy/shorter wavelength excitation sources can lead to

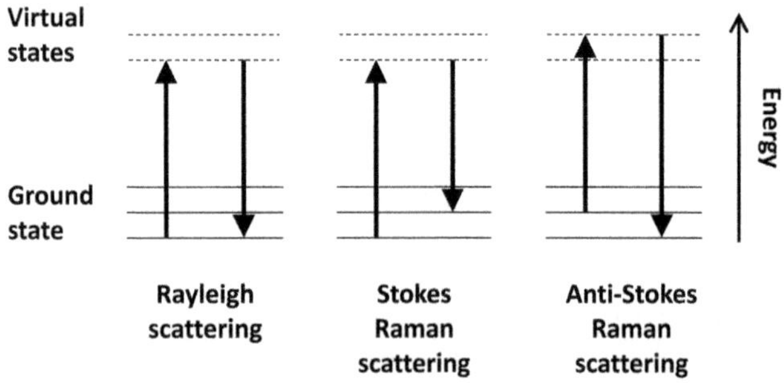

Figure 9.1. An illustration of Rayleigh (elastic) and Stokes and anti-Stokes Raman (inelastic) scattering.

issues such as sample degradation and fluorescence. In particular, fluorescence is a phenomenon characterised by an intrinsically higher probability than Raman scattering, hence often resulting in a hindering of Raman signals. At the expense of decreasing the efficiency of the Raman scattering process, lower energy/longer wavelength laser sources are employed to overcome these issues. These lower energy lasers have also been implemented in handheld Raman spectrometers, with 784 and 1064 nm lasers now being readily available,[8] thus promoting their uses for the in-field application.

9.2. General Characterisation of NPS Using Raman Spectroscopy

The applicability of Raman spectroscopy for structural identification is afforded by means of scattered *light* resulting from polarisability changes ascribed to particular nuclear motions upon sample irradiation. This gives rise to a characteristic spectrum or fingerprint for each substance. With the exception of a few substances (*e.g.*, GHB), NPS samples are highly conjugated and theoretically strong Raman scatterers which facilitates the employment of low energy excitation sources (*vide supra*). In fact, functional groups which are ubiquitous in NPS are characterised by medium to high intensity Raman signals[10,12] such as aromatic ring quadrant stretching (*ca.* 1600 cm^{-1}), carbonyl stretching (*ca.* 1700 cm^{-1}), and C–N stretching (*ca.* 1380–1260 cm^{-1}). As such, the use of Raman spectroscopy has shown success in the challenging field of NPS chemical detection,[1–9] characterised by an increasingly large number of structurally diverse substances, yet exhibiting key structural similarities. Herein, the authors would like to bring to the attention of the reader a number of recent publications dedicated to the identification/detection of NPS by means of Raman spectroscopy (*i.e.*, *via* the scattering of *light*).

Stewart and co-workers reported on the examination of reference and seized samples of cathinone derivatives by means of Raman spectroscopy employing a 785 nm laser source.[4] In particular, this study evaluated the β-ketophenethylamine (β-KP) family of synthetic cathinones with the systematic structural changes on a common moiety resulting in distinct Raman spectral profiles as illustrated in Figure 9.2.

Clear spectral differences can be observed on comparing the mono-substituted (*N*-ethylcathinone), methyl substituted (4-MEC) and methylenedioxy (βK-MDMA) substituted cathinones with all these derivatives exhibiting high intensity bands at positions which are consistent with those of their aromatic model compounds, such as toluene. All the investigated compounds by Stewart *et al.* are characterised by strong bands at *ca.* 1600–1700 cm^{-1}, with the one at higher frequency (*i.e.*, higher Raman shift, cm^{-1}) associated to the carbonyl stretch. In turn, the shoulder on the second of these two high intensity bands is reported to play an

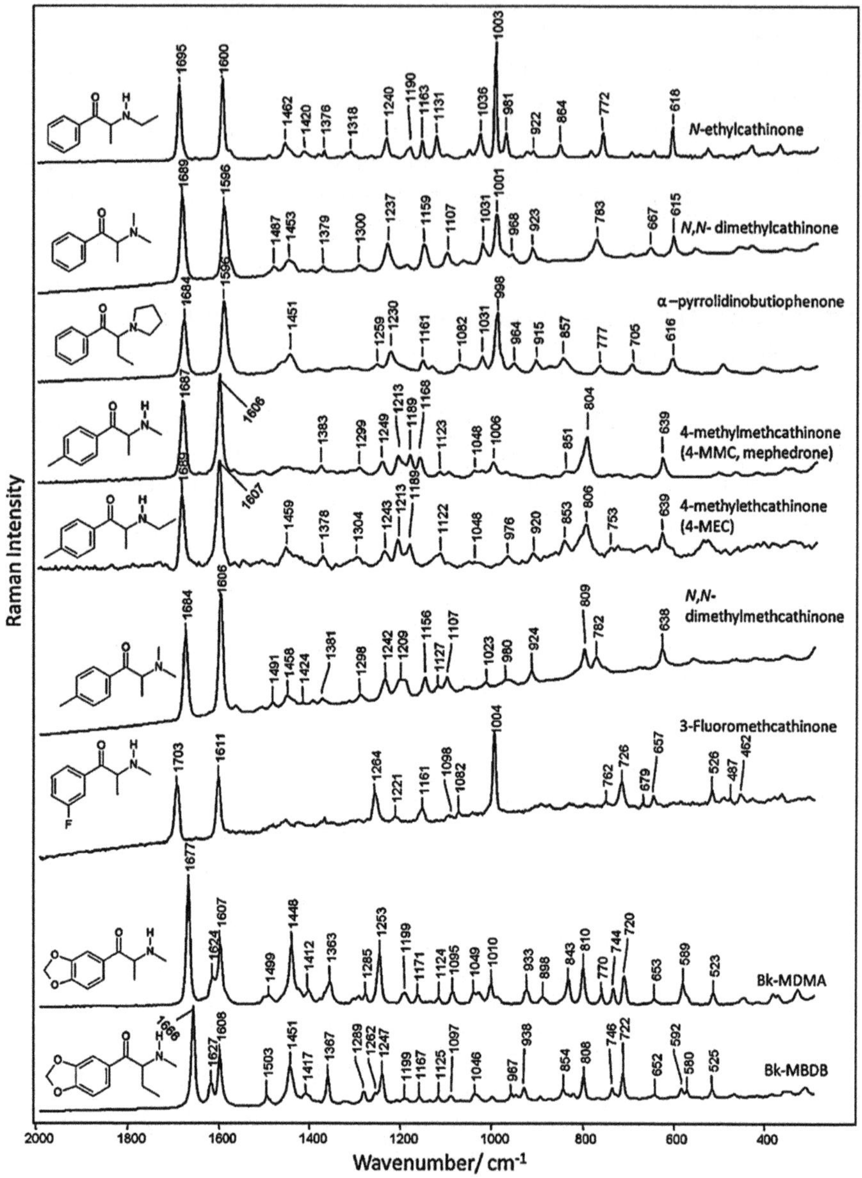

Figure 9.2. Chemical structures and associated Raman spectra of a representative series of β-KPs. Reprinted from *Analytica Chimica Acta*, **711**, S. P. Stewart, S. E. J. Bell, N. C. Fletcher, S. Bouazzaoui, Y. C. Ho, S. J. Speers and K. L. Peters, Raman spectroscopy for forensic examination of β-ketophenethylamine "legal highs": Reference and seized samples of cathinone derivatives, 1–6, Copyright (2012), with permission from Elsevier.

important role in the delineation of the 3,4-ethylenedioxy substituted β-KPs with respect to other analogues where this is not observed. Also, subtle spectral differences between structural analogues 4-MEC and 4-MMC (mephedrone), solely differing in a terminal methyl group, were observed in the region of *ca.* 1000 cm^{-1}.

Employing the same excitation wavelength (*i.e.*, 785 nm), a number of monosubstituted and methylenedioxy substituted cathinones were investigated in detail, with particular focus on regioisomers of mephedrone and flephedrone.[7] In this study, four spectral regions are identified that can be used to delineate between the positional isomers of these two closely related analogues. The first and second regions are associated to intense vibrational frequency bands ascribed to aromatic C=C and C–H stretching and in-plane vibrations at *ca.* 1600 and 1300–1240 cm^{-1} respectively, with consistent shifts to lower frequencies on progression from ortho to meta and para substituted analogues. The authors also observed two other regions at lower frequency, *ca.* 690 and 540–490 cm^{-1} respectively, that can be used to discriminate between these regioisomers. Bands in the 540–490 cm^{-1} spectral region were associated to vibrational frequencies from aromatic ring breathing vibrational motions coupled to C–F (flephedrone) and C–CH_3 (mephedrone), exhibiting significant spectral shifts and hence allowing the delineation of the different positional isomers.

Cutting agents/adulterants, which are often used to mask the side effects of psychoactive substances, can also induce complications regarding the positive identification of NPS given their highly-conjugated structures, close structural similarity to NPS in some cases (*e.g.*, benzocaine and dimethocaine, adulterant and NPS respectively, Figure 9.3) and thus associated spectral similitudes. More recently, model mixtures containing different NPS (*e.g.*, 2-aminoindane (2-AI)) as well as commonly used cutting agents/adulterants such as benzocaine, microcrystalline cellulose, paracetamol, procaine, talc, caffeine, *etc.* as well as a number of internet products containing NPS were chemically analysed using various handheld spectrometers by Assi *et al.*[8] In this study, the authors reported that the employed spectroscopic techniques were affected differently by the presence of cutting agents/adulterants. In short, both NIR and ATR-FT-IR (Attenuated Total Reflectance – Fourier Transform – Infrared) spectroscopy were observed to be more sensitive to the carbohydrate cutting agents present than Raman spectroscopy. Consequently, Raman outperformed the other investigated spectroscopic

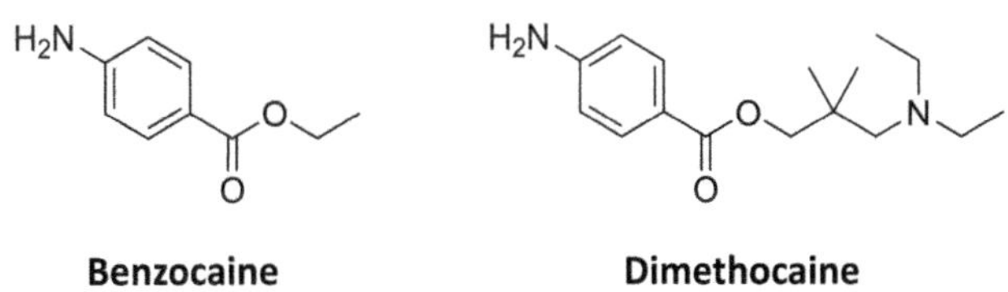

Figure 9.3. Chemical structures of benzocaine and dimethocaine.

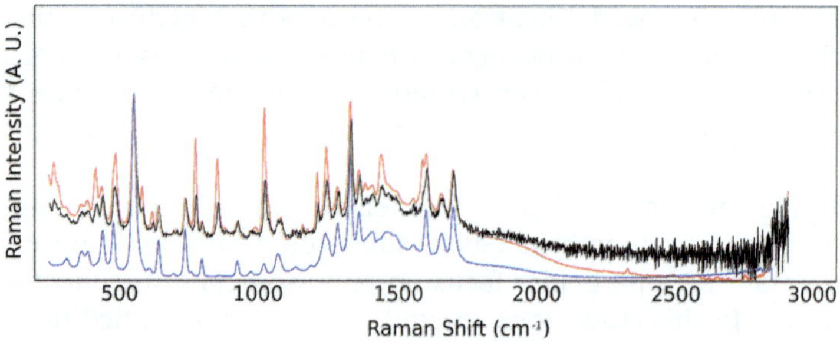

Figure 9.4. Raman spectra for a mixture of 2-AI/caffeine 20:80% w/w (black solid line), 2-AI/caffeine 82% (red solid line) and caffeine 18% (blue solid line). Reproduced from ref. 8 with permission from The Royal Society of Chemistry.

techniques for the identification of the NPS in the presence of common cutting agents.

Figure 9.4 illustrates the Raman spectra of a mixture of 2-AI and caffeine containing solely 20% w/w of the NPS. Application of the in-built algorithm in the handheld instrument employed resulted in a Hit Quality Index (HQI) correlation to library spectra for 2-AI/caffeine (82% HQI) and caffeine (18% HQI), illustrated as red and blue solid lines respectively. In addition, results obtained employing the Raman spectrometer for the analysis of street samples revealed successful identification of the claimed NPS on the product label in three out of seven samples. In turn, in the remaining street samples the identification was hindered by strong fluorescence signals thus illustrating one of the major drawbacks of Raman spectroscopy as mentioned above. Accordingly, Section 9.3 is devoted to the impact of and methods used to reduce fluorescence background signals during chemical characterisation of NPS related samples using Raman spectroscopy.

However, in order to overcome the challenges of identifying an NPS in a chemically complex product, Elie and co-workers reported on the absolute identification of mephedrone and methylenedioxyaminoindane (MDAI) employing a novel combination of microcrystalline testing coupled to Raman micro-spectroscopy.[5] Specific microcrystals of each of the two NPS investigated were produced by the addition of mercury chloride and the approach showed successful delineation through the analysis of associated regioisomers. Grown microcrystals were subsequently analysed by Raman micro-spectroscopy employing a visible 532 nm excitation wavelength, yielding a number of peak shifts when compared to their associated salt form and an additional peak in the low frequency region associated to the Hg–Cl stretching vibrational motion. Interestingly, the robustness of the proposed method was successfully proven in the presence of cutting agents (as well as on street samples), where the NPS was effectively extracted from the cutting agents during the microcrystalline procedure before analysis *via* Raman spectroscopy.

9.3. Fluorescence Interference When Characterising NPS Using Raman Spectroscopy

As mentioned in the earlier section, the generation of a fluorescence signal during Raman spectroscopy can hinder the identification of NPS products by masking the scattered signal produced by the drug of interest. Fluorescence emission denotes a photochemical process by which there is a radiative de-excitation from the lowest vibrational energy level of the first singlet excited state onto a vibrational energy level of the singlet ground state.[19] Consequently, absorption must occur in order for a substance to exhibit fluorescence. As anticipated in previous sections, the Raman scattering phenomenon is *ca.* 10^6 times less probable than fluorescence emission. By means of the fourth power relationship (λ^{-4}),[10] it can be readily understood that high frequency (short wavelength) excitation sources are sought after in order to enhance the intrinsically weak Raman scattering phenomenon. However, the use of relatively short wavelength lasers, often in the visible region, can also cause sample degradation and strong fluorescence backgrounds that consequently hinders Raman signals and identification of substances such as NPS. Typical wavelengths used for Raman spectroscopy include 532, 785 and 1064 nm.

Sythacaine, known as the 'legal cocaine' as well as other NPS such as MPA and 2-AI were chemically analysed with views of developing an in-field screening method employing a Raman spectrometer with a visible 514.3 nm excitation wavelength.[20] Whilst the 2-AI spectrum was successfully determined with a number of intense vibrational bands similar to those of its structural analogue amphetamine, the Raman spectra of MPA exhibited a prominent fluorescence background signal precluding its identification in mixture samples containing different weight ratios of 2-AI and methiopropamine (MPA). A similar scenario was observed for synthacaine which led the authors to explore alternative methods such as their reported electrochemical protocol.

Due to the issues associated with the use of visible wavelength excitations, the use of lower energy radiation such as NIR excitation sources is in fact warranted.[10] NPS reference standard materials are in most cases white powders and therefore NIR excitation wavelengths are red shifted with respect to the main absorption band. However, given the much greater probability of fluorescence emission with respect to that of Raman scattering (*vide supra*), fluorescence generated from impurities at very low concentration can result in precluding the observation of Raman signals even for NIR excitation. For example, Figure 9.5 shows the Raman spectra, using a NIR excitation of 785 nm, of two NPS reference standards, methylone and alpha-pyrrolidinobutiothiophenone (alpha-PBT), which exhibit large fluorescence background signals despite their high purity (>98%). Striking improvement of these Raman spectra was achieved upon smoothing and baseline subtraction in order to reduce the impact of shot/residual noise and the fluorescence background signal respectively. Figure 9.5 illustrates the potential of employing data

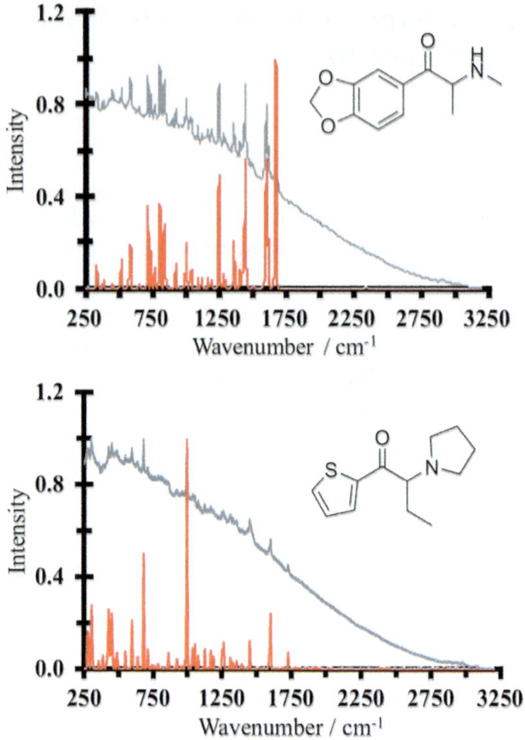

Figure 9.5. Raw (grey solid line) and baseline corrected (red solid line) Raman spectra of methylone (top) and alpha-PBT (bottom).

processing methodologies to improve spectral signatures when the use of longer wavelength excitation may not be available or there is intrinsically low Raman activity of the substance.

In a different study,[9] Raman spectroscopy (785 nm) was also investigated in the identification of a large number of seized samples containing NPS, with an increase of 41–76% of the samples fully initially identified following the updating of the reference libraries with relevant standard compounds. Jones *et al.* emphasise the large variability of the spectral quality between samples, particularly in relation to the fluorescence background. As such, the authors classified the samples in four groups according to the quality of the spectral data and the level of processing required (Figure 9.6), with substances classified as grade 1 requiring no processing to give positive matching and those classified as grade 4 characterised by the absence of spectral information even upon extensive processing. More than 75% of the samples in the reported study were classified as grade 1 or 2.

Recent studies[21,22] report on the use of longer wavelength excitation sources (*i.e.*, 1064 nm) in order to minimise the possibility of fluorescence backgrounds when using handheld Raman instruments for NPS characterisation, which resulted in improved Raman spectra ascribed to a reduction

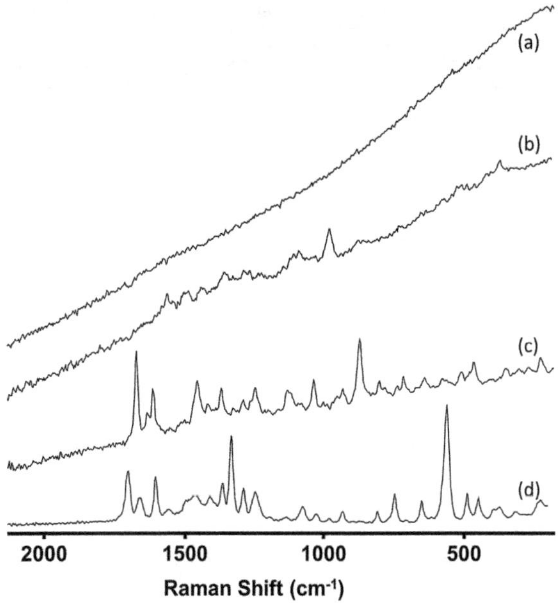

Figure 9.6. Typical Raman spectra of a grade 4 (a), 3 (b), 2 (c) and 1 (d) sample, with the spectra offset for ease of visualization. Reproduced from ref. 9 with permission from The Royal Society of Chemistry.

in the fluorescence background. For example, Harkai *et al.* compared three handheld Raman instruments to evaluate cannabinoid and cathinone samples where the longer wavelength (*i.e.*, 1064 nm) improved spectral quality. Another study focused on identification of 60 different NPS products, containing NPS from different classes such as the aminoindanes and benzodiazepines, using Raman spectroscopy with two different NIR excitation wavelengths, 785 and 1064 nm.[21] On progression from lower (785 nm) to higher (1064 nm) excitation wavelength, an improved identification (*i.e.*, 18–48% HQI positive correlation) was observed. This was primarily associated with a reduction in the fluorescence background for lower frequency excitations as shown in Figure 9.7 for a variety of NPS products.

9.4. Surface-Enhanced Raman Spectroscopy (SERS) in NPS Investigations

One of the key challenges for the detection of NPS in-field and in biological matrices is the limited amount of sample available and/or complex matrices of the investigated samples. In both cases, the Raman signal of the molecule of interest has to have an adequate signal with respect to the instrument noise as well as to be distinguished from the signals of the other non-NPS components. The use of longer laser wavelengths has shown excellent promise in

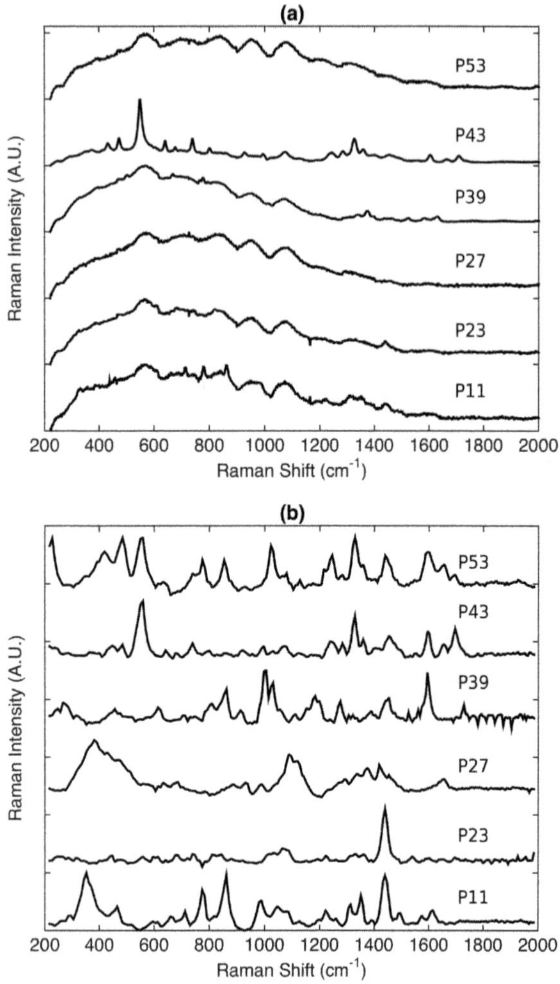

Figure 9.7. Raman spectra of selected NPS internet products using the 785 nm (a) and 1064 nm (b) excitation wavelength. The spectra presented have been normalised to the maximum peak. Reprinted from *Forensic Science International*, **273**, A. Guirguis, S. Girotto, B. Berti, and J. L. Stair, Identification of new psychoactive substances (NPS) using handheld Raman spectroscopy employing both 785 and 1064 nm laser sources, 113–123, Copyright (2017), with permission from Elsevier.

overcoming fluorescence issues as mentioned in the previous section, however a significant decrease in sensitivity can be observed when using NIR excitation. Again, the Raman scattering efficiency is proportional to λ^{-4}. In the case of using an excitation laser at 1064 nm, Raman scattering is over three times less efficient than that when using a laser excitation at 785 nm. Some of those limitations are overcome by improved detector sensitivity in new instruments and increased measurement times as well as utilising other phenomena that lead to an increase in the Raman signal. This is particularly evident in the case of enhancements of the electromagnetic fields around molecules of interest that are optically excited near an intense and sharp plasmon resonance. The interaction of light with metal surfaces generate collective oscillations of conductive electrons and localised secondary fields. These fields generated by surface plasmons are coupled into Raman scattered light and can result in significant increases in intensity. This Surface-Enhanced Raman Scattering (SERS) has been utilised for various applications since its discovery over forty years ago,[23] and the key concepts, some of the applications and a tutorial on SERS can be found in a recent review on this subject.[24]

In order to achieve the SERS phenomena, the analyte of interest has to be transferred from bulk onto nanostructure metal surfaces (Figure 9.8a) and thus ideally resulting in Raman signal enhancements of several orders of magnitude improving the signal to noise (S/N) ratio of the analyte measurement (Figure 9.8b).[25] The increase of Raman signal intensity can arise from two separate multiplicative effects, electromagnetic enhancement (EM) and chemical enhancement (CE) (Figure 9.8c). The recent evidence indicates that charge transfer resonance between the analyte and metal surface contributes to SERS,[26] although it is still unclear how much each of the above effects contribute to overall SERS.[27] However, the enhancement factors of 10^{10} that can be achieved in the context of single molecule SERS[28] provide exciting opportunities for detection and identification of drugs of abuse such as NPS.

The emergence of NPS combined with the increasing number of visits to emergency rooms due to drug overdose has amplified the need for rapid identification of these new offending drugs. The first application of SERS in analysing drugs of abuse demonstrated the use of gold colloids in enhancement of Raman signals of cocaine and methamphetamine,[29] which has opened opportunities for developing methods for rapid screening of illicit drug detection. More recently, a factorial design approach was used to optimise the conditions for the acquisition of SERS spectra for the cathinone, mephedrone, using a gold colloid (Figure 9.9). The specific Raman bands of mephedrone were identified despite obscuring signals from the nanoparticle stabilising agent, citrate, where the lowest limit of detection for mephedrone was determined to be 9.07×10^{-6} M at the Raman shift of 1606 cm^{-1}.[30] In another study, careful optimisation of the concentrations of silver colloids and salt were used to enable identification of other NPS, achieving limits of detection of 5.4×10^{-5} M for the aminoindane, MDAI.[31]

However, other media have been used as a substrate for promoting SERS and obtaining spectral information for NPS. The silver coated surface of a roughened copper coin soaked in a methanolic solution of sample yielded vast information in SERS spectra with accuracy of identification of mephedrone of 98%.[30] SERS active films prepared by addition of hydroxyethylcellulose polymer to a solution of aggregated metal nanoparticles can form an inexpensive SERS substrate that can be easily analysed by portable handheld Raman

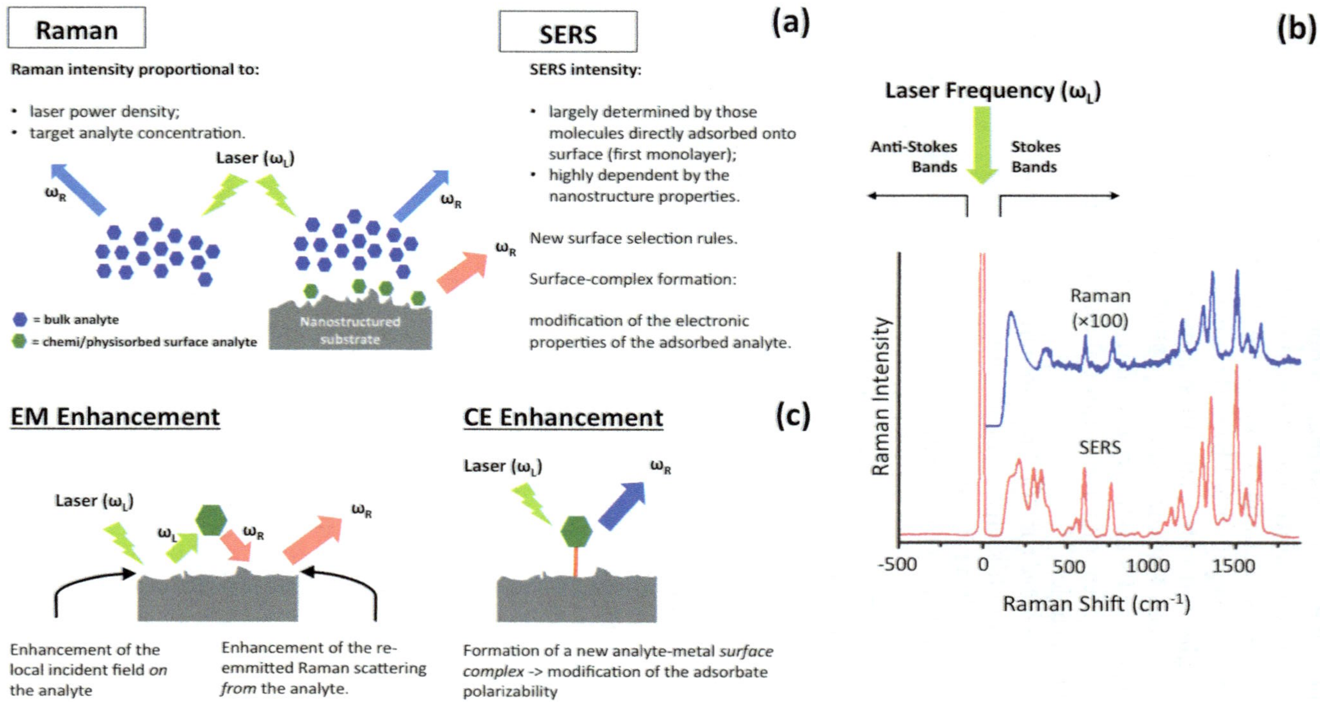

Figure 9.8. (a) Schematic comparative visualization of the Raman and SERS phenomena. (b) SERS spectrum of Rhodamine 6G 10^{-7} M in silver hydroxylamine colloid (red line) and normal Raman spectrum of Rhodamine 6G 10^{-3} M in milli-Q water (blue line) multiplied by a factor of 100. The spectra were acquired using a 633 nm laser line under the same experimental conditions. (c) Schematic outline of the electromagnetic and chemical enhancements in SERS. Adapted from ref. 25 with permission from The Royal Society of Chemistry.

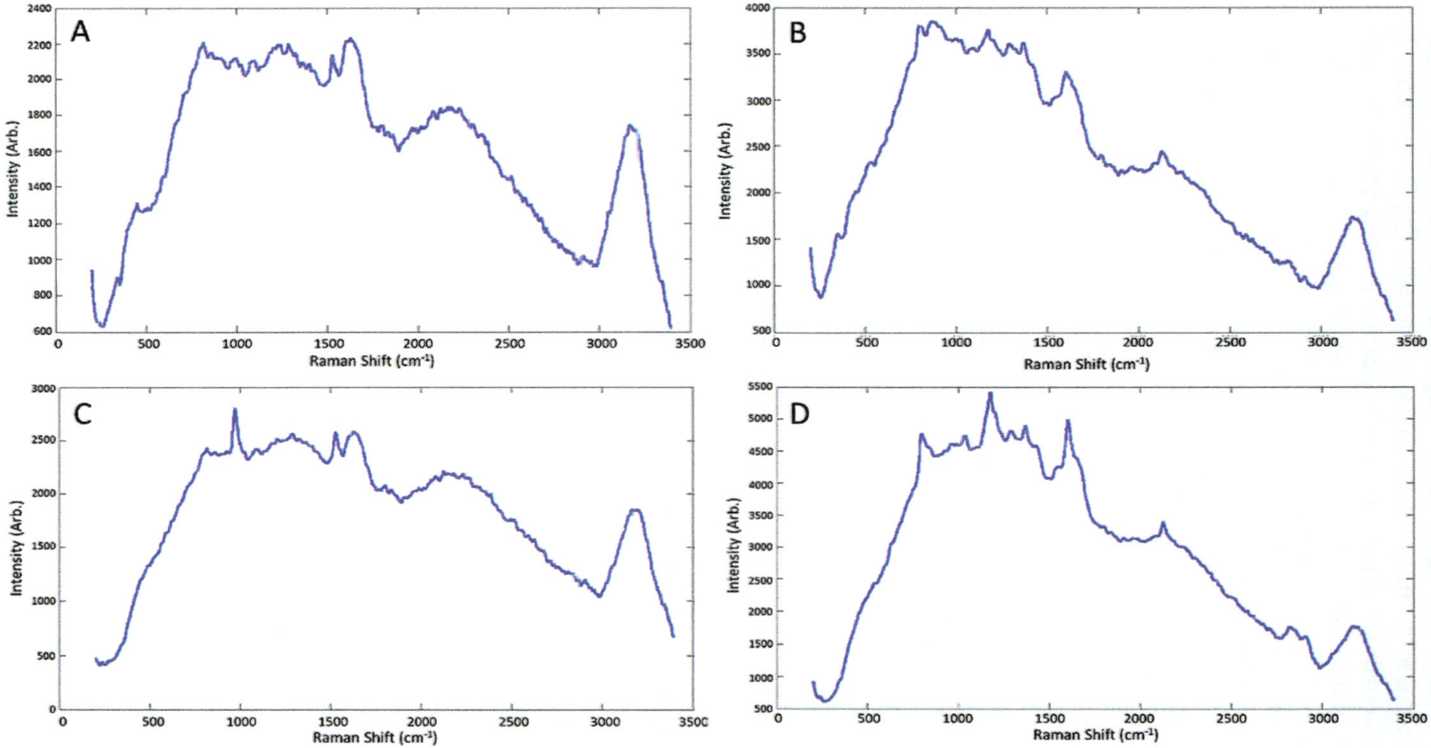

Figure 9.9. Example of mephedrone SERS spectra generated using gold colloid and the factorial design. (A) 40% Colloid (200 μL), no salt, 10% analyte (50 μL), and 50% water (250 μL) at pH 3. (B) 70% Colloid (350 μL), no salt, 10% analyte (50 μL), and 20% water (100 μL) at pH 3. (C) 40% Colloid (200 μL), 10% K2SO4 (50 μL), 10% analyte (50 μL) at pH 3 and 40% water (200 μL). (D) 70% Colloid (350 μL), 2% KNO3 (10 μL), 10% analyte (50 μL), and 18% water (90 μL) at pH 7. Adapted with permission from S. Mabbott, E. Correa, D. P. Cowcher, J. W. Allwood and R. Goodacre, *Analytical Chemistry*, 2013, **85**, 923–931. Copyright 2013 American Chemical Society.

spectrometers. SERS spectra of street samples of MPA recorded using a Raman microscope and citrate reduced silver colloid (CRSC) poly-SERS films, were very similar to those of high-purity reference standards (Figure 9.10). The use of SERS removed the obscuring fluorescence signals from components of the street samples in addition to enhancing the signal of MPA.[2]

This substrate system can be used in criminal investigations where no bulk powder is available, but there are items linked to a criminal investigation with traces of prohibited samples. The collection of samples on swabs and transferring them onto silver (CRSC) poly-SERS films can provide a suitable system for NPS or other drugs of abuse identification, as demonstrated by investigating model evidence samples with different amounts of methcathinone. A minimum of 23 μg was required to obtain a good quality methcathinone SERS spectrum with sufficiently high S/N for positive compound

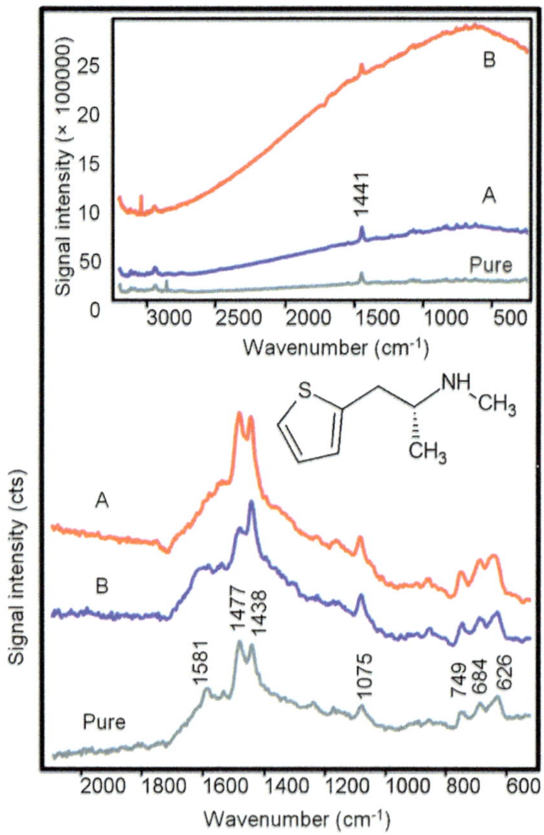

Figure 9.10. SERS spectra of seized samples A and B (15 mL aqueous droplets on Ag (CRSC) Poly-SERS films), 785 nm excitation. Insets show the molecular structure of methiopropamine and the Raman spectra of methiopropamine along with sample powders A and B on the same vertical axis. Reproduced from ref. 2 with permission from The Royal Society of Chemistry.

identification (Figure 9.11). Furthermore, the most prominent peak (1002 cm^{-1}) of methcathinone was identified in the spectrum using just 2.3 μg, however a confident identification was not possible as some of the weaker bands were not visible. A simple swabbing method has shown potential as a cost-effective way to detect trace NPS on evidence.[2]

SERS provides an exciting possibility to enable quicker response by legal authorities as well as medical professionals in emergency, especially as it is possible to utilise handheld Raman instruments in-field. The enhancement ratios that are achieved by use of SERS when compared to traditional Raman spectroscopy improves the detection of NPS by lowering the limit

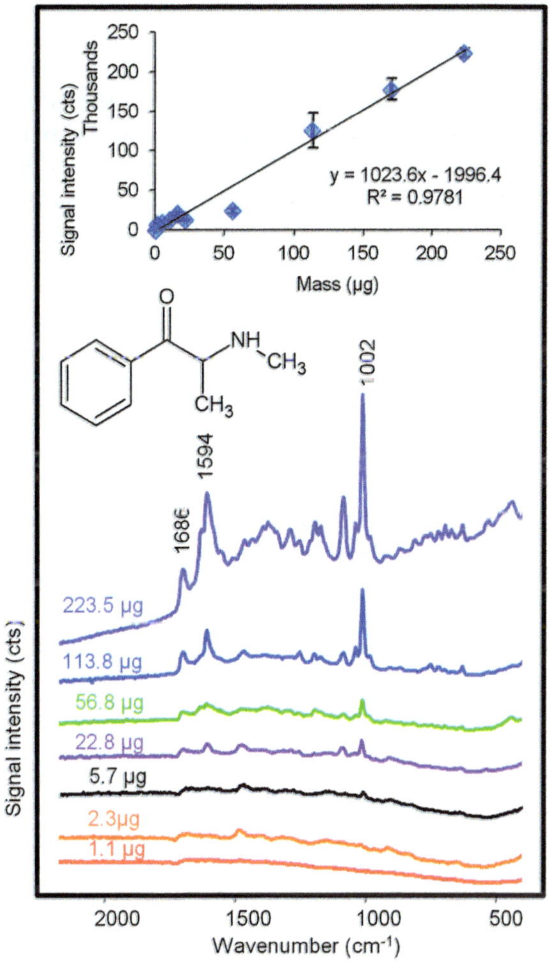

Figure 9.11. SERS spectra of methcathinone obtained by swabbing substrates which carried the stated mass of drug with a wet cotton bud and then swiping the bud over an Ag (HRSC) Poly-SERS film. Reproduced from ref. 2 with permission from The Royal Society of Chemistry.

of detection. However, further optimisation of SERS substrates and evaluation of the use of 1064 nm laser in handheld devices to obtain SERS spectra would be needed to ensure method robustness and strengthen the efforts in crime scene investigations.

9.5. Protocol for the Analysis of NPS Using Handheld Raman Spectroscopy

As extensively discussed in this chapter, the application of Raman spectroscopy for the detection of NPS has seen a large surge of interest in recent years, with special emphasis on in-field applications facilitating forensic investigations. However, it is often that the forensic scientist must develop robust and reproducible approaches regarding sample preparation and data acquisition in order to ensure subsequent accurate identification or quantification of unknown substances. Accordingly, an experimental protocol is outlined below which can be applied to the acquisition of Raman spectral data for NPS samples in different forms. Firstly, the focus is given to the use of benchtop instrumentation and then more extensively to the detailed particularities associated with sample preparation for in-field analysis such as the use of small (<3 mg) sample quantities.

Analysis of NPS in powder form using Raman benchtop instrumentation does not often represent any particular issue regarding sample preparation. However, in light of the observed complexity and heterogeneity of NPS samples, it is advised that a number of replicates must be taken for each sample by focusing on different areas of the sample surface. The latter can result in the identification of different components within the sample mixture, especially when using a Raman microscope. A number of neat NPS are inherently in oil form and available in small volumes (<10 μL) which can be a challenge to analyse. In such cases, it is advised that the oil sample is retrieved from the supplied container upon addition of a suitable volatile solvent (*e.g.*, ethanol) onto a small and previously indented (*ca.* 2 × 3 mm) area of an aluminium plate. This is done in a controlled manner by depositing small volumes and letting the solvent evaporate prior to adding subsequent aliquots.

Analysis of NPS in powder form using Raman handheld instrumentation does have some challenges regarding sample preparation. Whilst large (>150 mg) amounts of powder can be easily analysed using handheld devices by means of vials and widely available vial holders, experimental examination of small (<3 mg) amounts of powder require the development of a protocol detailing the optimum way of presenting the sample to the instrument. Nonetheless, as in the case of benchtop instrumentation (*vide supra*), the measurement of a number of replicates remains crucial in order to account for the intrinsic heterogeneity of NPS samples. Thus, samples presented in vials must be vortexed in between replicates so that irradiation of distinctly different sample areas is ensured. This approach is often limited to samples present at 150 mg or more to allow the laser to optimally interact with the sample.

Figure 9.12 illustrates a developed method for sample preparation of small quantities of NPS (<3 mg) prior to analysis on handheld devices. This is of particular use when obtaining NPS library spectra with reference standards which are often available in only small quantities (*ca.* 5–10 mg). The reader should note that the protocol herein described has been validated using the Rigaku Progeny and Xantus-1 handheld devices, although may be adapted for use with other instrumentation. Firstly, the powder is weighed directly onto a previously cut 2 × 2 cm aluminium plate to avoid any loses (**A**). By means of a plastic-coated spatula, the sample is compacted forming a macroscopically uniform flat surface (**B**). Subsequently, a glass slide is carefully placed covering the sample as illustrated in Figure 9.12 above (**C**).

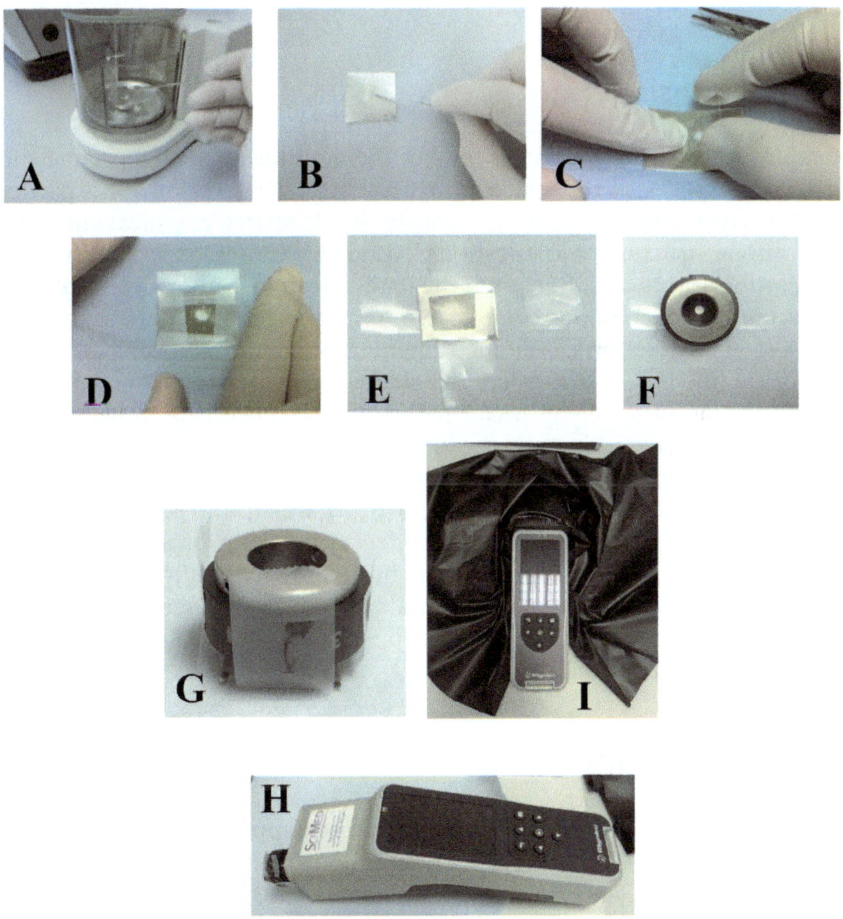

Figure 9.12. Illustration of steps A-H in the preparation of small quantities of NPS for analysis *via* handheld Raman instrumentation.

The position of the glass slide with respect to the aluminium plate is then fixed by designated tape (**D**). It is of note in this last step that the tape does not cover the sample and more importantly sample movement between the glass slide and the aluminium plate must be avoided. The tape is then folded under the aluminium plate and any remaining removed. The sample is then placed onto a cross made of tape with the sticky side facing up (**E**) and the instrument nozzle is faced against it (**F**). By sticking the remaining tape onto the nozzle the position of the sample with respect to it is fixed (**G**) and the nozzle can be attached to the instrument (**H**). To avoid any interference from ambient light, it is recommended to cover it with a blackout piece of clothing (**I**). Lastly, the data acquisition parameters such as focal distance, laser power, exposure time and number of accumulations must be optimised. In the case for a specific NPS product, this will be dictated by the method developed using associated reference standards when possible.

9.6. Conclusions

Raman spectroscopy is a useful tool for the monitoring of NPS products for forensic purposes as the spectrum represents a 'fingerprint' for a particular NPS. The technique is able to identify closely related chemical analogues with minor structural modifications. Although fluorescence from impurities/cutting agents can often mask the Raman response from an NPS of interest, a number of methods were shown to minimise this interference. This required consideration of the instrumentation (*e.g.*, excitation wavelength) and/or sample (*i.e.*, formulation and sample size) available. For example, a longer excitation wavelength can reduce fluorescence from a sample and improve NPS detection, while the application of SERS can produce greater signal enhancements with improved limits of detection. A number of practical considerations need to be made with regards to the NPS product; the sample form and heterogeneity can affect acquisition of a representative spectrum. Thus, further improvements in NPS reference libraries, matching algorithms, and SERS substrates will make Raman spectroscopy a very promising tool indeed for forensic investigations of NPS products in various crime scene investigation settings.

Acknowledgements

This chapter was partially carried out with the support of the University of Hertfordshire and the European Commission (EC) *via* the EU MADNESS (JUST/2013/DPIP/AG/4823) and EPS-NPS (JUST/2013/ISEC/DRUGS/AG/6429) projects. This publication has been produced with the financial support of the Drug Prevention and Information Programme and the ISEC Programme of the European Union. The contents of this publication are the sole responsibility of the authors and can in no way be taken to reflect the views of the European Commission. The authors would like to thank Dr Ian van der Linde for assistance with Figure 9.1.

References

1. J. Lobo Vicente, H. Chassaigne, M. V. Holland, F. Reniero, K. Kolář, S. Tirendi, I. Vandecasteele, I. Vinckier and C. Guillou, *Forensic Sci. Int.*, 2016, **265**, 107.
2. W. W. Y. Lee, V. A. D. Silverson, L. E. Jones, Y. C. Ho, N. C. Fletcher, M. McNaul, K. L. Peters, S. J. Speers and S. E. J. Bell, *Chem. Commun.*, 2016, **52**, 493.
3. S. E. J. Bell, D. T. Burns, A. C. Dennis and J. S. Speers, *Analyst*, 2000, **125**, 541.
4. S. P. Stewart, S. E. J. Bell, N. C. Fletcher, S. Bouazzaoui, Y. C. Ho, S. J. Speers and K. L. Peters, *Anal. Chim. Acta*, 2012, **711**, 1.
5. L. Elie, M. Elie, G. Cave, M. Vetter, R. Croxton and M. Baron, *J. Raman Spectrosc.*, 2016, **47**, 1343.
6. B. Li, A. Calvet, Y. Casamayou-Boucau, C. Morris and A. G. Ryder, *Anal. Chem.*, 2015, **87**, 3419.
7. R. Christie, E. Horan, J. Fox, C. O'Donnell, H. J. Byrne, S. McDermott, J. Power and P. Kavanagh, *Drug Test. Anal.*, 2014, **6**, 651.
8. S. Assi, A. Guirguis, S. Halsey, S. Fergus and J. L. Stair, *Anal. Methods*, 2015, **7**, 736.
9. L. E. Jones, A. Stewart, K. L. Peters, M. McNaul, S. J. Speers, N. C. Fletcher and S. E. J. Bell, *Analyst*, 2016, **141**, 902.
10. E. Smith and G. Dent, *Modern Raman Spectroscopy. A Practical Approach*, John Wiley & Sons, Ltd., Chichester, 2005.
11. R. L. McCreery, *Raman Spectroscopy for Chemical Analysis*, John Wiley & Sons, Ltd., New York, 2000.
12. P. Larkin, *Infrared and Raman Spectroscopy. Principles and Spectral Interpretation*, Elsevier Academic Press, Oxford, 2011.
13. G. Turrel and J. Corset, *Raman Microscopy. Developments and Applications*, Elsevier Academic Press, Oxford, 1996.
14. J. M. Chalmers, H. G. Edwards and M. D. Hargreaves, *Vibrational Spectroscopy Sampling Techniques*, John Wiley & Sons, Ltd., Chichester, 2012.
15. M. D. Hargreaves, *Drugs of Abuse–Application of Handheld FT-IR and Raman Spectrometers*, John Wiley & Sons, Ltd., Chichester, 2012.
16. S. Christesen, B. Maciver, L. Procell, D. Sorrick, M. Carrabba and J. Bello, *Appl. Spectrosc.*, 1999, **53**, 850.
17. S. Coticone, B. Morgan and N. Ludwigsen, *Internet J. Forensic Sci.*, 2011, **4**, 1.
18. B. Diehl, C. S. Chen, B. Grout, J. Hernandez, S. O'Neill, C. McSweeney, J. M. Alvarado and M. Smith, *Eur. Pharm. Rev.*, 2012, **17**, 1.
19. J. R. Lakowicz, *Principles of Fluorescence Spectroscopy*, Springer, New York, 3rd edn, 2006.
20. L. R. Cumba, A. V. Kolliopoulos, J. P. Smith, P. D. Thompson, P. R. Evans, O. B. Sutcliffe, D. R. do Carmo and C. E. Banks, *Analyst*, 2015, **140**, 5536.
21. A. Guirguis, S. Girotto, B. Berti and J. L. Stair, *Forensic Sci. Int.*, 2017, **273**, 113.
22. S. r. Harkai and M. Pütz, *Toxichem Krimtech*, 2015, **82**, 229.
23. M. Fleischmann, P. J. Hendra and A. McQuillan, *Chem. Phys. Lett.*, 1974, **26**, 163.
24. S. Schlücker, *Angew. Chem., Int. Ed.*, 2014, **53**, 4756.
25. L. Guerrini and D. Graham, *Chem. Soc. Rev.*, 2012, **41**, 7085.
26. M. Galperin, M. A. Ratner and A. Nitzan, *J. Chem. Phys.*, 2009, **130**, 144109.

27. P. S. Londero, M. Leona and J. R. Lombardi, *Appl. Phys. Lett.*, 2013, **102**, 111101.
28. E. C. Le Ru, E. Blackie, M. Meyer and P. G. Etchegoin, *J. Phys. Chem. C*, 2007, **111**, 13794.
29. S. M. Angel, J. N. Roe, B. D. Andresen, M. L. Myrick and F. P. Milanovich, *Optical Fibers in Medicine V*, 1990, vol. 1201, p. 469.
30. S. Mabbott, E. Correa, D. P. Cowcher, J. W. Allwood and R. Goodacre, *Anal. Chem.*, 2013, **85**, 923.
31. S. Mabbott, O. Alharbi, K. Groves and R. Goodacre, *Analyst*, 2015, **140**, 4399.

Chapter 10

Light in Forensic Science: Microcrystalline Tests of New Psychoactive Substances Using Light Microscopy

M. Baron*, L. Elie, M. Elie and M. Vetter

University of Lincoln, School of Chemistry, JBL, Green Lane, Lincoln, LN6 7DL, UK

*E-mail: mbaron@lincoln.ac.uk

Table of Contents

Light in Forensic Science: Issues and Applications
Edited by Giorgia Miolo, Jacqueline L. Stair and Mire Zloh

Published by the Royal Society of Chemistry, www.rsc.org

10.1. Introduction

The historical development of microcrystalline tests can be traced back to the early 1800s when the formation of crystals under the light microscope was observed following the mixing of chemical reagents with a range of analytes in drugs, pharmaceuticals and foods.[1] The term 'chemical microscopy' was used for this new form of testing in which chemistry at the microscale could be observed through a microscope indicating that the targeted analyte was present in the sample. Identification of the analyte is determined by observing the morphology (shape, size, colour and spatial arrangement) of the microcrystals with comparison against the microcrystals obtained from known reference standards.

In its simplest form, a test consists of mixing a drop of analyte solution and a drop of reagent solution together on a microscope slide and then observing and recording the formation of microcrystals within the drop under a microscope (Figure 10.1). Within seconds to minutes, microcrystals grow spontaneously within the periphery of a droplet deposited on the microscope slide. Microcrystals grow in size until an equilibrium is reached; after that residual analyte/reagent crystallises indicating the end of the test. Finally, the remaining solvent dries leaving the test unreadable in most cases due to excess reagent and/or drug crystallising and covering the microcrystals of interest. The ideal time to observe the test and record images of the result is normally the window between formation of fully formed analyte-reagent microcrystals and excessive crystallisation of reagent and analyte and/or complete drying of the drop.

Microcrystalline tests are cost effective, quick, conserve evidential material and can be performed competently with minimal training. Photomicrographs or videos of the crystals are permanent records that provide an objective test result requiring little interpretation. As stated by Fulton: "It is the sort of record that with a little explanation can be understood by anyone who can see."[2]

The recommendation to use microcrystalline tests seems to divide forensic scientists and drug analysis chemists. While the first edition of Clarke's

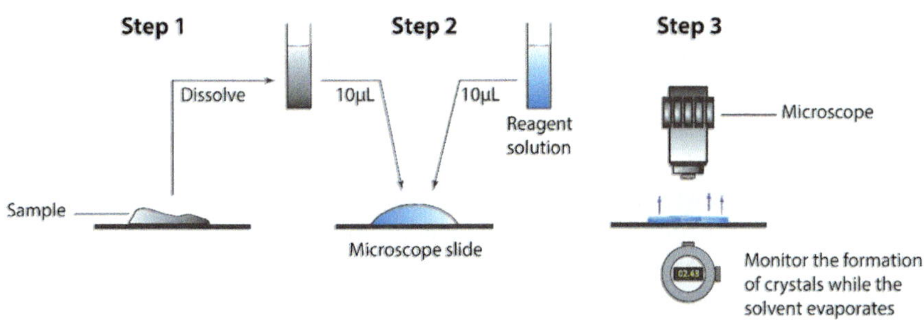

Figure 10.1. Microcrystalline test procedure.

'Isolation and identification of drugs in pharmaceuticals, body fluids and post-mortem material' from 1969 contained microcrystalline tests in the analysis portfolio, they were entirely removed in the second edition from 1986. In contrast, forensic textbooks discuss microcrystal tests although often described wrongly as presumptive tests.

Much of the groundwork for modern microcrystalline tests for drugs of abuse was made by Fulton who published the first textbook[2] in 1969 on their use for the identification of numerous drugs of abuse. Many of the tests in current use for these substances still follow Fulton's procedures. Fulton recorded his observations of microcrystals as hand drawn sketches and this remains an acceptable method of recording and presenting test results. However, it is now standard practice to attach a digital camera to the microscope and obtain both photomicrographs and microvideo of the test. The latter offers a temporal dimension to the test that can be incorporated into the identification criteria.

Clarke reported microcrystal tests for 45 atropine-like alkaloid drugs.[3] He developed and used the hanging drop technique in which the volatility of the free base form of the drug is utilised to separate the drug from its matrix. Clarke's description of crystal morphology used in this early work is still used today with 6 individual crystal shapes described as rods, needles, plates, blades, tablets and prisms and 10 aggregates and agglomerates described as rosettes, burrs, tufts, fans, stars, crosses, sheafs, clusters, bundles and dendrites. Further terms that resemble everyday objects such as wheat and swallow tails[4] have also been used to describe crystal habits that can't be fully described using only Clarke's morphology descriptors. Figure 10.2 shows a selection of different crystal morphologies along with their 'Clarke type' descriptions.

The shape of individual crystals and the arrangement of crystals are both identifying characteristics for a particular test along with crystal colour and size, spatial and temporal formation of the crystals. A compendium of microcrystalline tests for 18 illicit drugs has been reported.[4] Details of the tests and microcrystal observations are provided including photomicrographs, illustrations and optical properties; the latter determined using polarised light microscopy (PLM) include birefringence, extinction position, sign of elongation and optical axes. PLM uses transmitted white light, parallel polarised light, crossed polarised light and crossed polarised light with various retardation filters such as the Red-1 and quarter-wave compensators. Interference effects introduced by the crystal's birefringence can be used to determine the birefringence, sign of elongation and direction of the optic axis.

The need for analyte separation is an important consideration when developing new tests as drugs in real samples can be combined with different excipients and the drug may also be present in different salt forms. This may affect the presence of crystals and crystal morphology in comparison to the test performed on 'pure' reference standards when a direct analysis test is used. For a heterogeneous solid sample, it may be possible to select particles of

Figure 10.2. Examples of microcrystal structures' designation based on the morphology system used by Fulton and Clarke showing definition of individual crystal shapes, aggregates and agglomerates. (**a**) Rosettes of needles; (**b**) serrated plates; (**c**) rods; (**d**) blades; (**e**) feather combs; (**f**) cluster of dendritic serrated rods; (**g**) skeleton of wire with tufted ends; (**h**) dendritic rosettes of needles; (**i**) bundles of blades; (**j**) prisms; (**k**) skeletons of blades; (**l**) hairs and tablets.

active ingredient from excipients through sample observation with a stereomicroscope. If mechanical separation is not possible then a simple micro liquid–liquid extraction can be used in which the drug is extracted into a suitable organic solvent that is then evaporated to dryness on the microscope slide before proceeding with the test.[4] The hanging drop technique developed by Clarke[3] is also a separation technique in which basic drugs are converted into the more volatile free base form by increasing pH. The requirement is that the free base is sufficiently volatile to transfer from the sample solution to the suspended reagent drop.

In comparison to modern instrumental techniques the microcrystalline test may appear somewhat out-dated however they provide a simple, cost effective method of confirming the presence of a drug when used in an appropriate analytical scheme. The laboratory only requires a light microscope and trained microscopist to be able to perform the tests. Microcrystalline tests should not be confused with presumptive screening tests used in unknown analysis such as colour tests in which a positive result may be obtained by a number of different substances. In fact, microcrystalline tests would have little value being used in this way. In most cases, a fully validated microcrystalline test will give an unequivocal identification for the target drug through interpretation of microcrystal morphology.

Microcrystalline tests now exist for most illicit drugs and are well established for substances such as cocaine, amphetamine, methamphetamine and phencyclidine.[5–7] Microcrystalline tests for 18 of the 25 most frequently encountered controlled substances in forensic laboratories across the United States have been published.[4] Forty-two tests using 24 different reagents are described with most reagents being simple binary compounds of the transition metals platinum, gold and mercury either used alone in aqueous solution or combined with an acid. The microchemistry taking place is mainly precipitation of coordination compounds formed between metal ions and organic ligands. Factors such as pH, temperature, humidity and sample solubility can influence the chemistry and crystal formation and so these factors must be taken into account when carrying out tests and in developing new tests. Also, the salt form of the drug and the presence of cutting agents and adulterants can lead to distortions in the expected crystal form requiring a higher level of chemical microscopy skill than when comparing clearly formed microcrystals with the expected crystal morphology.[2,10] Table 10.1 gives a selection of common reagents used in microcrystalline tests.

Wielbo and Tebbett[8] were the first to combine a spectroscopic technique with microcrystalline tests. The subjectivity associated with test results of distorted microcrystals can be reduced if a spectroscopic measurement can confirm the presence of the drug within the crystal. Micro-Fourier Transform infrared spectroscopy (micro-FTIR) was used to obtain the infrared spectrum of individual microcrystals. The technique requires that crystals of sufficient size are formed, the drop dries quickly to remove residual water and dried reagent crystals do not mask the microcrystals. Not all microcrystalline tests

Table 10.1. Microcrystalline test reagents and their common target drugs, adapted from Fulton.[2]

	Complex metal halides		**Organic acids**
$HAuCl_4$	Cocaine, amphetamine, cathinones, phenethylamines	Picric acid	Nicotine, mescaline, procaine, cocaine
$HAuBr_4$	Heroin, amphetamine	5-Nitrobarbituric acid (dilituric acid)	Methamphetamine, methylphenidate, brucine, strychnine
H_2PtCl_6	Amphetamine, cocaine, clonazepam		
	Neutral/basic reagent		**Simple oxygen acid**
$HgCl_2$	Diacetylmorphine, methadone, methylmethcathinone, benzylpiperazine, caffeine	$KMnO_4$	Cocaine, phencyclidine

meet these criteria however, it was shown that infrared spectra could be obtained for a number of common drugs such as cocaine, heroin, phencyclidine and methamphetamine. The microcrystal spectra showed some differences from the pure drug spectra but were sufficiently similar to confirm the presence of the drug within the microcrystals. Infrared spectra have also been obtained for a number of other microcrystalline tests.[4]

Micro-Raman spectroscopy offers advantages over micro-FTIR in not requiring the same sample pre-treatment while still providing the specificity of vibrational spectra. A Raman spectrum can be obtained from a microcrystal *in situ* within the micro-drop. This technique has been successfully demonstrated with selected new psychoactive substances (NPS).[9] When used in this way the microcrystalline test has a dual function; as an identification technique and as a selective extraction technique for either micro-Raman or micro-FTIR. This potentially removes the need for a separation technique when analysing mixtures particularly when there is temporal and/or spatial separation of microcrystals within the micro-drop.

The emergence of NPS has challenged drug laboratories to develop and validate suitable methods for their analysis. It is often possible to adapt existing methods to include NPS however validation of the method requires the use of reference standards and the need to be able to distinguish between the target drug molecular structure and closely related molecular structures such as structural isomers. Microcrystalline tests for relatively few NPS have been published. These include benzylpiperazine (BZP), mephedrone (4-MMC), 3,4-methylenedioxy-2-aminoindane (MDAI), and 3,4-methylenedioxypyrovalerone (MDPV).[10,11] The United Nations Office on Drugs and Crime include microcrystalline tests for benzylpiperazine and mephedrone in their recommended methods publications.[12,13]

The chemical structures of selected NPS are shown in Figure 10.3. NPS largely fall into 5 groups: stimulants, cannabinoids, psychedelics, dissociatives and sedatives. 4-MMC, BZP and MDAI are all examples of stimulants. Caffeine is included as it is often added as a cutting agent to NPS products. When developing a test for a new substance, it is essential that consideration is given to the other substances that are likely to be present in street samples. Method validation requires that the selectivity of the test and matrix effects are evaluated.[14] It is known, for example, that the presence of caffeine mixed with cocaine distorts and inhibits crystal formation in the cocaine test with a gold chloride/acetic acid reagent.[4]

It is relatively straightforward to develop a microcrystalline test for an NPS. Only a few milligrams are required that can be either as a reference standard, if available, or by purification of the NPS from a product. The NPS is then screened using a bank of test reagents such as those in Table 10.1 and those that produce microcrystals further investigated as a possible test. Once developed and validated the test is quick, inexpensive and easy to carry out requiring no more than some basic manipulative and microscopy skills. The tests are portable and when combined with other techniques can provide a robust analytical scheme. Often more than one reagent will produce

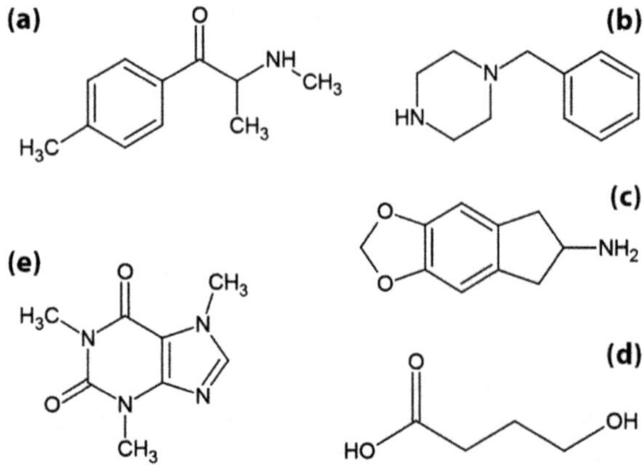

Figure 10.3. Molecular structures of selected NPS and cutting agent. (**a**) mephedrone (4-methylmethcathinone, 4-MMC, 2-(methylamino)-1-(4-methylphenyl)propan-1-one), (**b**) BZP (1-benzylpiperazine), (**c**) MDAI (6,7-dihydro-5*H*-indeno[5,6-*d*][1,3]dioxol-6-amine), (**d**) GHB (γ-hydroxybutyrate), (**e**) caffeine (1,3,7-trimethyl-3,7-dihydro-1*H*-purine-2,6-dione).

microcrystals for an NPS and so more than one test can be developed further adding to the discrimination available. A comprehensive identification can be achieved with screening drug tests such as the Marquis reagent followed by confirmation using two independent microcrystalline tests. Microcrystalline tests are classified as a category B technique in the SWGDRUG recommendations for the analysis of seized drugs.[14]

It is known that microcrystalline tests can be used to distinguish between isomeric forms of drugs. The methamphetamine test[6] can be used to distinguish between the racemic mixture and the pure *d*- and *l*-forms and so provides an analytical test for determining the enantiomeric form of the seized drug if enantiomeric standards are available. The emergence of NPS has seen the proliferation of chemical structures differing only by minor modifications such as that seen between positional isomers. Unequivocal identification is a challenge for forensic drug laboratories as routine methods such as gas chromatography-mass spectrometry may have difficulty in distinguishing between these forms without extensive method development and validation. A microcrystalline test may be able to distinguish between the different isomeric forms through differences in microcrystal morphology and/or when combined with micro-Raman or micro-FTIR.

This chapter presents microcrystalline tests for selected NPS showing how established test reagents can be used to provide unequivocal identification. The versatility of the reagent mercury(II) chloride is demonstrated by showing how it can discriminate between several different NPS. Crystal morphology descriptions are given along with optical properties obtained by PLM showing the value of incorporating PLM into standard laboratory procedures. The ability to distinguish between isomeric chemical structures is demonstrated by coupling microcrystalline tests with micro-Raman spectroscopy.

10.2. Experimental

One of the advantages of the microcrystalline test is its simplicity requiring only that analyte and reagent chemically react within a solvent micro-drop leading to the formation of microcrystals. Seized drug substances tend to be in the solid form and so the test requires that the substance and the reagent are brought together in a micro-drop of solvent. One of three methods is commonly used to achieve this.

Solution mixing: The solid sample is first dissolved in a suitable solvent and then micro-drops of the sample and reagent solutions are mixed together on a microscope slide. Classically, the applied volume was simply a drop. However, it is now recommended to use automatic pipettes to improve repeatability.[1] Typical volumes to perform a microcrystalline test are between 5–20 μL. Most commonly, the drug solution is applied first onto the slide, swiftly followed by the reagent solution using the same volume. Generally, there is no need to follow a certain order, however, some reagents can be sensitive to ambient conditions and by applying it last the exposure time is greatly reduced. Most tests are carried out as an exposed micro-drop however a coverslip can be added after microcrystal growth if needed to improve the quality of photomicrographs.

Nucleation can be assisted by gently swirling the pipette tip inside the drop. This movement has a positive effect in two ways: it mixes both solutions which can be slow when different viscosities are involved (*e.g.* either drug or reagent are dissolved in concentrated acid), and it provides crystallisation sites generally speeding up crystal formation. It is not recommended to perform assisted nucleation categorically as it can influence the habit of some crystal tests. Moreover, some microcrystals develop rapidly upon the addition of the reagent so any foreign object introduced into the test drop would destroy those newly formed crystals. Therefore, a test should always be performed with and without assisted nucleation, and a newly developed test protocol must include the effects assisted nucleation has on the test outcome.

Direct addition: A small amount of the solid sample is placed on a microscope slide and a micro-drop of reagent applied adjacent to it. The drop and the sample are then carefully brought into contact and the formation of microcrystals observed within the reagent micro-drop.

Hanging drop method: A micro-drop of reagent is supported over the analyte solution and then, after waiting for a period of time, microcrystals are observed in the reagent drop. As the free base is often more soluble than the salt form, the drug solution is made alkaline to convert any salt into the free base.

Universally, each test should be accompanied by three control samples. The first one is the drug solution deposited onto a slide and left to dry, the second one is the reagent solution left to dry on its own; and the third one is the microcrystalline test performed with a certified reference standard of the suspected drug. If the drug or reagent solution form crystals similar to those obtained with the chosen microcrystal test, the test is invalid and cannot be used. Positive identification of a substance is made by comparing habit, shape and colour to microcrystals obtained under the same conditions using reference standards of known purity.

Using a high-power microscope, the test slide can be observed easily and instantly following the application of drug and reagent solution. While light microscopy can be sufficient for unequivocal identification, further characterisation of crystals forming can be achieved by using polarising light microscopy.[2] Further chemical analysis of the formed microcrystals can then be carried out by a micro-spectroscopic technique such as micro-FTIR or micro-Raman.

In developing a microcrystalline test for NPS casework a full validation should be carried out consistent with that required for any qualitative analysis method.[14] This includes establishing the following for the method: selectivity, matrix effects, repeatability and reproducibility, limit of detection, robustness and ruggedness.

The microcrystalline results presented in this chapter for selected NPS were obtained by mixing together 10 μL of NPS aqueous solution (10 mg mL^{-1} or 1 mg mL^{-1}) with aqueous silver/lanthanum nitrate (for GHB) or 10 μL of mercury(II) chloride solution (10 mg mL^{-1}) for all other NPS using the solution mixing technique unless otherwise stated. If assisted nucleation was used then this is stated. Mercury(II) chloride (>99.5%), silver nitrate (>99%) and caffeine were obtained from Sigma-Aldrich (Gillingham, UK). NPS certified standards were obtained from LGC Standards (Teddington, UK).

Light microscopy was carried out using a Meji ML5000 microscope under low and high magnification using calibrated transmitted light microscopy and phase contrast microscopy set for Kohler illumination at 400× overall magnification. Photomicrographs were captured with a Canon EOS Rebel T2i digital SLR fitted with a Meji Techno adapter ring, camera attachment and 1.9× magnification photo-eyepiece. For 3-dimensional image capture, multiple partially focussed images taken of different focal planes were imported into Helicon Focus software and merged into a single image.

Polarised light microscopy was carried out using a Meiji MT9300 polarising light microscope fitted with a Meiji Techno adapter ring camera attachment and 1.9 magnification photo eye piece and a Canon EOS Rebel T2i digital SLR. Images were captured in unpolarised light, crossed polars and crossed polars with either a Red-1 or a quarter-wave compensator. The Red-1 and quarter-wave compensators produce retardations equivalent to about 550 nm and 125 nm respectively.

Micro-Raman spectra were captured using a Horiba Jobin Yvon LabRam 300 equipped with an Olympus BX41 confocal microscope and fitted with a 532 nm solid state laser (Laser Quantum). The confocal hole was set at 300 μm and the slit at 100 μm. The exposure time, laser power at the sample and the spectral scan range were optimised for each microcrystal type. Spectra were baseline corrected and normalised using the LabSPEC software (version 5.64.15, 2008).

A FEI Inspect 50 instrument fitted with a secondary electron detector was used to acquire the SEM images. Images were obtained under high vacuum (≈67 Pa) and beam power of 12.5 keV with a spot size of 4.5, a working distance of 12.3 mm, and an overall magnification of 111×. Close-ups were captured with a working distance of 7.7 mm and an overall magnification of 254×.

10.3. Results and Discussion

10.3.1. Gamma-hydroxybutyrate (GHB)

One of the earliest reported microcrystalline tests for NPS was that developed for the identification of gamma-hydroxybutyrate (GHB). GHB presents a number of problems in analysis such as its hygroscopic nature hindering infrared analysis and the need for derivatisation to obtain identification by gas chromatography-mass spectrometry. Often GHB is presented to the lab as an aqueous solution. The microcrystalline test can be applied to the sample in this form and requires no more than 5 minutes to complete an identification. The original test reagent consisted of a mixture of 0.1 g silver nitrate and 0.1 g copper(II) nitrate dissolved in 10 mL water.[15] GHB forms rectangular crystals in the region of 100 μm length at the drop periphery within 5 minutes down to GHB concentrations of 2 mg mL^{-1}. These crystals are birefringent and this can be used to distinguish the crystals from the reagent crystals that also grow from the periphery of the drop. It was reported that the silver nitrate reagent alone was unable to produce these crystals and that it was necessary to have both silver (Ag^+) and cupric (Cu^{2+}) ions present. However, it was later shown that rectangular microcrystals are formed between GHB and high purity silver nitrate alone and that the presence of other metal ions may play a role in either enhancing or inhibiting crystal growth.[16]

An optimised reagent combination of silver nitrate and lanthanum nitrate was established[16] that optimised the quality of the microcrystals in terms of shape and size. These crystals were analysed by single crystal X-ray diffraction and were found to be monoclinic consisting of only GHB molecules and silver ions.[17] Lanthanum nitrate therefore appears to play a role in enhancing the crystallisation process possibly by controlling the rate of crystal formation. The enhanced silver nitrate/lanthanum nitrate test has a limit of detection of 0.5 mg mL^{-1}; about 4× lower than the cupric nitrate/silver nitrate test reagent.[15]

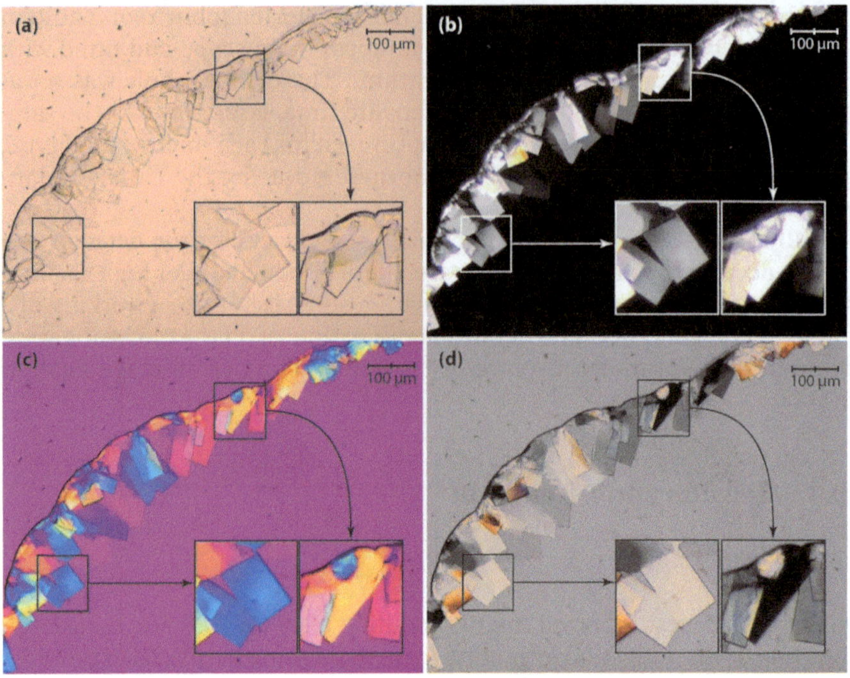

Figure 10.4. GHB microcrystals at 10 mg mL^{-1} with silver nitrate/lanthanum nitrate reagent. **(a)** Original, **(b)** crossed polar, **(c)** crossed polar and Red-1 compensator, **(d)** crossed polar and quarter-wave compensator.

Figure 10.4a shows microcrystals obtained using the silver nitrate/lanthanum nitrate reagent. Rectangular shaped crystals growing individually can be seen clearly growing from the edge of the drop under unpolarised white light transmission illumination.

When GHB microcrystals were observed using scanning electron microscopy (SEM), the enhanced resolution provided by SEM enabled the direct observation of the faces constituting the thickness of the crystals (crystal depth) which revealed that the orthogonality observed between length and width appears to extend to depth, *i.e.* the crystals are cuboids (Figure 10.5). Crystals presented a different morphology when grown on the SEM carbon stub surface than when grown on an untreated microscope slide. They did not preferentially grow from the edge of the test drop but covered the entire test area. This can be explained by the quasi-spherical shape that the droplet adopts when deposited onto a hydrophobic surface such as carbon contrasting with the flattened semi-ovoid shape adopted when placed onto a hydrophilic surface such as an untreated glass microscope slide. On the hydrophobic surface, crystals have space to grow in three dimensions as the droplet gradually loses its volume from all directions at the same time, as compared to the hydrophilic surface where the thickness of the liquid phase is reduced and evaporation occurs faster on the edge of the flattened droplet. Consequently,

Figure 10.5. (**a**) Image of [GHB Ag] micro-cocrystals obtained using a SEM. (**b**) Examples of clusters of blades growing from the same nucleation point and forming '*paddle wheels* of cuboid blades'. (**c**) and (**d**) Close up of the *paddle wheel* crystalline arrangement.

instead of systematically growing inwardly flat and starting from the edges, *i.e.* following the glass slide surface plane, many of them appeared to grow from shared nucleation points (localised all over the test area) to form '*paddle-wheels* of cuboid blades' (Figure 10.5b–d). This phenomenon allowed for the detailed observation of the edges of the crystals and helped in establishing their orthogonality and demonstrates that the substrate surface properties is another factor to consider when developing a microcrystalline test.

The same crystals viewed under cross polars (Figure 10.4b) exhibit the double refraction expected by monoclinic crystals with the brightest crystals seen in a 45° geometry between the cross polars. Extinction is observed when the crystal length of elongation is parallel to the direction of the polariser indicating parallel extinction. The large number of crystals visible in the field of view aligned in different orientations relative to the crossed polars gives a range of interference colours consistent with Newton's scale expected for crystals showing birefringence. Full extinction occurs when crystals are aligned parallel to either the polariser or analyser (North-South or East-West looking at the photomicrographs). A range of colours through iron grey, blue grey, pale grey is seen up to white as crystal orientation moves to 45° relative to the crossed polars; the orientation giving greatest retardation. Where crystals overlap with each other retardation is increased due to thickness giving the yellow and orange colours observed.

The maximum retardation seen under cross polars can be used to estimate birefringence. For GHB crystals as viewed in Figure 10.4b this is first order white light; a retardation of about 250 nm. Assuming that the crystals have a thickness of 10 μm[17] gives a moderate birefringence of 0.025.

When observed through the Red-1 compensator (Figure 10.4c) the crystals in the NE-SW direction are yellow. The Red-1 plate gives a retardation of about 550 nm which is reduced to 300 nm by the GHB crystals indicating that the retardation produced by the birefringence of the crystals is about 250 nm which agrees with the observation under crossed polars.

The Red-1 compensator image shows that the fast direction (low refractive index) is along the direction of crystal elongation. The quarter-wave plate (Figure 10.4d) also shows this with full extinction of the crystals aligned NE-SW. The sign of elongation for the crystals is therefore negative.

10.3.2. *Benzylpiperazine (BZP)*

1-Benzylpiperazine (BZP) is an NPS that emerged in the early 2000s.[12] It was used in party pills along with 3-trifluoromethylphenylpiperazine (3-TFMPP) as an ecstasy substitute. It has since been controlled in many countries. BZP was controlled in the UK in 2009 as a Class C drug.

A microcrystalline test using a 5% aqueous platinum bromide (H_2PtBr_6) reagent gives yellow square plates with BZP.[4] The presence of 3-TFMPP does not affect the BZP crystals when BZP is in mass excess to 3-TFMPP. However, some distortion is observed when 3-TFMPP is in excess of BZP. Interestingly these microcrystals are described as almost isotropic. An alternative test using a solution of platinic chloride ($H_2PtCl_2 \cdot 6H_2O$) dissolved in 40% hydrobromic acid further diluted by approximately 40% sulphuric acid gives rectangular shaped crystals with indented ends.[12]

A BZP test using a 10 mg mL^{-1} mercury(II) chloride ($HgCl_2$) aqueous reagent has been reported.[10] The drug concentration was 1 mg mL^{-1} and 10 μL

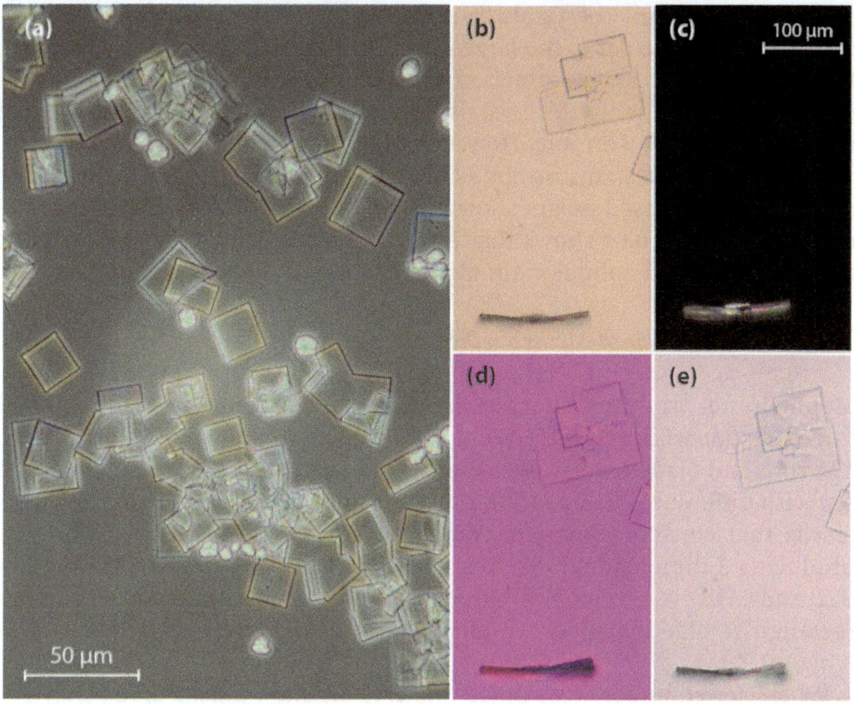

Figure 10.6. BZP microcrystals at 1 mg mL^{-1} with mercury(II) chloride reagent. **(a)** Under phase contrast, **(b)** original, **(c)** crossed polar, **(d)** crossed polar and Red-1 compensator, **(e)** crossed polar and quarter-wave compensator.

volumes of both solutions were combined. Assisted nucleation was used by gentle swirling of a plastic pipette within the micro-drop. Indication of BZP is given by the appearance of transparent flat square-cut plates (Figure 10.6a and b) within approximately 30 s of mixing along the line of assisted nucleation and after a few minutes anywhere within the drop. The limit of detection of the test is 0.2 mg mL^{-1}.

When viewed under phase contrast (Figure 10.6a) the characteristic Becke line is seen around the edge of the crystals expected when the crystals and the surrounding medium have different refractive indices. BZP-platinum bromide crystals are reported to have a refractive index greater than 1.700 [4] and so the contrast shown in Figure 10.5a is expected if the BZP-$HgCl_2$ crystals have a similar refractive index.

Some distortion of crystals occurs where crystal growth is impaired by the presence of other crystals. This is seen in Figure 10.6a with clusters of crystals overlaying each other and showing clear distortion from the square planar shape. More specific distortions such as twinning where crystals grow sharing a common face are also seen with this BZP test. This is a feature of the test and can be used as an additional identification characteristic when comparing with a reference standard.

The BZP-$HgCl_2$ microcrystals do not appear to show anisotropy when illuminated perpendicular to the large square face. All the crystals viewed at different orientations show extinction when viewed with cross polarisers. This is presumably similar to the crystals formed with platinum bromide that are reported as almost isotropic. However, when viewed parallel to this face *i.e.* along the thin faces, anisotropy is observed as shown by the interference colours under crossed polars shown in Figure 10.6c. The orientation of the crystal in Figure 10.6c shows that extinction is not parallel, however, when the stage is rotated oblique extinction is observed.

10.3.3. Mephedrone (4-Methylmethcathinone, 4-MMC)

Mephedrone produces characteristic rod or rectangular blade shaped microcrystals with mercury(II) chloride reagent as shown in Figure 10.6a. The rods and blades can grow to greater than several hundred μm in length but with differences in size from the same nucleation site. A feature of this test is that clusters (burrs) of rods grow from the same nucleation site producing 3-dimensional rosettes. This clearly distinguishes 4-MMC from BZP and GHB even though all three drugs give similar transparent square/rectangular blade shaped crystals. The limit of detection of the test is 3 mg mL^{-1}.

When viewed under cross polars the crystals show anisotropic behaviour with parallel extinction (Figure 10.6b). The Red-1 and quarter-wave compensators show that the fast direction of the crystal is along the length of elongation (Figure 10.6c and d).

Another factor to consider is the rate of crystal formation. For 4-MMC it was seen that crystals could be observed between 1 and 30 minutes depending on drug concentration. Assisted nucleation speeds up crystal growth along the pipette giving trail chains of blades resembling barbed wire (Figure 10.7a). When nucleation is unassisted, microcrystals form more slowly producing dense aggregates of blades appearing to grow from a central nucleation site.

The rate of evaporation of the solvent can also be used to control the formation of microcrystals. Microcrystals must be observed and recorded within a window of time when micro-drops are exposed allowing the solvent to evaporate. In Figure 10.7a the centre of each burr is very dense making it difficult to observe detail in this region. This can be overcome by growing microcrystals within capillary tubes[18] as shown in Figure 10.8. This extension to the solution mixing technique restricts solvent evaporation and provides a 3-dimensional medium for crystal growth. This works for tests where spontaneous crystallisation occurs at the concentrations following mixing of the two solutions. Assisted nucleation was possible using a thin nichrome wire.

Three-dimensional crystal morphology cannot be captured with a single 2-D photomicrograph. Multiple partially focussed 2-D photomicrographs of different focal planes within the capillary tube can be combined into a single

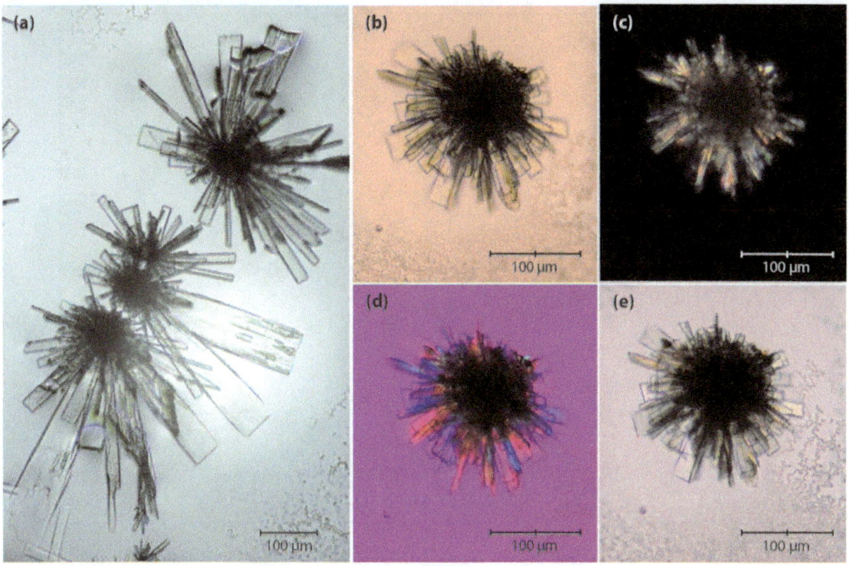

Figure 10.7. Mephedrone microcrystals at 10 mg mL^{-1} with mercury(II) chloride. (**a**) Original, (**b**) cluster close-up, (**c**) crossed polar, (**d**) crossed polar and Red-1 compensator, (**e**) crossed polar and quarter-wave compensator.

image using software such as Helicon Focus. When viewed in this way it can be seen that crystal morphology of the 4-MMC-$HgCl_2$ microcrystals is best described as rosettes of blades rather than burrs of blades; the description given for the exposed micro-drop test.

4-MMC exists in two other regioisomeric forms, 2-methylmethcathinone (2-MMC) and 3-methylmethcathinone (3-MMC) (Figure 10.9) varying in the position of the methyl group on the aromatic ring. The mercury(II) chloride test distinguishes between these isomeric forms through microcrystal observation (Figure 10.8) although absolute identification is only possible for 4-MMC due to the poor quality and lack of distinctiveness of the microcrystals for the other two isomers. Micro-Raman analysis of the crystals however, is able to identify the individual isomer demonstrating that microcrystalline tests can be used within an analytical scheme to provide identification to the level of structural isomer through crystal morphology, crystal optical properties and spectroscopic analysis. For the latter, microcrystals provide a method of separation by extracting the NPS into a crystalline form that can be viewed and analysed by micro-Raman.

10.3.4. 3,4-Methylenedioxy-2-aminoindane (MDAI)

3,4-Methylenedioxy-2-aminoindane (MDAI) was classed as a 'second generation legal high' and appeared around 2010. It has been found in single substance NPS products as well as mixed with other substances in other

Figure 10.8. Mephedrone microcrystals at 10 mg mL^{-1} with mercury(II) chloride. Combination of partially focussed focal planes **(a-f)** resulting in one reconstructed fully focussed image **(g)** using Helicon Focus™. Reprinted from L. E. Elie, M. G. Baron, R. S. Croxton and M. P. Elie, *Drug Testing and Analysis*,[18] John Wiley and Sons. Copyright © 2012 John Wiley & Sons, Ltd.

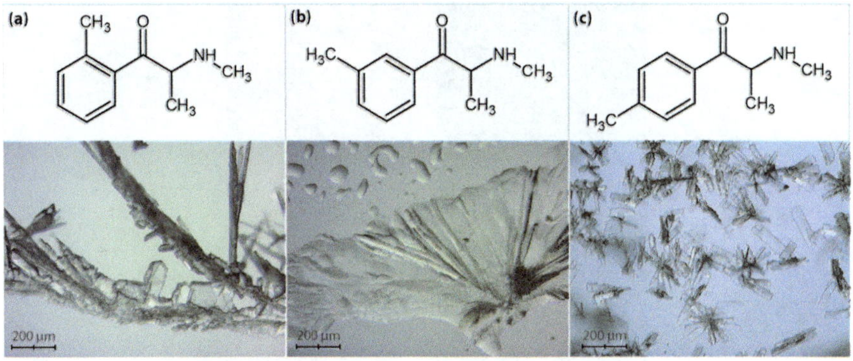

Figure 10.9. Mephedrone isomer microcrystals at 10 mg mL^{-1} with mercury(II) chloride. **(a)** 2-methylmethcathinone, **(b)** 3-methylmethcathinone, **(c)** 4-methylmethcathinone (mephedrone). Reprinted from L. Elie, M. Elie, G. Cave, M. Vetter, R. Croxton and M. Baron, *Raman Spectroscopy*,[9] John Wiley and Sons. Copyright © 2016 John Wiley & Sons, Ltd.

NPS products. As MDAI was never controlled in the UK it was still appearing in products up until the introduction of the Psychoactive Substances Act in 2016.

MDAI free base and hydrochloride salt both give characteristic blade microcrystals with mercury(II) chloride. For the free base solubility is improved by using a 10% aqueous methanol solvent for both the drug and the reagent. The free base forms serrated blades growing either individually or as 2-dimensional clusters. With assisted nucleation microcrystals are observed within 60 s along the pipette trail. Crystals also develop elsewhere within the drop but more slowly.

MDAI hydrochloride trapezoidal blades develop quickly with assisted nucleation along the pipette trail with more crystals developing throughout the drop after about 2 minutes (Figure 10.10a). As MDAI hydrochloride is more soluble in water than the freebase, aqueous solutions can be used for both the drug and test reagent. The limit of detection of the test is 0.5 mg mL^{-1}. The blades show anisotropy with parallel extinction (Figure 10.10b) and have a positive sign of elongation determined with the Red-1 (Figure 10.10c) and quarter-wave (Figure 10.10d) compensators.

10.3.5. *Mixtures and NPS Products*

Most NPS products are mixtures of substances often consisting of the main NPS active ingredient(s) such as 4-MMC, MDAI *etc*, cutting and bulking agents. Caffeine is often found in NPS products. It is a legal stimulant and added as a cutting agent. It can often be the major constituent and so it is important that any test for an NPS is not masked by the presence of other substances such as caffeine.

Figure 10.10. MDAI and caffeine microcrystals at 10 mg mL^{-1} with mercury(II) chloride. **(a-d)** MDAI, **(e-h)** caffeine, **(i-l)** mixture. **(a, e, i)** Original, **(b, f, g)** crossed polar, **(c, g, k)** crossed polar and Red-1 compensator, **(d, h, l)** crossed polar and quarter-wave compensator.

Caffeine gives characteristic needle shaped microcrystals when mixed with mercury(II) chloride reagent that grow from the edge of the drop inwards (Figure 10.10e). The orientation of the crystals within the drop is 3-dimensional hence the range of interference colours seen along the length of many of the crystals. However, some crystals appear aligned normal to the vibrational direction of illumination as shown by the uniform colour of the crystal seen with crossed polars (Figure 10.10f), the Red-1 and the quarter-wave compensators (Figure 10.10g and h). The alignment of these crystals with the compensators indicate caffeine-$HgCl_2$ crystals have a positive sign of elongation with parallel extinction.

A mixture of caffeine and MDAI gives needle shape crystals of caffeine-$HgCl_2$ along with trapezoidal blades of MDAI-$HgCl_2$ showing that the test is able to detect both MDAI and the cutting agent caffeine (Figure 10.10i).

The characteristic interference colours of both crystals can be seen under crossed polars with the Red-1 and quarter wave compensators (Figure 10.10j, k and l) giving additional characteristics for confirming a match with the reference standards.

The identification of caffeine is a bonus of this test as the main validation requirement is that the test can detect MDAI in the presence of other substances such as caffeine that are likely to be found in NPS street samples. For NPS/caffeine mixtures, MDAI can be detected in mixtures ≥40%(w/w), BZP ≥20%(w/w) and 4-MMC ≥50%(w/w) when tested with the mercury(II) reagent. BZP and caffeine were identified in NPS products in which these substances were not advertised as being present[10] which was then confirmed by GC-MS.[19]

10.4. Conclusion

Microcrystalline tests for the analysis of standard drugs of abuse are well established and although they may seem somewhat out-dated in comparison to modern instrumental techniques, this chapter has shown that they still have a role to play in modern drug analysis. It is relatively straightforward to develop a test for a new substance by screening through the relatively small group of reagents that are available and so, with the proliferation of new designer drugs, the microcrystalline test offers a cost-effective alternative to putting in placc a validated method of analysis. The results for selected NPS presented show how unequivocal identification can be achieved using optimised tests. The example of GHB shows that it is relatively straightforward to establish a test that works but that further development may be needed to obtain a fully optimised test. Thc use of techniques such as X-ray Diffraction is helping to more clearly understand the nature of microcrystals and their formation giving rise to a more systematic approach to developing new tests.

The tests presented have also shown how the same reagent, mercury(II) chloride can give rise to highly distinctive unique microcrystals for different substances. This shows that a potentially unlimited number of microcrystalline tests are possible from the small group of reagents that are commonly used. The search for new reagents will be an interesting area of future research. It has also been shown that control of factors such as rate of crystal growth may enhance the test. Enclosing the test within capillary tubes achieves this by restricting solvent evaporation producing superb 3-D rosettes of 4-MMC-$HgCl_2$ microcrystals. When combined with digital image capture and image processing software this represents a modernising of the photomicrograph procedure. The capillary tube platform also offers a portable, standardised approach for at scene analysis.

While the use of polarising light microscopy is not a modernising of the technique, the use of PLM with microcrystalline tests is a powerful combination offering additional specificity. This has been demonstrated for all the tests presented in the chapter. The crystal optical measurements can be added to the match criteria used in determining an identification. Combining microcrystalline tests with suitable modern analytical techniques is a modernising of the technique and has been shown using Raman Spectroscopy. Not only does this confirm the presence of the drug within the crystal using an *in situ* measurement but also gives identification of the NPS to the level of structural isomer as shown by the identification of the three structural isomers 2-MMC, 3-MMC and 4-MMC.

References

1. M. P. Elie and L. E. Elie, in *Encyclopedia of Analytical Chemistry*, ed. R. A. Meyers, John Wiley & Sons Ltd, 2009.
2. C. C. Fulton, *Modern Microcrystal Tests for Drugs*, John Wiley & Sons, Inc., New York, 1969.
3. E. G. C. Clarke, *Isolation and Identification of Drugs*, The Pharmaceutical Press, London, 1st edn, 1969.
4. K. M. Brinsko, D. Golemis, M. B. King, G. J. Laughlin and S. B. Sparenga, *A Modern Compendium of Microcrystal Tests for Illicit Drugs and Diverted Pharmaceuticals*, McCrone Research Institute, Inc., Chicago, 2016.
5. ASTM E1968-11, *Standard Guide for Microcrystal Testing in the Forensic Analysis of Cocaine*, ASTM International, West Conshohocken, PA, 2011.
6. ASTM E1969-11, *Standard Guide for Microcrystal Testing in the Forensic Analysis of Methamphetamine and Amphetamine*, ASTM International, West Conshohocken, PA, 2011.
7. ASTM E2125-07, *Standard Guide for Microcrystal Testing in the Forensic Analysis of Phencyclidine and Its Analogues*, ASTM International, West Conshohocken, PA, 2007.
8. D. Wielbo and I. R. Tebbett, *J. Forensic Sci.*, 1992, **37**(4), 1134.
9. L. Elie, M. Elie, G. Cave, M. Vetter, R. Croxton and M. Baron, *J. Raman Spectrosc.*, 2016, **47**(11), 1343.
10. L. Elie, M. Baron, R. Croxton and M. Elie, *Forensic Sci. Int.*, 2012, **214**, 182.
11. M. R. Wood, R. A. Lalancette and I. Bernal, *Acta Crystallogr., Sect. C: Struct. Chem.*, 2015, **71**, 32.
12. United Nations Office On Drugs and Crime, *Recommended Methods for the Identification and Analysis of Piperazines in Siezed Materials*, United Nations, New York, 2013, p. 18.
13. United Nations Office On Drugs and Crime, *Recommended Methods for the Identification and Analysis of Synthetic Cathinones in Siezed Materials*, United Nations, New York, 2015, p. 17.
14. *Scientific Working Group for the Analysis of Seized Drugs (SWGDRUG) Recommendations*, 2016.
15. K. M. Andera, H. K. Evans and C. M. Wojcik, *J. Forensic Sci.*, 2000, **45**(3), 665.
16. M. P. Elie, M. G. Baron and J. W. Birkett, *J. Forensic Sci.*, 2008, **53**(1), 147.
17. M. P. Elie, Development of Analytical Methods for the Analysis of γ-Hydroxybutyrate (GHB), PhD Thesis, University of Lincoln, UK, 2015.
18. L. E. Elie, M. G. Baron, R. S. Croxton and M. P. Elie, *Drug Test. Anal.*, 2013, **5**(7), 573.
19. M. Baron, M. Elie and L. Elie, *Drug Test. Anal.*, 2011, **3**, 576.

Chapter 11

Optical Detection of NPS Internet Products *via* HPLC-DAD Systems: A Selective Review

M. G. Schmid

Department of Pharmaceutical Chemistry, Institute of Pharmaceutical Sciences, University of Graz, Universitätsplatz 1, A-8010 Graz, Austria
*E-mail: martin.schmid@uni-graz.at

Table of Contents

Light in Forensic Science: Issues and Applications
Edited by Giorgia Miolo, Jacqueline L. Stair and Mire Zloh

Published by the Royal Society of Chemistry, www.rsc.org

11.1. Introduction

Over the last decade, a significant change in recreational drug use has been monitored. Novel psychoactive substances (NPSs) compete with former and well known drugs of misuse such as cannabis, opium, cocaine or amphetamine on the worldwide illicit drug market. In the last six years, no less than 345 non-scheduled synthetic new compounds have been reported to the European Monitoring Centre for Drugs and Drug Addiction (EMCDDA) (2015: 100, 2014: 101, 2013: 81, 2012: 73, 2011: 49, 2010: 41).[1]

These compounds are spread *via* the internet as "legal highs", "research compounds", "leisure drugs", "plant food", "bird cage cleaner" or "room odorizers"; due to their chemical structure, they are intended to fall outside of drug legislation. These substances belong structurally to several compound classes including cathinones, synthetic tryptamines, phenetylamines, synthetic cannabinoids, arylamines or piperazines and mimic the psychoactive effects of scheduled compounds.[2]

Novel recreational drugs are mostly synthesized in Asian countries and East Europe and are traded *via* Internet shops with a disclaimer "Not for human consumption". In many cases, online vendors even publish product analyses by mass spectroscopy or nuclear magnetic resonance (NMR) data on their websites in order to certify the purity of their offered products. However, the correctness of the data is doubtful as well as the fact that the provided information may not be understandable for consumers.

Many of the NPSs contain a stereogenic centre. Therefore it is assumed that their enantiomers exhibit different pharmacological effects, as it is already reported *e.g.* for methcathinone[1,3] and amphetamine.[4,5]

As non-enantioselective synthesis is easier and cheaper to perform, new designer drugs are mainly traded as racemic mixtures.[6–11]

A considerable challenge of these NPSs is that due to their novelty only a few analytical methods for their identification and separation are available. Quantitative analysis is often very difficult or even impossible since there are no or very expensive certified standards available.[12] Moreover, there is little knowledge and data about their pharmacology and toxicology as well as the long-term damage they may cause. Hence, there is great forensic and medical interest to develop chromatographic and electrophoretic methods for the analysis of NPSs.

These new compounds are to be analyzed *e.g.* as powder samples or pills, which can be purchased *via* the internet. However, in a bioanalytical context, they have to be analysed in physiological samples, such as in urine, hair, saliva, blood as well as serum.[2,13-16]

This chapter is intended to give an overview of methods to detect NPSs traded *via* the Internet by UV detection in combination with a high-performance separation technique, mainly HPLC, but also capillary electrophoresis (CE), capillary electrochromatography (CEC) and supercritical fluid chromatography (SFC) are to be taken into account. For UV-detection, the number of papers is rather limited, since in most cases structure elucidation of new compounds is of main interest, which has to be accomplished by other detection techniques, mainly MS due to low UV significance. In the following chapters, NPSs will be presented according to their chemical substance class. There, it can be assumed that each classic illicit drug can be replaced by a NPS alternative.[2]

11.2. Achiral and Chiral Separation of Novel Psychoactive Substances by HPLC or by Other High Performance Separation Techniques with UV-detection

Generally, there are two analytical approaches for illicit drugs: on the one hand immunoassay screening, which is advantageous because of high throughput and high speed, but which has drawbacks because of limited specifity and false positive responses.[17] On the other hand, method development for HPLC, GC, CE, CEC and SFC is important to confirm the identity of NPSs or their quantification. Since preliminary immunoassays mostly fail to produce correct results for NPSs, the latter techniques have attracted more interest. The following subchapters will give an overview of the state of the art of analysis of NPSs by chemical compound classes.[17]

11.2.1. Synthetic Cannabinoids

During the last 10 years, the role of cannabis, mainly as marijuana, hashish or hash-oil has changed for consumers due to the upcoming availability of partially legal alternatives. A prominent example was "Spice", which appeared in Europe in 2008 and was sold *via* the Internet or in bricks and mortar shops. It was claimed to be a natural substitute for cannabis; later it turned out to be a cheap herbal mixture, to which a synthetic cannabinoid was added. Since homogenisation of the drugs in the matrix was poor, in some cases drug overdose occurred.

Later, a solution of the active ingredient was sprayed on the herbal matrix prior to the drying procedure. As a consequence, a huge number of these synthetic cannabinoid agonists were used as recreational drugs in the scene. These compounds are also available as liquids for e-cigarettes and shishas. Table 11.1 gives a few examples of different chemical compound classes. They can be classified as synthetic cannabinoids, non-classical cannabinoids, aminoalkylindoles, aminoalkylindazoles, eicosanoides, fatty acid amide hydrolase inhibitors and others.[18] Meanwhile the variety of synthetic cannabinoids has become vast: some of which were discovered only 2–4 years ago and have not yet been scheduled. For these newest compounds, to date no certified standards are available. Often, they are labelled with misleading names as can be verified on various platforms. Similar to delta-9-tetrahydrocannabinol, these substances are rather lipophilic, however, they can be analysed by normal-phase or reversed-phase systems

Table 11.1. Examples for new synthetic cannabinoids, which are available *via* the Internet.

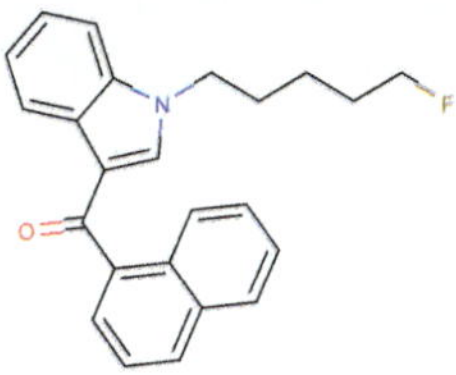	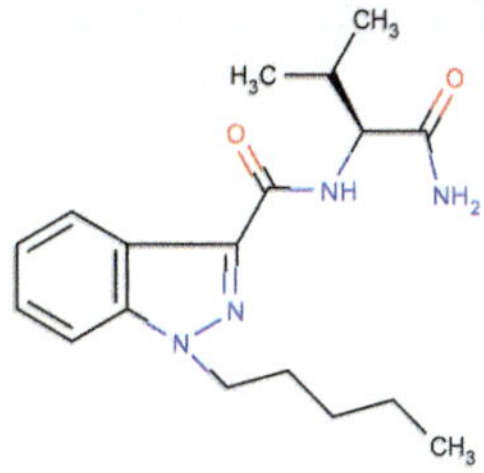
AM 2201 Chemical name: 1-[(5-Fluoropentyl)-1*H*-indol-3-yl]-(naphthalen-1-yl) methanone	**AB-PINACA** Chemical name: *N*-[(1*S*)-1-(aminocarbonyl)-2-methylpropyl]-1-pentyl-1*H*-indazole-3-carboxamide
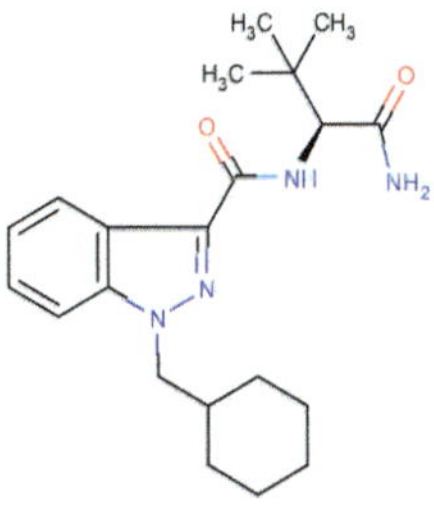	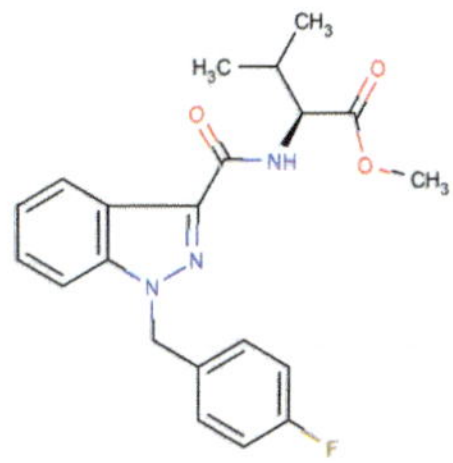
MAB-CHMINACA Chemical name: *N*-[(2*S*)-1-amino-3,3-dimethyl-1-oxobutan-2-yl]-1-(cyclohexylmethyl) indazole-3-carboxamide	**AMB-FUBINACA** Chemical name: Methyl (2*S*)-2-{[1-[(4-fluorophenyl)methyl]indazole-3-carbonyl] amino}-3-methylbutanoate

containing a high content of organic components, such as methanol or acetonitrile.

As an example, Figure 11.1 demonstrates the determination of a sample of AMB-FUBINACA by HPLC-DAD on a simple reversed-phase column. This technique has mainly been reported for the separation of new synthetic cannabinoids of purchased or seized samples, however, identification has to be done by mass spectrometry or NMR. For example, Lee and co-workers report on the identification of APINAC as a new synthetic cannabinoid using GC-TOF-MS, IR and NMR. Additionally, they used HPLC-UV after having monitored the appropriate UV spectra.[19] Another novel compound, MDMB-CHMICA, was investigated for its absolute configuration and abundance in illegal products. The authors used normal-phase preparative chromatography and a UV-VIS spectrophotometer for its isolation from a herbal blend sample.[20] Additionally, chiral separation of racemic MDMB-CHMICA was achieved using a Chiralpak IA-3 column and a wavelength of 290 nm. Different synthetic cannabinoids such as AM- HU- JWH- and RCS-derivatives were quantified in plant materials by HPLC-UV by Ciolino.[21] To emphasize the risk of inhomogeneities in herbal mixtures, Auwärter's group published a determination method of synthetic cannabinoids by means of HPLC-DAD.[22] After their extraction, they were quantified using a C18 column and detection

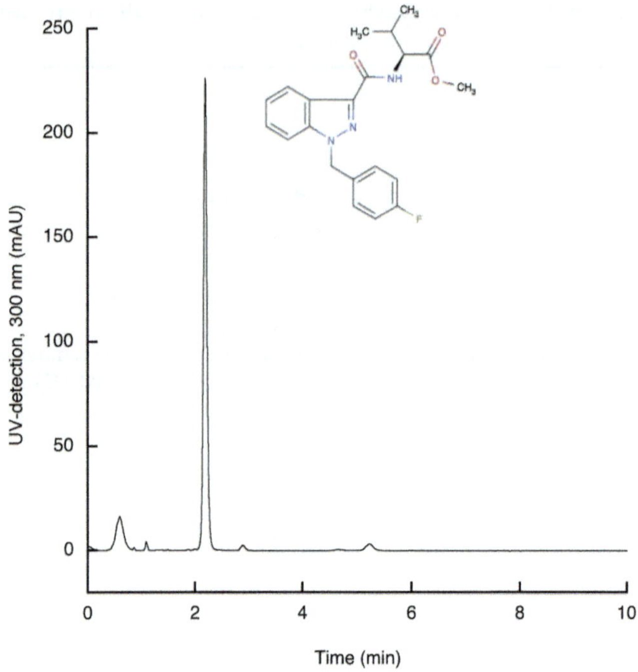

Figure 11.1. Determination of the cannabinoid AMB-FUBINACA by HPLC-DAD. Conditions: Stationary phase: LiChrospher 100 RP-18 (5 μm) LiChroCART 125-4. Mobile phase: 25 mM triethylammonium phosphate buffer pH 3.0/acetonitrile (36 : 64), injection: 3 μl, flow rate: 1.6 ml min^{-1}, detection wavelength: 300 nm (M. G. Schmid, unpublished work).

wavelengths of 209 or 217 nm. Another attempt for the use of preparative HPLC was done by Park *et al.* for the isolation of 1-butyl-3-(2methoxybenzoyl)indole prior to identification as a new synthetic cannabinoid.[23] Similarly, the synthetic cannabinoid JWH-018 was extracted from commercially available herbal marijuana alternatives by Dunham.[24] In 2012, Uchiyama *et al.* identified two new-type synthetic cannabinoids, *N*-(1-adamantyl)-1-pentyl-1*H*-indole-3-carboxamide (APICA) and *N*-(1-adamantyl)-1-pentyl-1*H*-indazole-3-carboxamide (APINACA) *via* HPLC using ESI mass spectra, but also UV spectra, since both compounds could also be distinguished by their UV spectra.[25] Furthermore, they succeeded in detecting five synthetic cannabinoids, AM-1220, AM-2233, AM-1241, CB-13 (CRA-13), and AM-1248, as designer drugs in illegal products. Finally, a manual for use by national drug analysis laboratories was published by The United Nations Office on Drugs and Crime presenting recommended approaches for the identification and analysis of synthetic cannabinoids in seized materials.[26] Herein, a UHPLC-DAD method for a broad spectrum of compounds is presented along with suitable detection wavelengths. Furthermore, a scheme for the general

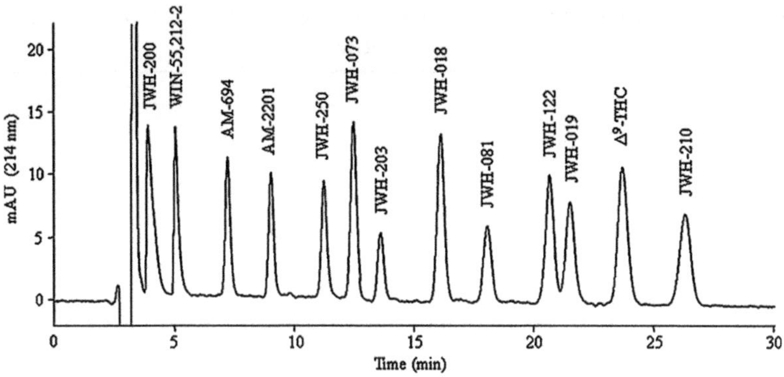

Figure 11.2. Nano-LC separation of synthetic cannabinoids and Delta-9-THC in the optimum conditions: mobile phase acetonitrile/methanol/water/formic acid 69/5/25/1 (v/v/v/v); flow rate 500 nl min^{-1}; concentration of all analytes 25 μg ml^{-1} and Delta-9-THC 50 μg mL^{-1}. Reprinted from *Journal of Pharmaceutical and Biomedical Analysis*, **71**, G. Merola, Z. Aturki, G. D'Orazio, R. Gottardo, T. Macchia, F. Tagliaro and S. Fanali, Analysis of synthetic cannabinoids in herbal blends by means of nano-liquid chromatography, 45–53, Copyright (2012), with permission from Elsevier.

isolation and chemical characterization of new synthetic cannabinoids is described.[26]

In 2012, Fanali's group reported on a simultaneous separation of 12 synthetic cannabinoids and delta-9-tetrahydrocannabinol in herbal blends.[27] Nano-liquid chromatography was performed in a 25 cm capillary column (100 μm ID) packed with RP-18 silica particles. Figure 11.2 shows the separation of the compounds by isocratic elution mode in less than 30 min. Recently, SFC was also found to be a powerful alternative to HPLC.

An appropriate method for the systematic toxicology analysis of cannabinoids and their metabolites was developed by Geryk *et al.* performing urine spiking experiments.[28]

11.2.2. New Benzodiazepines

Since chlordiazepoxide was discovered in 1955, dozens of derivatives bearing a benzene ring and a diazepine ring were used as sedative and hypnotic medication. This compound class replaced barbiturates in the 1970s and nowadays represents a frequently prescribed psychoactive drug class. Also in this field, there is abuse, since many benzodiazepines can be purchased illegally *via* the Internet. Here, it is necessary to distinguish between Internet pharmacy shops, where fakes of common formulations can be purchased, and new, chemically modified benzodiazepines, which are traded in Internet shops, but not in common drug stores. Table 11.2 shows a choice of benzodiazepines being abused as recreational drugs.

Table 11.2. Examples for new benzodiazepines, which can be purchased *via* the Internet.

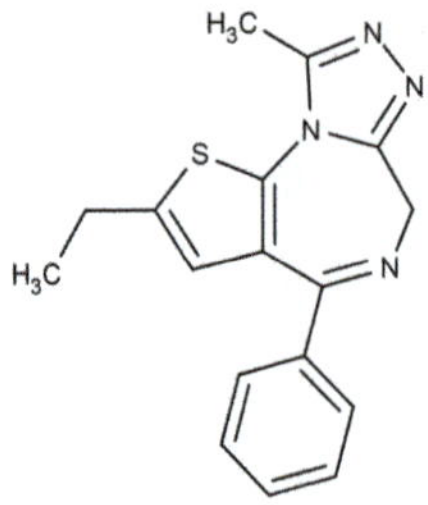 **Deschloroetizolam** Chemical name: 2-Ethyl-9-methyl-4-phenyl-6*H*-thieno[3,2-f][1,2,4]triazolo[4,3-a][1,4]diazepine	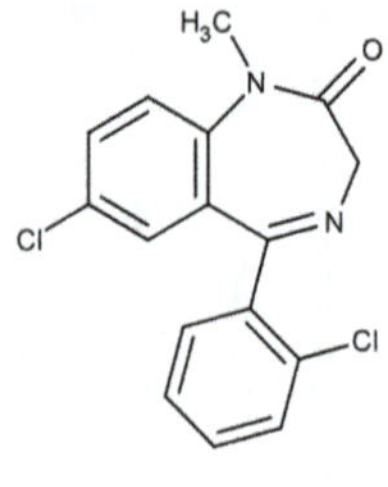 **Diclazepam** Chemical name: 7-Chloro-5-(2-chlorophenyl)-1-methyl-1,3-dihydro-2*H*-1,4-benzodiazepin-2-one
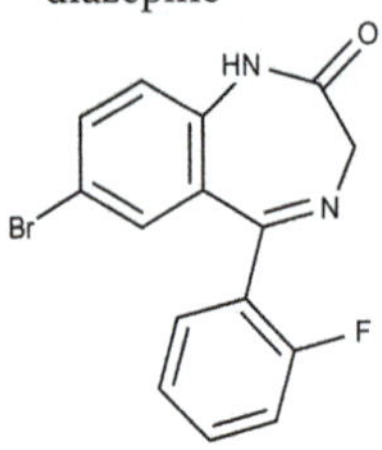 **Flubromazepam** Chemical name: 7-Bromo-5-(2-fluorophenyl)-1,3-dihydro-1,4-benzodiazepin-2-one	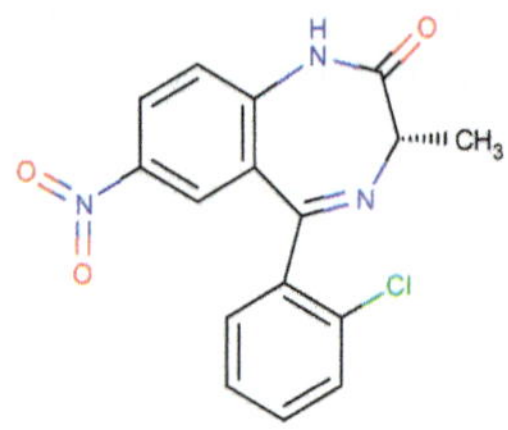 **Meclonazepam** Chemical name: (3*S*)-5-(2-chlorophenyl)-3-methyl-7-nitro-1,3-dihydro-1,4-benzodiazepin-2-one
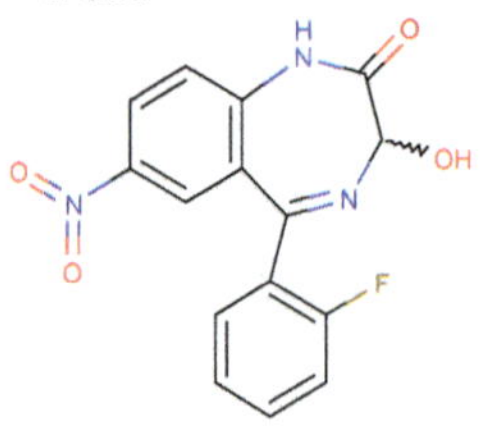 **Nifoxipam** Chemical name: 5-(2-Fluorophenyl)-3-hydroxy-7-nitro-2,3-dihydro-1*H*-1,4-benzodiazepin-2-one	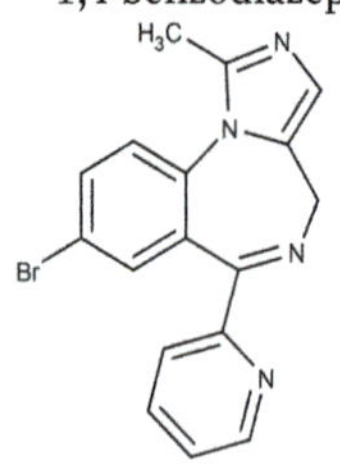 **Pyrazolam** Chemical name: 8-Bromo-1-methyl-6-(pyridin-2-yl)-4*H*-[1,2,4]triazolo[4,3-a][1,4]benzodiazepine

Since benzodiazepines in general show good UV absorption, seizures can easily be analyzed by HPLC-DAD. Figure 11.3 shows the HPLC-analysis of a pill of clonazepam purchased at www.officialbenzofury.com using the reversed-phase mode.

Mashayekhi and Khalilian report on an approach for the simultaneous determination of oxazepam, lorazepam and diazepam in human urine and plasma by HPLC-UV.[29] To gain sufficient sensitivity, samples are enriched by solid-phase extraction coupled with dispersive liquid–liquid microextraction. A supramolecular solvent-based hollow fiber liquid phase microextraction of five different classic benzodiazepines was developed

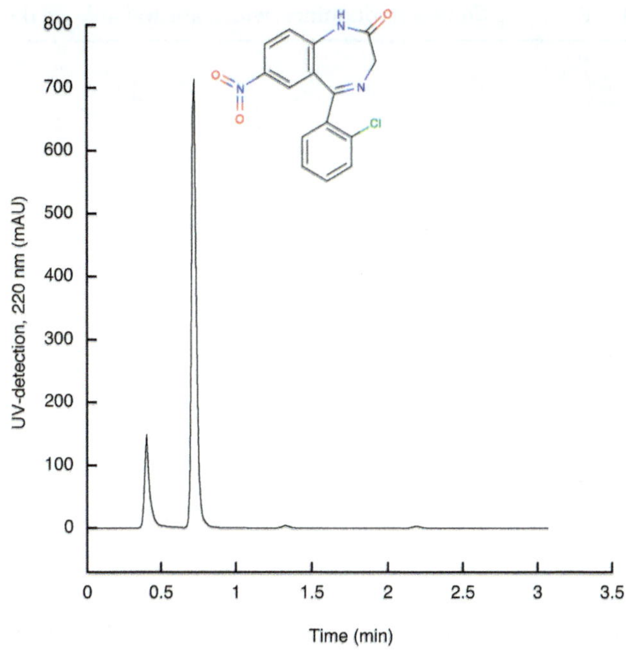

Figure 11.3. Analysis of clonazepam by HPLC-UV. Conditions: Stationary phase: LiChrospher 100 RP-18 (5 μm) LiChroCART 125-4. Mobile phase: 25 mM triethylammoniumphosphate buffer pH 3.0/acetonitrile (36 : 64), injection: 3 μl, flow rate: 1.6 ml min^{-1}, detection wavelength: 220 nm (M. G. Schmid, unpublished work).

for extraction of aqueous contaminated solutions. Ingredients were separated and detected by HPLC-UV.[30] Quantification of some real life samples is also shown. Another group described the determination of some benzodiazepines besides tricyclic antidepressants in saliva by HPLC along with LOQ and LOD values.[31] To increase concentration of the samples, sequential solid phase extraction was performed. In future, all these afore presented methods could be transferred for separation and detection of new benzodiazepines.

11.2.3. *New Tryptamines*

Synthetic tryptamines, such as LSD or AMT have been used by consumers for decades as hallucinogenic drugs. Natural representatives, such as dimethyltryptamine or ingredients of hallucinogenic mushrooms have been used in South America during religious ceremonies for centuries.[32] Alexander Shulgin dedicated an entire book to tryptamines and named it TIKAL, which stands for "Tryptamines I Have Known and Loved".[33] Published in 1997, he described 55 tryptamine derivatives. Some of them are prohibited in many countries, such as DMT and LSD, but on the other hand,

Table 11.3. Examples for new tryptamines, which are available *via* the Internet.

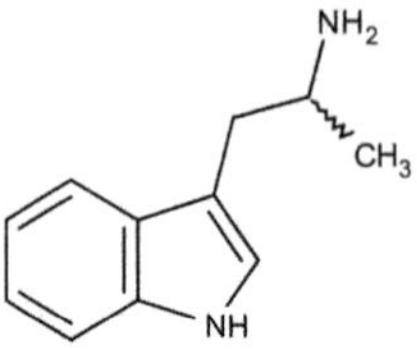 **AMT**, α-Methyltryptamine Chemical name: 2-(1*H*-indole-3-yl)-1-methyl-ethylamine	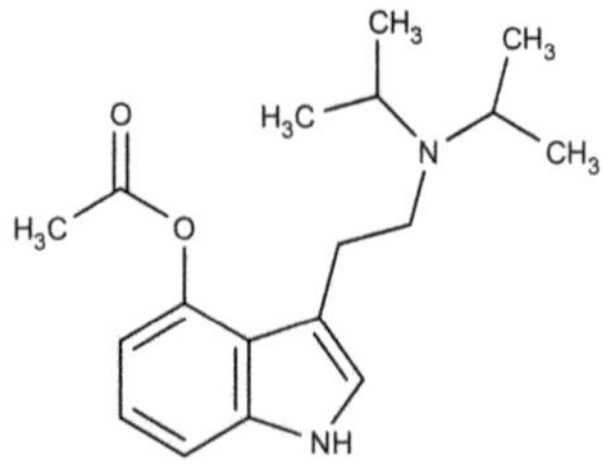 **4-AcO-DiPT**, 4-acetoxy-*N*,*N*-diisopropyltryptamine Chemical name: 3-[2-(Diisopropylamino)-ethyl]-1*H*-indole-4-yl acetate
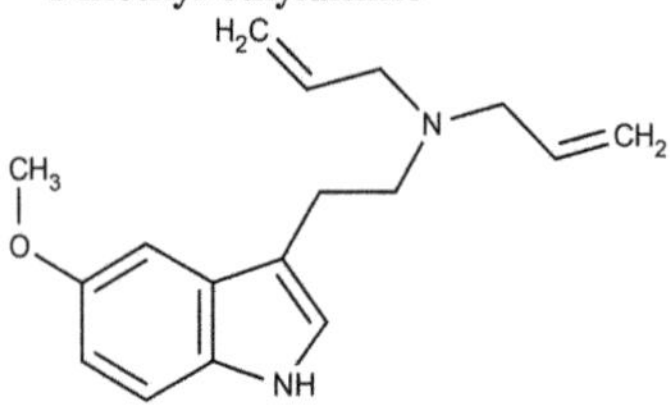 **5-MeO-DALT**, 5-methoxy-*N*,*N*-diallyl-tryptamine Chemical name: *N*-Allyl-*N*-[2-(5-methoxy-1*H*-indole-3-yl)-ethyl]-prop-2-en-1-amine	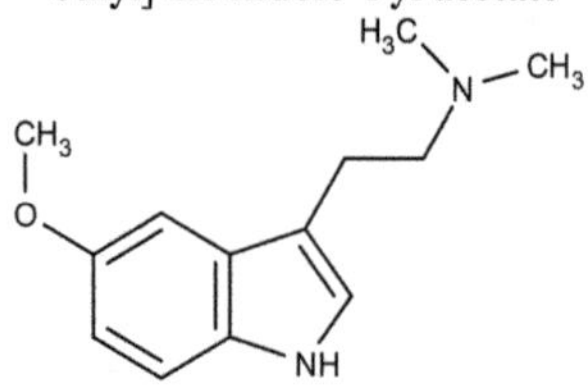 **5-MeO-DMT**, 5-methoxy-*N*,*N*-dimethyl-tryptamine Chemical name: 2-(5-Methoxy-1*H*-indole-3-yl)-*N*,*N*-dimethyl-ethanamine
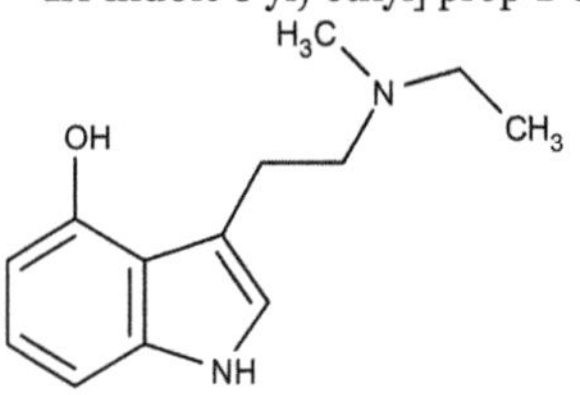 **4-OH-MET**, 4-Hydroxy-*N*-methyl-*N*-ethyltryptamine Chemical name: 3-(2-(Ethyl(methyl)amino) ethyl)-1*H*-indole-4-ole	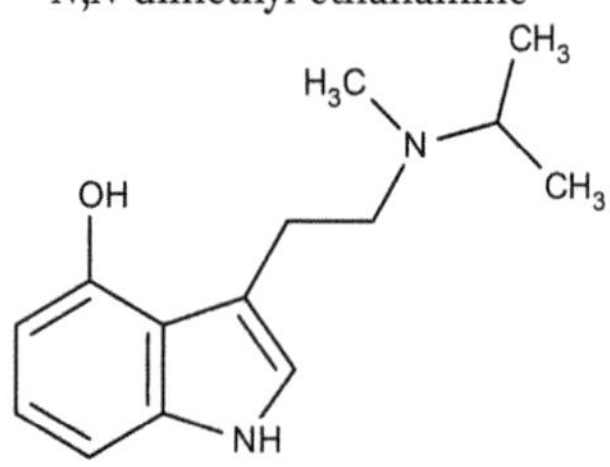 **4-OH-MiPT**, 4-Hydroxy-*N*-methyl-*N*-isopropyltryptamine Chemical name: 3-(2-[Isopropyl(methyl) amino]ethyl)-1*H*-indole-4-ol

many legal alternatives of LSD are traded *e.g.* as 150 μg blotters *via* the Internet. A set of tryptamines available at online stores can be found in Table 11.3.

A return of the lysergamides was claimed by Brandt *et al.* in two papers dealing with analytical and behavioural characterization of 1P-LSD, AL-LAD and LSZ, which are traded *via* the Internet.[34,35] Structure elucidation was accomplished by NMR, LC-MS and GC-MS. Since UV data was given as well, it can be assumed that HPLC-UV methods developed for the analysis of LSD should also work for these new alternatives. Figure 11.4 shows quantification of a classic LSD blotter by RP-HPLC-UV using a wavelength of 310 nm.

In 2012, Elliott *et al.* published analytical data of AMT (3-(2-aminopropyl) indole) and 5-IT (5-(2-aminopropyl)indole) by means of NMR, GC-EI/CI and LC-MS.[36] Besides this, also an attempt to separate these two compounds by

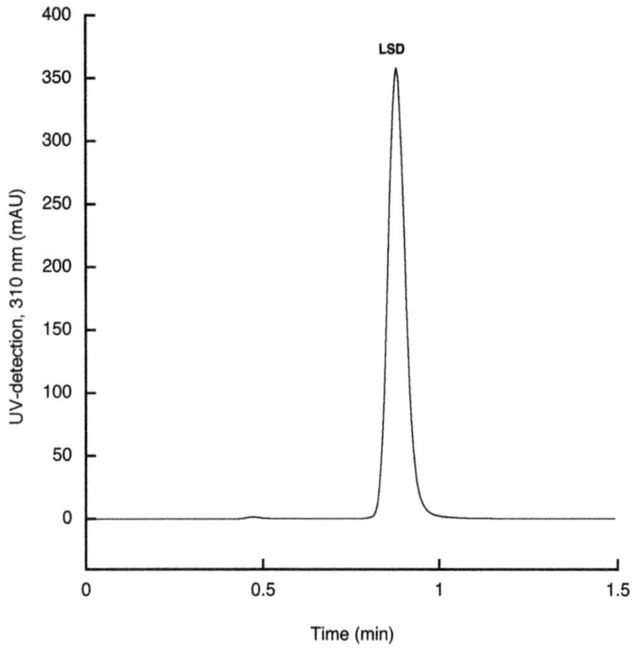

Figure 11.4. Determination of a LSD blotter extract by HPLC-UV Conditions: Stationary phase: LiChrospher 100 RP-8 (5 μm) LiChroCART 125-4. Mobile phase: Methanol/acetonitrile/TEA (50 : 50 : 0.1), injection: 3 μl, flow rate: 1.5 ml min^{-1}, detection wavelength: 310 nm (M. G. Schmid, unpublished work).

HPLC-DAD within four minutes was presented. Soh and Elliott presented stability studies of a broad spectrum of NPSs, among them 5-MeO-DALT and AMT. These compounds were spiked in blood and plasma for a time period at ambient temperature. Then, suspected breakdown products were identified and detected by HPLC-DAD.[37]

After test purchase, a set of 13 new tryptamines was checked for identity by GC-MS and separated by HPLC-UV. They were divided into five substance classes, among them tryptamines with an acetoxy group, methoxy group or hydroxy group on position 4 or 5 of the indol ring. Since some of them were chiral, successful enantioseparation of (±)-5-(2-aminopropyl)indole on an Astec Cyclobond I 2000 chiral column was the proof that this drug is traded as enantiomer.[38] In the following, CE-UV was also used to separate tryptamines. Figure 11.5 displays an electropherogram of the simultaneous separation of seven tryptamines.

11.2.4. New Phenethylamines

This compound class gained high popularity after the publication of Alexander Shulgin's book "PIKAL: A Chemical Love Story" in 1991. Herein, Shulgin describes 179 different phenethylamines being used by himself and his wife

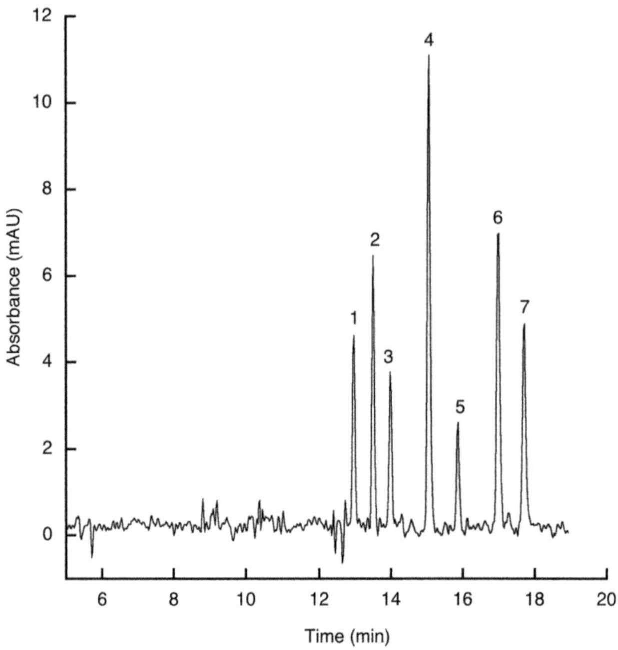

Figure 11.5. Electropherogram of the simultaneous separation of (1) tryptamine, (2) *N*-ω-methyltryptamine, (3) 5-IT, (4) MiPT, (5) 4-OH-MiPT, (6) 5-MeO-DiPT, (7) 4-AcO-DPT. Conditions: Capillary: 50 µm ID fused silica capillary 68.5 cm (effective length 60 cm); electrolyte: 30 mM ammonium phosphate solution pH 2.1, methanol (90 : 10); cassette temperature: 40 °C; UV: 208 nm; applied voltage: 20 kV; injection: 5 kV for 5 s (M. G. Schmid, unpublished work).

as psychedelics.[39] Meanwhile, many of them have been scheduled. Particularly the subgroup of 2C (2C-X) compounds (Table 11.4) became famous.

They contain on the 2 and 5 positions of a benzene ring two methoxy groups and various substituents at positions 3 and/or 4. Later, another very potent subfamily was discovered, when Ralf Heim prolonged 2004 at the Free University of Berlin the amino residue of 2C-B with a 2-methoxyphenylmethyl chain and published his work.[40] To date, the latter compound class represents an important replacement for LSD, since it requires similar small doses as LSD for the hallucinogenic effect. Table 11.4 shows two 2C compounds and two rather new phenethylamine analogues available *via* the internet.

25C-NBOMe was tested for its metabolites by Soh and Elliot by means of HPLC-DAD.[37] Lin *et al.* reported on the bioactivity of 2C-T-2 (2,5-dimethoxy-4-ethylthiophenethylamine). The group showed the detection of this phenethylamine in rat urine using capillary electrophoretic techniques combined with an on-line sample concentration technique.[41] A technique demonstrating how to analyze 2C-B in human urine by HPLC-UV after SPE was shown by Soares and coworkers.[42] In the chromatograms shown,

Table 11.4. Examples for new phenethylamines, which are offered *via* the Internet.

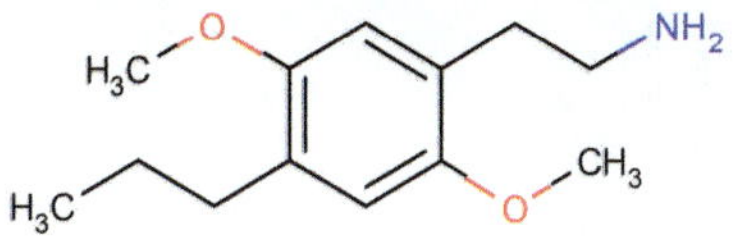 **2C-P** Chemical name: 2,5-Dimethoxy-4-propylphenylethylamine	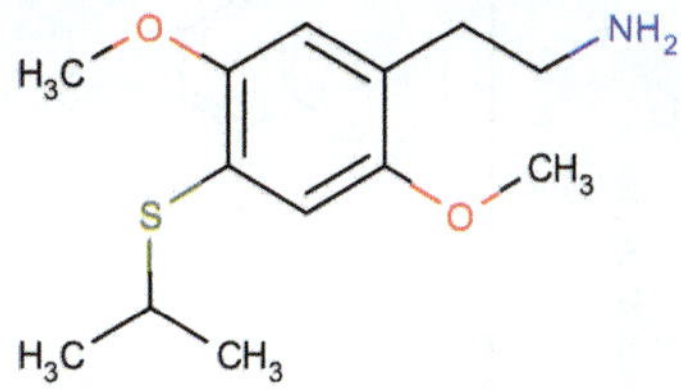 **2C-T4** Chemical name: 2-[4-(Isopropylthio)-2,5-dimethoxyphenyl]-ethanamine
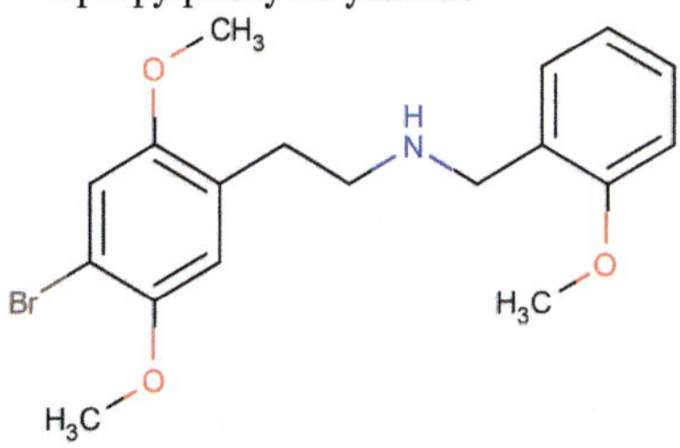 **25Br-NBOMe** Chemical name: 2-(4-Bromo-2,5-dimethoxy-phenyl)-*N*-[(2-methoxyphenyl)methyl] ethan-1-amine	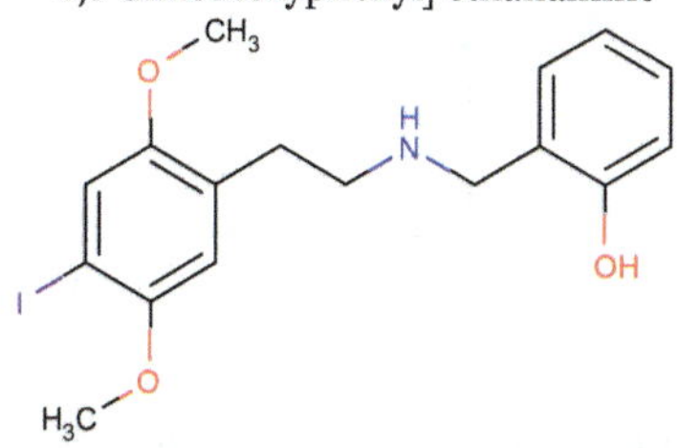 **25I-NBOH** Chemical name: 2-((2-(4-iodo-2,5-dimethoxyphenyl)ethylamino)methyl) phenole

2C-B eluted as the last compound compared to some other amphetamines. Another derivative, DOC, was successfully resolved into its enantiomers by a Lux Cellulose-2 column.[43] Figure 11.6 demonstrates the chiral separation of 2,5-dimethoxy-4-chloro-amphetamine on a Lux Cellulose-4 chiral column, which showed similar chiral separation ability.

11.2.5. New Amphetamines and Methamphetamines

Amphetamine was reported in 1887 for the first time and was administered as Benzedrone for decades. After the United Nations 1971 Convention on Psychotropic Substances, amphetamine became scheduled, but was still popular. After 2008, halogenated amphetamines appeared as legal alternatives on the global drug market, starting with 4-fluoroamphetamine (4-FA), later 3-fluoroamphetamine (3-FA) and 2-fluoroamphetamine (2-FA), but also some of their *N*-methamphetamine analogues.

Table 11.5 shows some amphetamines, which were purchased from various internet platforms and tested for identity. Contrary to phenethylamines, all amphetamines and *N*-methamphetamines possess a chiral centre leading to two possible enantiomers. It can be expected that there is a difference in the effect of the *R*- and the corresponding *S*-enantiomer, however, to date only a few studies have been published. Taking into consideration that *e.g.* fluoroamphetamine is available as ortho- meta- and para-isomer with two enantiomers each, we have to distinguish between six

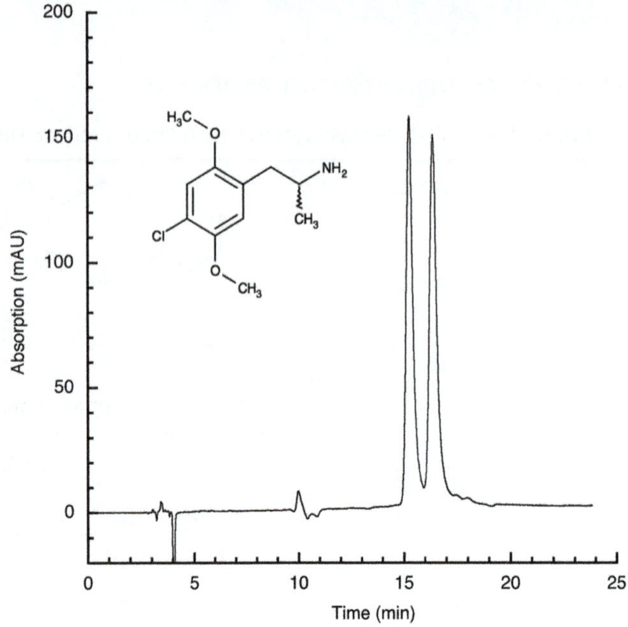

Figure 11.6. Enantioseparation of 2,5-dimethoxy-4-chloro-amphetamine. Conditions: Column: Lux Cellulose-4, 250 × 4.6 mm, 3 µm, mobile phase: ACN: isopropanol:DEA:FA (95:5:0.1:0.1), ambient temperature, flow: 1 ml min^{-1}, UV: 254 nm, injection: 5 µl (M. G. Schmid, unpublished work).

Table 11.5. Examples for amphetamines, which are available *via* the Internet.

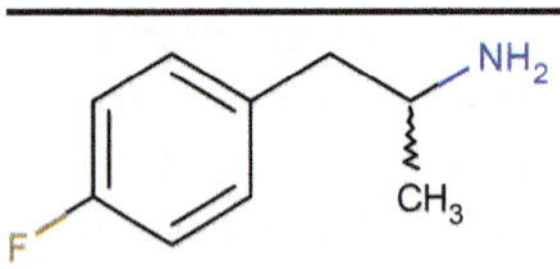 **4-FA,** 4-fluoroamphetamine Chemical name: 1-(4-Fluorophenyl) propan-2-amine	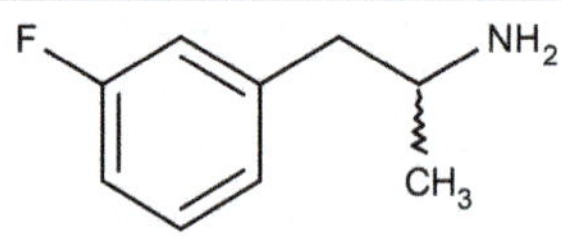 **3-FA,** 3-fluoroamphetamine Chemical name: 1-(3-Fluorophenyl) propan-2-amine
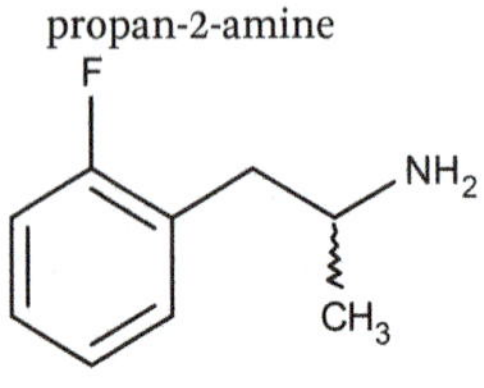 **2-FA,** 2-fluoroamphetamine Chemical name: 1-(2-Fluorophenyl) propan-2-amine	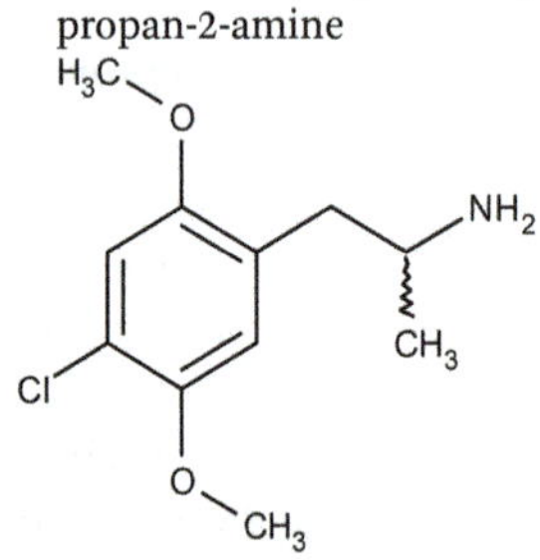 **DOC,** 2,5-dimethoxy-4-chloro-amphetamine Chemical name: 1-(4-Chloro-2,5-dimethoxy-phenyl)propan-2-amine
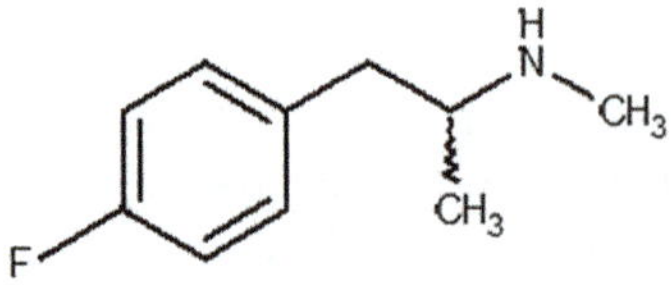 **4-FMA,** 4-Fluoromethamphetamine Chemical name: 1-(4-Fluorophenyl)-*N*-methylpropan-2-amine	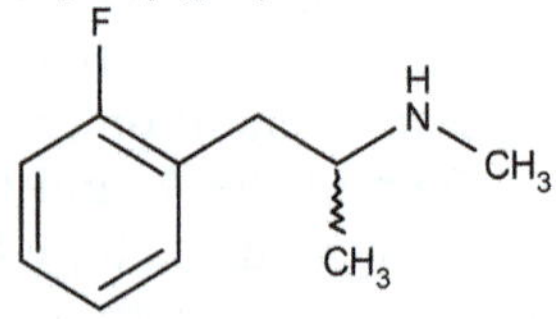 **2-FMA,** 2-Fluoromethamphetamine Chemical name: 1-(2-Fluorophenyl)-*N*-methylpropan-2-amine

isomers. Different amphetamine derivatives available on the internet were resolved into their enantiomers by HPLC-UV on a Lux-Cellulose 2 chiral column.[43] Some amphetamine derivatives were resolved in their enantiomers when the polar organic mode was chosen. Another attempt was the enantioseparation of amphetamines using a C-18 column and sulfated β-cyclodextrin as a mobile phase additive.[44] Moreover, SFC was carried out on a commercially available CHIRAL ART Amylose SA column for fluorinated amphetamines.[45] Mangeling's group compared chiral separation of 4-FA and 4-FMA using different chiral columns by means of CEC, SFC and three LC-modes. Detection wavelengths were set at 220 or 208 nm.[56] All tested amphetamines from the internet were found to be present as racemic mixtures.

Iwata *et al.* applied chemically modified capillaries in chiral capillary electrophoresis for methamphetamine profiling. In this work, chiral separations were also shown.[46] A general approach for enantioseparation of primary amines in human urine was shown by Chung's group.[47] They developed a single drop microextraction in-line coupled to capillary electrophoresis. In the following, electropherograms of amphetamine, *p*-chloro-amphetamine (4-CA) and 2-amino-1,2-diphenylethanol, one in buffer and one in urine are shown in this work.

11.2.6. Cathinones

This compound class is very similar to amphetamines, the difference is a beta keto group in the chemical structure. The basic molecule, S-cathinone, is a naturally occurring monoamine alkaloid, which can be found in *Catha edulis*, the khat shrub. Since stimulant effects are very similar to those of amphetamines, this substance class gained enormous popularity in the last 10 years. Since both parent compounds, cathinone and methcathinone are scheduled by the United Nations 1971 Convention on Psychotropic Substances, synthesis of derivatives, mainly of methcathinones increased dramatically. EMCDDA stated that cathinone derivatives represent a considerable percentage of synthetic NPSs.[1] More than 100 new substituted cathinones emerged, which act as stimulant and entactogenic NPSs.

Table 11.6 gives an overview of some substituted cathinones, starting with para-methylated methcathinone (4-MMC), which was named mephedrone and gained high popularity in the years 2008–2012.[48] After its ban, it was replaced by its meta- and ortho- analogues. Beside this, also halogenated methcathinones, such as flephedrone (4-FMC), brephedrone (4-BMC), and about 2 years ago, clephedrone (4-CMC) entered the market. Monitoring the availability of cathinone derivatives in Internet shops, during the last two years ethcathinones also appeared (Table 11.7). After the mephedrone ban in 2010, a well known substitute called 4-MEC was introduced in the European community. During 2016 and 2017, particularly pyrovalerones were sold *via* the Internet as recent NPSs, as illustrated in Table 11.8.

Table 11.6. Examples for methcathinones, which are traded *via* the Internet.

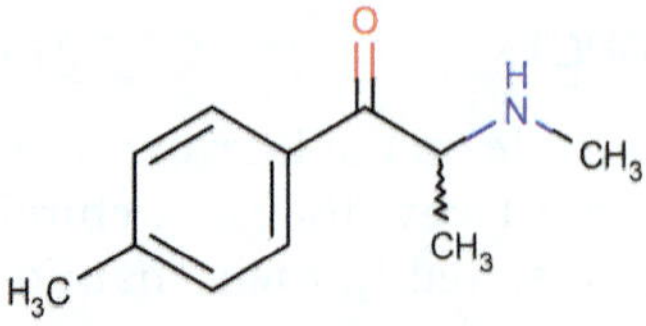 **4-MMC**, 4-Methylmethcathinon Chemical name: 2-(Methylamino)-1-(4-methylphenyl)-1-propan-1-one	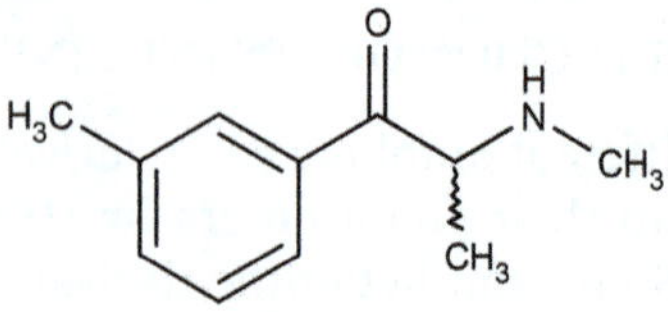 **3-MMC**, 3-Methylmethcathinon Chemical name: 2-(Methylamino)-1-(3-methylphenyl)-1-propan-1-one
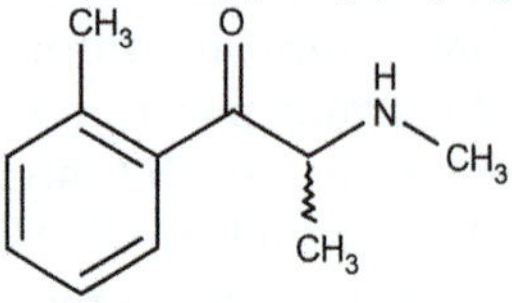 **2-MMC**, 2-Methylmethcathinone Chemical name: 2-(Methylamino)-1-(2-methylphenyl)-1-propan-1-one	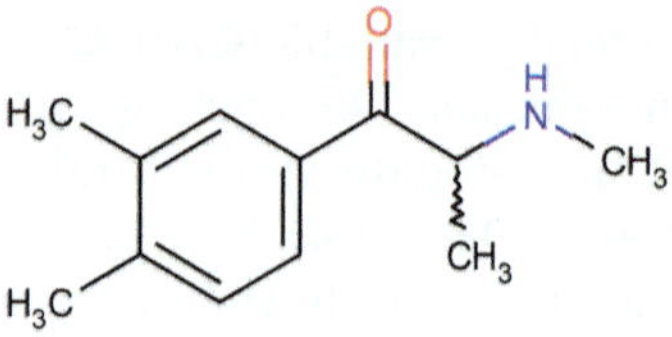 **3,4-DMMC**, 3,4-Dimethylmethcathinone Chemical name: 1-(3,4-Dimethylphenyl)-2-(methylamino)-propan-1-one
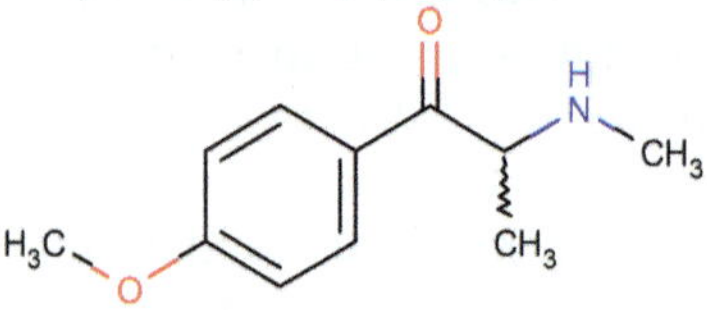 **Methedron**, 4-Methoxymethcathinone Chemical name: 1-(4-Methoxyphenyl)-2-(methylamino)-propan-1-one	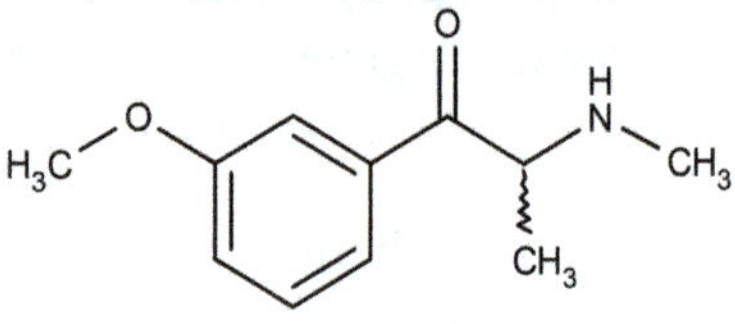 **3-MeOMC**, 3-Methoxymethcathinone Chemical name: 1-(3-Methoxyphenyl)-2-(methylamino)-propan-1-one
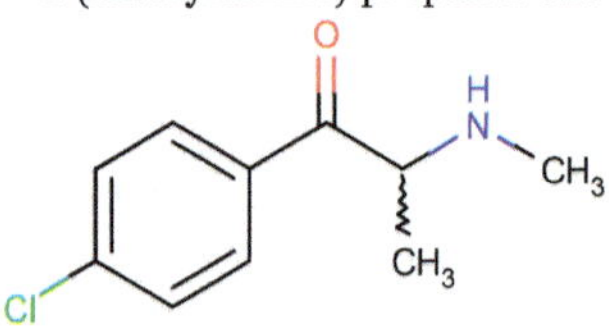 **4-CMC**, 4-Chloromethcathinone, clephedrone Chemical name: 1-(4-Chlorophenyl)-2-(methylamino)-1-propan-1-one	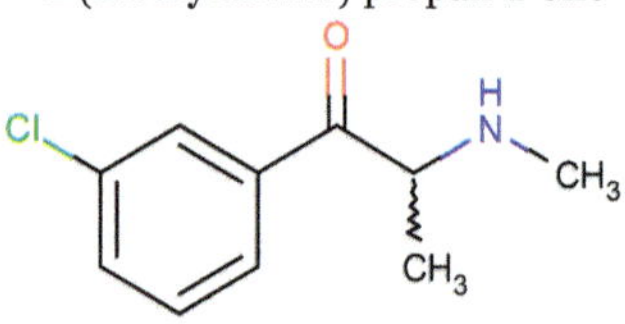 **3-CMC**, 3-Chloromethcathinone Chemical name: 1-(3-Chlorophenyl)-2-(methylamino)-1-propan-1-one
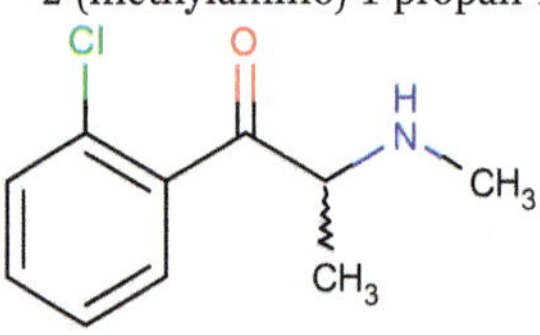 **2-CMC**, 2-Chloromethcathinone Chemical name: 1-(2-Chlorophenyl)-2-(methylamino)-1-propan-1-one	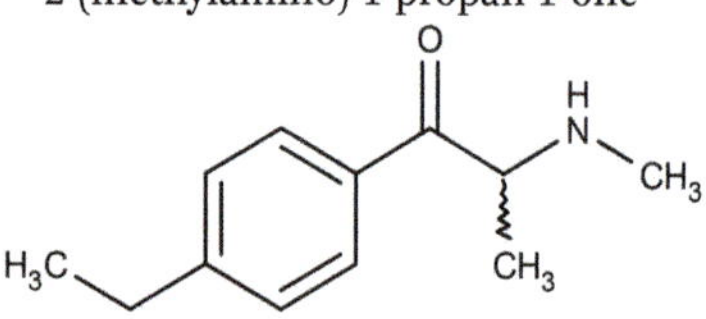 **4-EMC**, 4-Ethylmethcathinone Chemical name: 1-(4-Ethylphenyl)-2-(methylamino)propan-1-one
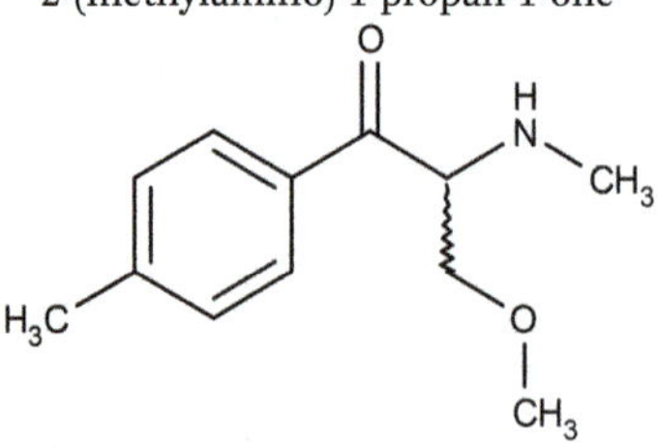 **Mexedrone** Chemical name: 3-Methoxy-2-(methyl-amino)-1-(4-methyl-phenyl) propan-1-one	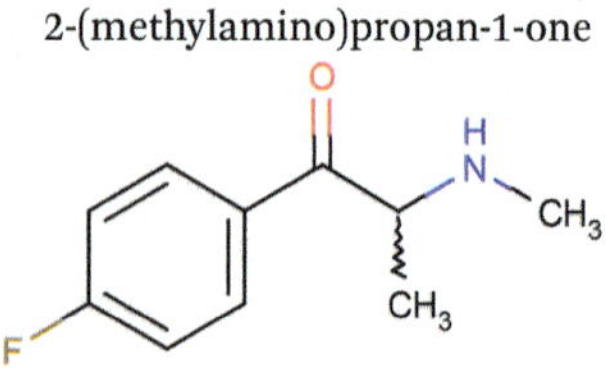 **4-FMC**, flephedrone Chemical name: 1-(4-Fluorophenyl)-2-(methylamino)propan-1-one

(continued)

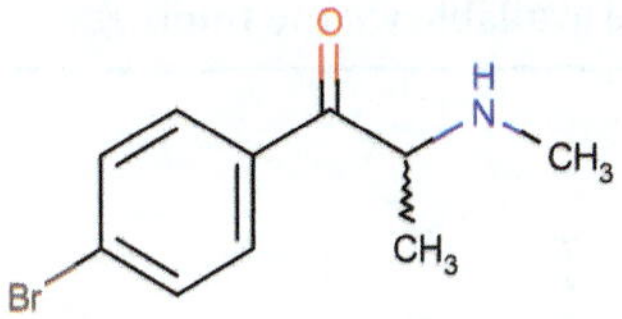

4-BMC, brephedrone
Chemical name: 1-(4-Bromphenyl)-2-(methylamino)propan-1-one

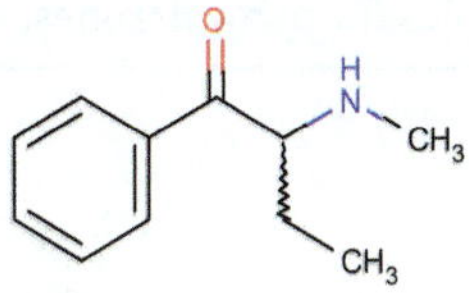

Buphedrone
Chemical name: 2-(Methylamino)-1-phenylbutan-1-one

Table 11.7. Examples for ethcathinones, which are available *via* the Internet.

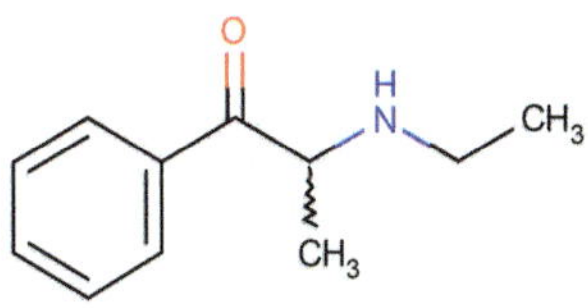

Ethcathinone (parent compound)
Chemical name: 2-Ethylamino-1-phenyl-propan-1-one

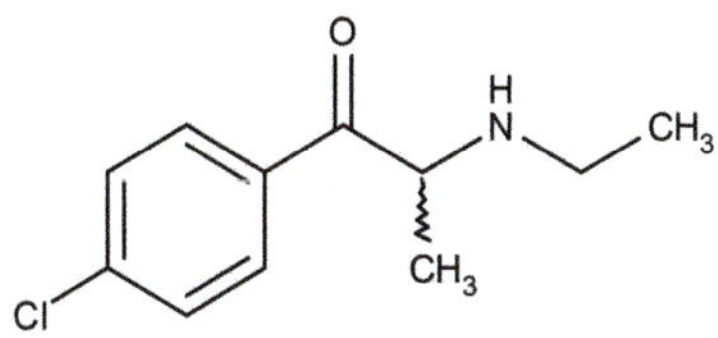

4-CEC, 4-Chloroethcathinone
Chemical name: 1-(4-Chlorphenyl)-2-(ethylamino)1-propan-1-one

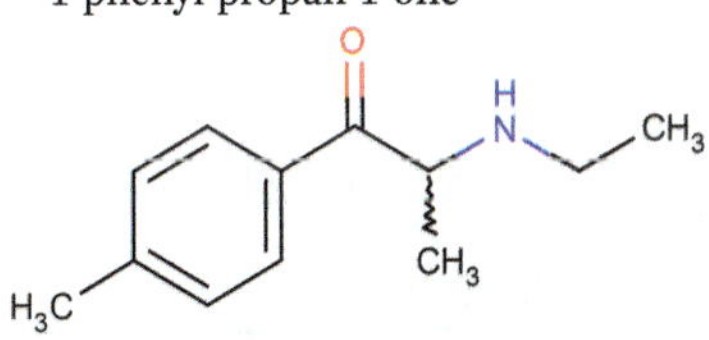

4-MEC, 4-Methylethcathinone
Chemical name: 2-(Ethylamino)-1-(4-methylphenyl)-1-propan-1-one

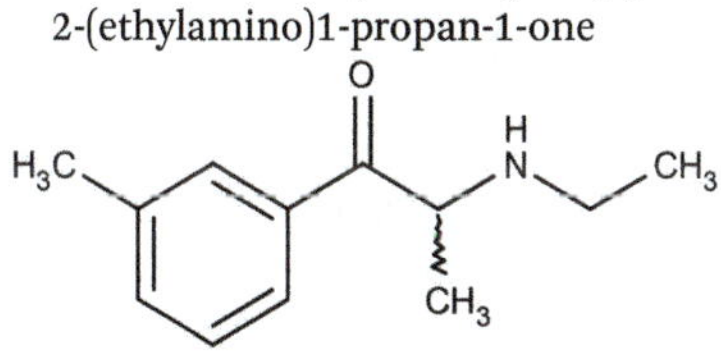

3-MEC, 3-Methylethcathinone
Chemical name: 2-(Ethylamino)-1-(3-methylphenyl)-1-propan-1-one

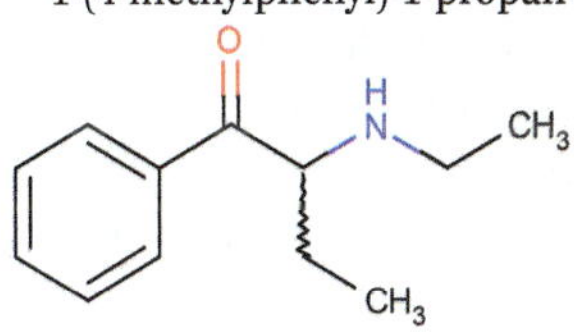

***N*-Ethyl-Buphedrone**
Chemical name: 2-(Ethylamino)-1-phenylbutan-1-one

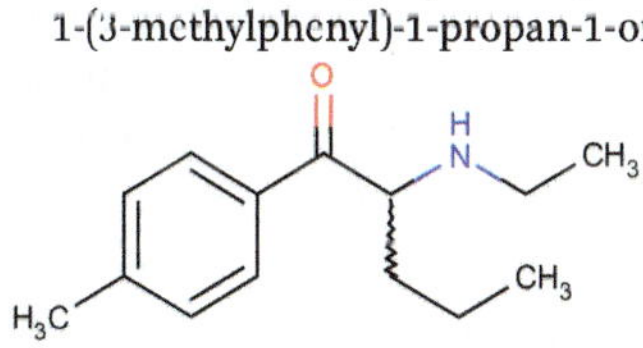

4 MPD
Chemical name: 1-(4-Methylphenyl)-2-methylamino-pentan-1-one

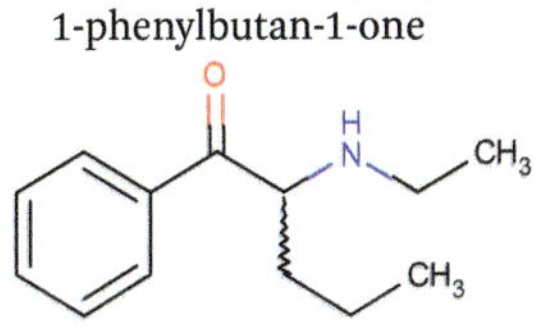

***N*-Ethyl-Pentedrone**
Chemical name: 2-(Ethylamino)-1-phenylpentan-1-one

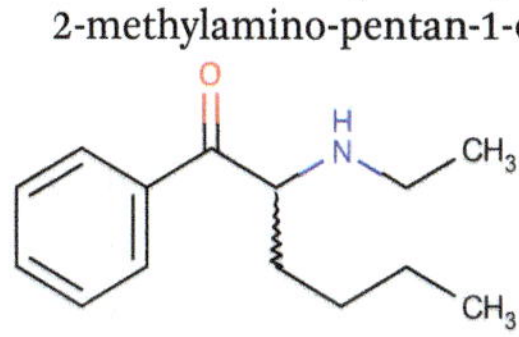

***N*-Ethyl-Hexedrone**
Chemical name: 2-(Ethylamino)-1-phenylhexan-1-one

Table 11.8. Examples for pyrovalerones, which are available *via* the Internet.

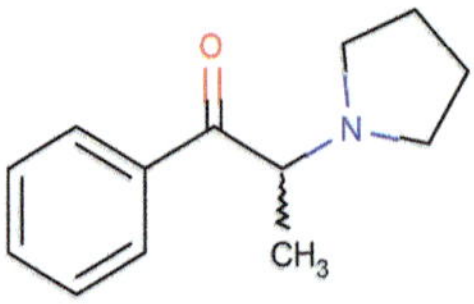 **Alpha-PPP** Chemical name: 1-Phenyl-2-(1-pyrrolidinyl)-1-propanone	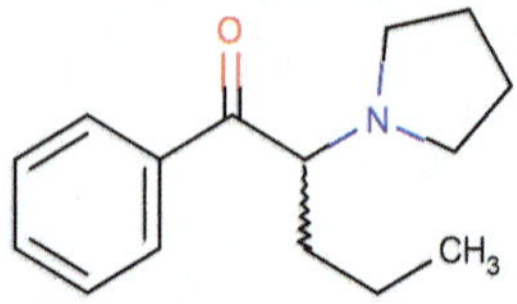 **Alpha-PVP** Chemical name: 1-Phenyl-2-(1-pyrrolidinyl)-1-pentanone
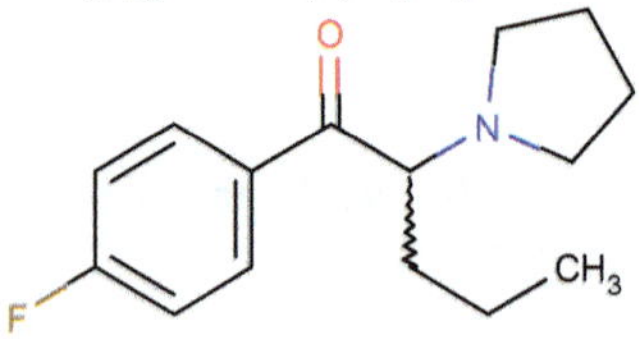 **4 F-PVP** Chemical name: 1-(4-Fluorphenyl)-2-(pyrrolidin-1-yl)pentan-1-one	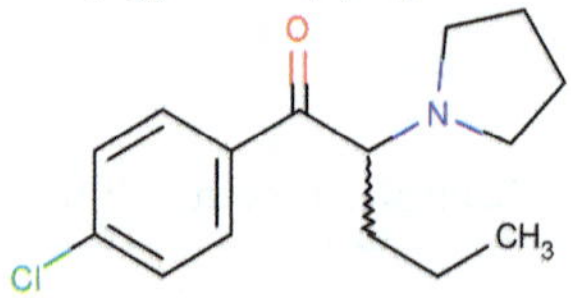 **4-Cl-PVP** Chemical name: 1-(4-Chlorophenyl)-2-(pyrrolidin-1-yl)pentan-1-one
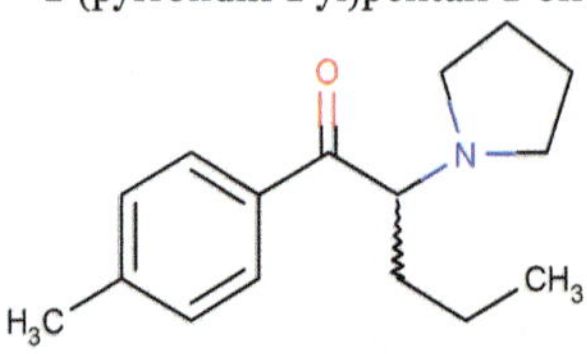 **4-MPrC** Chemical name: 1-(4-Methyl-phenyl)-2-pyrrolidin-1-yl-pentan-1-one	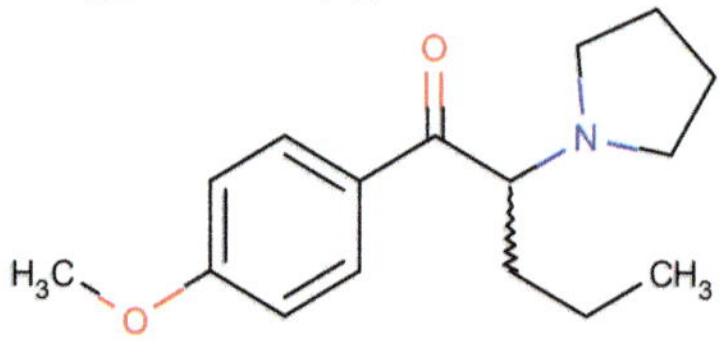 **4-MeO-α-PVP,** 4-methoxy-α-pyrrolidinovalerophenon Chemical name: 1-(4-Methoxyphenyl)-2-(1-pyrrolidinyl)-1-pentan-1-one

A new subfamily of pyrovalerones emerged from alpha-PVP as parent compound, which was chlorinated, fluorinated, methylated and methoxylated to date (all in para-form). It can be expected that meta- and ortho- analogues will appear soon. In the same way that *N*-methamphetamine was modified to 3,4-methylenedioxy-*N*-methamphetamine (MDMA, Ecstasy), a 3,4-methylenedioxy group was added to many compounds of the three afore mentioned substance subclasses. In the following, again, new compound subclasses emerged as shown in Table 11.9.

Mostly, investigations of new cathinones were done by GC-MS and HPLC-MS for determination and verification of structures and quantification. Atlabachew *et al.* investigated different khat chewing leaves and determined naturally occurring cathinone using HPLC-UV.[49] Forensic analysis of hallucinogenic mushrooms and khat plants using cation-exchange liquid chromatography was introduced by Laussmann and Meier-Giebing.[50] Cathinone was monitored by UV-detection at 257 nm. Recently, 4-fluoro-PV9 and α-PHP were identified and characterized as new designer drugs in seized materials by a Polish group.[51] Besides LC-MS, they used HPLC-DAD and a Hypersil ODS column. In 2011, Santali *et al.* reported on synthesis, full chemical characterization and development of validated methods for the quantification of mephedrone.[52] They presented the use of HPLC-UV at a wavelength

Table 11.9. Examples for two 3,4-methylendioxy-analogues, which are available *via* the Internet. On the left side the parent compound is shown.

Parent compound	3,4-Methylendioxy-analogue
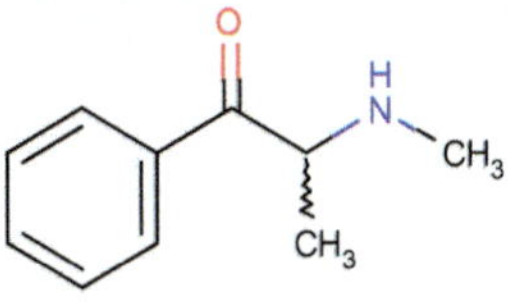 **Methcathinone** Chemical name: 2-(Methylamino)-1-phenyl-propan-1-one	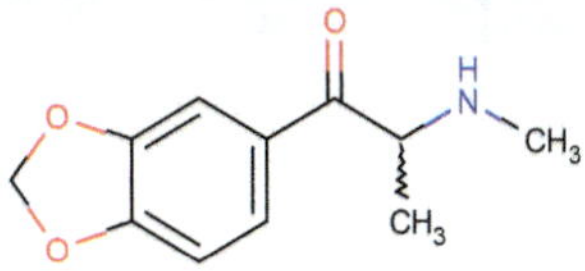 **Methylone,** 3,4-Methylendioxymethcathinone Chemical name: 1-(1,3-Benzodioxol-5-yl)-2-(methylamino)propan-1-one
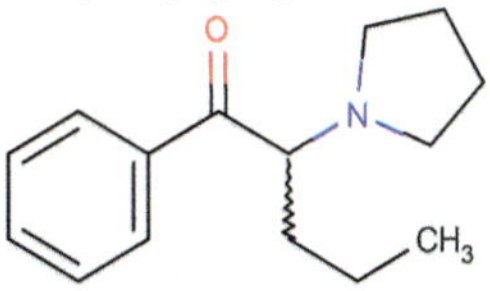 **Alpha-PVP** Chemical name: 1-Phenyl-2-(1-pyrrolidinyl)-1-pentanone	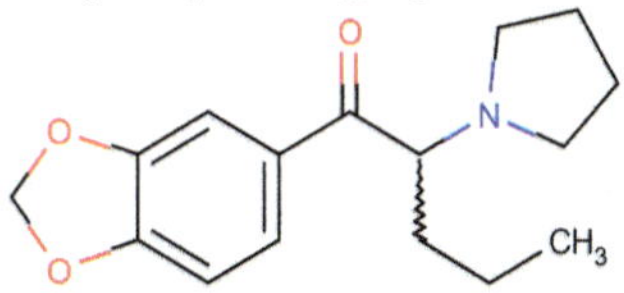 **MDPV,** 3,4-Methylendioxy-pyrovalerone Chemical name: 1-(1,3-Benzodioxol-5-yl)-2-(pyrrolidin-1-yl)pentan-1-one

of 258 nm for UV-detection showing also UV-spectra of mephedrone in different solvents. Gil *et al.* determined 4-MEC and mephedrone in powder samples using HPLC-UV.[53] Mephedrone-related deaths were investigated by Maskell's group[54] using HPLC-UV for 4-MMC after recording the appropriate UV-spectrum.

Also in terms of enantioseparation of new cathinones, some publications came up during the last 5 years: Mohr *et al.* showed enantioseparations of cathinones on a Chiralpak AS-H column using the normal phase mode and UV detection. They succeeded in resolving 19 out of 24 compounds, which were available in the Internet as powders at that time.[55] To circumvent the purchase of a chiral column, the use of a common RP-18 HPLC column and addition of sulfated β-cyclodextrin to the mobile phase was demonstrated by the same group in 2014. Herein, 17 cathinones were resolved in their enantiomers[44] such as halogenated cathinones, methedron and naphyrone. Albals *et al.* compared the enantioseparation of cathinone derivatives by means of HPLC with normal phase, polar organic and RP mode as well as CEC and SFC.[56] Geryk *et al.* compared four immobilized polysaccharide-based chiral HPLC phases for their chiral recognition ability of various drugs, among them 2-FMC.[57] Recently, Taschwer *et al.* showed enantioseparation of 19 cathinone derivatives on a Lux Cellulose-2 column and UV-detection.[43] With some of these chiral separation techniques, it is easy to differentiate between positional isomers, *e.g.* of 2-MMC, 3-MMC and 4-MMC or 3-MeOMC and 4-MeOMC.[43] Separation of the enantiomers of three positional isomers of fluoromethcathinone was shown simultaneously on a Lux Cellulose-2 column (Figure 11.7). Moreover, other chiral selectors such as cyclofructan-based, Cyclobond I 2000 RSP[58] or chiral ion-exchange type stationary phases[59] were applied successfully for enantioseparation of various cathinone derivatives. HPLC-UV was also used besides structure elucidation techniques to describe

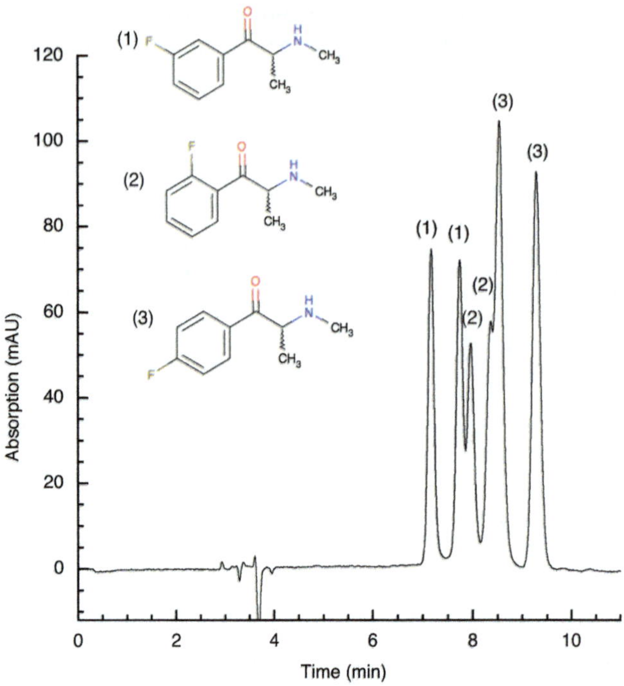

Figure 11.7. Separation of the enantiomers of three positional isomers of FMC. Conditions: Column: Lux Cellulose-2, 250 4.6 mm, 5 μm, mobile phase: ACN: isopropanol:DEA:FA (95:5:0.1:0.1), ambient temperature, flow: 1 ml min^{-1}, UV: 254 nm, injection: 5 μl (M. G. Schmid, unpublished work).

novel upcoming cathinone derivatives as *e.g.* DL-4662 for the first time.[12] A clinical approach was presented by Mayer *et al.*[60]: Mephedrone, 4-MEC and 4-FMC were determined in human urine by HPLC–DAD.

Besides HPLC, racemates of cathinone derivatives were separated by other high performance separation techniques, such as CEC[61] on a packed capillary containing an amylose derivative as chiral selector and SFC on a commercially available CHIRAL ART Amylose SA column.[45] Both techniques were coupled with UV-detection and for the latter approach, *e.g.* detection wavelengths of 230 or 254 nm were used. Also, CE-UV was used for enantioseparation of this compound class. Mohr *et al.* resolved a broad spectrum of cathinones by CE using sulfated-β-cyclodextrin as an additive for the background electrolyte.[62] Merola *et al.* employed native β-cyclodextrin or highly sulfated β-cyclodextrin for successful enantioseparation of 12 cathinones by CE-UV or CE-MS.[63] Some of these samples were seized in Florida packaged as 'bath salts'.

Methods for the analysis of clephedrone, which has been available on the drug market since 2013, were presented by Taschwer *et al.* using GC-MS, NMR and CE-UV.[64] A sample of crystals was resolved in its enantiomers by CE-UV and sulfated β-cyclodextrin as chiral additive. Geryk *et al.* presented chiral separation of biologically active basic compounds in ultra-performance supercritical fluid chromatography, among them mephedrone, brephedrone and flephedrone; detection was performed at 254 nm as shown in Figure 11.8.[45]

A remarkable approach was presented by Baciu *et al.*[65]: They determined enantiomers of mephedrone and MDPV in human hair by CE-UV. For this purpose, a combined in-line with solid-phase extraction process was necessary

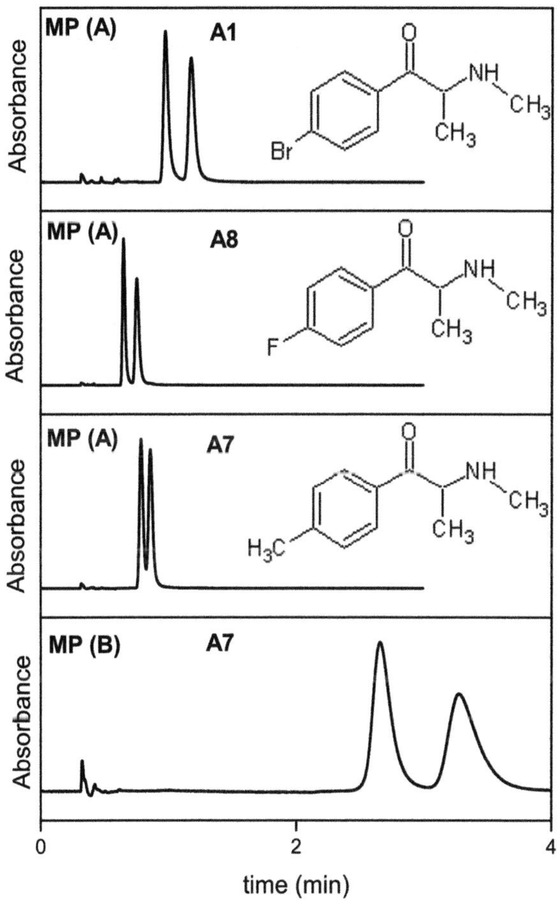

Figure 11.8. Chiral separation of mephedrone, brephedrone and flephedrone by ultra-performance supercritical fluid chromatography. Conditions: Chiral column: CHIRAL ART Amylose SA (150 mm 3.0 mm i.d., 3 μm); MP (A): CO_2/PrOH/TFA/IPA 90/10/0.05/0.05 (v/v/v/v); MP (B): CO_2/PrOH/TFA/IPA 95/5/0.05/0.05 (v/v/v/v); flow rate 2.5 mL min^{-1}; column temperature 35 °C; injection volume: 1 μl. Reprinted from *Analytica Chimica Acta*, **932**, R. Geryk, K. Kalikova, M. G. Schmid and E. Tesarova, Enantioselective separation of biologically active basic compounds in ultra-performance supercritical fluid chromatography, 98–105, Copyright (2016), with permission from Elsevier.

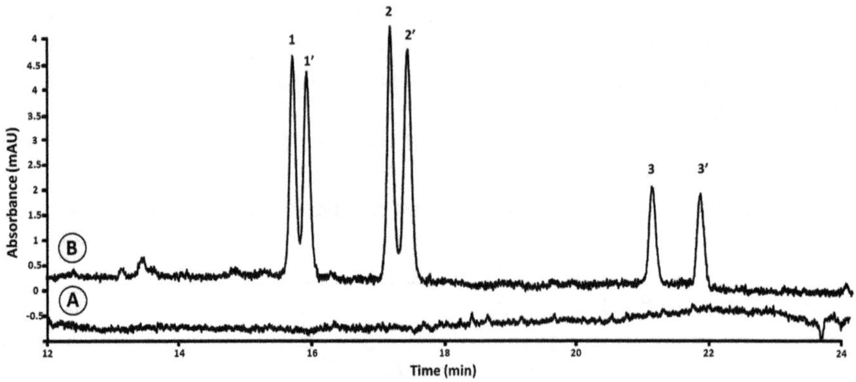

Figure 11.9. Electropherograms obtained under the optimal PLE/in-line SPE-CE conditions of pooled clean hair (A) and pooled clean hair spiked with the studied compounds at a concentration of 0.15 ng mg^{-1} for R,S-mephedrone (1,1′) and its metabolite (2, 2′), and 0.50 ng mg^{-1} for R,S-MDPV (3, 3′) (B). CE conditions: 80 mM disodium phosphate solution adjusted with phosphoric acid to pH 2.5 as the BGE, hydrodynamic injection applying 50 mbar for 5 s, cassette temperature of 25 °C, separation voltage of +30 kV and a detection wavelength of 200 nm. Reprinted with permission from T. Baciu, F. Borrull, M. Calull and C. Aguilar, *Electrophoresis*, 2016, 37, 2352–2362. © 2016 WILEY-VCH Verlag GmbH & Co. KGaA, Weinheim.

prior to analysis. Figure 11.9 displays electropherograms obtained under the optimal conditions of pooled non-contaminated hair and pooled blank hair spiked with the studied compounds.

A presumptive color spot test method for the detection of synthetic cathinones in seized illicit materials was introduced by Philp *et al.*[66] using Cu(ɪɪ)-neocuproine as color reagent. In this approach, a positive result for 39 out of 44 cathinone substances tested was achieved.

11.2.7. *Substituted Benzofurans*

Another substance class, which is closely related to amphetamines represents 2-aminopropyl modified benzofurans.[67] While studying non-neurotoxic analogs of MDMA, Nichols *et al.* synthesized the first compound of this substance class, namely "Benzofury", also known as 6-APB, in 1993.[68]

Table 11.10 gives a few examples of substituted benzofurans, which can be purchased *via* the Internet. Two of them, namely 6-APB and 5-MAPB, were tested for breakdown products in blood and plasma samples by HPLC-DAD using a 25 mM TEAP buffer and an acetonitrile gradient up to 70%.[37]

These compounds are chiral and were separated into their enantiomers *e.g.* by a Lux-Cellulose 2 [43] or a Lux-Cellulose 4 column and a mobile phase consisting of acetonitrile/methanol/TEAA. Enantioseparation of 6-APB was studied by four different polysaccharide-based chiral stationary phases as chiral selectors.[56] In a second step, a comparative study between CEC, SFC

Table 11.10. Examples for substituted benzofurans, which are available *via* the Internet.

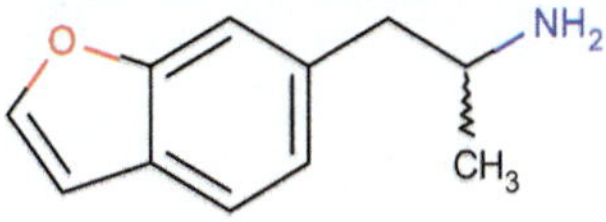 **"Benzofury", 6-APB, 6-(2-Aminopropyl) benzofuran** Chemical name: 1-(1-Benzofuran-6-yl) propan-2-amine	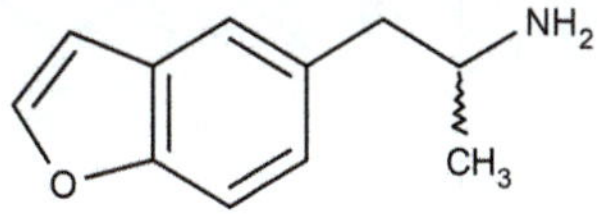 **5-APB, 5-(2-Aminopropyl)benzofuran** Chemical name: 1-(1-Benzofuran-5-yl) propan-2-amine
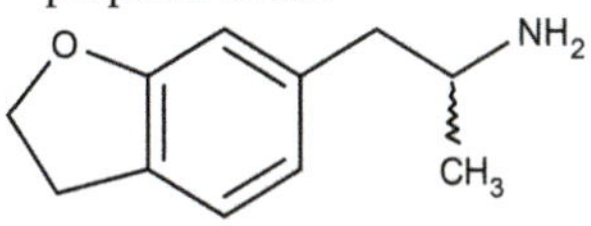 **6-APDB** Chemical name: 1-(2,3-Dihydro-1-benzofuran-6-yl)propan-2-amine	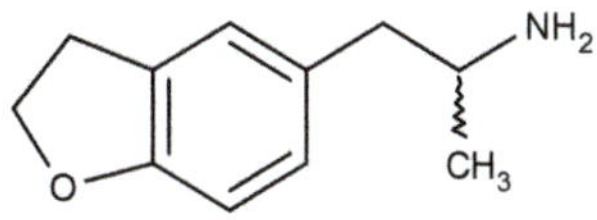 **5-APDB** Chemical name: 1-(2,3-Dihydro-1-benzofuran-5-yl)propan-2-amine
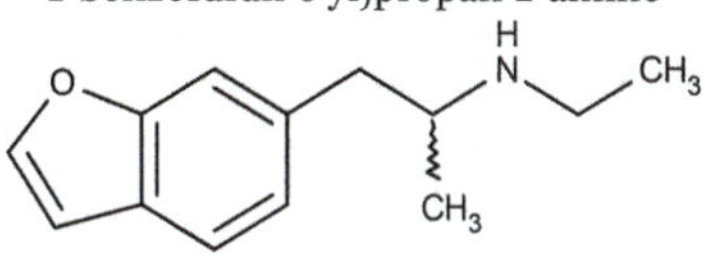 **6-EAPB** Chemical name: 1-(Benzofuran-6-yl)-*N*-ethylpropan-2-amine	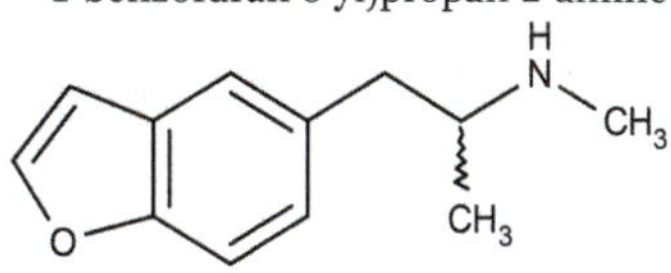 **5-MAPB** Chemical name: 1-(Benzofuran-5-yl)-*N*-methylpropan-2-amine

and three liquid chromatographic modes was performed. Also in CE-UV, enantioseparation of benzofurans was achieved successfully.[69] Figure 11.10 illustrates the chiral separation of 2-chlor-amphetamine and 4-APDB by CE using (+)-18-crown-6-tetracarboxylic acid as chiral selector. Figure 11.11 shows an electropherogram of the separation of the isomers of 4-APB, 5-APB and 7-APB in one run. With this method, it is easy to distinguish between positional isomers of benzofurans.

11.2.8. Ketamine and Derivatives

Introduced as an anaesthetic drug in the 1960s, ketamine and some of its derivatives meanwhile also turned out to act as recreational drugs, since they may cause hallucinations. Some typical examples are given in Table 11.11. These compounds are chiral, but it is only from ketamine that (S)-ketamine shows a stronger effect than the (R)-enantiomer. Methoxetamine has been offered *via* the Internet for almost 7 years as the first derivative of this compound class. Successful enantioseparation was achieved on a Lux-Cellulose 2[43] or a Lux Cellulose 4 column and a mobile phase at polar organic mode. Figure 11.12 shows a chromatogram of the chiral separation of a powder sample of methoxetamine purchased *via* the internet by HPLC-UV on a Lux-Cellulose 4 column.

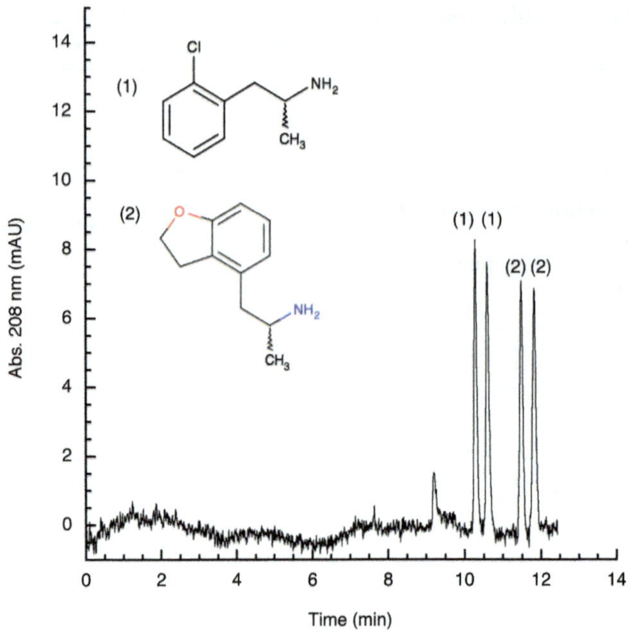

Figure 11.10. Enantioseparation of 2-chlor-amphetamine and 4-APDB by CE using (+)-18-crown-6-tetracarboxylic acid as chiral selector. Conditions: 20 mM (+)-18-crown-6-tetracarboxylic acid, 10 mM Tris, 30 mM citric acid, pH 2,1, cassette temperature: 20 °C, applied voltage: 30 kV, injection: 10 mbar for 5 s (M. G. Schmid, unpublished work).

11.2.9. Morpholines

Within this compound class, *e.g.* derivatives of phencyclidine (angel dust) can be found (Table 11.12). Methoxy-derivatives of phencyclidine are available as recreational drugs. Furthermore, derivatives of the scheduled stimulant phenmetrazine have also gained a certain attraction as substitutes. Since these methylated phenylmorpholines are chiral (Table 11.12), more knowledge about the effects of the single enantiomers would be desirable. Attempts at enantioseparation were made using a Lux-Cellulose 2 [43] column and a mobile phase at polar organic mode.

11.2.10. Phenidines

The first documented recreational use of methoxyphenidine was in 2013.[70] Since then, at least two derivatives were introduced as recreational drugs (Table 11.13). To date, diphenidine, methoxyphenidine and ephenidine are available. Also, these compounds possess a chiral centre; some of them were resolved into their enantiomers by HPLC-UV on a Lux-Cellulose 2 [43] column. Another successful approach was to separate diphenidine and methoxyphenidine by CE-UV into their enantiomers.[69]

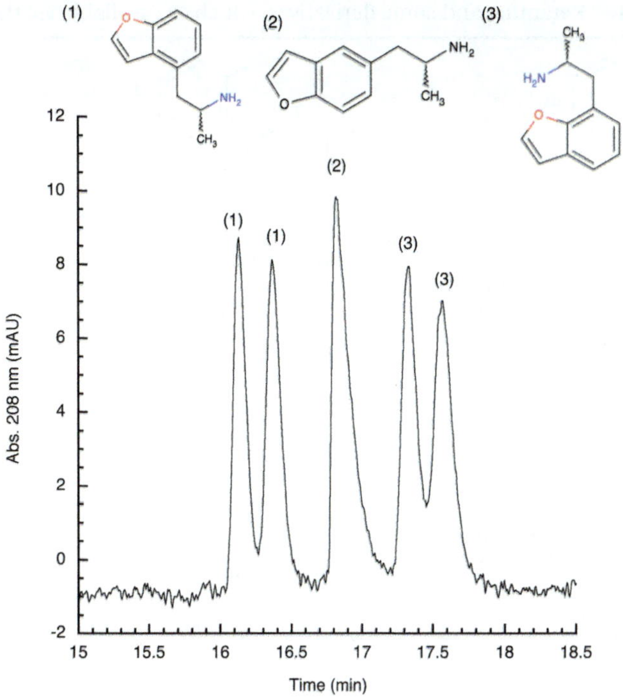

Figure 11.11. Separation of the positional isomers 4-APB, 5-APB and 7-APB by CE. Conditions: 20 mM (+)-18-crown-6-tetracarboxylic acid, 10 mM Tris, 30 mM citric acid, pH 2,1, cassette temperature: 20 °C, applied voltage: 30 kV, injection: 10 mbar for 5 s (M. G. Schmid, unpublished work).

11.2.11. Phenidates

A popular phenidate is represented by methylphenidate, which is widely used under the name Ritalin to treat attention deficit hyperactivity order (ADHD). Within the last years, derivatives appeared as designer drugs, some of them are shown in Table 11.14. Subsequently, the methyl group was replaced by an ethyl or propyl side chain or the phenyl ring underwent substitution. These compounds are chiral and were separated into their enantiomers by a Lux-Cellulose 2 [43] or a Lux-Cellulose 4 column and a mobile phase consisting of acetonitrile/methanol/TEAA. Figure 11.13 shows a chromatogram of the chiral separation of a recently available compound, 4-fluoro-ethylphenidate, on a Lux-Cellulose 4 column. Moreover, enantioseparation of ethylphenidate was performed successfully using CE-UV.[44]

11.2.12. Thiophenes

The first thiophene acting as a recreational drug was methiopropamine, which came up as a "legal high" in 2010. The idea was to circumvent the law by replacing the phenyl ring of *N*-methamphetamine by a thiophene

Table 11.11. Ketamine and some derivatives, which are available *via* the Internet.

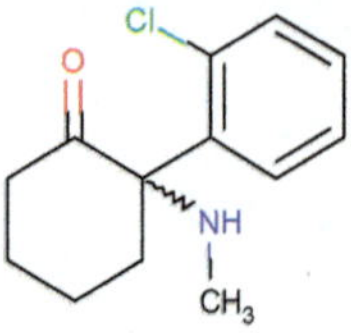 **Ketamine** Chemical name: 2-(2-Chlorophenyl)-2-(methylamino)cyclohexanone	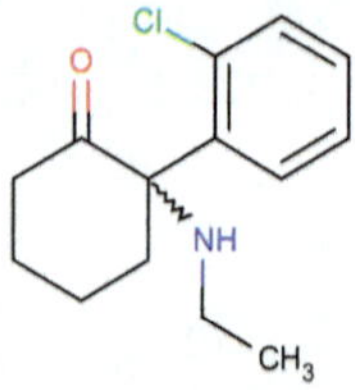 ***N*-Ethyl-ketamine** Chemical name: 2-(2-Chlorophenyl)-2-(ethylamino)cyclohexan-1-one
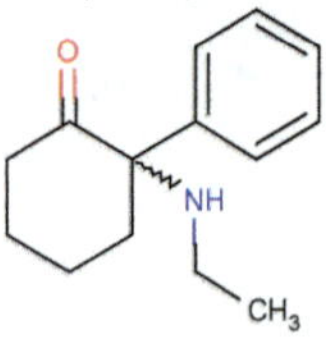 **2-Oxo-PCE**, O-PCE, Eticyclidone, Deschloro-*N*-ethyl-ketamine Chemical name: 2-(Ethylamino)-2-phenylcyclohexan-1-one	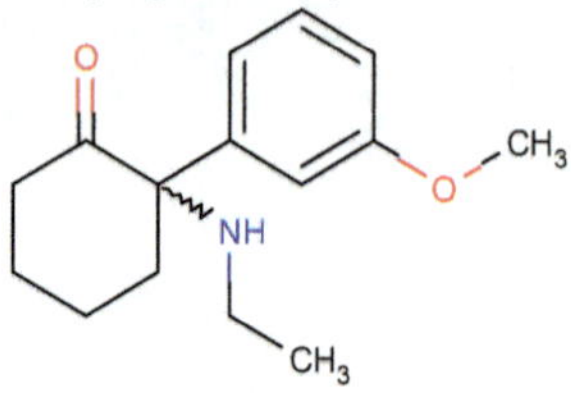 **Methoxetamine** Chemical name: 2-(3-Methoxyphenyl)-2-(ethylamino)cyclohexanone

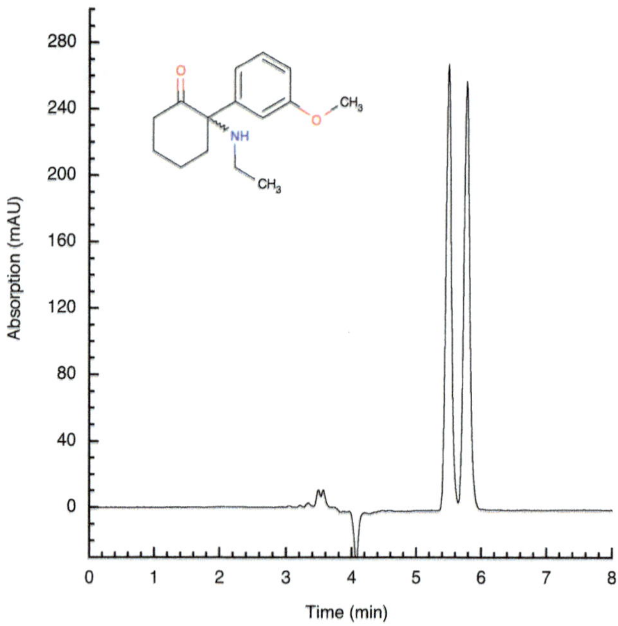

Figure 11.12. Enantioseparation of methoxetamine by HPLC-UV. Conditions: Column: Lux Cellulose-4, 250 × 4.6 mm, 5 µm, mobile phase: ACN: isopropanol:DEA : FA (95 : 5:0.1 : 0.1), ambient temperature, flow: 1 ml min^{-1}, UV: 254 nm, injection: 5 µl (M. G. Schmid, unpublished work).

Table 11.12. Examples for morpholines, which are available *via* the Internet.

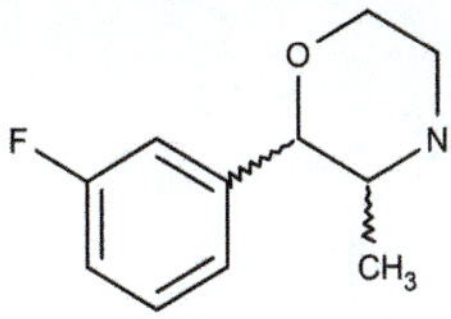 **3-FPM,** 3-Fluorophenmetrazine Chemical name: 2-(3-Fluorophenyl)-3-methylmorpholine	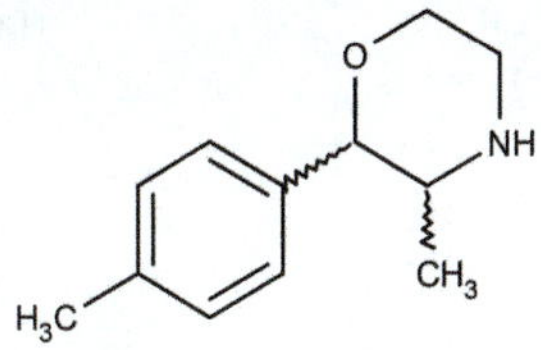 **4-MPM,** 4-Methylphenmetrazine Chemical name: 2-(4-Methylphenyl)-3-methylmorpholine
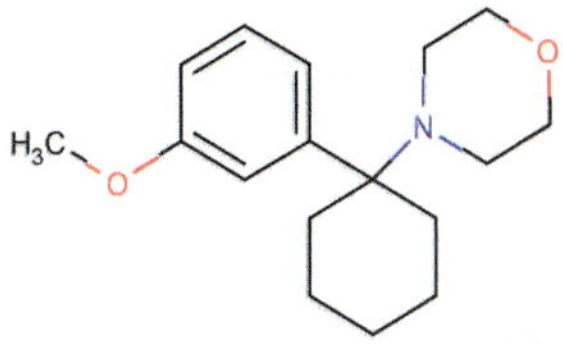 **3-MeO-PCP** Chemical name: 4-[1-(3-Methoxyphenyl) cyclohexyl]morpholine	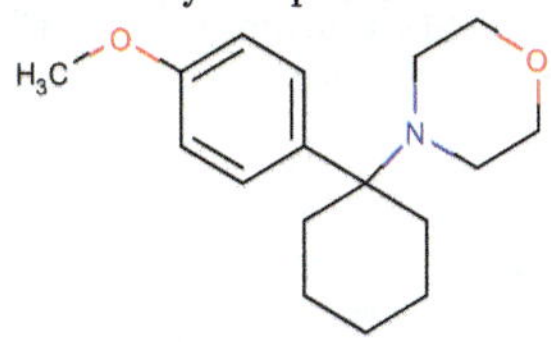 **4-MeO-PCP** Chemical name: 4-[1-(4-Methoxyphenyl)-cyclohexyl]morpholine

Table 11.13. Examples for phenidines, which are available *via* the Internet.

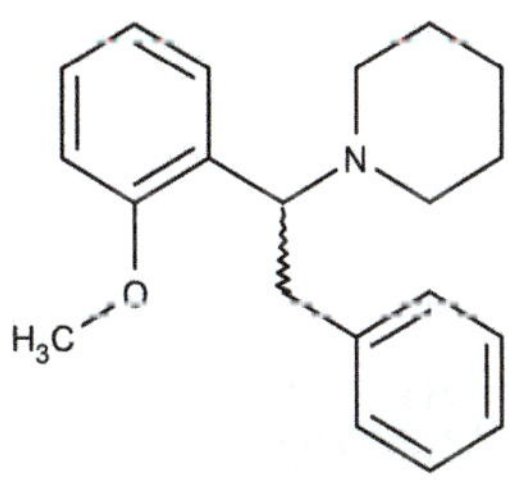 **Methoxphenidine**, MXP Chemical name: 1-[1-(2-Methoxy-phenyl)-2-phenylethyl]piperidine	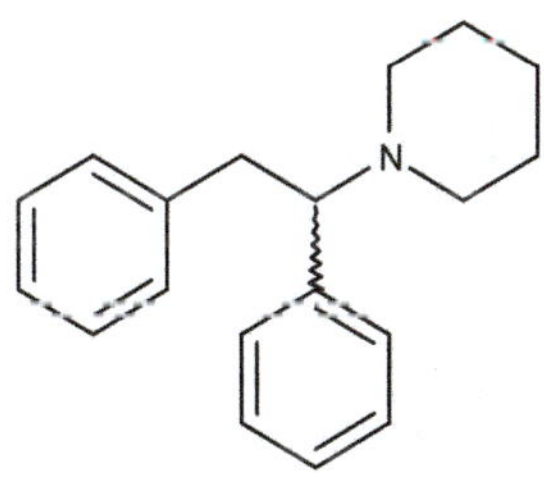 **Diphenidine** Chemical name: 1-(1,2-Diphenylethyl) piperidine
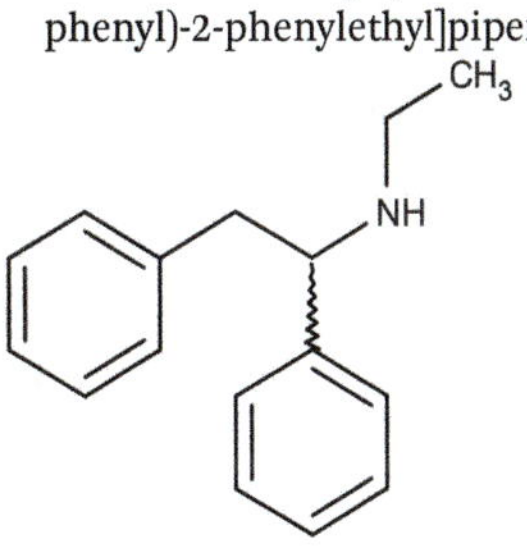 **Ephenidine** Chemical name: *N*-Ethyl-1,2-diphenylethylamine	

Table 11.14. Examples for phenidates, which are available *via* the Internet.

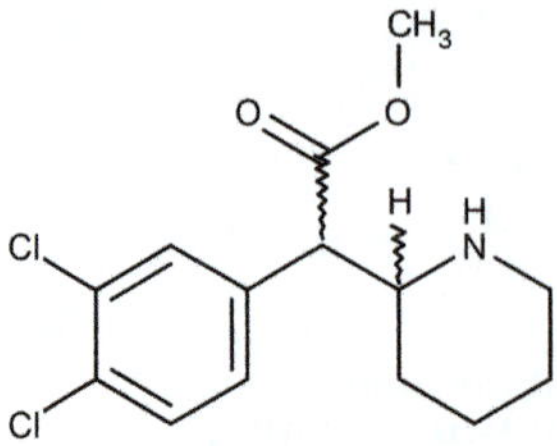 **3,4-CTMP**, 3,4-Dichloromethylphenidate Chemical name: Methyl-2-(3,4-dichloro-phenyl)-2-(piperidin-2-yl)acetate	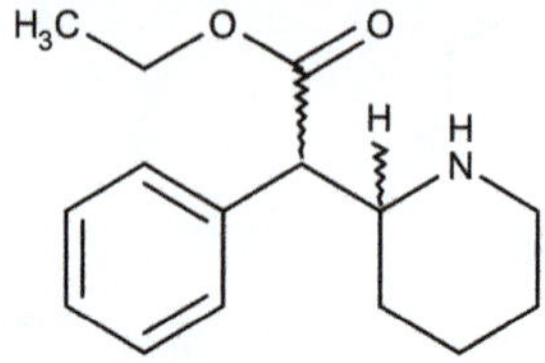 **Ethylphenidate** Chemical name: Ethyl 2-phenyl-2-piperidin-2-yl-acetate
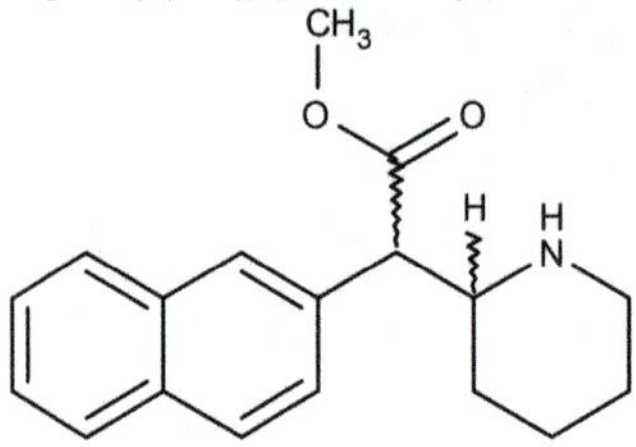 **Methylnaphtidate**, HDMP-28 Chemical name: Methyl-2-(naphthalen-2-yl)-2-(piperidin-2-yl)acetate	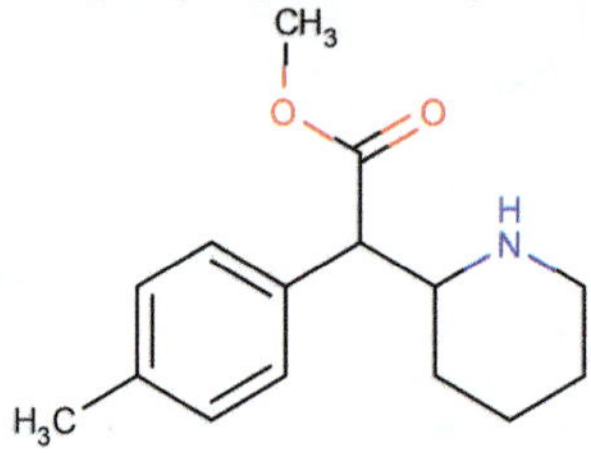 **4-Methyl-Methylphenidate (4 Me TMP)** Chemical name: Methyl (2*R*)-2-(4-methyl-phenyl)-2-[(2*R*)-piperidin-2-yl]acetate
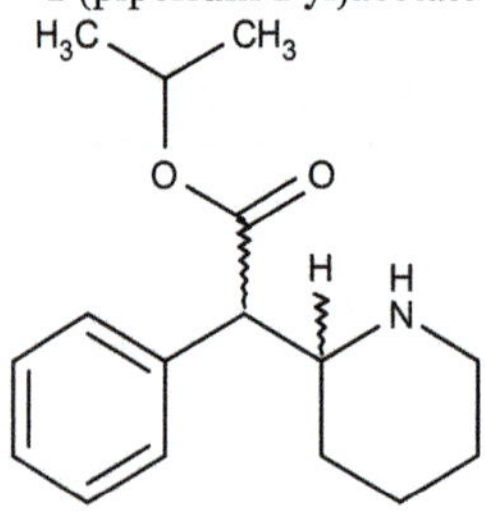 **Isopropylphenidate**, IPPD Chemical name: Propan-2-yl 2-phenyl-2-(piperidin-2-yl)acetate	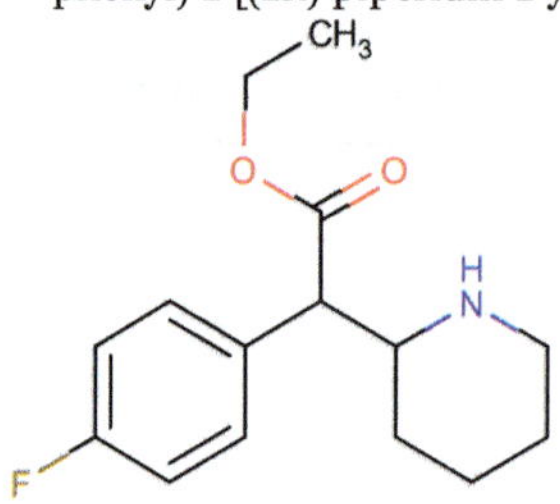 **4-Fluoroethylphenidate (4-FEPH)** Chemical name: Ethyl 2-(4-fluoro-phenyl)-2-(piperidin-2-yl)acetate

ring (Table 11.15). Later, thiopropamine and thiothinone were traded as analogues to amphetamine and methcathinone, respectively. Methiopropamine was tested for its metabolites in blood and plasma samples by HPLC-DAD.[37] Some of these "research chemicals" bear a chiral centre and exist in two enantiomeric forms. An approach for enantioseparation was carried out on a Lux Cellulose 2 column[43] and a mobile phase at polar organic mode as described previously.

11.2.13. Other Compounds

Besides the previously described NPSs, there are a few compounds that do not fit into any of the subgroups. 3-Fluorophenmetrazine, for instance, represents a fluorinated derivative of phenmetrazine and is being sold as a designer drug. Other examples can be found in Erowid.[70]

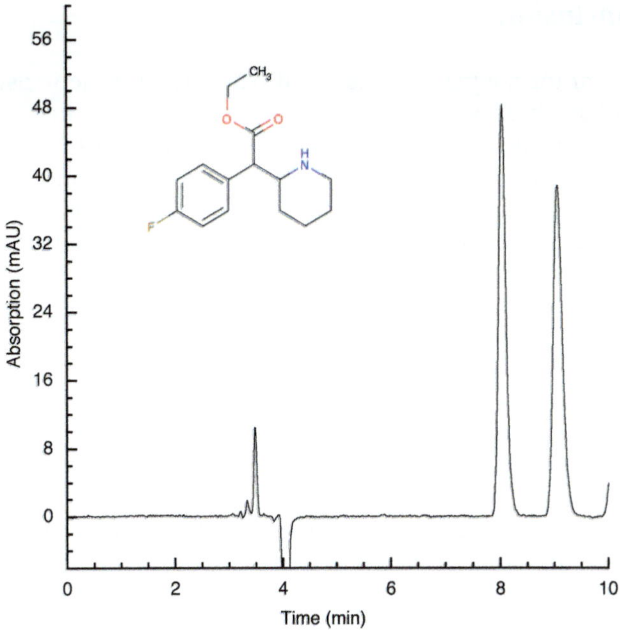

Figure 11.13. Enantioseparation of 4-fluoro-ethylphenidate (4-FEPH) as a sample purchased *via* the Internet by HPLC-UV. Conditions: Column: Lux Cellulose-4, 250 × 4.6 mm, 3 μm, mobile phase: ACN: isopropanol:DEA : FA (95 : 5:0.1 : 0.1), ambient temperature, flow: 1 ml min^{-1}, UV: 254 nm, injection: 5 μl (M. G. Schmid, unpublished work).

Table 11.15. Examples for thiophenes, which are available *via* the Internet.

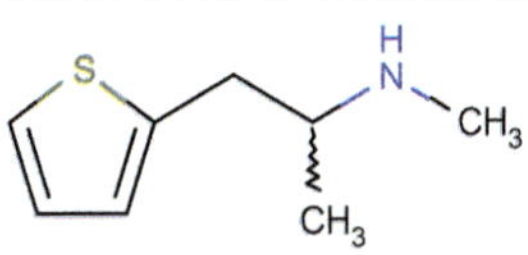 **Methiopropamine** Chemical name: 1-(Thiophen-2-yl)-2-methylaminopropane	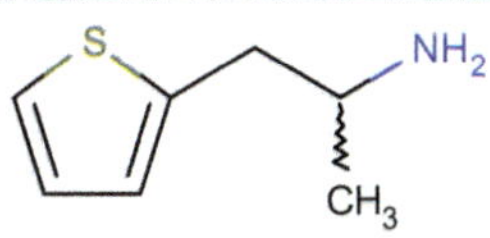 **Thiopropamine** Chemical name: 1-(Thiophen-2-yl)-2-methylaminopropane
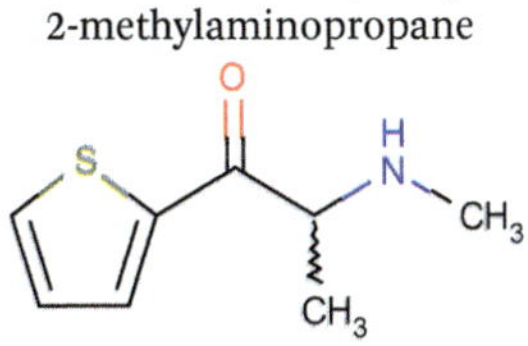 **Thiothinone**, βk-MPA Chemical name: 2-(Methylamino)-1-(thiophen-2-yl)propan-1-one	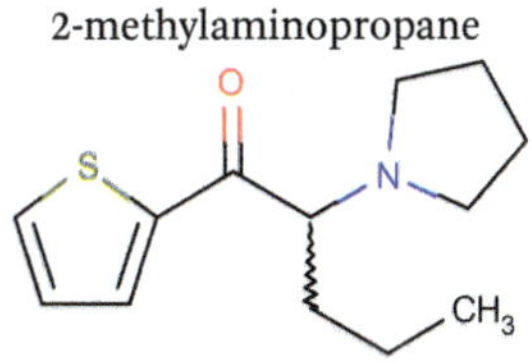 **α-PVT**, α-Pyrrolidinopentiothiophenone Chemical name: 2-(Pyrrolidin-1-yl)-1-(thiophen-2-yl)pentan-1one

11.3. Conclusion

Since every year for the past one hundred years, new synthetic psychoactive compounds have been registered *via* EMCDDA, there is a need to develop methods for their analysis. Since they are traded on the Internet, there is no quality assurance. For proof of identity, GC-MS, LC-MS and NMR are the first choice. For the testing of powder samples, HPLC-UV was shown to be a robust and less expensive alternative, particularly for quantification and enantioseparation experiments. The present review focused on the progress in development of analytical methods for compounds, where little pharmacological and clinical data are available.

Furthermore, it has been shown that many of these NPSs are chiral and that there is a need to develop methods to check enantiomeric purity. It has been shown that chiral NPSs are traded as racemic mixtures, however, in most cases it is unclear which enantiomer is responsible for the psychotropic effect.

For the future, it will be a challenge to elucidate the chemical structure of upcoming new compounds as well as to develop further methods for purity checks.

References

1. European Monitoring Centre for Drugs and Drug Addiction, http://www.emcdda.europa.eu/publications/edr/trends-developments/2016, last accessed December 2016.
2. P. I. Dargan and D. M. Wood, *Novel Psychoactive Substances: Classification, Pharmacology and Toxicology*, Academic Press, 2013.
3. R. A. Glennon, R. Young, B. R. Martin and T. A. Dal Cason, *Pharmacol., Biochem. Behav.*, 1995, **50**, 601–606.
4. R. A. Glennon, R. Young, B. R. Martin and T. A. Dalcason, *Pharmacol., Biochem. Behav.*, 1995, **50**, 601–606.
5. D. Jirovsky, K. Lemr, J. Sevcik, B. Smysl and Z. Stransky, *Forensic Sci. Int.*, 1998, **96**, 61–70.
6. L. B. Rasmussen, K. H. Olsen and S. S. Johansen, *J. Chromatogr. B*, 2006, **842**, 136–141.
7. S. Mohr, M. Taschwer and M. G. Schmid, *Chirality*, 2012, **24**, 486–492.
8. M. Taschwer, J. A. Weiss, O. Kunert and M. G. Schmid, *Forensic Sci. Int.*, 2014, **244**, e56–e59.
9. M. Taschwer, Y. Seidl, S. Mohr and M. G. Schmid, *Chirality*, 2014, **26**, 411–418.
10. J. A. Weiss, S. Mohr and M. G. Schmid, *Chirality*, 2015, **27**, 211–215.
11. J. A. Weiss, M. Taschwer, O. Kunert and M. G. Schmid, *J. Sep. Sci.*, 2015, **38**, 825–828.
12. F. T. Peters and M. R. Meyer, in *Novel Psychoactive Substances: Classification, Pharmacology and Toxicology*, ed. P. I. Dargan and D. M. Wood, Academic Press, 2013, pp. 131–150.
13. J. R. Archer, S. Hudson, D. M. Wood and P. I. Dargan, *Curr. Drug Abuse Rev.*, 2013, **6**, 86–90.

14. M. H. Tang, C. K. Ching, C. Y. Lee, Y. H. Lam and T. W. Mak, *J. Chromatogr. B: Anal. Technol. Biomed. Life Sci.*, 2014, **969**, 272–284.
15. M. Concheiro, M. Castaneto, R. Kronstrand and M. A. Huestis, *J. Chromatogr. A*, 2015, **1397**, 32–42.
16. A. J. Barnes, E. Spinelli, S. Young, T. M. Martin, K. L. Kleete and M. A. Huestis, *Ther. Drug Monit.*, 2015, **37**, 661–669.
17. Z. Aturki, A. Rocco and S. Fanali, *LCGC North Am.*, 2015, **28**, 18–25.
18. V. Auwärter, P. I. Dargan and D. M. Wood, in *Novel Psychoactive Substances: Classification, Pharmacology and Toxicology*, ed. P. I. Dargan and D. M. Wood, Academic Press, 2013, pp. 317–338.
19. H. J. Lee, H. N. Park, T.-S. Leem, J.-h. Jeon, S. Cho, J. Lee and Y. B. Baek, *Forensic Toxicol.*, 2017, **35**, 45–55.
20. L. Andernach, S. Pusch, C. Weber, D. Schollmeyer, S. Munster-Muller, M. Putz and T. Opatz, *Forensic Toxicol.*, 2016, **34**, 344–352.
21. L. A. Ciolino, *J. Forensic Sci.*, 2015, **60**, 1171–1181.
22. B. Moosmann, V. Angerer and V. Auwärter, *Forensic Toxicol.*, 2015, **33**, 54–60.
23. Y. Park, C. Lee, H. Lee, J. Pyo, J. Jo, J. Lee, H. Choi, S. Kim, R. S. Hong, Y. Park, B. Y. Hwang, S. Choe and J. H. Jung, *Forensic Toxicol.*, 2013, **31**, 187–196.
24. S. J. B. Dunham, P. D. Hooker and R. M. Hyde, *Forensic Sci. Int.*, 2012, **223**, 241–244.
25. N. Uchiyama, M. Kawamura, R. Kikura-Hanajiri and Y. Goda, *Forensic Toxicol.*, 2012, **30**, 114–125.
26. *Recommended Methods for the Identification and Analysis of Synthetic Cannabinoid Receptor Agonists in Seized Materials*, United Nations Office on Drugs and Crime, New York, 2013, https://www.unodc.org/documents/scientific/STNAR48_Synthetic_Cannabinoids_ENG.pdf.
27. G. Merola, Z. Aturki, G. D'Orazio, R. Gottardo, T. Macchia, F. Tagliaro and S. Fanali, *J. Pharm. Biomed.*, 2012, **71**, 45–53.
28. R. Geryk, M. Svidrnoch, A. Pribylka, K. Kalikova, V. Maier and E. Tesarova, *Anal. Methods*, 2015, 7, 6056–6059.
29. H. A. Mashayekhi and F. Khalilian, *J. Chromatogr. Sci.*, 2016, **54**, 1068–1073.
30. F. Rezaei, Y. Yamini, M. Moradi and B. Daraei, *Anal. Chim. Acta*, 2013, **804**, 135–142.
31. M. N. Uddin, V. F. Samanidou and I. N. Papadoyannis, *J. Chin. Chem. Soc.*, 2011, **58**, 142–154.
32. S. L. Greene, in *Novel Psychoactive Substances: Classification, Pharmacology and Toxicology*, ed. P. I. Dargan and D. M. Wood, Academic Press, 2013, pp. 363–378.
33. A. Shulgin and A. Shulgin, *TIHKAL: The Continuation*, Transform Press, 1997.
34. S. D. Brandt, P. V. Kavanagh, F. Westphal, S. P. Elliott, J. Wallach, T. Colestock, T. E. Burrow, S. J. Chapman, A. Stratford, D. E. Nicols and A. L. Halberstadt, *Drug Test. Anal.*, 2017, **9**, 38–50.
35. S. D. Brandt, P. V. Kavanagh, F. Westphal, A. Stratford, S. P. Elliott, K. Hoang, J. Wallach and A. L. Halberstadt, *Drug Test. Anal.*, 2016, **8**, 891–902.
36. S. P. Elliott, S. D. Brandt, S. Freeman and R. P. Archer, *Drug Test. Anal.*, 2013, **5**, 196–202.
37. Y. N. A. Soh and S. P. Elliott, *Drug Test. Anal.*, 2014, **6**, 696–704.
38. M. Taschwer, E. Ebner and M. G. Schmid, *J. Appl. Pharm. Sci.*, 2016, **6**, 28–34.
39. A. Shulgin and A. Shulgin, *PIHKAL: A Chemical Love Story*, Transform Press, 1991.
40. R. Heim, Doctoral Thesis, 2003.
41. Y. C. Chiu, S. H. Chou, J. T. Liu and C. H. Lin, *J. Chromatogr. B*, 2004, **811**, 127–133.
42. M. E. Soares, M. Carvalho, H. Carmo, F. Remiao, F. Carvalho and M. L. Bastos, *Biomed. Chromatogr.*, 2004, **18**, 125–131.

43. M. Taschwer, J. Grascher and M. G. Schmid, *Forensic Sci. Int.*, 2017, **270**, 232–240.
44. M. Taschwer, Y. Seidl, S. Mohr and M. G. Schmid, *Chirality*, 2014, **26**, 411–418.
45. R. Geryk, K. Kalikova, M. G. Schmid and E. Tesarova, *Anal. Chim. Acta*, 2016, **932**, 98–105.
46. Y. T. Iwata, T. Mikuma, K. Kuwayama, K. Tsujikawa, H. Miyaguchi, T. Kanamori and H. Inoue, *Forensic Sci. Int.*, 2013, **226**, 235–239.
47. K. Choi, J. Kim, Y. O. Jang and D. S. Chung, *Electrophoresis*, 2009, **30**, 2905–2911.
48. D. M. Wood and P. I. Dargan, in *Novel Psychoactive Substances: Classification, Pharmacology and Toxicology*, ed. P. I. Dargan and D. M. Wood, Academic Press, 2013, pp. 211–228.
49. M. Atlabachew, N. Torto, B. S. Chandravanshi and M. Redi, *Chromatographia*, 2013, **76**, 401–408.
50. T. Laussmann and S. Meier-Giebing, *Forensic Sci. Int.*, 2010, **195**, 160–164.
51. M. Majchrzak, M. Rojkiewicz, R. Celinski, P. Kus and M. Sajewicz, *Forensic Toxicol.*, 2016, **34**, 115–124.
52. E. Y. Santali, A. K. Cadogan, N. N. Daeid, K. A. Savage and O. B. Sutcliffe, *J. Pharm. Biomed.*, 2011, **56**, 246–255.
53. D. Gil, P. Adamowicz, A. Skulska, B. Tokarczyk and R. Stanaszek, *Forensic Sci. Int.*, 2013, **228**, E11–E15.
54. P. D. Maskell, G. De Paoli, C. Seneviratne and D. J. Pounder, *J. Anal. Toxicol.*, 2011, **35**, 188–191.
55. S. Mohr, M. Taschwer and M. G. Schmid, *Chirality*, 2012, **24**, 486–492.
56. D. Albals, Y. Vander Heyden, M. G. Schmid, B. Chankvetadze and D. Mangelings, *J. Pharm. Biomed.*, 2016, **121**, 232–243.
57. R. Geryk, K. Kalikova, J. Vozka, D. Plecita, M. G. Schmid and E. Tesarova, *J. Chromatogr. A*, 2014, **1363**, 155–161.
58. N. L. T. Padivitage, E. Dodbiba, Z. S. Breitbach and D. W. Armstrong, *Drug Test. Anal.*, 2014, **6**, 542–551.
59. D. Wolrab, P. Frühauf, A. Moulisova, M. Kuchar, C. Gerner, W. Lindner and M. Kohout, *J Pharm. Biomed.*, 2016, **120**, 306–315.
60. M. Mayer, A. Benkö, A. Huszar, K. Sipos, A. Lajtai, A. Lakatos and Z. Porpaczy, *J. Chromatogr. Sci.*, 2013, **51**, 861–866.
61. Z. Aturki, M. G. Schmid, B. Chankvetadze and S. Fanali, *Electrophoresis*, 2014, **35**, 3242–3249.
62. S. Mohr, S. Pilaj and M. G. Schmid, *Electrophoresis*, 2012, **33**, 1624–1630.
63. G. Merola, H. Z. Fu, F. Tagliaro, T. Macchia and B. R. McCord, *Electrophoresis*, 2014, **35**, 3231–3241.
64. M. Taschwer, J. A. Weiss, O. Kunert and M. G. Schmid, *Forensic Sci. Int.*, 2014, **244**, 56–E59.
65. T. Baciu, F. Borrull, M. Calull and C. Aguilar, *Electrophoresis*, 2016, **37**, 2352–2362.
66. M. Philp, R. Shimmon, N. Stojanovska, M. Tahtouh and S. L. Fu, *Anal. Methods*, 2013, **5**, 5402–5410.
67. S. L. Greene, in *Novel Psychoactive Substances: Classification, Pharmacology and Toxicology*, ed. P. I. Dargan and D. M. Wood, Academic Press, 2013, pp. 383–390.
68. A. P. Monte, D. Marona-Lewicka, N. V. Cozzi and D. E. Nichols, *J. Med. Chem.*, 1993, **36**, 3700–3706.
69. M. Taschwer, M. G. Hofer and M. G. Schmid, *Electrophoresis*, 2014, **35**, 2793–2799.
70. Erowid, https://www.erowid.org/, last accessed December 2016.

Chapter 12

Supramolecular Approach in Detecting Drugs of Abuse: Optical Sensors

Karl J. Wallace* and J. Hugh Broome

The Department of Chemistry and Biochemistry, The University of Southern Mississippi, Hattiesburg MS 39406, United States
*E-mail: karl.wallace@usm.edu

Table of Contents

Light in Forensic Science: Issues and Applications
Edited by Giorgia Miolo, Jacqueline L. Stair and Mire Zloh

Published by the Royal Society of Chemistry, www.rsc.org

12.1. Introduction

Supramolecular chemistry is now an established discipline, evolving rapidly, transcending many fields of chemistry and redefining major areas of science.[1] Traditionally, supramolecular chemistry can be split into two main branches: self-assembly and host-guest recognition. These two branches have given rise to complex chemical systems, for example molecular machines,[2] molecular knots[3] and stimuli responsive self-assembled structures and materials.[4,5] The field of supramolecular chemistry has expanded well beyond its historical scope and affected almost every other scientific discipline: for instance, medicine,[6] biomimetics,[7] data storage and processing.[8] These areas use our understanding of the intermolecular interactions, defined by Lehn as "*chemistry beyond the molecule*,"[9] to solve complex problems facing humanity. The focus of these solutions are on non-covalent interactions, *i.e.*, electrostatic forces (ion-pairs), hydrogen bonding, dipole interactions, π-π interactions and van der Waals forces, which can occur between host molecules and guest ions or molecules.[10] More recently, reversible covalent bonds have now been accepted in the lexicon of bonding interactions between separate molecules.[11,12]

A lot of the work in the field of supramolecular chemistry has been inspired by Mother Nature and has concentrated on the interactions between receptors and analytes. Much of the work that falls into the realms of host-guest recognition is sensing reagents, whereby the analyte produces a signal change upon the binding event. An elegant perspective written by Anslyn shows how supramolecular chemistry has infiltrated a more traditional area of chemistry: analytical chemistry.[13] Bringing this into the context of supramolecular chemistry, Anslyn introduces the reader to the concept of "supramolecular analytical chemistry".[14] Anslyn gives a broad review of the field and the approaches used. From single analyte sensing to the development of differential array and sensor technology, he highlights the many scientists that have influenced the field and gives his own perspective on the future direction. There is no doubt that the field is addressing the challenges of sensing specific analytes in complex mixtures; for example, biological systems (blood, urine, saliva/oral fluid) and environmental settings (industrial water, streams, oceans). Supramolecular chemistry has also infiltrated society: for instance, many cellular imaging agents utilize host-guest concepts that have been developed to specifically target and probe a desired target analyte, which produce an optical response upon binding.

Some classical examples of host-guest systems are the calcium indicators used for calcium mapping in cells, developed by the late Roger Tsien, which utilized two fluorescent calcium indictors Fura-2 (compound **1**) and Indo-1 (compound **2**) based on a calcium specific chelator 1,2-bis(*o*-aminophenoxy) ethane-*N*,*N*,*N'*,*N'*-tetraacetic acid (BAPTA).[15–17]

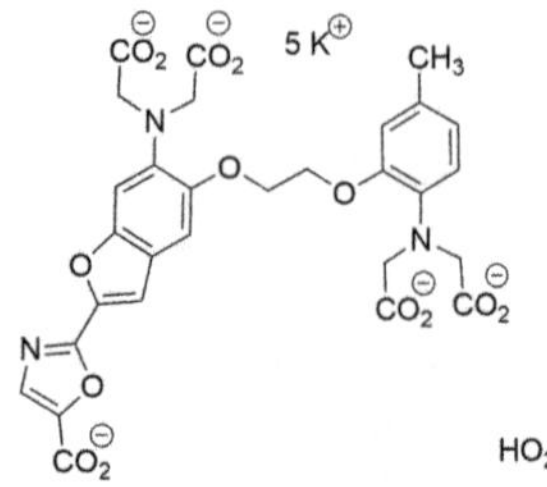

(1)

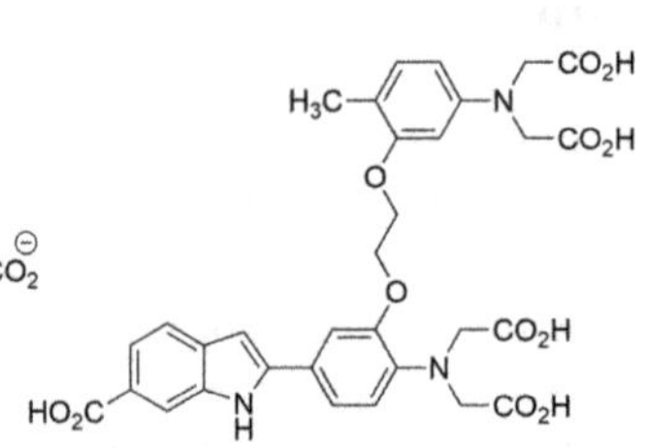

(2)

Other calcium indicators in this family include, fluo-3, fluo-4 and calcium green-1. All of these have been used in calcium imaging; however, monitoring Ca^{2+} ions in plant cells is often challenging, due to the hydrophobicity of the Ca^{2+} chelators and the hydrophilic nature of the plant cell. Although, Ueda *et al.* showed that calcium green-1 can be tethered to a protein ligand by a linker and bound to the cell wall (compound **3**). Compound **3** was used to monitor cytosolic concentration of Ca^{2+} ions, under physiological conditions, in *Nicotiana tabacum*, and in real time, Figure 12.1.[18]

Recent advances in host-guest and self-assembly chemistry are trending towards the area of soft materials, specifically in bioapplications. It is well known that bowl-shaped macrocycles, such as cyclodextrins, cucurbit[n]urils and calix[n]arenes have good bio-compatibility. Bowl-shaped molecules have been used in supramolecular biomaterials, alongside nanoparticles, vesicles, micelles and hydrogels, as delivery carriers.[19] One elaborate drug delivery example, was developed by Tan and Zhao, who utilized the core concepts of supramolecular chemistry: host-guest recognition and self-assembly, to target tumors with nanoparticle functionalized polyacrylates.[20] Their approach was to prepare a nanoparticle containing the chemotherapy drug (doxorubicin-Dox), which acted as a target-specific drug delivery system. They utilized the inclusion chemistry of β-cyclodextrins (β-CD) and a derivatized adamantine (AD) molecule, which is known to form a host-guest system. Both the β-CD and AD were synthetically modified by attaching polyacrylic acid and polyethylene glycol functional groups, Figure 12.2. Additionally, other AD molecules contained a fluorescent tag for imaging purposes. In order to target the specific cancer line (Human breast cell and a mouse melanoma cell), they took advantage of the folate-receptor mediated endocytosis of cancerous cells by incorporating folic acid in their system. The idea was to carry the Doxorubicin to the target cell and unload the chemotherapy drug into the intracellular compartment to achieve target-specific cancer therapy, Figure 12.2.

Another example of biocompatibility is the use of an amphiphilic polymer to inhibit the proliferation of HeLa cells *via* Photodynamic Therapy. Photodynamic Therapy (PDT), uses light to photosensitize a chemical molecule, to form the reactive triplet oxygen species (Figure 12.3A), which is known to kill cells.[21] Zhu and Zhou prepared a supramolecular amphiphilic polymer, which used the host-guest nature of a hydrophilic calix[4]arene derivatized with PEG groups.[22] The hydrophobic chlorin e6 was bound inside the calix[4]arene cavity. These molecules self-assembled into a micelle structure with

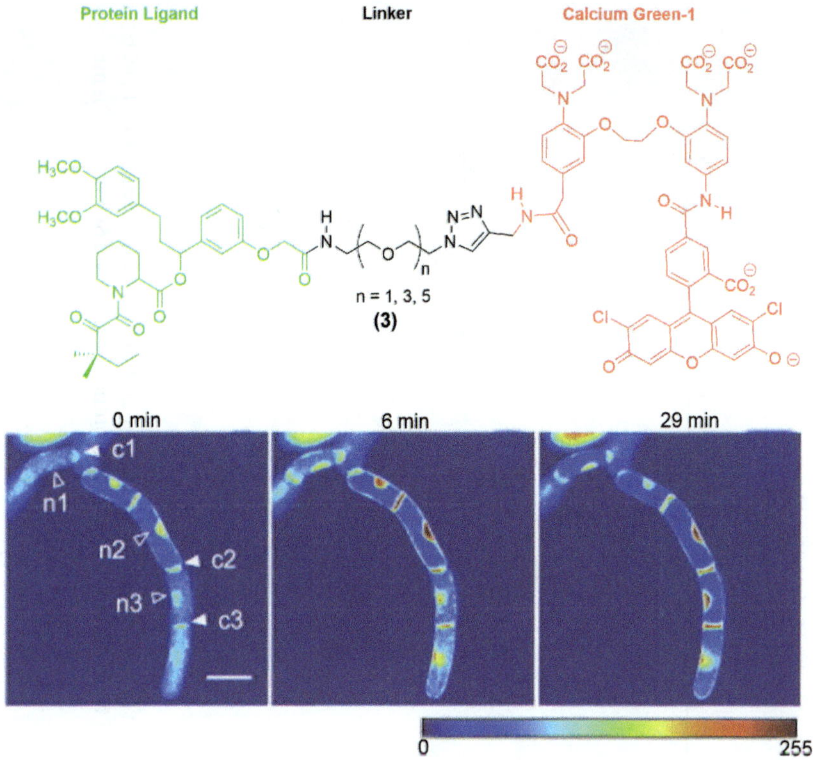

Figure 12.1. Time-lapse imaging of X-stained BY-2 cells expressing hFKBP12-mRFP monitoring the mobility of Ca^{2+} ions in plant cells. The increased concentration is shown by the red color. Reprinted from *Bioorganic & Medicinal Chemistry Letters*, **26**, Y. Takaoka, M. Shigenaga, M. Imai, Y. Nukadzuka, Y. Ishimaru, K. Saito, Y. Yokoyama, K. Nishitani and M. Ueda, Protein ligand-tethered synthetic calcium indicator for localization control and spatiotemporal calcium imaging in plant cells, 9–14, Copyright (2016), with permission from Elsevier.

approximately 12% loading of chlorin e6, Figure 12.3. They showed that the amphiphilic structures inhibit the proliferation of HeLa cells more successfully than using free chlorin e6 as a sensitizers for PDT.[22]

The elegant examples mentioned above only scratch the surface on how supramolecular chemistry has influenced society. Two recent reviews by Li and Loh and Hasan *et al.* have written perspectives on other drug delivery systems[19] and photodynamic therapy,[23] respectively. Additionally molecular storage devices, known as molecular organic frameworks (MOF's) are now being used to store hydrogen and other gases.[24,25] The examples above have demonstrated how the concepts of supramolecular chemistry have been used in real life applications throughout society.

One area of society on which supramolecular chemistry has not had a significant impact is the field of forensic science. Webster's dictionary

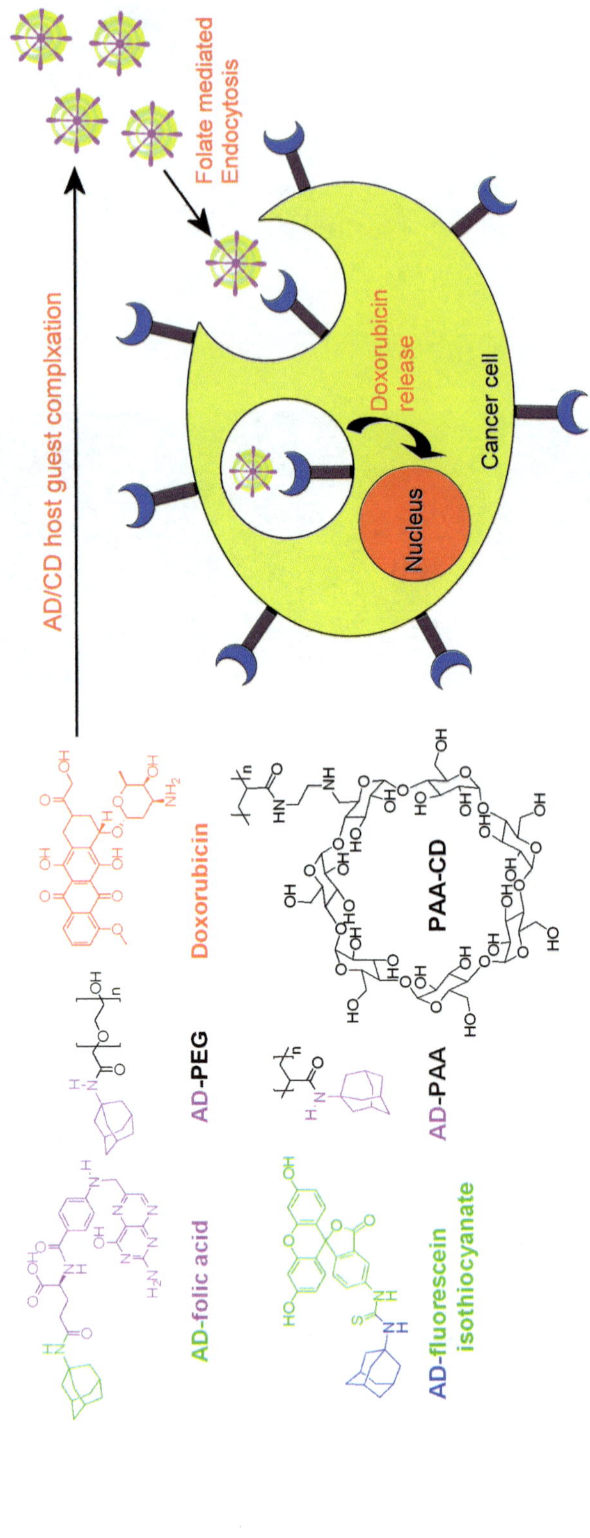

Figure 12.2. A cartoon representation of the chemotherapy drug doxorubicin being transported and delivered to a cancer cell *via* host guest mediated self-assembly using a target-specific therapy. Adapted from ref. 20 with permission from The Royal Society of Chemistry.

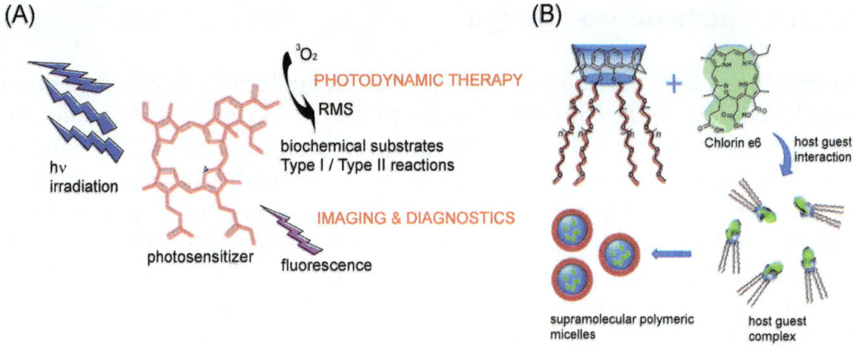

Figure 12.3. (A) A photosensitizer compound is being excited by irradiation, which generates cytotoxic $^1O_2$ and RMS from ground state $^3O_2$. Adapted and reproduced from ref. 23 with permission from The Royal Society of Chemistry. (B) A cartoon representation of the calix[4]arene chlorin e6 host guest molecule which then self-assembles into polymeric micelles, for PDT. Adapted from ref. 22 with permission from The Royal Society of Chemistry.

defines forensic science as "the application of scientific principles and techniques to matters of criminal justice especially as relating to the collection, examination, and analysis of physical evidence".[26] Forensic science, like supramolecular chemistry transcends many disciplines. Broadly speaking, forensic science means any discipline of science that can be used for the purpose of law ("the science of the public"). A scientific discipline can be used to resolve disputes, upholding laws, government regulation and to help public safety: for example, environmental and public health. The evidence needs to be upheld using rigorous scientific standards and ethical values, with little interpretation and minimizing bias. Additionally, it is important that evidence is preserved for deliberations and potential disputes. Therefore, detecting trace analytes at low concentrations in a non-destructive and reversible manner is imperative in the forensic filed.

The traits required for forensic chemistry can be addressed using the core concepts of supramolecular chemistry. Anslyn has elegantly said that supramolecular chemistry "opens up completely new molecular designs and approaches for the use of synthetic receptors in analytical sciences, giving supramolecular chemists new and fertile ground to explore that has truly practical and useful ramifications. Supramolecular analytical chemistry is destined for a bright future."[13] One would have thought that sensor research and technology, from a supramolecular approach, would have a high profile in the field of forensic science; however, this is not true. The vast majority of the published supramolecular literature is dominated by medical and environmental applications. This chapter discusses how supramolecular approaches can be used to detect drugs of abuse, using chemosensors, specifically optical sensors, namely UV-Vis and fluorescence spectroscopy.

12.2. Chemosensor Design

In order for a host molecule to bind to a particular guest, considerable attention is required in the initial design of a molecular receptor. Supramolecular sensors generally have three common components, a scaffold, the binding functional groups and a reporting moiety. Additionally, the chemist needs to take into account other design principles; complimentarity,[27] preorganization[28] and convergence; these will influence the thermodynamics of the host-guest interaction, specifically the enthalpy-entropy compensation.[29] Despite the immense body of research on host design, there is no holy grail for designing the perfect host molecule. However, a chemist can design a reasonable host molecule from scratch using their chemical intuition, pen and paper and computational calculations. As the emphasis of this chapter is on optical sensors, the focus will be on the reporting group of the receptor. If the molecular host contains a reporting functional group, for example, an electrochemical, chromophore, or fluorophore moiety, the chemical environment will change in the event of a host-guest interaction, which stimulates a change in the reporting group. A change in the reporting group is converted to the change in a measurable spectroscopic signal.

These molecules are broadly defined as chemosensors (also known as chemoreactors).[30] Chemosensors originally evolved as macrocyclic hosts targeting cationic guests; however, it has now expanded to include anions and small molecular guests. It also includes podands, which are acyclic molecules, molecular cavities and a plethora of other types of molecular hosts. Crown ethers, calixarenes, thiourea motifs and many other heteroatom moieties that can participate in intermolecular bonding are incorporated into host molecules.[31] Generally speaking, the relatively weak intermolecular forces allow for a reversible interaction between the host and guest; therefore, the chemosensor can be reused multiple times, making it an attractive technique for forensic science.

Two different processes occur during an analyte detection with a chemosensor: molecular recognition and signal transduction. There have been a plethora of literature reviews on chemosensor design and signal transduction in the presence of target analyte.[30] Chemosensors are usually designed based on five broad signal transduction approaches represented in Figure 12.4.

- **on-OFF sensors:** A chromophore or fluorophore changes its electronic properties upon the coordination of an analyte. If the chemosensor is a colorimetric sensor the "on-off" mechanism is indicated by a drastic spectroscopic color change *i.e.*, colored to colorless. If the signaling moiety is a fluorophore, then the fluorescence signal is quenched upon the binding of the guest.
- **off-ON sensors**: This mechanism is the opposite of the "on-off" sensors, whereby a drastic spectroscopic change is seen from colorless to colored

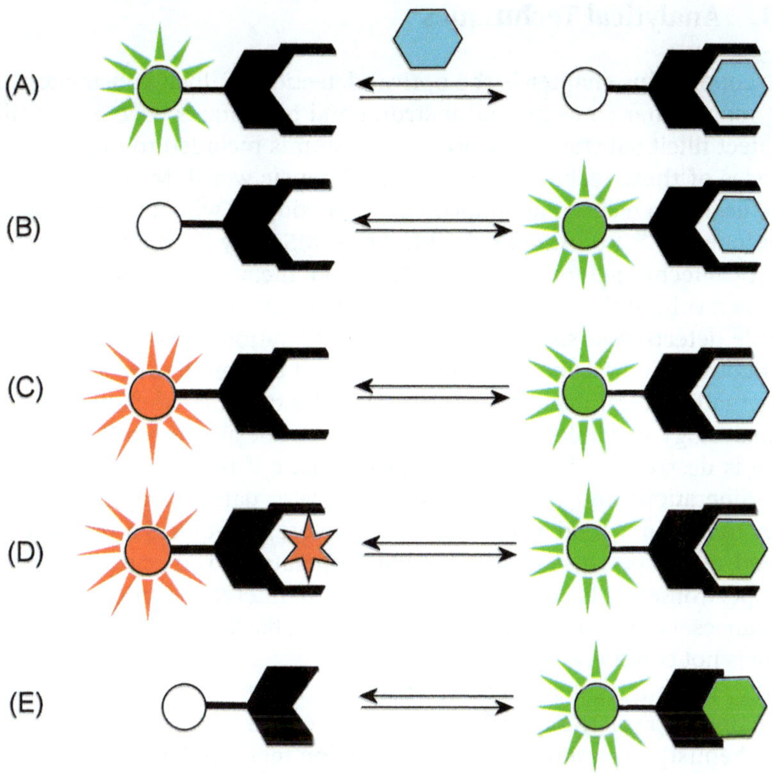

Figure 12.4. Cartoon representation of the five common mechanism used in sensor design (A) on-off (B) off-on (C) ratiometric (D) displacement assay and (E) chemoreactand.

is observed upon on the binding of a guest or if the host contains a fluorophore the fluorescence signal is increased.

- **Ratiometric sensors:** The binding of a guest molecule will disturb the original electronic state of the free chemosensor, with the appearance of a new absorbance or emission band appearing at a different wavelength. Often through an isobestic or an isoemissive point.
- **Displacement Sensors:** This method uses a competition between a dye molecule that has a spectroscopic signature at one wavelength and a specific analyte, having a greater binding affinity. The specific analyte will then displace the dye molecule, inducing a colorimetric or fluorometric spectroscopic change.
- **Chemodosimeter and Chemoreactand:** When an analyte is bound to a chemosensor *via* a non-reversible covalent bond (chemodosimeter) or reversible covalent bonds (chemoreactands) the optical properties are different from the bound and unbound chemical adducts.

12.3. Analytical Techniques

The scope of this chapter is the optical detection of illicit substances; however, many other conventional instrumental techniques have been utilized to detect illicit substances. A brief discussion is included to highlight what the uses of these techniques are in the forensic world. Mass spectroscopy (MS), used as a technique on its own[32–40] or coupled with a chromatography technique,[35,41–46] for example GC-MS or LC-MS is by far the most exploited analytical technique for illicit drug detection. It can certainly be argued that MS has revolutionized forensic labs, as this tandem approach provides fast, reliable detection at significantly low concentration limits. Therefore, why would other methods be used and exploited if this method is "king"? There are actually several disadvantages of MS. The medium of choice for forensic toxicology is whole blood, a specimen not desirable for MS. The technique is destructive, which can be problematic if the evidence is required for deliberations and potential disputes at later dates. Sample preparation is key to obtain a good MS spectrum; therefore, it requires a professionally trained mass spectroscopist to prepare, operate and interpret the results. The spectrometer also needs to be tailored to different settings for different substances. Finally, the samples have to be sent back to the lab, and this technique is not conducive for onsite analysis.

Other techniques that have been used for detecting drugs are NMR,[47] IR[48] and Raman.[49] Even though NMR is arguably one of the most important tools for a chemist, it is seldom used as a detection method. The technique is not sensitive; pre-treatment of the sample is often needed. In addition, high field strengths are required (greater than 600 Mz) for good resolution. It is very difficult to analyze biological samples; and just like MS, this technique requires an expert to interpret the spectrum. However, NMR has been successfully used in detecting GHB, known as liquid "X" or liquid ecstasy, which is often referred to as the date rape drug in both oral fluid and beer.[50] Again, NMR is not favorable for onsite usage.

Two other spectroscopic methods that are becoming very popular in illicit drug detection are techniques that are complimentary to one another, IR (NIR)[48] and Raman.[51] Both of them can be miniaturized easily and used as hand held spectrometers for onsite detection; they are both non-destructive techniques. Depending on the drug, Raman can detect concentrations of drugs in the μg L^{-1} range, but IR requires a greater concentration, 400 mg L^{-1}. Unlike MS and NMR, minimal pretreat is required, but IR spectroscopy requires drug extraction before analysis.[52] A major disadvantage of IR spectroscopy is a significant amount of spectral overlap, from sugars and proteins in the sample; therefore, background substitution is needed.

There are other analytical techniques that have played significant roles in forensic science.[5,53,54] For example, biosensors are used in bioanalytical techniques in the forensic field. A biosensor is defined by IUPAC as "...a self-containing integrated device which is capable of providing specific quantitative or semi-quantitative analytical information using biological recognition element".[55] This has led to a plethora of methodologies, which include immunoassays and other point-of-care (POC) testing technology, and broadly incorporates nanosensors and biochips. The explosion of this technology really owes much of its success to the human genome project. Many bioanalytical techniques are commercially available and as computer power and software sophistication has improved, the biosensors are becoming increasingly complex and generate large volumes of data. Now, biosensors have a marked increase in their sensitivity and a drastic reduction in response time. An advantage of the biosensor approach is the specificity due to the antibody–antigen interaction. Despite the advantages, there are a couple of disadvantages; for example, biosensors often produce false positives and negatives. There is a lack of quantitative application, but this is improving due to better computer power. Biosensor usage is limited to urine and semen samples, mostly. Many biosensors are also electrochemically active; electrochemical techniques have been utilized in the forensic field. There are a few reported electrochemical sensors for cocaine,[56–59] amphetamines,[60,61] cannabis and New Psychoactive Substances (NPS).[62] Although, electrochemical sensors have extensively been utilized in the supramolecular community for many years for the detection of anions,[63] cations,[64] and neutral species.[65] However, the approach for the electrochemical detection of illicit substances is from a more traditional analytical standpoint and not necessary from a supramolecular perspective (see Section 12.8).

The techniques described above have played, and will continue to play, important roles in the forensic science field. Not one of these methods is more important than any other; they all have equal parts to play. In fact, many of these methods are now coupled together to help in the identification of illicit drugs. The remaining sections of this chapter will elaborate on optical spectroscopic sensors; in particular, colorimetric and fluorescence spectroscopy. Chromophores and fluorophores are being incorporated into molecular sensors for illicit drug detection; some examples are also coupled to the detection methods mentioned in this current section.

12.4. Optical Techniques

Optical spectroscopy is a group of spectroscopic analytical techniques that measure the intensity and wavelength of radiation absorbed and/or emitted by a sample. Optical spectroscopy includes both absorption and luminescence spectroscopy (photoluminescence and chemiluminescence). Luminescence, which covers both fluorescence and phosphorescence involves the emission of light; however, there are varies mechanisms in which this can happen (*vide infra*). The theory behind optical spectroscopy can be highlighted with a Jablonski diagram (Figure 12.5). Once a molecule absorbs energy in the form of electromagnetic radiation, the electron is excited from ground state (S_0) and populates one of the higher electronically excited states (S_1, S_2, S_3, *etc.*). The absorption process is fast and occurs in the range of

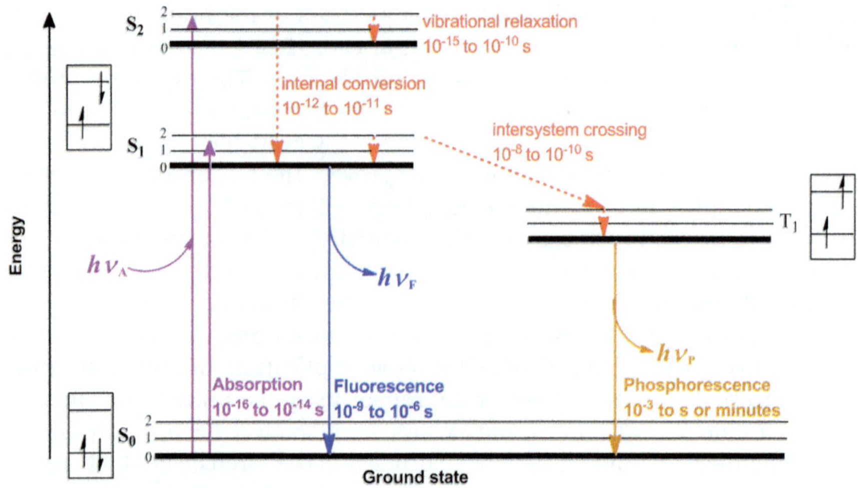

Figure 12.5. A simplified Jablonski diagram. The sequence of events leading to fluorescence and phosphorescence are shown (S_0 is the ground state, S_1 and S_2 are excited singlet states; T_1 is an excited triplet state). From the T_1 state photochemical reactions take place for imaging, diagnostics and photodynamic therapy.

10^{-14} to 10^{-16} s. The excited state molecule quickly loses its excess vibrational energy in the form of heat to the surroundings, and the electrons reach the lowest vibration level of the first electronic state (S_1). This process is known as "internal conversion". This occurs between the electronic states that have the same spin multiplicity. Internal conversion, from S_1 back to the ground state (S_0), is seldom possible because of the large energy gap. Once electrons have populated the first electronic state, they can relax to the ground state in various mechanisms. The emission of photons from the first electronic state (S_1) to ground state (S_0) is called fluorescence. It generally occurs in the range of 10^{-8}–10^{-10} s and emits at a higher wavelength than absorption spectrum due to the loss in energy by internal conversion and vibrational relaxation. The difference between the maximum of the first absorbance band to the maximum of the fluorescence is called the Stokes Shift. When the states are relatively close in energy, the first electronic state can change to a triplet excited (T_1) state through intersystem crossing. The emission of light from this triplet excited state to ground state is called phosphorescence. The relaxation time for phosphorescence is in the second time period, significantly slower than the fluorescence, milliseconds, seconds and even minutes!

Optical spectroscopy, for example, UV-Vis and fluorescence, are of particular interest to the sensor community. Their advantages and disadvantages for both techniques are summarized in Table 12.1. Fluorescence is particularly attractive as the scientist can manipulate different fluorescent mechanisms (*vide infra*). When developing molecular probes for imaging, the probes require a binding site, a reporter group and a molecular scaffold (see Section 12.2). The design and synthesis of molecular receptors that can coordinate guest species for detection and can produce changes in fluorescence intensity

Table 12.1. Advantages and disadvantages of optical techniques.

Technique	Advantages	Disadvantages
UV-Vis spectroscopy	Simple, fast, portable, inexpensive Many organic molecules absorb UV-vis light Quantitative Non-destructive Naked eye detection	Overlap of absorption bands Spectra is not specific for particular compounds Solvatochromic effects Low sensitivity Must obey Beers Law Light scatting can limit the precision of the spectrometer
Fluorescence spectroscopy	Simple, fast, portable, inexpensive	Samples containing mixture of substances cannot be analyzed
	Non-destructive	Very few molecules are naturally luminescent
	Highly sensitive (pico/femto molar)	More complicated equipment required compared to UV-Vis absorption
	Applicable to remote analysis using fiber optics or laser sources (*e.g.*, atmospheric remote sensing) Selection of emission and excitation wavelength allows selectivity control Information content of measurement is high (Polarization, decay times as well as spectral data)	Highly susceptible to multiplicative and additive interferences unless specialized techniques are used

upon coordination, is a topic of intense current interest.[66] Over the past 25 years, there has been a surplus of reviews on various aspects of fluorescent sensors.[67] A molecular receptor that is able to detect analytes often has common characteristic parts in the molecular design, *i.e.*, a spacer, binding site and reporting groups. The fluorophore can be in the form of organic dyes,[68] luminescent metal complexes, (such as lanthanides as a binding unit)[69–71] and quantum dots[69,72] (due to their size-dependent specific spectroscopic characteristics). Furthermore, the photophysical properties of a fluorophore can be easily tuned using a range of mechanisms. These mechanisms include: photo-induced electron transfer (PET), intramolecular charge transfer (ICT), the influence of heavy metal ions, the destabilization of nonemissive n–n* excited states, metal-to-ligand charge transfer (MLCT), chelation-enhanced quenching (CHEQ), Dexter (exchange mechanism) and Förster (coulombic) resonance energy transfer (FRET).[70,71] As these mechanisms have been used as molecular fluorescent sensors in the supramolecular chemistry community, they can be tuned to prepare as supramolecular probes for illicit drug detection.

12.5. Optical Detection for Illicit Drugs

The "traditional" drugs of abuse make up the vast majority of illegal substances on the market, and the EU drug market is estimated to be worth at least 24 billion euros a year. The cannabis market is the largest (38%), followed by heroin (28%) and cocaine (24%). The remaining 10% is made up of new synthetic drugs, which includes the amphetamine and NPS family. There is a real need and a current problem regarding rapid chemical testing for the identification of new drugs of abuse (illicit compounds).

Cocaine has been used recreationally for many years by a vast number of people. The estimated retail market value is in the range of 4.5 to 7.0 billion euros and is the most common illicit drug in Western and Southern Europe. The use of cocaine has stayed relatively steady; however, the use of maritime containers used at ports throughout the World is causing concern, as cocaine is being hidden using elaborate concealment methods.

Synthetic stimulants, such as 3,4-methylenedioxymethamphetamine (MDMA or "ecstasy"), amphetamine and methamphetamine have an estimated value of 2.5 billion euros, and they are consumed by both recreational and marginal drug users. Even though MDMA was synthesized in early part of the 20th century, the drug was really used and reported as recreational purposes until the 1970's. Since 1990, the "World Drug Report 2016" estimates that up to 30 million people are using the ecstasy class of compounds.[73] A concerning aspect of the rise of these synthetic compounds is not only the health risks, but also the increasing amount being dumped into the environment.

The cathinone family of drugs contains two widely used substances; mephedrone and flephedrone.[74] Due to their lipophilic nature, they can easily pass through the blood brain barrier and stimulate the central nervous system by releasing dopamine and inhibiting the re-uptake of epinephrine, norepinephrine and serotonin.[75] Mephedrone is of particular concern due to its negative health implications and continued popularity, despite efforts to control the substance.[74] The surge in popularity of cathinones has resulted in broad restrictive legislation from the US and EU governments seeking to ban as many derivatives as possible.

Cathinone derivatives were classified as Class B, Schedule I substances through generic wording in legislation in the UK.[76] The European Council followed suit in December of 2010. The DEA (Drug Enforcement Agency) and many US states introduced similar measures in 2011. A troubling aspect of NPS is the ease in which supply chains are easily moved around the globe, as they are often packaged as 'legal highs', 'research chemicals' and 'dietary supplements' and sold through 'brick and mortar' and online shops. With increased availability, the number of acute, and even fatal, health issues has increased poisonings and injuries associated with injecting cathinones are also on the rise.

Other drugs of abuse, such as diazepam, ketamine and GHB, are commonly used drugs, known for their anticonvulsant, anesthetic, anti-depressive, hypnotic, tranquilizing and sedative properties. Although these drugs have been used in the medical world, their use has steadily fallen since 1970. However, it is their recreational use that is of public concern. The following sections highlight some of the most recent advances in drug sensing for the previously mentioned classes of drugs. The discussion will focus on how supramolecular concepts have been utilized in the detection of illicit drugs, using optical approaches.

12.6. Optical Sensors Using a Bioinspired Approach: Aptameric and Immunoassay Assays

Much of supramolecular chemistry has been inspired by Mother Nature. Therefore, it is appropriate to discuss systems that utilize a bioinspired optical response approach upon the addition of illicit drugs. A popular bioinspired method for the detection of cocaine utilizes aptamers. An aptamer is a small nucleic acid ligand comprised of RNA or single-stranded DNA oligonucleotides, which have high specificity and affinity for their targets, in the following examples the target drug is cocaine.

Lin *et al.* developed a system that utilized a single stand probe (5′-CCC CGG CGT CAA GAA CGA A-3′) and a 'smart' strand (5′-**FAM**-CGG CGT CCT TCG TTC TTC AAT GAA GTG GGA CGC CGGGG-3′), which contains a fluorescein moiety (FAM) that acts as the optical antenna.[77] Two half sites on the aptamer have a very low binding affinity in the absence of the target drug. However, in the presence of a cocaine molecule the binding affinity between the two half-sites is drastically enhanced; for example, the signal intensity increase was 113% in PBS buffer (pH 7.7). Initially, the fluorophore signal is turned off, quenched by a PET mechanism from a GGG-triplet state on the 5′-terminus. The addition of cocaine induced structural changes in the hairpin probe resulting in an increase in fluorescence emission. The smart probe can be reused by introducing the enzyme exonuclease III, which exo-cleaves the sequences and dismantles the complex, releasing cocaine. This again produces the two half-sites; however, there was partial cleavage of bases. Nevertheless, another cocaine molecule can be bound again and released by adding more enzyme, Figure 12.6. Their smart hairpin probe was found to be selective for cocaine in the presence of caffeine and heroin, which didn't bind to the smart probe.

Detection chip technology, shortened to DETECHIP® has been developed by Holmes.[78] This system contains molecular sensors that can discriminate a large number of analytes utilizing colorimetric responses or fluorometric changes. The analysis was performed by quantifying the RGB (red, green, blue) response based on optical changes upon increasing drug concentration.[78] One of Holmes' systems used oligonucleotide-aptamers conjugated with different organic dyes (8-hydroxypyrene, Allura Red, Blue Litmus, cobalt thiocyanate, Erythrosine B, Eosin Y, Mandeline Reagent, Marquis Test, Neutral Red, Phloxine B, Red Litmus, Simon's Reagent, Sunset Yellow and Universal pH indicator dyes). The fabrication of the DETECHIP utilizes a 96 well-plate reader containing an optical bottom. The colorimetric or fluorescent dye was placed in each row and each well contains an oligonucleotide-aptamer and an organic dye, which was buffered and immobilized on a sol–gel support. The optical response was assigned a '0' or a '1'; this produces a specific binary sequence for each well upon the addition of a particular drug. The combinatorial approach allowed for high selectivity; in other words, only a specific drug would have a specific binary sequence, which improved the reliability of the test. This technology has been shown to be a simple, sensitive and selective method for abusive narcotics and has been used for different

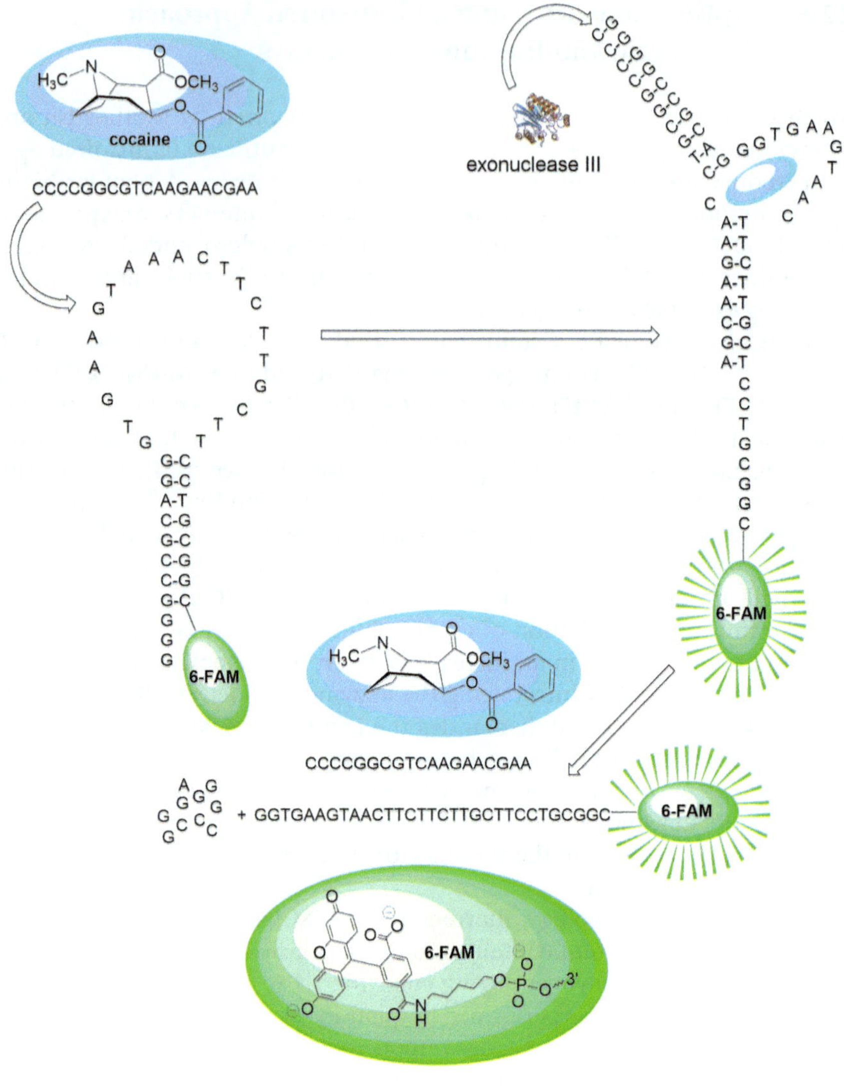

Figure 12.6. Cartoon representation of the aptamer fluorescent smart probe for the detection of cocaine.

sensing applications.[79] An elegant example that uses this technology in compounds of abuse was developed by Sikich and Holmes who used DETECHIP's response towards ketamine and phenylalanine detection. They utilized a photo flatbed scanner to analyze the RGB image changes pre- and post-exposure to the analyte.[80] The higher the concentration of ketamine the more '1' values were seen when a color response was observed. For example, 16 values of '1' appear at the 10 mM concentration and 25 values of '1' are seen when the concentration is 25 mM, see Table 12.2.

Table 12.2. The addition of increasing concentration of ketamine produces a unique 48-digit reading.

[Ketamine] mM	48-digit RGB code (red, green and blue)	N° color changes
10	000-011-011-001-001-011-000-001-011-001-000-000-011-011-000-001	16
25	011-011-011-001-011-111-000-001-011-011-101-000-011-011-011-000	25
50	011-011-011-001-111-111-001-111-011-011-101-001-011-011-011-000	30
60.5	011-011-011-001-111-111-111-111-011-011-101-011-011-011-011-000	33
80	011-011-011-001-111-111-111-111-011-011-101-111-011-011-111-000	35
100	011-011-011-001-111-111-111-111-011-011-111-111-011-011-111-000	36

From the RGB data obtained for ketamine (Table 12.2) and phenylalanine, a calibration curve can be formed based on the green and blue colorimetric values from six experimental trials. This allowed for the calculation of a threshold value of 1000 difference in RGB values in order for a reliable binary signal change. The correlation coefficient (r) were calculated to be between 0.97 and 0.99, however, the authors did not report a limit of detection using the DETECHIP approach.

The pyrene fluorophore has also been tethered to aptamers and used as a signaling unit. The photophysics of pyrene is well understood.[81] The two most important aspects associated with pyrene fluorophore are the wavelength associated with the excimer emission typically located between 410 and 480 nm and the intensity ratio of the excimer and monomer (IE/IM). The monomer emission shows three or four transitions, between 380 to 420 nm. Both of these are extremely convenient methods to be utilized for understanding conformational changes in host–guest recognition. When investigating the use of pyrene sensors in solution, another factor that needs to be taken into consideration is concentration, especially if one is investigating conformational changes induced by guest binding. The excimer formation must come from the formation of intramolecular excimer formation as opposed to intermolecular formation between pyrene units of different molecules. If the concentration of the solution is significantly high, a false positive can be observed as the excimer emission band, due to the interaction between two adjacent molecules in solution rather than the host–guest complex.

There have been a number of prominent examples, whereby a single pyrene unit is used as a signaling motif and many others that contain two pyrene units, whereby a conformational change is seen upon the addition of an analyte, bringing the pyrene units into close proximity. An excellent review published by de Silva in the late 1990s has reported many sophisticated examples of pyrene-containing receptors.[82] There have also been examples reported in the literature of pyrene groups incorporated into lipid membranes that undergo structural changes upon the addition of Cu^{2+} ions and our own group recently reported how the pyrene excimer signal is generated upon the ion-induced self-assembly process and other host-guest systems.[83] This signaling mechanism has been employed in aptameric sensors, too. A pyrene motif was attached to each terminus on complementary aptamer fragments (CAF);

this was developed by Yang *et al.*[84] Upon the addition of cocaine, the two fragments combine bringing the two pyrene units close together to produce the excimer signal. The two halves of the aptamer sequence were spliced from a single aptamer, and each of the two fragments were labelled with a pyrene fluorophore. One of the aptamer fragments was labelled on the 5′ terminus, while the other strand was labelled on the 3′ terminus, Figure 12.7A. The development of the cocaine–aptamer complex induced the formation of a double helix between the two fragments; as a consequence, the pyrene motifs were brought into close proximity, causing excimer emission signal to switch on. The three mono bands were seen between 375 to 425 nm, and a featureless excimer band was observed at 485 nm. This is typical for the pyrene fluorophore, Figure 12.7B. To confirm the selectivity, one of the fragments was replaced by another fragment that had a completely random DNA sequence. Upon the addition of cocaine to the non-complimentary solution, no excimer emission was observed. They also showed that their bioengineered system could also detect cocaine in a complex biological sample, urine. It is also known that excimer emission is long lived (100 ns). Therefore, fluorescence decay experiments were used to differentiate the excimer signal from the aptamer fragments from the background fluorescence after 40 ns after the excitation pulse. Using the aptamer approach allows for high selectivity, due to the complementarity and inherent selectivity of the aptamer fragments for cocaine. Whereas the fluorescence decay experiment allows for high sensitivity in biological samples (5 μM in urine). As a specific wavelength is used, this avoided interference from other contaminants.

Recent advances in nanomaterials have produced a new class of fluorescent labels called quantum dots. Quantum dots (QD) are semiconducting nanocrystals composed of groups of II-VI, III-V, or IV-VI group elements.

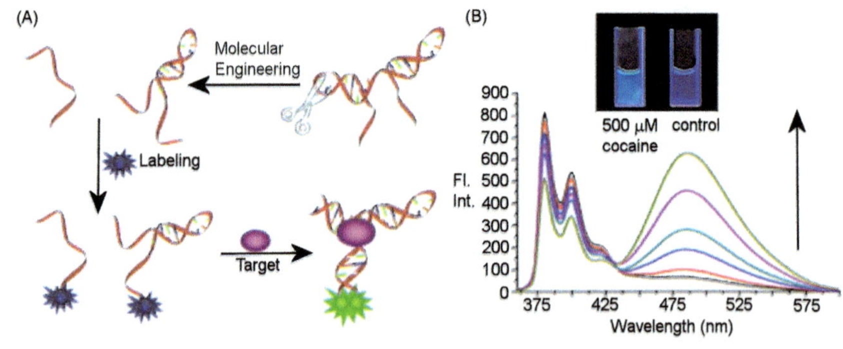

Figure 12.7. (A) Excimer formation is turned "on" when two complementary apatamer fragments (CAP) combine in the presence of the target cocaine (B) Fluorescence spectra showing the excimer band increase at 485 nm on the addition of cocaine (buffer triethyl amine acetate pH 7.0). Reprinted from *Biosensors and Bioelectronics*, **25**, C. Wua, L. Yana, C. Wanga, H. Lina, C. Wanga, X. Chen, and C. J. Yanga, A general excimer signaling approach for aptamer sensors, 2232–2237, Copyright (2010), with permission from Elsevier.

One type of QD commonly used in biological and medical science is the cadmium selenide/zinc sulfide (CdSe/ZnS core/shell) crystal, which form stable aqueous suspensions and is biocompatible.[85–88] The unique optical properties of QD make them ideal for optical sensing applications. Unlike organic fluorophores, QD exhibit broad excitation or absorption spectra, narrow emission spectra and do not photobleach. The CdSe/ZnS QD exhibit a size-dependent, tunable and bright photoluminescence that spans the visible spectrum. Coskunol *et al.* developed a covalently-linked QD-aptamer cocaine biosensor that signals detection *via* a quenching in the fluorescence signal of the QD.[89] They developed a combinatorial platform with carboxyl functionalized CdSe/ZnS QDs, which are covalently bound to micro-well surfaces (a 96 well plate). They tethered the cocaine specific aptamer, which also has a gold nanoparticle attached to the 3′ terminus of the aptamer to the QD *via* a peptide bond. In the presence of cocaine, the aptamer folds and brings the Au-nanoparticle into proximity of the QD, causing a quench in the fluorescence emission at 655 nm. Using this principle, they could determine the limit of detection to be 0.14 nM. Coskunol *et al.* tested their approach in urine; they fabricated a synthetic urine sample and demonstrated that cocaine can be detected in the nM concentration range.

Another system was developed by Lu *et al.*, who used an aptameric system for the detection of adenosine and cocaine.[90] Gold nanoparticle tagged aptamers contained a 3′-single stranded "adenosine aptamer." The sequence of the adenosine aptamer was complementary for the recognition of adenosine. In the presence of adenosine, the purple-colored nanoparticle aggregates produced a red colored solution; upon disassembly of the aggregates, the absorbance at 522 nm increased, while the absorbance at 700 nm decreased. Using a similar technique, a cocaine complementary aptamer was appended to a similar system and behaved as a colorimetric sensor in the presence of cocaine. Lu *et al.* carried out absolute kinetic studies and could quantify the rate of color change of the aptamer probe for cocaine between 50–500 μM, which was found to occur within a two minute period.

12.7. Optical Molecular Imprinted Polymers (MIPS)

Molecular-Imprinted Polymers (MIPS) are highly cross-linked polymers that form a matrix in the presence of a template.[91] The template is removed after the synthesis of the MIP and forms a cavity that is a complimentary size and binding site for the target/template. This is a highly attractive approach in sensor design as this method introduces specificity into the system. Consequently, MIPS are now being used in many sensor array applications, now termed molecular imprinted technology (MIT).[92]

Molecular imprinted polymers can also be attached to surfaces. One example of a MIP system that has been incorporated onto a fiber optic system has been developed by Grattan *et al.* for the detection of cocaine.[93] They attached a acrylamidofluorescein (AAF) *via* a covalent bond to the distal end on the

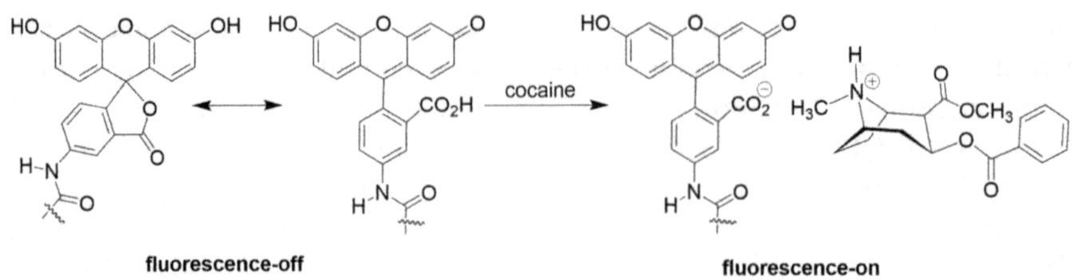

Scheme 12.1. The fluorescein dye is attached to MIP, the acid base nature of the dye and cocaine forms the ion–pair interaction required to turn on the fluorescence signal.

fiber optic cable, containing the LED light source and mini-spectrometer to measure the UV-Vis response to the cocaine binding event. The sensing mechanism was highly dependent on the frontier orbitals of the fluorescein fluorophore, whereby the anionic form was highly fluorescent and the neutral lactone form had low fluorescence intensity (the protonated form is not fluorescent). As cocaine is basic, this would accept a proton from the MIP host forming an ion–pair interaction. This proton abstraction ring opened the lactam ring, Scheme 12.1. This formed the fluorescein anion, which turns on the fluorescent signal. This is a common approach in sensing with the xanthene class of molecules are used as reporting units in molecular probes.[94,95] The MIP-cocaine interaction was further stabilized by the intermolecular molecular interactions between cocaine's amine motif and the carboxylic acid functionality on the fluorophore. Their system could observe a 300% increase in fluorescence intensity for cocaine in a 9:1 H_2O:MeCN solution, which was significantly better than the other drugs tested (ketamine, codeine, amphetamine sulfate, ecgonine methyl ester and buprenorphine HCl). Their system was also shown to be reversible; by washing the surface with methanol, the cocaine was removed. The sensing surface could be used repeatedly for up to one month.

Moreda-Piñeiro *et al.* designed a MIP that also incorporated a QD (Mn-doped ZnS QD), Figure 12.8.[96] In Moreda-Piñeiro's system, the fluorescence emission was tuned to emit in the IR range, which is a desirable wavelength for sensing applications in biological systems. Just like the aptamer-QD based systems described in Section 12.6, a reduction in the fluorescence signal was observed. The quenching effect is a consequence of an electron-transfer induced mechanism in the presence of cocaine. Their system was tested against two cocaine metabolites, benzoylecgonine and ecgonine methyl ester. The Stern–Volmer (*K*sv) constants for the metabolites was calculated to be half that of cocaine, implying the MIP had a preference for cocaine in urine sample in the mg L^{-1} range.

Another approach is to form MIP containing colloidal crystals. The nanocrystals form highly ordered macroporous structures within a MIP polymer. These crystals can either directly detect an analyte in the traditional host-guest manner, or the target molecule can perturb the crystals in the chemical

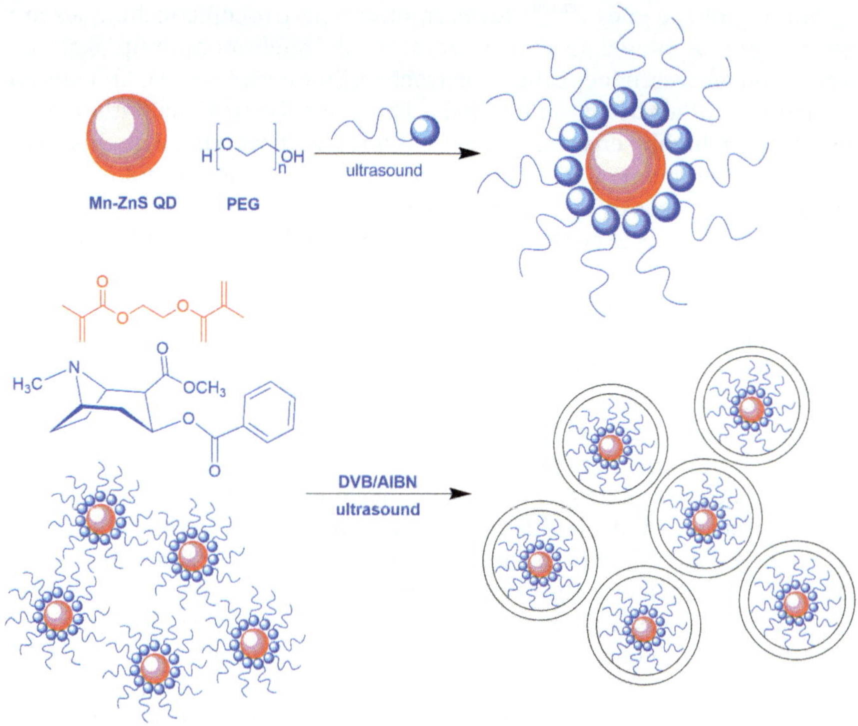

Figure 12.8. The preparation of MIP coated PEG-QD in the detection of cocaine. Adapted from *Biosensors & Bioelectronics*, **75**, M. P. Chantada-Vázquez, J. Sánchez-González, E. Peña-Vázquez, M. J. Tabernero, A. M. Bermejo, P. Bermejo-Barrera and A. Moreda-Piñeiro, Synthesis and characterization of novel molecularly imprinted polymer-coated Mn-doped ZnS quantum dots for specific fluorescent recognition of cocaine, 213–221, Copyright (2016), with permission from Elsevier.

environment. The latter approach has been used in liquid crystal technology.[97] As a consequence, the crystal lattice will change, thus the volume will be altered. This makes the Bragg diffraction angle different; and, drastically alters the optical signatures. Wang *et al.* investigated a novel molecularly imprinted hydrogels (MIPHs).[98] The MIP was prepared from methacrylic acid (MAA) monomer and an ethylene glycol dimethylacrylate (EGDMA), templated by ketamine. Due to the strong electrostatic and hydrophobic interactions with the analytes in the polar environment, the MIPHs showed varying degree of Bragg diffraction angles with and without ketamine. Other targets were also studied: morphine, amphetamine, phencyclidine (PCP) and tiletamine, to ensure the MIPH's selectivity bound the intended target, ketamine.

Many early MIP systems did not contain a way to spectroscopically monitor the uptake of the desired target. There are a few examples that utilized an assay displacement approach, whereby a dye molecule would be released upon the binding of the guest.[99,100] However, other spectroscopic techniques are now being incorporated into MIP systems to accurately monitor uptake. One electro-optical technique, called electrochemiluminescence (ECL), is often coupled with other detection methods. The low noise-to-signal ratio makes this approach highly sensitive, but there is one significant flaw with this technique as a standalone detection approach. The ECL signal requires a co-reactant and this has to be removed from the electrochemical set up to make sure that the drug is generating the light and not the co-reactants itself, *i.e.* remove any false positives. As a consequence, the ECL approach is combined with other detection methods. The most common ECL-type sensors are the ECL-immunosensors sensors, due to their specificity (see Section 12.6). However, molecular imprinting technology (MIT) is another tandem approach. One early example has been developed by Song *et al.*, who used a MIT-based ECL sensor for the detection of drugs.[101] In order to fabricate the ECL-MIP sensor, a multi walled carbon nanotube(MWCNT)-Nafion-Ru(bpy)$_3^{2+}$ modified electrode was initially fabricated, while a methamphetamine templated MIP was attached to the modified electrode using sol–gel method. Elution of the methamphetamine template molecules left behind an ECL-MIP sensor. The sensor's response in the presence of methamphetamine in a PBS buffer system at pH 8.0 was studied. They found there was an increase in ECL emission and could still observe an ECL response at concentrations as low as 4.0 fM! The stability of their system was shown to last for six weeks. They argue that their ECL-MIP sensor approach exhibited the high selectivity of the MIP and the sensitivity of an ECL system. Their system could be easily adapted to other biologically significant amines by altering the template molecule during the formation of the MIP.

12.8. Fluorescent Molecular Receptors

The definition of Drug-Facilitated Sexual Assault (DFSA) is "voluntary or involuntary ingestion of a drug by a victim that results in an act of sexual activity without consent." These drug compounds are known as "date-rape" drugs and include gamma hydroxbutyic acid (GHB), gamma—butyrolactone (GBL), Rohypnol, Ketamine and Soma. The date rape drug GHB and GBL are often used, as they are easily expelled from the body. This makes them difficult to detect, even though there are commercially available detection kits for GHB and GBL. There has been interest in fluorescence systems based on the development of Diversity-Orientated Fluorescence Approach (DOFLA).[102] One early example of a fluorescent probe for GBL was reported by Chang *et al.*, who used a library of BODIPY dyes and monitored how their optical response changes by interacting with both GBL and GHB.[103] A two fold fluorescence increase was seen in the presence of 10 mg mL^{-1} of GBL, and the emission showed a linear increase up to a concentration of 100 mg mL^{-1}, Figure 12.9B. The detection limit was determined to be 3 mg mL^{-1}. They also showed that solvent conditions representing alcoholic beverages do

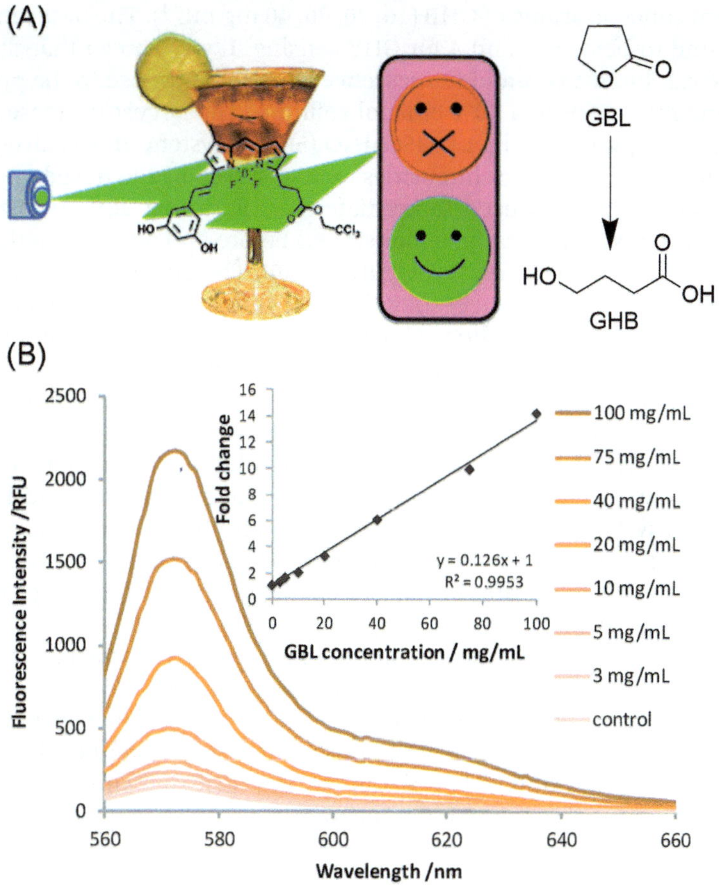

Figure 12.9. The fluorescent BODIPY dye (Green date) turns on its fluorescence signal on aggregation with GBL (A) A cartoon representation of the fluorescence signal can be seen by shining a green laser pointer through a "spiked" alcoholic drink (B) the fluorescence spectra of Green dye on the increasing concentration of GBL. Adapted from ref. 103 with permission from The Royal Society of Chemistry.

not produce a change in the optical response. The sensor could also be used under various pH conditions and in solutions containing up to 10% ethanol. Interestingly, the visible detection of GBL in an alcoholic beverage can be detected with their green laser pointer, using the Green Date molecule as the fluorescent probe, Figure 12.9A. The increase in fluorescence intensity was due to aggregation of the dye molecule and de-aggregation in the presence of GBL, influencing the optical response.

Chang *et al.* reported a second similar system for the detection of GHB with another BODIPY dye, which they call GHB Orange (compound **4**).[104] They screened 5500 dyes to see how their spectroscopic signature changed upon the addition of GHB. The best 17 hits were selected and analyzed with different concentrations of GHB (10, 20, 30, 40 mg mL^{-1}). The best candidate was found to be compound **4** for GHB sensing. It was shown that the GHB Orange exhibited a two-fold fluorescence emission decrease in the presence of 10 mg mL^{-1} GHB in a 50% ethanol solution. The largest decrease, a five-fold decrease, was seen in a DMSO : H_2O (50 : 50) system. It was also shown that different fluorescence intensities were observed by the naked eye, when a hand-held lamp was used to irradiate the solution at 365 nm. Different optical responses are clearly demonstrated before and after the addition of GBH in different spiked drinks, Figure 12.10B. The optical response can be seen within 30 seconds, suggesting that this system can be implemented as "naked eye" detection method and can be used for on on-site analysis.

The amine functional group is ubiquitous in illicit drugs, like methamphetamines (MAPA). Therefore, using this functional group as a binding moiety in sensor design is certainly of interest. Many molecular receptors that target illicit drug detection are utilized in the solution phase; however, there is interest in vapor detection because this phase offers significant advantages over liquid systems. For example, the potential for highly sensitive, selective sensors with rapid response times is high, as there are no competing solvents. That said, this approach can be a challenge as the chemical properties of the analyte can be different in the gas phase. The handling of gases is often difficult. Gas phase detection can produce false positives, which can be problematic. To overcome the sensitivity approach, one avenue is to use fluorescence, conductive polymers (CP). Cheng *et al.* developed an elegant system; whereby, they doped their CP with known quenchers tetraphenylporphyrin (TPP) or metallotetraphenylporphyrin (MTPP). Utilizing simple Lewis acid base chemistry, they doped a fluorochromic metaloporphyrin into the CP matrix.[105] They proposed that their poly(9,9-octylfluorene) polyfluorene was efficiently quenched *via* FRET (Förster Resonance Energy Transfer) in the solid state. Upon the vapor addition of MAPA, the amine functional group coordinated to the M(II) ion (Zn, Cu, Ni, Pd and Pt), and there was a drastic increase in the fluorescence. It was shown that exposing their PdTTP/polyfluorene films to a 215 ppm of MAPA vapor causes the fluorescence signal to increase rapidly within 15 s, and the signal continues to increase successively for the next 600 s. The increase in the fluorescence signal resulted from the amine vapor reaching the PdTPP coordination sites deep within the film. They tested their MTPP/polyfluorene films against other vaporous amines, such as aniline, *N*,*N*-dimethylaniline, benzylamine and *N*,*N*-dimethylbenzylamine. Except for aniline, each of the analytes produced a fluorescence recovery (Figure 12.11); therefore, more work is required to tackle selectivity to overcome false positives, nevertheless this study is a promising start towards vapor sensing systems.

The synthesis of artificial molecular receptors is often difficult for neutral targets. Orti *et al.* synthesized a diazatetraester 1*H*-pyrazole crowns for the fluorescent detection of amphetamine, methamphetamine, ecstasy and dopamine.[106] Their molecular design was based on a pseudo-bibracchial lariat ether (BiBLE). The two pendant arms incorporated either two naphtha-2-yl

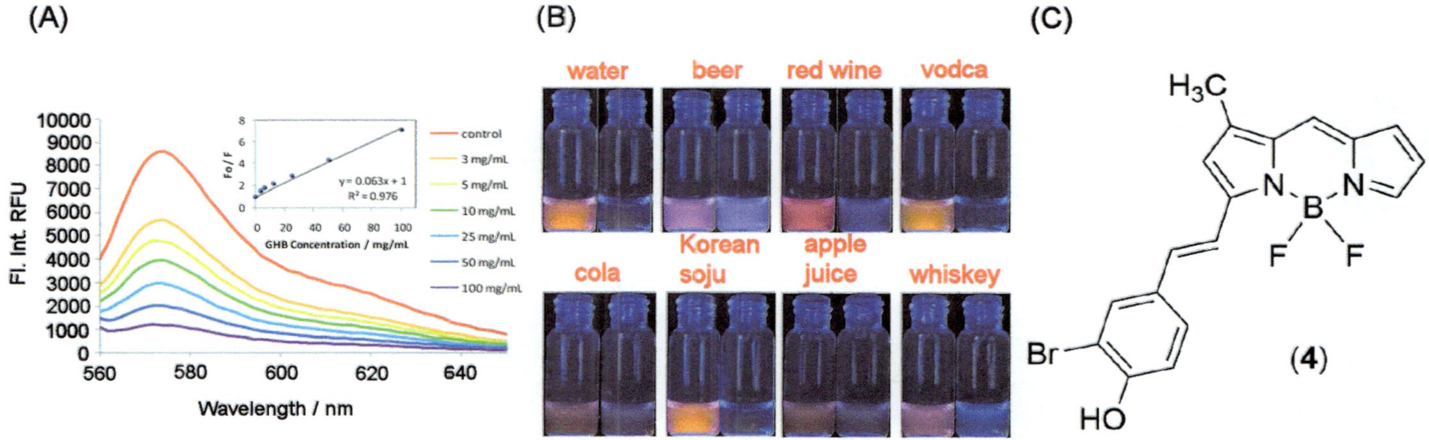

Figure 12.10. (A) Fluorescent spectra of GHB orange (A BODIY molecule) upon the addition of different concentrations of GHB (B) The optical response observed in different beverages with GBH orange and GBH (50 μM and 5 mg mL^{-1}, respectively) in DMSO : H_2O (50 : 50) and (C) chemical structure of GHB Orange. Adapted from ref. 104 with permission from the Royal Society of Chemistry.

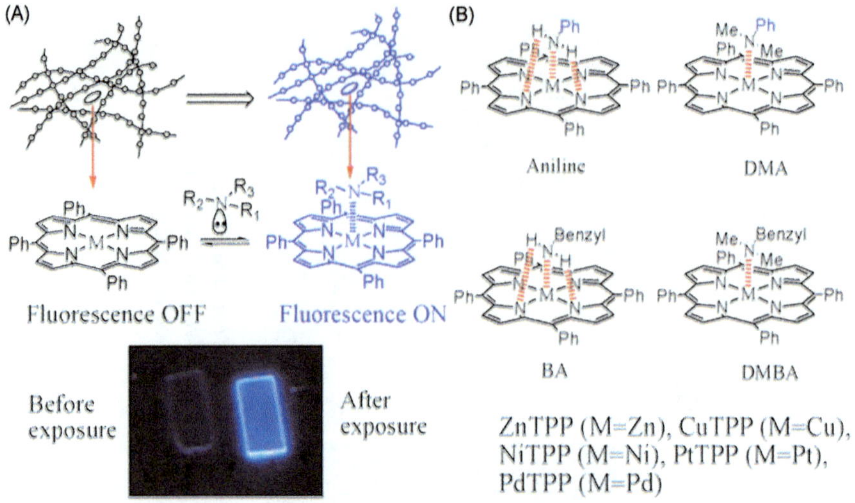

Figure 12.11. (A) Schematic representation of a "turn-on" of the fluorescence signal in the vapor phase, upon the coordination of the amine group and (B) High-lighting different binding modes. Reproduced from ref. 105 with permission from The Royal Society of Chemistry.

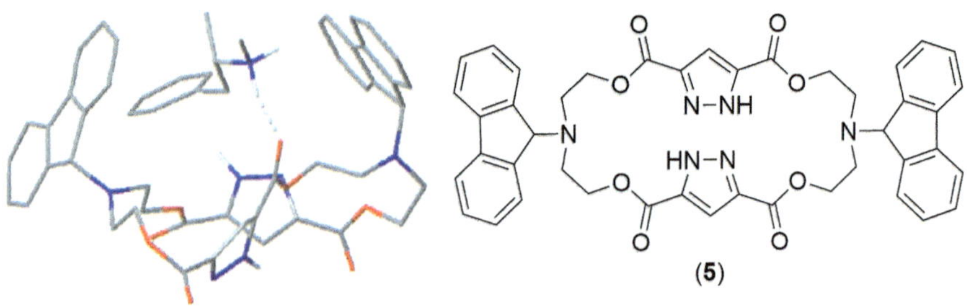

Figure 12.12. Molecular model of compound **5** and amphetamine. Reprinted with permission from ref. 106. Copyright 2016 American Chemical Society.

groups or two fluoren-9-yl moieties, compound **5**. They showed that an intense fluorescence signal was seen upon the addition of methamphetamine in a H_2O : EtOH (70 : 30) mixture, Figure 12.12. The optical response was seen at pH levels below six; whereby, both tertiary amines were protonated, due to fluoren-9-yl fragments. Probe **5** did not fluoresce at higher pH values. In contrast, the naphtha-2-yl system did not fluoresce at all in the studied pH range (2.0–11.0), suggesting that a PET fluorescence mechanism played a role in their molecular sensor design, which was supported by computational studies.

Compound **5** showed a clear hyperchromic shift in fluorescence intensity upon the addition of the psychoactive compounds (AMPH, METH, MDMA and DA). From this data, they were able to calculate K_a values, Table 12.3. However,

Table 12.3. Binding constants (K_a) between compound **5** and AMPH, METH, MDMA and DA determined by spectrofluorometric titration in 70 : 30 H_2O : EtOH, theoretical Ec values and calculated limit of detection (ppb).

Amine	K_a (M^{-1})	Ec (kcal mol^{-1})	LoD (ppb)
(+)-AMPH	2.20×10^5	−12.03	7.4
(+)-METH	1.80×10^5	−6.97	1.3
(±)-MDMA	2.20×10^5	−9.95	8.0
DA	5.60×10^5	−13.71	67

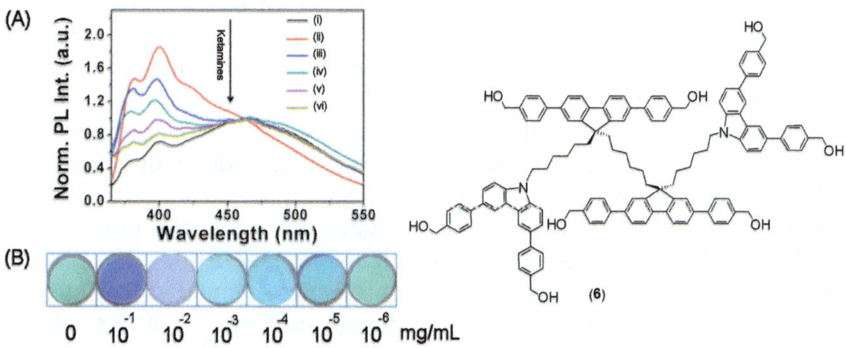

Figure 12.13. (A) The emission spectra of compound **6** doped filter paper after the addition of different concentrations of ketamines (i) no ketamine (ii) 1×10^{-1} mg mL^{-1} (iii) 1×10^{-2} mg mL^{-1} (iv) 1×10^{-3} mg mL^{-1} (v) 1×10^{-4} mg mL^{-1} (vi) 1×10^{-5} mg mL^{-1}. (B) Photograph under UV Lamp (253 nm). Adapted with permission from ref. 107. Copyright (2016) American Chemical Society.

the binding constants that were obtained were very similar for all of the amine groups studied. This indicates that there was very little preference between the drugs and the molecular probe. That said, computational studies agreed reasonably well with the experimental association constants. As the fluorescence intensity was significant, limits of detection could also be calculated.

Cheng *et al.* designed solid-state fluorescent probes for the naked-eye detection of picogram amounts of ketamine and methylamphetamine.[107] They utilized a multi-branched three-dimensional structure; fluorenyl motifs were used as the signaling motif in their system. To improve the system and reduce false positives from other structural similar compounds and ions, they developed a vapor phase detection approach to alleviate cross-reactivity. Compound **6** was prepared and exposed to ketamine or methylamphetamine, which resulted in a visible color change, Figure 12.13B. Even a 50 pg cm^{-2} amount of ketamine could be detected by the naked-eye. Theoretical calculations suggested that the binding of ketamine with compound **6** was occurring *via* hydrogen bonding interactions between the amine group on the drug and the benzyl alcohol moiety on **6**. This resulted in significantly enhanced, blue-shifted fluorescence emission at 398 nm, but no significant intensity change of the 463 nm band was seen, Figure 12.13A.

Carbon nanotubes, alternatively known as carbon quantum dots or carbon nanodots, are often functionalized with hydroxyl or carboxyl groups. These compounds have been shown to have unique optical properties and have been used for biological imaging and light harvesting applications.[108] Recently, carbon nanodots are now being used in sensing applications for neutral molecules, such as the organic precursors for illicit drugs. One early example of their sensing applications has been towards drug precursors and was developed by Lin *et al.*[109] They synthesized an amine functionalized carbon nanodot (A-CD). The A-CD's were found to be colloidal in nature, they did not aggregate, nor form particulates that might settle out of solution. The photoluminescence properties were also obtained, and they were found to be stable for one year. The carbon nanodots exhibited an absorption band at 350 nm with a shoulder at 310 nm, which they attributed to the n–π* and π–π* transitions corresponding to the carbonyl and the sp^2 carbon core. Fluorescence emission indicated an almost excitation-independent emission with the maximum emission observed at 420 nm, when excited at 320 nm. The optical properties of the CD's were studied towards methamphetamine precursors (phenylpropane-1,2-diol and phenylpropane-2-one). The addition of these compounds resulted in a 50% fluorescence quenching. The quenching was due to a charge-transfer-quenching mechanism, which was believed to occur as a consequence of the hydrogen-bonding interaction between the carbonyl groups on the drug precursor and the amine and carboxyl groups on the surface of the CD.

Recently, Wallace and Stair synthesized a molecular receptor (compound 7) to target cathinones.[110] It is known that amphetamines can interact with proteins *via* non-covalent interactions, in particular, π–π stacking and hydrogen bonding interactions.[111] Therefore, an examination of the Brookhaven Protein DataBank[112] was carried out to identify high-quality protein-ligand complexes between receptors, drugs of abuse, common adulterants and endogenous psychoactive substances (*i.e.*, dopamine and serotonin) that are similar to mephedrone. These were used to develop a consensus pharmacophore of mephedrone-receptor binding to support host molecule selection. The validation criterion was obtained by investigating protein-ligand binding interactions of nine proteins from the PDB and five ligands (caffeine, methamphetamine, MDMA, paracetamol and L-dopamine). From this study, a molecular tweezer, containing thiourea groups, was designed and synthesized. Due to their strong hydrogen bonding properties as both donors and acceptors, the thiourea moieties were chosen to interact with the carboxyl motif on mephedrone. Anthracene was chosen as the aromatic group for π-stacking, due in part to its strong fluorescent properties. Additionally, the spatial orientation of the binding interactions was imperative to develop a consensus probe design, compound 7. It was anticipated that the molecular probe would re-organize around mephedrone. The rigid anthracene scaffold was used to enable π–π stacking with aromatic functionalities on the mephedrone as well as to provide an organic backbone and incorporate hydrogen-bonding motifs

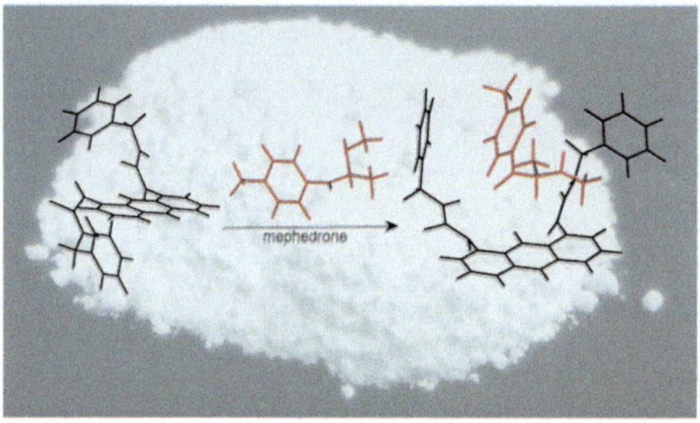

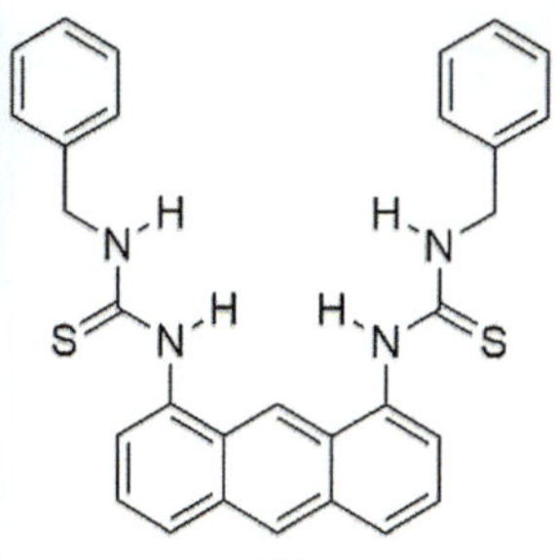

(7)

Figure 12.14. DFT fully optimized structure of compound **8** showing the conformational change on the binding with mephedrone.

(thiourea moieties).[113] It was shown through extensive NMR studies that the molecular probe interacts preferentially with cathinone-like structures (*i.e.*, requiring both a carbonyl and amine on the guest). The adulterants tested (lidocaine, acetaminophen, benzocaine and caffeine) resulted in no proton-NMR shift changes on the host, further supporting the need for a β-ketoamine arrangement to induce interactions. The DFT calculations confirmed that the lowest energy conformations of each cathinone are truly indicative of the way they bind to probe **7** and were in good agreement with the experimentally observed data. Unfortunately, the optical property was modest. Moreover, the addition of common cutting agents did not affect interaction between mephedrone and probe **7**, which is promising for use with "street" samples (Figure 12.14).

Many illicit drugs are chiral; for example, the aminoindane compounds that fall under the class of NPS substances. Chiral recognition is an attractive area of research but can be challenging; that said chiral recognition is an important area in host-guest recognition. There are some excellent examples in the literature on chromophoric, chiral recognition events; however, colorimetric chiral sensors geared towards drugs of abuse are scarce.[114] An advantage of this system is that there is the potential to observe distinctive color changes *via* the naked-eye, upon the addition of different enantiomers. Cho *et al.* synthesised a colorimetric chiral sensor, compound **8**. They modified binaphthyl unit by preparing a crown ether derivative in the 3 and 3′ positions on the ring unit, thereby acting as a chiral barrier. The signaling unit was tethered to their architecture (diazophenol and dinitrobenzene). They showed that their molecular probe produced different colors from several biologically important primary amines or alcohols enantiomers (phenylalaninol, phenylglycinol and 1-aminoindane). For example, compound **8** in CH_3CN was green in color, upon the addition of *R* or *S* phenylalaninol produced a purple and a royal blue color, respectively. They were able to calculate binding constants, which showed a modest preference for the (*S*)-enantiomer over the (*R*)-enantiomer.

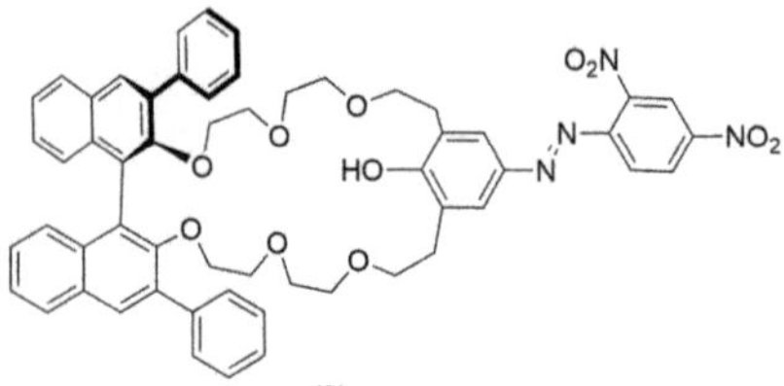

(8)

12.9. Surface Detection Methods

In addition to the supramolecular analytical approach mentioned in the previous section, there has been interest in the detection of psychoactive compounds utilizing surface chemistry. The most common method is dominated by electrochemical techniques (*vide infra*).[115,116] Detection of illicit substances at the surfaces is an alternative approach utilizing supramolecular concepts, however, the analyte is often bound to the surface by non-covalent interactions, which can provide valuable insight to another approach in illicit drug detection. In addition, many of the subsequent surface techniques are designed for portability and are cost effective, like host-guest style sensors.

Wael *et al.* applied electrochemical methods for the rapid on-site screening of cocaine.[117] They incorporated a conductive gel electrolyte into the fingertips of gloves, Figure 12.15. Whereby solid cocaine and other cutting agents can be directly detected at your fingertips! This type of technology can certainly be useful for law enforcement agencies and hospitals needing to rapidly identify an unknown powder.

Sutcliffe and Banks have developed electrochemical techniques for drug screening.[62,118] They recently designed a low cost, disposable screen-printed electrode, for the detection of "legal highs." By investigating the redox properties of the "legal highs" themselves, they utilized various commercially available electrodes, *i.e.*, boron doped diamond and glassy carbon, to analyze their effect on the oxidation-reduction of the amine moiety on methamphetamine.[62] These were then compared to other electrochemical responses of the commercially available electrodes to their screen-printed electrodes. The oxidation of (±)-methcathinone was possible with glassy carbon, boron-doped and their screen-printed electrodes, but there was overlap in the redox potentials of the drug. Therefore, it was difficult to determine the selectivity between (±)-methcathinone, (±)-mephedrone and 4′-methyl-*N*-ethyl-cathinone at pH 2 and pH 12. The common cutting agents also showed similar redox potentials, rendering this technique difficult to selectively detect a specific legal high or their common agents.

Another electrochemical approach that Sutcliffe used was a graphite screen-printed electrodes for the indirect detection of synthacaine (synthetic cocaine), a "legal high" of cocaine.[119] Although synthacaine can contain many cutting agents and other psychoactive adulterants, it is principally known to be composed of methiopropamine and 2-aminoindane in the sample. The indirect

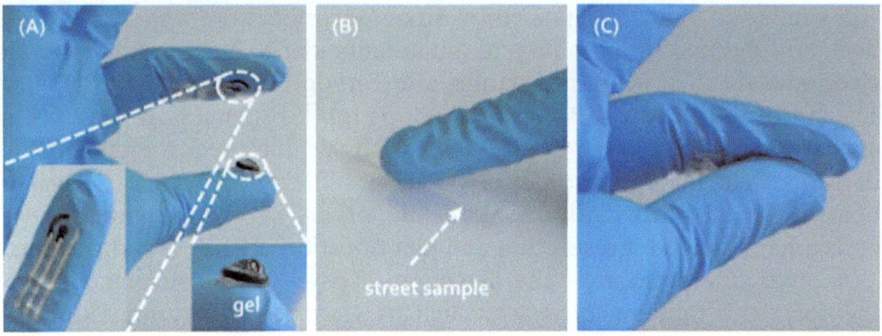

Figure 12.15. A wearable fingertip device. (A) highlights the screen-printed electrode onto a flexible nitrile finger cot (bottom left inset), (B) Applying the swiping method from a street sample, directly to the electrode (C) The electrochemical circuit is complete by joining the index finger with electrodes to the thumb coated with the conductive gel electrolyte. Reproduced from ref. 117 with permission from The Royal Society of Chemistry.

electrochemical approach with a graphite screen-printed electrode and *N*,*N*′-(1,4-phenylene)dibenzenesulfonamide, as a mediator, can quantify the components of the binary mixture of methiopropamine and 2-aminoindane using linear sweep voltammetry. This technique is in agreement with traditional HPLC techniques; whereby, the indirect electrochemical approach determined that the binary mixture contained 58% and 42% methiopropamine and 2-aminoindane, respectively. The HPLC technique resulted in 57% methiopropamine and 42% 2-aminoindane. Methiopropamine contains a secondary amine functional group, while 2-aminoindane has a primary amine moiety. Therefore, the electrochemical response was significantly difficult to resolve the redox potential due to the two different amine groups. Another study by Sutcliffe *et al.* used disposable screen-printed electrodes for the detection of new psychoactive substances (NPS), in particular cathinones.[62] They also demonstrated that the formation of an *in situ* mercury and bismuth film modified graphite screen-printed electrodes could detect mephedrone, 4′-methyl-*N*-ethylcathinone and mephedrone in the μg mL^{-1} range for mephedrone and the *in situ* formed bismuth film and 38 μg mL^{-1} for mephedrone and the *in situ* formed mercury film. This demonstrated that the *in situ* formed films were not necessary for the detection of mephedrone and 4-methylecathinone. Using an acetate buffer at pH 4.3, they explored the direct electrochemical reduction of mephedrone with a range of commercially available carbon-based electrodes. The limits of detection were calculated to be approximately 12 μg mL^{-1} for both mephedrone and 4-methylecathinone, using graphite screen-printed electrodes. They could also demonstrate the quantification of synthetic cathinone in a seized street sample of NRG-2 (a brand name given to legal highs to bypass legislative control), which was independently verified with HPLC. It is anticipated that the future direction of this validated protocol could be used as a portable "legal high" electrochemical sensor.

A similar system was reported by Arcos-Martinez *et al.*, who developed a detection method for cocaine and adulterants using screen-printed carbon electrodes (SPCEs) containing multi wall carbon nanotubes (MWCNTs).[116] They incorporated the MWCNTs *via* a 3 mg mL^{-1} dispersion in DMF. Using square wave voltammetry, they could exploit cocaine's redox activity for its detection. The incorporation of MWCNTs into their electrochemical system helped separate the oxidation peaks of adulterants, *i.e.* codeine, paracetamol and caffeine for better resolution and identification of sample components.

Another way to improve the detection of psychoactive substances is to incorporate a tandem approach, such as electrochemistry and luminescence. Chen *et al.* developed an electrochemiluminescent methamphetamine HCl sensor based on a composite electrode with MWCNTs and $BMIMPF_6$ (1-butyl-3-methylimidazolium hexafluorophosphate, an ionic liquid).[120] Methamphetamine-HCl was found to enhance the ECL emission of immobilized $Ru(bpy)_3^{2+}$, significantly. Using a three-electrode, system, with $Ru(bpy)_3^{2+}$, an ECL emitter, Chen *et al.* developed a composite electrode containing MWCNTs, $[Ru(bpy)_3]$ Cl_2 and 1-butyl-3-methylimidazolium hexafluorophosphate. It was shown that methamphetamine-HCl could be monitored in a pH range of 7.0 to 9.0 with a detection limit of 8.0×10^{-9} mol L^{-1}, which is significantly lower than the direct redox techniques utilized with the screen-printed electrodes, Figure 12.16.

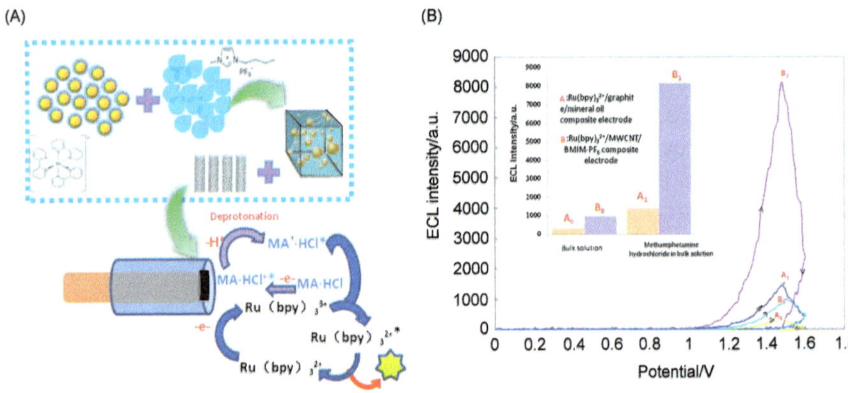

Figure 12.16. (A) Fabrication and mechanism for the electrochemiluminescent composite sensor, generating light at 610 nm (B) ECL signal generated of different luminescent composite sensors in the solution with and without methamphetamine hydrochloride, $Ru(bpy)_3^{2+}$–graphite–mineral oil composite electrode, (B) $Ru(bpy)_3^{2+}$–MWCNT–BMIM·PF_6 composite electrode. Reprinted from *Biosens. Bioelectron.*, **24**, H. Dai, Y. Wang, X. Wu, L. Zhang and G. Chen, An electrochemiluminescent sensor for methamphetamine hydrochloride based on multiwall carbon nanotube/ionic liquid composite electrode, 1230–1234, Copyright (2009), with permission from Elsevier.

12.10. Conclusion and Future Perspective

Historically, supramolecular chemistry has made significant contributions to a number of disciplines of science as highlighted in the introduction of this chapter. Over the last several years, the supramolecular approach has been used to detect drugs of abuse. The focus of this chapter was to highlight how light can be used, in a friendly way, for the "war on drugs" in the streets. The optical response, whether UV-vis or fluorescent, is either turned on or turned off upon the binding of a specific target molecule. We have tried to focus our attention on the most recent work and how host-guest recognition can be used in the forensic field, by using aptamer based technology, polymer (specifically MIPs) and solution based systems. Examples of common instrumental techniques utilized in drug detection have also been presented. Mass spectrometry, chromatography and electrochemical examples are all common alternatives to the host-guest approach, but have significant disadvantages. Many of the examples that have been presented in this chapter deal with a single analyte and a single molecular probe. This is certainly a traditional approach on host-guest recognition; however, many illicit drugs are now being cut with other drugs and other cutting agents. For example, caffeine, lidocaine, and benzocaine produces a complex mixture of compounds. These complex mixtures of structurally similar cutting agents can produce false positives. Additionally, the number of new legal highs is ever increasing, due to minor modifications in their functionalities. In order to keep track and monitor these structurally similar compounds, the future will have to lie in sensor array technology, for example lab-on-a-chip or molecular tongue, linear discrimination analysis (LDA) and principle component analysis (PCA). This will allow for multiple probes to be used for monitoring multiple drugs in a complex system using statistical analysis to differentiate between the drugs. Whereby there are many responses, providing high selectivity and multiple data points for identification. Development of advanced sensors that can provide quick, efficient and a cost effective solution for in-field detection of unknown illicit substances is the future of a new field that should be entitled, "Supramolecular Forensic Chemistry".

Acknowledgements

KJW and JHB would like to thank the editors Giorgia Miolo, Jacqueline L. Stair and Mire Zloh for the invitation to contribute to *Light in Forensic Science*. KJW and JHB would also like to thank the department of Chemistry and Biochemistry at the University of Southern Mississippi for their continued support and financial help and the NSF (OCE-0963064 and GK-12-0947944) by financing the research carried out in the Wallace lab over the years.

References

1. *Supramolecular Chemistry: From Molecules to Nanomaterials*, ed. J. W. Steed and P. A. Gale, John Wiley & Sons, Chichester, 1st edn, 2012.
2. C. Cheng and J. F. Stoddart, *ChemPhysChem*, 2016, **17**, 1780.
3. N. C. H. Lim and S. E. Jackson, *J. Phys.: Condens. Matter*, 2015, **27**, 35.
4. R. Dong, Y. Zhou and X. Zhu, *Acc. Chem. Res.*, 2014, **47**, 2006.
5. J. Alenus, P. Galar, A. Ethirajan, F. Horemans, A. Weustenraed, T. J. Cleij and P. Wagner, *Phys. Status Solidi A*, 2012, **209**, 905.
6. A. Hirst, B. Escuder, J. F. Miravet and D. Smith, *Angew. Chem., Int. Ed.*, 2008, **47**, 8002.
7. T. R. Simmons, G. Berggren, M. Bacchi, M. Fontecave and V. Artero, *Coord. Chem. Rev.*, 2014, **270**, 127.
8. A. Tozzi, *Cogn. Neurodyn.*, 2015, **9**, 463.
9. J.-M. Lehn, *presented in Part at Royal Swedish Academy of Sciences; Nobel Lecture, "Supramolecular Chemistry-scope and Perspectives Molecules-supramolecules-molecular Devices" Stockholm*, 8th December, 1987.
10. I. Mati and S. Cockroft, *Chem. Soc. Rev.*, 2010, **39**, 4195.
11. G. J. Mohr, *Chem.–Eur. J.*, 2004, **10**, 1082.
12. S. Rowan, S. Cantrill, G. Cousins, J. Sanders and J. F. Stoddart, *Angew. Chem., Int. Ed.*, 2002, **41**, 898.
13. E. V. Anslyn, *Supramol. Chem.*, 2016, **28**, 339.
14. E. V. Anslyn, *J. Org. Chem.*, 2006, **72**, 687.
15. G. Grynkiewicz, M. Poenie and R. Y. Tsien, *J. Biol. Chem.*, 1985, **260**, 3440.
16. A. Miyawaki, O. Griesbeck, R. Heim and R. Y. Tsien, *Proc. Natl. Acad. Sci. U. S. A.*, 1999, **96**, 2135.
17. R. Y. Tsien, *presented in Part at Royal Swedish Academy of Sciences; Nobel Lecture, "Constructing and Exploiting the Fluorescent Protein Paintbox" Stockholm*, 8th December, 2008.
18. Y. Takaoka, M. Shigenaga, M. Imai, Y. Nukadzuka, Y. Ishimaru, K. Saito, Y. Yokoyama, K. Nishitani and M. Ueda, *Bioorg. Med. Chem. Lett.*, 2016, **26**, 9.
19. A. A. Karim, Q. L. Dou, Z. Li and X. J. Loh, *Chem.–Asian J.*, 2016, **11**, 1300.
20. C. Y. Ang, S. Y. Tan, X. Wang, Q. Zhang, M. Khan, L. Bai, S. T. Selvan, X. Ma, L. Zhu, K. T. Nguyen, N. S. Tan and Y. Zhao, *J. Mater. Chem. B*, 2014, **2**, 1879.
21. R. Bonnett, *Chem. Soc. Rev.*, 1995, **24**, 19.
22. C. Tu, L. Zhu, P. Li, Y. Chen, Y. Su, D. Yan, X. Zhu and G. Zhou, *Chem. Commun.*, 2011, **47**, 6063.
23. G. Obaid, M. Broekgaarden, A.-L. Bulin, H.-C. Huang, J. Kuriakose, J. Liu and T. Hasan, *Nanoscale*, 2016, **8**, 12471.
24. A. M. Ullman, J. W. Brown, M. E. Foster, F. Léonard, K. Leong, V. Stavila and M. D. Allendorf, *Inorg. Chem.*, 2016, **55**, 7233.
25. P. Falcaro, R. Ricco, A. Yazdi, I. Imaz, S. Furukawa, D. Maspoch, R. Ameloote, J. D. Evans and C. J. Doonan, *Coord. Chem. Rev.*, 2016, **307**, 23.
26. Merriam-Websters Colegiate and Dictionary, *Forensic Science*, https://www.merriam-webster.com/medical/forensic%20science, 2016.
27. A. Nangia and G. R. Desiraju, *Acta Crystallogr., Sect. A: Found. Crystallogr.*, 1998, **54**, 934.
28. C.-E. Chang and M. K. Gilson, *J. Am. Chem. Soc.*, 2004, **126**, 13156.
29. J. D. Dunitz, *Chem. Biol.*, 1995, **2**, 709.

30. B. Wang and E. V. Anslyn, *Chemosensors: Principles, Strategies, and Applications*, John Wiley & Sons, Hoboken, New Jersey, 2011.
31. J. W. Steed, D. R. Turner and K. J. Wallace, *Core Concepts in Supramolecular Chemistry and Nanochemistry*, John Wiley & Sons, Chichester, 2007.
32. V. Shevyrin, V. Melkozerov, O. Eltsov, Y. Shafran and M. Morzherin, *Forensic Sci. Int.*, 2016, **259**, 95.
33. I. González-Mariño, E. Gracia-Lor, R. Bagnati, P. P. B. Martins, E. Zuccato and S. Castiglioni, *Anal. Bioanal. Chem.*, 2016, **408**, 4297.
34. T. Guinan, M. Ronci, R. Vasani, H. Kobus and N. H. Voelcker, *Talanta*, 2015, **132**, 494.
35. E. Bertol, F. Vaiano, M. G. Di Milia and F. Mari, *Forensic Sci. Int.*, 2015, **256**, 21.
36. C. R. Wilson, C. C. Mulligan, K. D. Strueh, G. W. Stevenson and S. B. Hooser, *J. Vet. Diagn. Invest.*, 2014, **26**, 428.
37. S. Saha, M. K. Mandal and K. Hiraoka, *Anal. Methods*, 2013, **5**, 4731.
38. M. Morelato, A. Beavis, P. Kirkbride and C. Rouxa, *Forensic Sci. Int.*, 2013, **226**, 10.
39. T. Guinan, M. Ronci, H. Kobusb and N. H. Voelcker, *Talanta*, 2012, **99**, 791.
40. R. Gottardo, I. Mikšík, Z. Aturki, D. Sorio, C. Seri, S. Fanali and F. Tagliaro, *Electrophoresis*, 2012, **33**, 599.
41. I. Ojanperä, S. Mesihää, I. Rasanen, A. Pelander and R. A. Ketola, *Anal. Bioanal. Chem.*, 2016, **408**, 3395.
42. K. Y. Zuway, J. P. Smith, C. W. Foster, N. Kapur, C. E. Banks and O. B. Sutcliffe, *Analyst*, 2015, **140**, 6283.
43. C. S. D. Wink, J. A. Michely, J. Jacobsen-Bauer, J. Zapp and H. H. Maurer, *Drug Test. Anal.*, 2016, **8**, 1005–1014.
44. J. A. Michcly, A. G. Helfer, S. D. Brandt, M. R. Meyer and H. H. Maurer, *Anal. Bioanal. Chem.*, 2015, **407**, 7831.
45. M. R. Meyer, A. Holderbaum, P. Kavanagh and H. H. Maurer, *J. Mass Spectrom.*, 2015, **50**, 1163.
46. E. Lendoiro, C. Jiménez-Morigosa, A. Cruz, M. Páramo, M. López-Rivadulla and A. de Castroa, *Drug Test. Anal.*, 2017, **9**, 96–105.
47. M. Grootveld and C. J. L. Silwood, *Biochem. Biophys. Res. Commun.*, 2005, **329**, 1.
48. R. Risoluti, S. Materazzi, A. Gregori and L. Ripani, *Talanta*, 2016, **153**, 407.
49. C. Weyermanna, Y. Mimounea, F. Anglada, G. Massonneta, P. Esseiva and P. Buzzini, *Forensic Sci. Int.*, 2011, **209**, 21.
50. M. Grootveld, D. Algeo, C. J. L. Silwood, J. C. Blackburn and A. D. Clarl, *BioFactors*, 2006, **27**, 121.
51. C. Andreou, M. R. Hoonejani, M. R. Barmi, M. Moskovits and C. D. Meinhart, *ACS Nano*, 2013, **7**, 7157.
52. K. M.-C. Hans, M. Müller, T. Petrosyan and M. W. Sigrist, *Anal. Methods*, 2014, **6**, 666.
53. E. Biavardi, S. Federici, C. Tudisco, D. Menozzi, C. Massera, A. Sottini, G. G. Condorelli, P. Bergese and E. Dalcanale, *Angew. Chem., Int. Ed.*, 2014, **53**, 9183.
54. K. Kang, A. Sachan, M. Nilsen-Hamilton and P. Shrotriya, *Langmuir*, 2011, **27**, 14696.
55. D. R. Theâvenot, K. Toth, R. A. Durst and G. S. Wilson, *Pure Appl. Chem.*, 1999, **71**, 2333.
56. Y. Wen, H. Pei, Y. Wan, Y. Su, Q. Huang, S. Song and C. Fan, *Anal. Chem.*, 2011, 7418.

57. J.-L. He, Y.-F. Yang, G.-L. Shena and R.-Q. Yua, *Biosens. Bioelectron.*, 2011, **26**, 4222.
58. B. Jiang, M. Wang, Y. Chen, J. Xiea and Y. Xiang, *Biosens. Bioelectron.*, 2012, **32**, 305.
59. R. J. White, N. Phares, A. A. Lubin, Y. Xiao and K. W. Plaxco, *Langmuir*, 2008, **24**, 10513.
60. M. Ebrahimi, H. Hamzeiy, J. Barar, A. Barzegari and Y. Omidi, *Sens. Lett.*, 2013, **11**, 566.
61. J. McGeehan and L. Dennany, *Forensic Sci. Int.*, 2016, **264**, 1.
62. J. P. Smith, J. P. Metters, O. I. G. Khreit, O. B. Sutcliffe and C. E. Banks, *Anal. Chem.*, 2014, **86**, 9985.
63. R. Saini, N. Kaur and S. Kumar, *Tetrahedron*, 2014, **70**, 4285.
64. N. H. Evans, H. Rahman, J. J. Davis and P. D. Beer, *Anal. Bioanal. Chem.*, 2012, **402**, 1739.
65. P. D. Beer, P. A. Gale and G. Z. Chen, *Coord. Chem. Rev.*, 1999, **185–186**, 3.
66. L. Tang, F. Li, M. Liu and R. Nandhakumar, *Bull. Korean Chem. Soc.*, 2011, **32**, 3400.
67. J. R. Lakowicz, *Principles of Fluorescence Spectroscopy*, Springer, New York, 3rd edn, 2006.
68. R. J. T. Houk, K. J. Wallace, H. S. Hewage and E. V. Anslyn, *Tetrahedron*, 2008, **64**, 8271.
69. J. M. Costa-Fernández, R. Pereiro and A. Sanz-Medel, *TrAC, Trends Anal. Chem.*, 2006, **25**, 207.
70. C. M. G. dos Santos, A. J. Harte, S. J. Quinn and T. Gunnlaugsson, *Coord. Chem. Rev.*, 2008, **252**, 2512.
71. E. J. O'Neil and B. D. Smith, *Coord. Chem. Rev.*, 2006, **250**, 3068.
72. X. Michalet, F. F. Pinaud, L. A. Bentolila, J. M. Tsay, S. Doose, J. J. Li, G. Sundaresan, A. M. Wu, S. S. Gambhir and S. Weiss, *Science*, 2005, **307**, 538.
73. United Nations Office on Drugs and Crime, *World Drug Report 2016*, United Nations Publications, Sales No. E.16XI.7, New York, 2016, http://www.unodc.org/wdr2016/.
74. European Monitoring Centre for Drugs and Drug Addiction and Europol, *EU Drug Markets Report In-depth Analysis*, Publications Office of the European Union, Luxembourg, 2016.
75. J. P. Kelly, *Drug Test. Anal.*, 2011, **3**, 439.
76. S. D. Brandt, H. R. Sumnall, F. Measham and J. Coled, *Drug Test. Anal.*, 2010, **2**, 377.
77. S. Niu, X. Lou, Y. Jiang and J. Lin, *Anal. Lett.*, 2012, **45**, 1919.
78. A. E. Holmes, United States Pat., US 2010/0197516 A1, 2010.
79. S. A. Bergman, *Anesth. Prog.*, 1999, **46**, 10.
80. H. Johnke, G. Batres, M. Wilson, A. E. Holmes and S. Sikich, *J. Sens. Technol.*, 2013, **3**, 94.
81. J. Birks, *Rep. Prog. Phys.*, 1975, **38**, 903.
82. A. P. de Silva, H. Q. N. Gunaratne, T. Gunnlaugsson, A. J. M. Huxley, C. P. McCoy, J. T. Rademacher and T. E. Rice, *Chem. Rev.*, 1997, **97**, 1515.
83. E. Manandhar and K. J. Wallace, *Inorg. Chim. Acta*, 2012, **381**, 15.
84. C. Wua, L. Yana, C. Wanga, H. Lina, C. Wanga, X. Chen and C. J. Yanga, *Biosens. Bioelectron.*, 2010, **25**, 2232.
85. S. Silvi, M. Baroncini, M. La Rosa and A. Credi, *Top. Curr. Chem.*, 2016, **374**, 65.
86. R. C. Somers, M. G. Bawendi and D. G. Nocera, *Chem. Soc. Rev.*, 2007, **36**, 579.

87. L. Basabe-Desmonts, D. N. Reinhoudt and M. Crego-Calama, *Chem. Soc. Rev.*, 2007, **36**, 993.
88. A. Burns, H. Ow and U. Wiesner, *Chem. Soc. Rev.*, 2006, **35**, 1028.
89. E. Guler, G. Bozokalfa, B. Demir, Z. P. Gumus, B. Guler, E. Aldemir, S. Timura and H. Coskunol, *Drug Test. Anal.*, 2017, **9**, 578–587.
90. J. Liu and Y. Lu, *Angew. Chem., Int. Ed.*, 2006, **45**, 90.
91. E. Turiel and A. Martín-Esteban, *Anal. Chim. Acta*, 2010, **668**, 87.
92. L. Figueiredo, G. Erny, L. Santos and A. Alves, *Talanta*, 2016, **146**, 754.
93. T. H. Nguyen, S. A. Hardwick, T. Sun and K. T. V. Grattan, *IEEE Sens. J.*, 2012, **12**, 255.
94. J. E. Whitaker, R. P. Haugland and F. G. Prendergast, *Anal. Biochem.*, 1991, **194**, 330.
95. O. Valdes-Aguilera and D. Neckers, *Acc. Chem. Res.*, 1989, **22**, 171.
96. M. P. Chantada-Vázquez, J. Sánchez-González, E. Peña-Vázquez, M. J. Tabernero, A. M. Bermejo, P. Bermejo-Barrera and A. Moreda-Piñeiro, *Biosens. Bioelectron.*, 2016, **75**, 213.
97. J. Prost, *The Physics of Liquid Crystals*, Oxford University Press, 1995.
98. L. Meng, P. Meng, Q. Zhang and Y. Wang, *Anal. Chim. Acta*, 2013, 77**1**, 86.
99. J. Tan, H.-F. Wang and X.-P. Yan, *Anal. Chem.*, 2009, **81**, 5273.
100. K. Ariga, G. J. Richards, S. Ishihara, H. Izawa and J. P. Hill, *Sensors*, 2010, **10**, 6796.
101. C. Han, Z. Shang, H. Zhang and Q. Song, *Anal. Methods*, 2013, **5**, 6064.
102. S.-W. Yun, N.-Y. Kang, S.-J. Park, H.-H. Ha, Y. K. Kim, J.-S. Lee and Y.-T. Chang, *Acc. Chem. Res.*, 2014, **47**, 1277.
103. D. Zhai, B. K. Agrawalla, P. S. F. Eng, S.-C. Lee, W. Xu and Y.-T. Chang, *Chem. Commun.*, 2013, **49**, 6170.
104. D. Zhai, Y. Q. E. Tan, W. Xu and Y.-T. Chang, *Chem. Commun.*, 2014, **50**, 2904.
105. C. He, Q. He, C. Deng, L. Shi, D. Zhu, Y. Fu, H. Caoa and J. Cheng, *Chem. Commun.*, 2010, **46**, 7536.
106. F. Reviriego, P. Navarro, E. García-España, M. T. Albelda, J. C. Frías, A. Domènech, M. J. R. Yunta, R. Costa and E. Ortí, *Org. Lett.*, 2008, **10**, 5099.
107. T. Fan, W. Xu, J. Yao, Z. Jiao, Y. Fu, D. Zhu, Q. He, H. Cao and J. Cheng, *ACS Sens.*, 2016, **1**, 312.
108. R. E. Smalley, M. S. Dresselhaus, G. Dresselhaus and P. Avouris, *Carbon Nanotubes: Synthesis, Structure, Properties, and Applications*, Springer-Verlag, Berlin Heidelberg, 2003.
109. T. H. Kim, H. W. Ho, C. L. Brown, S. L. Cresswell and Q. Li, *Anal. Methods*, 2015, **7**, 6869.
110. K. Kellett, J. H. Broome, M. Zloh, S. B. Kirton, S. Fergus, U. Gerhard, J. L. Stair and K. J. Wallace, *Chem. Commun.*, 2016, **52**, 7474.
111. S. Gibbons and M. Zloh, *Bioorg. Med. Chem. Lett.*, 2010, **20**, 4135.
112. H. M. Berman, J. Westbrook, Z. Feng, G. Gilliland, T. N. Bhat, H. Weissig, I. N. Shindyalov and P. E. Bourne, *Nucleic Acids Res.*, 2000, **28**, 235.
113. C. A. Hunter and J. K. Sanders, *J. Am. Chem. Soc.*, 1990, **112**, 5525.
114. F. Vögtle and P. Knops, *Angew. Chem., Int. Ed.*, 1991, **30**, 958.
115. P. Miao, K. Han, H. Sun, J. Yin, J. Zhao, B. Wang and Y. Tang, *ACS Appl. Mater. Interfaces*, 2014, **6**, 8667.
116. L. Asturias-Arribas, M. A. Alonso-Lomillo, O. Domínguez-Renedo and M. J. Arcos-Martínez, *Anal. Chim. Acta*, 2014, **834**, 30.

117. M. de Jong, N. Sleegers, J. Kim, F. van Durme, N. Samyn, J. Wang and K. de Wael, *Chem. Sci.*, 2016, 7, 2364.
118. J. P. Smith, O. B. Sutcliffe and C. E. Banks, *Analyst*, 2015, **140**, 4932.
119. L. R. Cumba, A. V. Kolliopoulos, J. P. Smith, P. D. Thompson, P. R. Evans, O. B. Sutcliffe, D. R. do Carmoa and C. E. Banks, *Analyst*, 2015, **140**, 5536.
120. H. Dai, Y. Wang, X. Wu, L. Zhang and G. Chen, *Biosens. Bioelectron.*, 2009, **24**, 1230.

Chapter 13

Photodegradation of Drugs/Illicit Drugs in Water

Paola Calza* and Davide Vione

Dipartimento di Chimica, Università degli Studi di Torino, via P. Giuria 5-10125, Torino, Italy

*E-mail: paola.calza@unito.it

Table of Contents

Light in Forensic Science: Issues and Applications
Edited by Giorgia Miolo, Jacqueline L. Stair and Mire Zloh

Published by the Royal Society of Chemistry, www.rsc.org

13.1. Introduction

Many drugs are sensitive to light,[1,2] and their formulated products may be degraded during manufacture, storage and administration. This may induce potency loss, changed efficacy, and adverse biological effects. Several monographs and reviews[3–5] have been published on the photochemistry, photostability, and related aspects of drugs and drug products. The biological consequences of the action of light on drugs are important as well. The photoproducts of a drug may be harmful and cause phototoxicity, photoallergy, photosensitization, and other effects.[6–9]

After their consumption, drugs are eliminated as the parent compounds and/or their metabolites through human excreta reaching sewage treatment plants (STPs). These plants are often unable to completely remove such xenobiotics, which may escape removal processes and reach the aquatic compartment. Here they often occur in the range of sub-ng L^{-1} to μg L^{-1}.

Light and oxygen are two of the most important abiotic factors involved in the environmental fate of drugs in aquatic systems and/or in the surface of soils. These factors have received particular attention over the last decade.[10,11] Direct or sensitized phototransformations in natural aquatic media can promote the complete degradation of the molecule, but in some cases they yield transformation products that can be more toxic and/or noxious than the parent compounds.[11,12] On the basis of this evidence, drugs, illicit drugs and their metabolites constitute a group of emerging organic contaminants of particular concern,[13] due to their worldwide occurrence in environmental waters.[14] Therefore, the evaluation of the potentially adverse effects of drugs and their derivatives is of increasing scientific concern.[15,16] The main sources of illicit drugs and metabolites into the aquatic environment are municipal wastewater treatment plant effluents, because of incomplete drug elimination with conventional treatment processes.[17]

Among the illicit drugs, cocaine (COC) is ubiquitous in natural surface waters and reaches levels of up to 115 ng L^{-1}.[18] However, it is also efficiently removed in conventional activated sludge processes, with elimination percentages usually above 90%.[19] Conversely, low removal is reported for both ketamine (a widely used club drug but also a general anesthetic and short-acting analgesic in veterinary practice, clinics and pediatrics), and methadone. Methadone is removed with low efficiency (<40%),[20] thus it reaches levels up to 735 ng L^{-1} and 18 ng L^{-1} in wastewater and river water, respectively.[21] Lin *et al.*[22] identified ketamine in Taiwan surface waters at concentrations of up to 341 ng L^{-1}.

Among drugs, anti-cancer drugs have shown potential to act as carcinogens, teratogens and/or mutagens,[23,24] and for this reason their environmental fate[25–27] and surface-water exposure[28] have been widely investigated. The anticonvulsant drug Gabapentin (GAB) exhibits teratogenic effects too. GAB is not metabolized, and it is poorly eliminated by STPs. Therefore, it has been detected in surface water and even in raw potable water.[29] Nonsteroidal anti-inflammatory drugs (NSAIDs) raise considerable concern due to their worldwide use and frequent environmental occurrence. Many of these compounds have shown a significant photoreactivity, often associated with phototoxic side effects.[30]

Although the environmental concentrations of drugs and illicit drugs are relatively low, the presence of these compounds and/or their transformation products in the aquatic environment may have negative effects on the aquatic organisms that are continuously exposed to them. Hence, knowledge about natural transformation is essential for an appropriate environmental risk assessment and to understand how drug fate and transport impact environmental and human health. There is some knowledge on the presence of these compounds in the aquatic environment, but rather limited knowledge on their effects on humans and wildlife once they enter the environment. Moreover, the effects of human metabolites and other transformation products (TPs) is much less known compared to the parent drugs. TPs can result from various treatment processes or from abiotic and biotic environmental processes such as biotransformation, hydrolysis or photolysis. The overall reaction pathways, together with possible identification strategies, are summarized in Scheme 13.1.

In the present chapter we focus on the photoinduced processes, with particular emphasis on the drug's phototransformation. Phototransformation is an important environmental depletion mechanism for drug removal in surface waters, because it may be a fast process under environmental conditions.[31–33] Phototransformation is particularly relevant to hydrophilic drugs with multiple chromophores.[34,35]

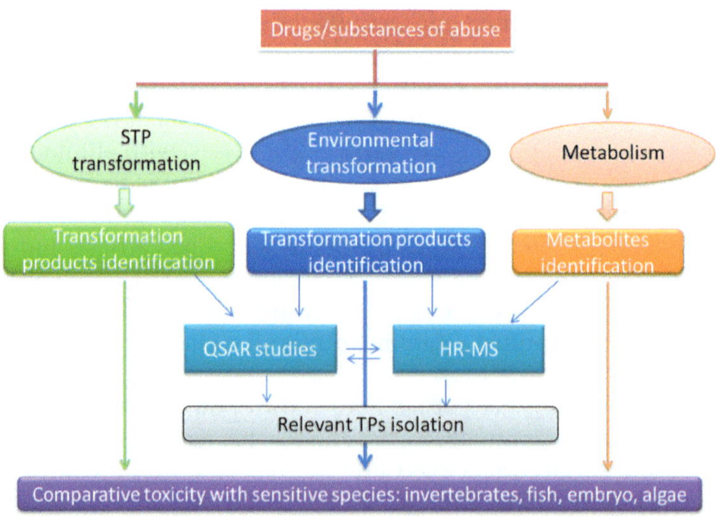

Scheme 13.1. Framework showing possible reaction pathways occurring on pharmaceutical and illicit drugs, and identification strategies.

13.2. Photochemical Reactions in Surface Waters

Photochemical reactions are important pathways by which several pollutants undergo natural attenuation in surface-water environments. These reactions are particular effective for biorecalcitrant compounds, which is the case for many emerging contaminants. The photoinduced processes that occur in sunlit surface waters can be divided into two main families, namely the direct photolysis and the indirect or sensitised photochemistry.[36] They will now be described in greater detail.

13.2.1. Direct Photolysis Processes

The direct photolysis is the transformation of a molecule as a consequence of radiation absorption, and in the environment it involves compounds that are able to absorb sunlight. For the direct photolysis process to be operational a molecule should absorb radiation above 280 nm, because shorter wavelengths do not occur in the sunlight spectrum at the ground. The absorption of sunlight (UV-vis radiation) by a molecule promotes an electron from the ground vibrational state of the ground electronic state to an excited vibrational state of an excited electronic state. Therefore, the molecule acquires both electronic and vibrational energy upon radiation absorption. In some cases, the absorbed radiation is so high as to abstract the electron from the molecule, *i.e.* to cause photoionisation. The probability of inducing photoionisation is higher in the case of the more energetic but environmentally insignificant UVC radiation ($\lambda < 280$ nm), particularly in the case of phenolic compounds, but for some molecules with extensive electron delocalisation the process can also take place effectively under environmental UV irradiation.[37]

If the absorbed radiation is not so energetic as to cause photoionisation, the excited electron may reach an excited vibrational state of an excited electronic state. In some cases the vibrational energy is so high as to cause bond breaking, which is usually a homolytic process that yields two radical species. For instance, the UVB or UVA irradiation of phenols often causes the homolytic rupture of the phenolic O–H bond with generation of a phenoxy radical and of a hydrogen atom ($H^{\bullet}$). In other cases, the excess vibrational energy is simply dissipated (for instance upon collision with the solvent), so that the electron reaches a ground vibrational level of an excited electronic state. The problem of *which* excited electronic state is reached at this point is a rather interesting issue. By labelling the electronic singlet states as S_0, S_1, S_2 and so on, where S_0 is the ground state, radiation absorption might shift the electron to, say, vibrationally excited S_3 (hereafter S_3*, where the '*' represents the excess of vibrational energy). One might expect that the loss of vibrational energy causes the electron to reach the ground vibrational state of S_3, but it is usually not so. According to Kasha's rule, the decay from a vibrationally excited state (*e.g.* S_1*, S_2*, S_3*, S_4* *etc.*) by loss of vibrational energy leads to S_1, which is the ground vibrational level of the first excited (singlet) electronic state. Such a state as well as the others mentioned previously are called "singlet" states because the electron, when promoted to an orbital with higher energy, maintains its spin orientation. There are some exceptions where a state S_n with $n > 1$ is reached by loss of vibrational energy, but in most cases Kasha's rule is followed.[36–38]

There are a number of possible evolution pathways from S_1, that is:[36–38]

(i) Loss of energy by internal conversion (*e.g.* by collisions). This process is favoured in the case of very large molecules that have several components that can interact with one another.
(ii) Emission of fluorescence radiation, which is favoured in the case of rigid multi-ring structures.
(iii) Chemical reactivity, which is usually inhibited by the short S_1 lifetimes. There are some exceptions, however, and a major one is 2-chlorophenol, that in its S_1 state undergoes a ring-contraction process to yield a C_5-ring derivative.[38]
(iv) Inter-system crossing (ISC), where the electron spin undergoes an inversion to reach the first excited triplet state (T_1). The ISC process is often a prerequisite for further reactions to take place, because of the longer lifetimes of the triplet states compared to the singlet ones. In fact, because the ground states of most molecules and particularly of organic ones are singlet states (S_0, the case of T_0 being quite rare with O_2 as a major exception), the spin-forbidden $T_1 \rightarrow S_0$ transition is less likely than $S_1 \rightarrow S_0$ and causes the T_1 lifetime to be longer than that of S_1.

Possible evolutions from the T_1 state are:[37,38]

(i) Internal conversion (again, loss of energy by *e.g.* collision with the solvent).
(ii) Emission of phosphorescence radiation. However, just a minority of molecules is able to emit phosphorescence at room temperature in aqueous solutions, and the standard measurements of phosphorescence are usually carried out in water-glycerol mixtures at the temperature of liquid N_2.
(iii) Reaction with molecular oxygen. Many triplet states react quickly with O_2, because the reaction is favoured by the triplet nature of the O_2 ground state. This energy-transfer process yields singlet oxygen ($^1O_2$) while the other molecule reaches its S_0 state.
(iv) Other chemical reactions. Some triplet states are oxidants and they can abstract e^- or H atoms from other dissolved molecules (including the same molecule in its ground state), and in some cases intramolecular redox processes can be triggered as well. Reactions with the solvent are also possible, and the T_1 states that undergo fast reaction with *e.g.* water are often unaffected by the presence of molecular oxygen in solution.

A general scheme of the processes induced by radiation absorption is provided in Scheme 13.2. Among these, the direct photolysis can be triggered by photoionisation, S_n*, S_1 and T_1 reactions that can cause transformation of the light-absorbing molecule. In contrast, photoluminescence and internal conversion quench the chemical reactivity.

As a transformation process, the direct photolysis is described by its reaction rate R and by the direct photolysis quantum yield Φ. The quantum yield is the ratio between the rate of photolysis and the photon flux absorbed by the molecule, P_a. Assume $A(\lambda)$ as the absorbance of the molecule in solution at the wavelength λ, and $p°(\lambda)$ as the incident spectral photon flux density from the radiation source (sunlight or a lamp in the laboratory). In this case the direct photolysis quantum yield can be expressed as follows:[36]

$$\Phi = \frac{R}{P_a} = \frac{R}{\int_{\lambda} p^{\circ}(\lambda)[1-10^{-A(\lambda)}]\mathrm{d}\lambda} \tag{a}$$

The above relationship works if the molecule is the only light-absorbing species in solution. In the presence of other absorbers with absorbance $A'(\lambda)$, the absorbed photon flux by the same molecule becomes:

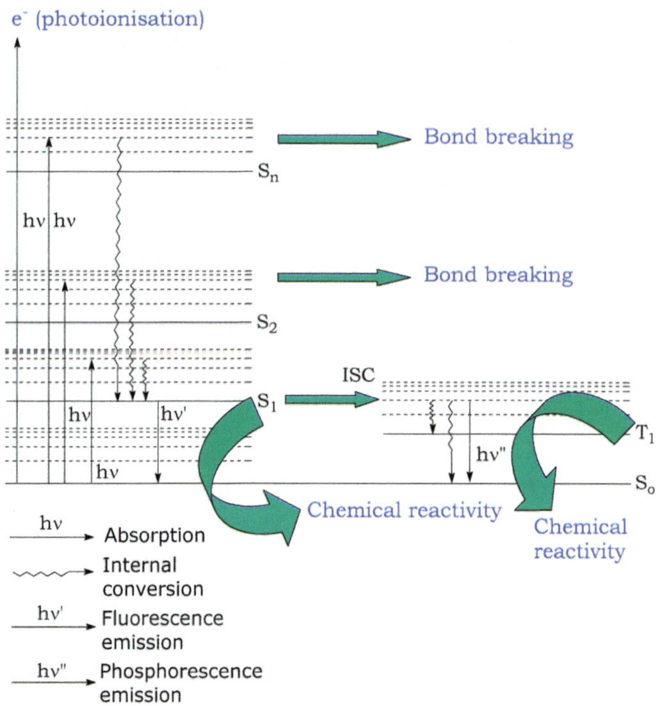

Scheme 13.2. Processes that follow radiation absorption by an organic molecule, originally occurring in its ground singlet state S_0. The processes highlighted in blue are those involved in transformation by direct photolysis.

$$P_a = \frac{A(\lambda)}{A(\lambda)+A'(\lambda)}\int_{\lambda} p^{\circ}(\lambda)\{1-10^{-[A(\lambda)+A'(\lambda)]}\}\mathrm{d}\lambda \quad \text{(b)}$$

In this case the expression for the quantum yield is modified as a consequence:[36]

$$\Phi = \frac{R}{P_a} = \frac{A(\lambda)+A'(\lambda)}{A(\lambda)}\frac{R}{\int_{\lambda} p^{\circ}(\lambda)\{1-10^{-[A(\lambda)+A'(\lambda)]}\}\mathrm{d}\lambda} \quad \text{(c)}$$

The main transitions of a molecule from S_0 to its excited states (S_n*, with $n \geq 1$) are characterised by bands in the absorption spectrum. Each band has peculiar values of the molar absorption coefficient $\varepsilon(\lambda)$ and thus, at equal optical path length and concentration, peculiar $A(\lambda)$ values that produce definite P_a values as a result. However, these light absorption properties are taken into account in the calculation of Φ. Independently of light absorption, after reaching the S_n* state most molecules will decay to S_1, from which further transformation processes are possible. The molecule has no memory of its state(s) before reaching S_1, thus the processes that follow will be the same independently of the initial S_n* state reached upon radiation absorption. This issue explains why many molecules have direct photolysis quantum yields that are independent of the excitation wavelength. A dependence of Φ on λ can occur in the case of photoionisation, of S_n* reactions (*e.g.* bond breaking), or with small molecules for which Kasha's rule (decay of S_n* into S_1) may not be respected.[37]

In the natural environment, the rate of direct photolysis can depend on several features including:[36]

(i) The irradiance of sunlight, because the direct photolysis is obviously enhanced if radiation is more intense.

(ii) The sunlight spectrum, which is connected to the irradiance. In fact, when the sun is low over the horizon (at sunrise, at sunset or in winter), in addition to the lower irradiance values (solar light is incident over a larger area if its elevation is lower, *i.e.* if the zenith angle is higher), radiation also has to travel through a thicker layer of the atmosphere. Atmospheric absorption depletes the shorter wavelengths (the UVB and, to a lesser extent, the UVA), thus sunlight in winter, at sunrise and at sunset is UVB-poor and the direct photolysis of UVB-absorbing molecules can be considerably inhibited.

(iii) The colour and turbidity of the natural water. If the water is particularly coloured (*i.e.* rich in chromophoric dissolved organic matter, CDOM, *vide infra*) or turbid, the incident sunlight is absorbed or scattered. As a consequence, the irradiance in the below-surface water layer gets considerably decreased.

(iv) The occurrence and concentration of some water components, including most notably the dissolved organic matter (DOM). While the previous factors affect P_a by modifying either $p°(\lambda)$ or $A'(\lambda)$, but by leaving Φ unchanged, DOM may modify Φ without affecting P_a. The rationale is that many molecules are photochemically transformed *via* their triplet states, and DOM can either react with the triplet states by reducing them (which is usually followed by back oxidation by dissolved O_2, yielding back the initial molecule) or, following triplet-induced reactions, cause reduction of partially oxidised species back to the parent molecule. Both effects are able to inhibit the direct photolysis process. By assuming M as a molecule that undergoes direct photolysis, the processes described above can take place as follows, where DOM_{ox} means oxidised DOM, M_{ox} represents an oxidised species (*e.g.* a radical) derived from M, and N is a reaction product. The formation of N is an index that the direct photolysis is operational, and any process that prevents N formation is expected to inhibit the direct photolysis.[39]

$$M(S_0) + h\nu \rightarrow M(S_n{}^*) \rightarrow M(S_1) \xrightarrow{ISC} M(T_1) \quad \text{(d)}$$

$$M(T_1) \rightarrow \rightarrow M_{ox} \rightarrow \rightarrow N \quad \text{(e)}$$

$$M(T_1) + DOM \rightarrow M^{\bullet -} + DOM^{\bullet +} \quad \text{(f)}$$

$$M^{\bullet -} + O_2 \rightarrow M + O_2{}^{\bullet -} \quad \text{(g)}$$

$$M_{ox} + DOM \rightarrow M + DOM_{ox} \quad \text{(h)}$$

The direct photolysis reactions do not usually cause the total mineralisation of a molecule, but they rather produce transformation intermediates. If a biorecalcitrant molecule yields biodegradable intermediates, mineralisation may occur following biological reactions instead of photochemistry. In some cases, the direct photolysis intermediates may be more harmful than the parent compound: this is for instance the case of the formation of 2,8-dichlorodibenzodioxin from triclosan, of acridine from carbamazepine, and of isobutylacetophenone from ibuprofen.[39]

13.2.2. *Indirect or Sensitised Photochemistry*

Differently from the direct photolysis, in the case of indirect photochemistry the reactions are triggered by the absorption of sunlight by dissolved molecules called photosensitisers. They are different from the molecule undergoing photodegradation. Common photosensitisers in surface waters are chromophoric dissolved organic matter (CDOM), nitrate and nitrite. CDOM is largely made up of humic and fulvic acids that derive from soil runoff and/or from the decomposition of organic material and vegetable litter.[39,40] It has a peculiar absorption spectrum that shows a featureless exponential decay of the absorbance with increasing wavelength. Therefore, the absorption spectrum of CDOM can be approximated with an equation of the form

$A(\lambda) = A_0 e^{-S\lambda}$, where S is the spectral slope. For most natural waters, S ranges between 10–20 μm^{-1}. In some cases, one would like to directly obtain the absorbance at a given reference wavelength ($A_{ref}(\lambda_{ref})$) from the numerical fit of the absorption spectrum, in which case it can be convenient to slightly modify the functional form of the fit function as follows: $A(\lambda) = [A_{ref}(\lambda_{ref})] e^{-S(\lambda-\lambda_{ref})}$.[41]

CDOM is by far the main sunlight absorber in natural waters in the 300–500 nm wavelength interval. Above 500 nm there is significant absorption by both chlorophylls and water itself, but between 300 and 500 nm the absorption of water is almost fully accounted for by CDOM.[41] The elevated absorbed photon flux, compared to other water components, makes CDOM the main photosensitiser in sunlit surface-water environments. CDOM is not yet precisely characterised, anyway it is thought to consist of aromatic carbonyls including acetophenones and benzophenones, and possibly also of quinones. The irradiation of CDOM yields a number of reactive species including most notably its triplet states (usually indicated as ^{3}CDOM*), which are rather effective oxidising agents towards a variety of pollutants including phenols, phenylurea herbicides and sulphonamide antibiotics.[42,43] The reaction between ^{3}CDOM* and dissolved O_2 yields $^1O_2$ as additional transient, which is for instance involved in the environmental attenuation of chlorophenolates.[39] Irradiated CDOM is the only significant source of $^1O_2$ in surface waters. Although not very reactive towards the majority of emerging pollutants, $^1O_2$ plays an important role in the sunlight-induced inactivation of pathogenic viruses.[36]

Irradiated CDOM also yields $^{\bullet}OH$, although the details of $^{\bullet}OH$ photoproduction by CDOM are still incompletely elucidated. Part of the process is linked to the photoproduction of H_2O_2, the importance of which is variable depending on the CDOM specimen. However, the H_2O_2-related pathway grossly counts on average for ~50% of $^{\bullet}OH$ photogeneration. CDOM-induced processes yield H_2O_2 *via* $O_2^{\bullet-}$ generation and finally produce $^{\bullet}OH$ through the Fenton reaction or H_2O_2 photolysis. This $^{\bullet}OH$ formation pathway can be inhibited upon addition to irradiated CDOM solutions of the H_2O_2-consuming enzyme catalase.[39,42,44] S–H in the reaction scheme below is a generic substrate undergoing oxidation, ISC = inter-system crossing.

$$\text{CDOM} + h\nu \rightarrow {}^1\text{CDOM}^* \xrightarrow{\text{ISC}} {}^3\text{CDOM}^* \quad \text{(i)}$$

$$^3\text{CDOM}^* + \text{S–H} \rightarrow \text{CDOM–H}^{\bullet} + \text{S}^{\bullet} \quad \text{(j)}$$

$$\text{CDOM–H}^{\bullet} + O_2 \rightarrow \text{CDOM} + HO_2^{\bullet}\ (O_2^{\bullet-} + H^+) \quad \text{(k)}$$

$$HO_2^{\bullet} + O_2^{\bullet-} + H^+ \rightarrow H_2O_2 + O_2 \quad \text{(l)}$$

$$Fe^{2+} + H_2O_2 \rightarrow FeOH^{2+} + {}^{\bullet}OH \quad \text{(m)}$$

$$H_2O_2 + h\nu \rightarrow 2\ {}^{\bullet}OH \quad \text{(n)}$$

There is also a H_2O_2-independent process of $^{\bullet}OH$ formation from irradiated CDOM, which is not inhibited by catalase.[44] Here the exact pathway of

$^{\bullet}OH$ photoproduction is fully unclear: some excited triplet states are able to oxidise H_2O and OH^- to $^{\bullet}OH$, but they are not necessarily representative of CDOM composition (an example is given by the triplet state of 1-nitronaphthalene, ^{3}1NN*). Another possibility is the photolysis of organic compounds having –OH or –OOH functional groups, which may occur in pristine CDOM and/or in CDOM after photoprocessing.[39,42,43]

As an additional photosensitiser, the nitrate anion (NO_3^-) absorbs both UVB and UVA radiation. It has an absorption maximum in the UVB region (~305 nm). Nitrate UV photolysis yields hydroxyl radicals *via* a pathway that, although apparently pH-dependent, is actually not expected to undergo significant pH effects below pH 12 (corresponding to the pK_a of the equilibrium $^{\bullet}OH \leftrightarrows H^+ + O^{\bullet-}$).[39,45]

$$NO_3^- + h\nu \rightleftarrows [^{\bullet}NO_2 + O^{\bullet}]_{\text{Solvent cage}} \rightarrow {}^{\bullet}NO_2 + O^{\bullet-} \xrightleftharpoons{H^+} {}^{\bullet}NO_2 + {}^{\bullet}OH \quad (o)$$

The photogenerated fragments ($^{\bullet}NO_2 + O^{\bullet-}$) are initially located within a cage of water molecules, where they can recombine back to nitrate (geminate recombination) or diffuse out of the cage into the solution bulk. There is evidence that only ~25% of the photogenerated fragments would diffuse out of the cage, the remainder undergoing geminate recombination. This situation can change if the solution contains radical scavengers at high enough concentration to react with cage $O^{\bullet-}$, and inhibit the recombination between $^{\bullet}NO_2$ and $O^{\bullet-}$.[46] The overall quantum yield of $^{\bullet}OH$ photoproduction *via* reaction (o) is ~1%.[45]

However, an unexpected pH trend is observed in $^{\bullet}OH$ photoproduction upon nitrate photolysis, the rate of which is about double at pH 6 compared to pH 8. A possible explanation for this finding might be the formation of peroxynitrous acid/peroxynitrite (HOONO/$ONOO^-$, pK_a = 7) upon nitrate photoisomerisation:[39]

$$NO_3^- + h\nu \rightarrow ONOO^- \xrightleftharpoons{H^+} HOONO \quad (p)$$

$$HOONO \rightarrow {}^{\bullet}OH + {}^{\bullet}NO_2 \quad (q)$$

The UVB absorption maximum of nitrate causes its photolysis to be considerably inhibited as the water depth increases, because UVB radiation is very effectively absorbed by CDOM and shows poor penetration into the water column.[36,41] Therefore, nitrate photolysis can only be important in shallow water bodies.

Nitrite absorbs environmental UV radiation with a maximum in the UVA region (~355 nm), and it yields $^{\bullet}OH$ *via* the following process:[41]

$$NO_2^- + h\nu \rightarrow {}^{\bullet}NO + O^{\bullet-} \xrightleftharpoons{H^+} {}^{\bullet}NO + {}^{\bullet}OH \quad (r)$$

The absorption of environmental UV radiation by nitrite involves two electronic transitions, one producing a shoulder in the absorption spectrum around 280 nm and the other causing the 355 nm absorption maximum.

Nitrite is small and its photolysis does not follow Kasha's rule: the UVB band has a photolysis quantum yield of approximately 6.5% and the UVA one of ~2.5%. Therefore, the quantum yield of $^{\bullet}OH$ photogeneration by nitrite decreases with increasing wavelength in the given interval.[47] The photoproduction of $^{\bullet}OH$ by irradiated nitrite has some pH dependence because nitrous acid (HNO_2, $pK_a = 3.3$) absorbs radiation to a comparable extent as nitrite but with higher photolysis quantum yield (~35%). However, the photoproduction of $^{\bullet}OH$ by HNO_2 *vs.* NO_2^- only becomes important at pH 5 or below, and it has limited significance for most surface-water environments. Compared to nitrate, nitrite is often less concentrated in surface waters (the nitrite concentration can be lower than that of nitrate by two orders of magnitude) but it absorbs sunlight to a higher extent and has higher photolysis quantum yield. At equal concentrations in shallow waters, the $^{\bullet}OH$ photoproduction rate by nitrite would be two orders of magnitude higher than for nitrate. This issue roughly compensates for the usually lower concentration of nitrite in natural waters, which makes nitrate and nitrite comparable $^{\bullet}OH$ sources in shallow water bodies. In deeper environments, the photochemistry of UVB-absorbing nitrate decreases faster with depth than that of UVA-absorbing nitrite (UVA radiation penetrates deeper in water columns than the UVB one). In this case, the photochemistry of nitrite is expected to prevail in the whole water column.[39]

To summarise, $^3CDOM^*$ and $^1O_2$ are produced by irradiated CDOM, while $^{\bullet}OH$ can be generated by CDOM, nitrate and nitrite. CDOM is often the main $^{\bullet}OH$ source in surface waters, but several exceptions may occur. The radical $^{\bullet}OH$ can react with several surface-water components, mainly with DOM but also with inorganic carbon (HCO_3^- and CO_3^{2-}). The reactions of $^{\bullet}OH$ with both HCO_3^- and CO_3^{2-} yield the carbonate radical, $CO_3^{\bullet-}$, which is a further photoinduced transient in surface waters. The radical $CO_3^{\bullet-}$ can also be produced upon oxidation of CO_3^{2-} by $^3CDOM^*$; this $CO_3^{\bullet-}$ formation process is usually minor compared to the $^{\bullet}OH$-induced reactions, but it can become competitive in high-DOC environments.[48]

The above discussion concerns the sources of photogenerated transients, but it is also important to take their sinks into account. The triplet states $^3CDOM^*$ undergo thermal inactivation (internal conversion) and reaction with O_2 to produce $^1O_2$. In oxygenated waters the reaction of $^3CDOM^*$ with O_2 can be an order of magnitude faster than the internal conversion.[43] Singlet oxygen ($^1O_2$) is mainly inactivated by collision with water molecules.[51] In freshwater the radical $^{\bullet}OH$ is mainly scavenged by DOM, and inorganic carbon (HCO_3^- and CO_3^{2-}) is usually the second most important $^{\bullet}OH$ sink.[39] In seawater, the bromide anion considerably overcomes DOM as $^{\bullet}OH$ scavenger. Interestingly the reaction between $^{\bullet}OH$ and bromide finally yields $Br_2^{\bullet-}$, which is a reactive transient and is involved into both oxidation and bromination reactions.[46]

$$^{\bullet}OH + Br^- \leftrightarrows HOBr^{\bullet-} \rightarrow OH^- + Br^{\bullet} \quad \text{(s)}$$

$$Br^{\bullet} + Br^- \leftrightarrows Br_2^{\bullet-} \quad \text{(t)}$$

Table 13.1. Overview of the main sources and sinks of photoinduced transient species in surface waters, and of their typical steady-state concentrations.[44]

	Source(s)	Sink(s)	$[x]$, mol L^{-1}
$^3CDOM^*$	CDOM	O_2, internal conversion	10^{-16}–10^{-14}
$^1O_2$	CDOM ($^3CDOM^* + O_2$)	Collisions with H_2O	10^{-16}–10^{-14}
$^{\bullet}OH$	CDOM, NO_2^-, NO_3^-	DOM, HCO_3^-, CO_3^{2-}, Br^- (seawater)	10^{-18}–10^{-16}
$CO_3^{\bullet-}$	$^{\bullet}OH + HCO_3^-/CO_3^{2-}$, $^3CDOM^* + CO_3^{2-}$	DOM	10^{-15}–10^{-13}

The radical $CO_3^{\bullet-}$ is almost exclusively scavenged by DOM. The scavenging of the cited transients is usually very fast, and the resulting lifetimes are typically in the range of μs to tens μs (somewhat longer lifetimes might apply to $CO_3^{\bullet-}$ only). Therefore, all the mentioned transients quickly reach a quite low ($<10^{-10}$ mol L^{-1}) steady-state concentration [x] in surface-water environments (with $x = {}^3CDOM^*$, $^1O_2$, $^{\bullet}OH$ or $CO_3^{\bullet-}$). Because the water chemistry does not change very quickly, the $[x]$ values mostly follow the irradiance of sunlight. Table 13.1 summarises the main sources and sinks of the described photoinduced transients, as well as typical steady-state concentrations values in sunlit surface waters.[39]

13.2.3. *Photochemical Reactions in Surface-water Environments*

The photoinduced degradation of compounds occurring in surface waters usually takes place by a mixture of direct and indirect photochemistry processes.[39,40,50] These processes are highly affected by water chemistry and depth and by seasonality. As far as water chemistry is concerned, elevated DOC is usually correlated with high values of both DOM and its chromophoric fraction CDOM. Elevated CDOM levels enhance the formation rates of both $^3CDOM^*$ and $^1O_2$, and DOM is not involved in their scavenging (see Table 13.1). Therefore, the reactions involving $^3CDOM^*$ and $^1O_2$ are usually favoured in high-DOC waters.[39] In contrast, DOM is a scavenger of both $^{\bullet}OH$ and $CO_3^{\bullet-}$ and, in the case of $CO_3^{\bullet-}$, it also inhibits the generation processes by scavenging $^{\bullet}OH$ (this effect is not compensated for by the reaction of $^3CDOM^*$ with CO_3^{2-} at elevated DOC). Therefore, $[^{\bullet}OH]$ and especially $[CO_3^{\bullet-}]$ tend to be lower in high-DOC waters. The formation of $CO_3^{\bullet-}$ is usually enhanced in the presence of elevated alkalinity (implying high levels of HCO_3^- and CO_3^{2-}), which in contrast has a negative effect on $[^{\bullet}OH]$ (inorganic carbon species scavenge the hydroxyl radical).[47] Anyway, high-DOC conditions affect $[^{\bullet}OH]$ to a much higher extent than elevated alkalinity. Moreover, the steady-state $[^{\bullet}OH]$ is usually very low in seawater because of scavenging by bromide.[39] The direct photolysis depends on the photon flux absorbed by a given molecule, and usually there is competition for sunlight irradiance between the molecule and CDOM. Therefore, direct photolysis is inhibited in high-DOC waters where, in some cases, elevated DOM has the additional effect of decreasing the direct photolysis quantum yields.[36]

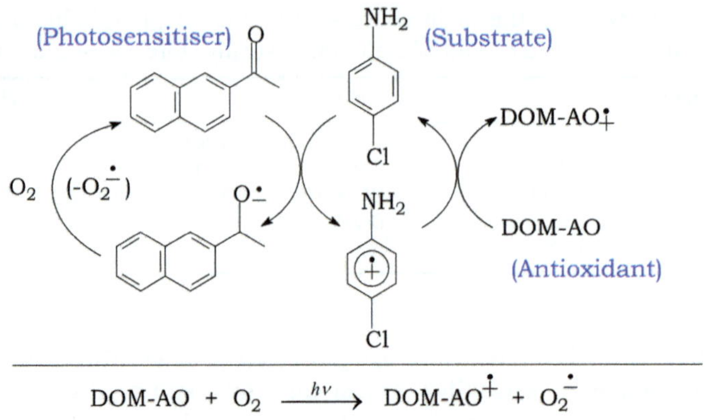

Scheme 13.3. Hypothesised pathways by which the antioxidant moieties occurring in DOM (DOM-AO) are supposed to inhibit the triplet sensitised transformation of a pollutant (4-chloroaniline). The process is at least partially catalytic in the photosensitiser and in the substrate, thus it might appear as a photocatalysed oxidation of DOM-AO by molecular oxygen.

A partial modification of the above scenario might be introduced in the case of $^3CDOM^*$ because, in some cases, the triplet-sensitised processes are inhibited by the antioxidant moieties occurring in DOM (*e.g.* phenolic compounds). These moieties could reduce back to the starting compounds the transient species (radicals or radical cations) that have been partially oxidised by $^3CDOM^*$. The hypothesised processes are depicted in Scheme 13.3. This phenomenon has been described in the case of pollutants such as aromatic amines and sulfonamides.[42,43] In these cases, it is still unclear whether the triplet-sensitised reactions are enhanced or inhibited in high-DOC waters. This topic clearly deserves additional research.

Depth and seasonality affect the sunlight irradiance in the water column, as well as the UVB/UVA intensity ratio. First of all, the photochemical reactions are enhanced in shallow waters that are thoroughly illuminated by sunlight. In contrast, only the surface layer of deep waters is well illuminated while the lower depths are often in the dark. Therefore, in deep water columns the elevated photoactivity of the well-irradiated surface layer is offset by the lack of photochemical reactivity in the lower depths.[39] The other issue is that CDOM absorbs UVB radiation to a higher extent than the UVA or visible one, thus the penetration of radiation in the water column increases with increasing wavelength. Therefore, in deep water bodies the sunlight radiation that reaches the sub-surface layers is considerably depleted in its UVB component, and the phenomenon is more marked as the CDOM content is higher.[41]

Seasonality, too, affects the irradiance and spectrum of sunlight. Obviously the irradiance is much higher in the summer months than in winter, and the photochemical reactions are correspondingly faster in summer. Moreover, winter sunlight has a considerable UVB deficit because the sun is lower over the horizon in winter, when radiation travels through a thicker atmospheric layer before reaching the ground. Because several atmospheric components and most notably ozone (with the main contribution from the stratospheric ozone layer compared to tropospheric ozone) are UVB absorbers, the absorption/depletion of UVB radiation is higher in winter. A similar issue also applies, all through the year, at sunrise and sunset compared to midday.[51]

Variations in sunlight irradiance affect all the photochemical reactions to a comparable extent. Therefore, these reactions are enhanced in shallow waters during the summer months at midday. In contrast the UVB deficit that can be found in deep waters and/or during winter, as well as at sunrise and sunset, does not affect all the processes in the same way. The photochemistry of UVB-absorbing nitrate is affected to a higher extent compared to UVA-absorbing nitrite and most notably to CDOM, which also absorbs radiation in the visible. Therefore, reactions involving $^{\bullet}OH$ and $CO_3^{\bullet-}$ are more inhibited in deep waters and/or during winter than the processes induced by $^3CDOM^*$ and $^1O_2$.[41] The effect on the direct photolysis depends on the absorption spectrum of the relevant molecule. Understandably, compounds that absorb sunlight only in the UVB region will be more affected than compounds that also absorb in the UVA and visible.[40] Scheme 13.4 gives insight

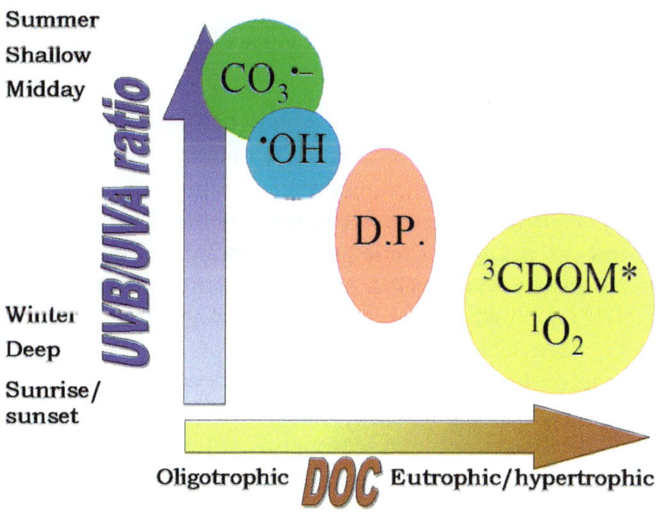

Scheme 13.4. Conditions (DOC and relative intensity of UVB *versus* UVA radiation) that favour or depress the main phototransformation processes occurring in surface-water environments. Indication is also provided of the conditions that favour the occurrence of high- or low-DOC conditions (trophic status of the water body) and that enhance or depress UVB radiation (water depth, seasonality, time of the day). Note that all photochemical processes are enhanced by an elevated irradiance and/or in shallow water, although not to the same extent.

Table 13.2. Overview of the compound classes that undergo main reactions with the listed photoinduced transients.[53,58,59]

Transients	Compound classes reacting with each transient
$^3CDOM^*$	Phenols, phenylurea herbicides, sulfonamide antibiotics
$^1O_2$	Chlorophenolates, aromatic aminoacids
$^{\bullet}OH$	Hardly oxidised compounds, hydrocarbons, recalcitrant pollutants (carbamazepine, nicotine, clofibric acid)
$CO_3^{\bullet-}$	Aromatic amines, sulphur-containing compounds (thiols, mercaptans)

into the impacts on photoreactions of the water DOC and of the UVB intensity compared to other regions of the sunlight spectrum. Indication is also provided of the conditions that favour the occurrence of high- or low-DOC conditions (trophic status) and that enhance or depress UVB radiation (water depth, seasonality, time of the day).

In addition to the environmental conditions, the prevailing photoreaction pathways also depend on the reactivity of a given substrate. The direct photolysis process is hardly predictable and should be assessed case by case, but the indirect photochemistry reactions tend to be favoured with particular compound classes. Table 13.2 summarises the types of compounds that undergo reaction with a particular photoinduced transient.[40,50,52]

13.2.4. *Daily, Seasonal and Long-term Trends of Water Chemistry: Implications for Photoreactions*

The water chemistry parameters of photochemical significance (DOC, nitrate, nitrite, carbonate and bicarbonate) undergo periodic fluctuations and longer-term trends as a consequence of several chemical and biological processes that occur in surface-water environments. During the day in the summer period the photosynthetic activity carried out by algae consumes CO_2 and increases the water pH. A pH increase induces an increase of the ratio $[CO_3^{2-}]\,[HCO_3^-]^{-1}$, which enhances the photogeneration of the carbonate radical $CO_3^{\bullet-}$ (see Scheme 13.5 below for the pH trend of the steady-state $[CO_3^{\bullet-}]$). The same phenomenon causes some decrease in the steady-state $[^{\bullet}OH]$ because CO_3^{2-} is a more efficient $^{\bullet}OH$ scavenger compared to HCO_3^-, although neither is usually able to scavenge as much $^{\bullet}OH$ as DOM (or as $Br_2^{\bullet-}$ in marine waters).[36]

Biological processes also cause some of the seasonal trends, such as the winter maxima (and summer minima) of nitrate and the summer maxima of the DOC. Nitrate is a nutrient for algal growth, and when the algal activity is most intense (*i.e.* during the summer season) NO_3^- is transformed into organic nitrogen. Intense algal activity in summer also means that these organisms proliferate, release extracellular organic matter during their metabolism and, even more importantly, release the whole cellular content into the water phase upon death and the subsequent cell lysis. Therefore, the water DOC is usually at its maximum during the summer season.[36,53]

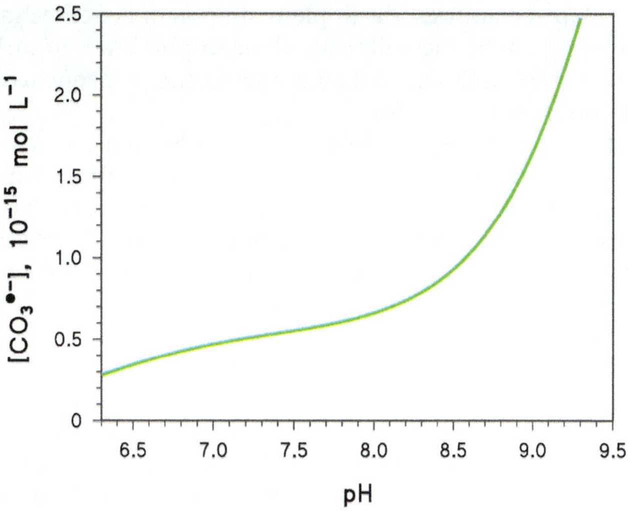

Scheme 13.5. Trend with pH of the steady-state $[CO_3^{\bullet -}]$ under 22 W m^{-2} sunlight UV irradiance (*i.e.*, 22 W m^{-2} in the 290–400 nm wavelength interval), as modelled with the APEX software.[49] Water conditions: 2 mM inorganic carbon, 5 mg C L^{-1} DOC, 5 m depth, 0.1 mM nitrate, 1 μM nitrite.

Chemical processes also affect the seasonal water chemistry: for instance, the occurrence of warmer water during summertime induces the precipitation of $CaCO_3$ that causes a decrease in water alkalinity.[36] The combined seasonal trends of nitrate, DOC and alkalinity affect the occurrence of $^3CDOM^*$, $^1O_2$, $^{\bullet}OH$ and $CO_3^{\bullet -}$. The main driver of the seasonal fluctuations is of course the irradiance of sunlight, but the photoinduced transients can undergo more or less marked seasonal changes than the irradiance, depending on water chemistry modifications.[53] The winter maxima and summer minima of nitrate and alkalinity and the opposite trend of the DOC cause the seasonal trends of $[^{\bullet}OH]$ and $[CO_3^{\bullet -}]$ to be smoother (*i.e.* they have less marked summer maxima and winter minima) compared to the irradiance. In contrast, the summer maximum of the DOC causes $^3CDOM^*$ and $^1O_2$ to undergo more marked seasonal fluctuations than the irradiance itself.[36,53]

Water chemistry can also vary in the long term, due to both direct human impact (*e.g.* variations in wastewater discharge loads) and indirect effects (most notably, climate change). The release of nutrients from wastewater discharges may induce eutrophication, which impacts the concentration values of the nitrogen species. Nitrate is usually low in oligotrophic environments, and human-induced eutrophication will likely cause the nitrate levels to increase over time. This increase is usually seen as a drift superposed on the seasonal fluctuations. However, nitrate levels can again become low in hypertrophic environments, where nitrogen may be a limiting factor for algal growth. The impact of human nutrients on nitrate would thus depend on the initial trophicity status and on the features of the human disturbance:

in many developed countries, the implementation of better technologies for wastewater treatment has actually caused the trophic levels of surface waters to decrease. In many, although not all the cases, the consequence has been a decrease in nitrate concentration.[36]

Differently from nitrate, the DOC consistently increases when passing from oligotrophic to eutrophic, and then to hypertrophic environments.[36] The DOC is a measure of DOM, which is the main scavenger of $^{\bullet}OH$ and $CO_3^{\bullet-}$ in natural waters. Nitrate is a direct $^{\bullet}OH$ source and an indirect $CO_3^{\bullet-}$ source, but the importance of DOM as sink of these radicals is usually higher than that of nitrate as source. Therefore, a parallel increase in both nitrate and DOC (as is for instance observed when passing from an oligotrophic to a eutrophic environment) will in most cases produce a decrease in both $[^{\bullet}OH]$ and $[CO_3^{\bullet-}]$.[39,53] *A fortiori*, the same holds when the DOC increases and nitrate decreases such as in the transition from eutrophic to hypertrophic waters. The waters with high DOC are usually rich in CDOM, which competes with pollutants for sunlight irradiance. Therefore, the rates of the direct photolysis reactions are affected by the trophic state and are expected to follow the order oligotrophic water bodies > eutrophic > hypertrophic.[36,39] In contrast, an increase in the DOC is expected to enhance the phototransformation processes induced by $^3CDOM^*$ and $^1O_2$. To summarise, the reactions induced by $^{\bullet}OH$ and $CO_3^{\bullet-}$, as well as the direct photolysis, would be enhanced in oligotrophic water bodies. In contrast, the $^3CDOM^*$ and $^1O_2$ processes would be favoured in hypertrophic environments.[36,39]

Climate change has a potential impact on several features of surface water bodies, including most notably chemistry and depth. Increasing water temperatures may enhance the solubility of minerals such as $CaSO_4$, where the Ca^{2+} ions are biologically stable but SO_4^{2-} may be transformed by microorganisms into organic sulphur species:[54,55]

$$SO_4^{2-} + 8\ H^+ + 6\ e^- + R\text{–}H \rightarrow R\text{–}SH + 4\ H_2O \qquad \text{(u)}$$

The above process causes an increase in both alkalinity and pH of surface waters, which is favourable to the $CO_3^{\bullet-}$-induced processes. However, the favourable effect towards $CO_3^{\bullet-}$ can be easily offset by a climate-change-induced increase of the DOC. In Nordic environments, an increase in precipitation can enhance both the organic matter content in soil and the leaching of organic compounds to surface waters. This phenomenon is usually termed "browning" because of the increased water coloration, and it involves an increase in the levels of both DOM and CDOM.[56] The expected consequence is an inhibition of the $^{\bullet}OH$- and $CO_3^{\bullet-}$-induced reactions (which in the latter case would totally overcome the increase in pH and alkalinity) and of the direct photolysis, as well as the enhancement of the $^3CDOM^*$- and $^1O_2$-assisted processes.[54]

In sub-tropical environments that are already affected by water scarcity, climate change might induce the onset of arid or extremely arid conditions. In this case, water bodies might become largely fluctuating, intermittent or even ephemeral as they would be characterised by large variations in the water column depth. Such variations might even produce complete water disappearance in some or most seasons in the case of intermittent and ephemeral environments. The reason is connected with the precipitation regime, which is highly variable around the year with a long dry season. During the dry period the water loss phenomena are insufficiently or not at all compensated for by precipitation, which causes a decrease in the water column depth. There are two extreme scenarios of water loss, namely outflow and evaporation (in the real world, part of the water is lost through outflow and part by evaporation). In the case of pure outflow, water simply flows away without changes in its chemical composition and just produces a shallower column. This might be an oversimplification, because microbiological processes could alter the scenario of constant water chemistry in an unpredictable fashion. In the case of pure evaporation, water is lost but the solutes are not (with the partial exception of dissolved gases), and an evaporative concentration phenomenon takes place. As a consequence, the same solutes (the same on a per-mole basis) occur in a shallower column. In the outflow scenario there would be only a decrease in the water column depth and one would observe an enhancement of the processes that are favoured in shallow waters ($^{\bullet}OH$- and $CO_3^{\bullet -}$-induced reactions, as well as the direct photolysis of UV-absorbing compounds). In the evaporative concentration scenario, the occurrence of the same amounts of solutes (albeit in a smaller volume) would not modify the absorbed photon fluxes. Therefore, the same amounts (moles) of $^{\bullet}OH$, $CO_3^{\bullet -}$, $^3CDOM^*$ and $^1O_2$ would be produced in a smaller volume. However, the scavengers of the transient species also undergo evaporative concentration, as is for instance the case of DOM that is the main quencher of $^{\bullet}OH$ and $CO_3^{\bullet -}$. Therefore, the two radicals are expected to undergo faster consumption in concentrated water, which offsets their being generated in a smaller volume. In contrast, $^1O_2$ is inactivated by collision with water that is not affected by evaporative concentration, while $^3CDOM^*$ is mostly scavenged by O_2. Dissolved oxygen often occurs very near saturation in surface waters, and the evaporative concentration in expected to induce O_2 outgassing from the evaporating water system (also note that the solubility of oxygen decreases as the water environment becomes more saline following evaporative concentration). Evaporative concentration is thus expected to enhance the reactions induced by $^3CDOM^*$ and $^1O_2$.[57]

Overall, climate change should modify water (photo)chemistry to a variable extent depending on the climatic zone (*e.g.* Nordic *versus* sub-tropical locations). In arid environments that largely occur in the subtropical ridge, the main effect is linked to the prevailing water loss process (outflow *versus* evaporation).[57] In contrast, in temperate water bodies the climate effects on water (photo)chemistry are probably much smaller. In some cases a "photochemical buffering" phenomenon has also been observed, with considerable changes in many water chemistry parameters but almost no changes in photochemistry, because of the compensation between opposite effects.[53]

13.3. Photoinduced Transformation of Drugs

After release into the receiving waters, drugs are subjected to environmental processes such as biodegradation/transformation, sorption and other chemical degradation processes that may contribute to their attenuation. Given the complexity and heterogeneity of drugs, it is difficult to predict or rationalize their behavior in the environment. The knowledge of factors (substituents, solvents) that affect the photochemical behavior of organic molecules are limited to relatively simple derivatives. Photolysis has been reported to significantly reduce the predicted environmental concentrations (based on consumption/usage rate) of many recalcitrant drugs[58–60] through both direct and indirect photochemistry. In addition to the photosensitizers described above, *i.e.* dissolved organic matter, nitrates and nitrites, the presence of bicarbonate has also been recently noted to greatly enhance the photolysis rates of *N*,*N*-dimethylaniline, cephalosporin antibiotics and 5-fluorouracil.[61] The cases of some peculiar drugs and illicit drugs are shown below, and a summary is provided in Table 13.3.

Table 13.3. List of discussed drugs together with their main phototransformation reactions.

Drug	Phototrans-formation reactions	Drug	Phototrans-formation reactions
Cocaine 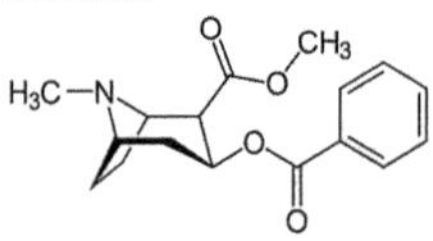	Ester bond cleavage Hydroxylation *N*-demethylation	**Gabapentine** 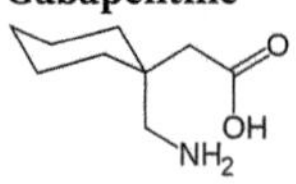	Hydroxylation Decarboxylation Ring-closure
Ketamine 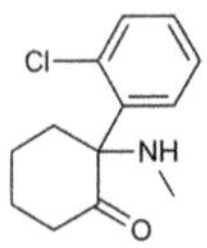	Photolysis *N*-demethylation	**Cyclophosphamide** 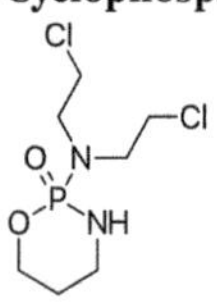	Hydroxylation De-chlorination Oxidation Ring-cleavage
Methadone 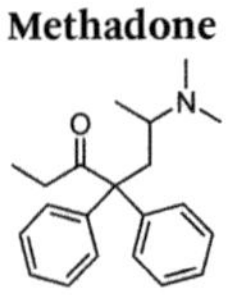	Hydrolysis Cyclization	**Tamoxifen** 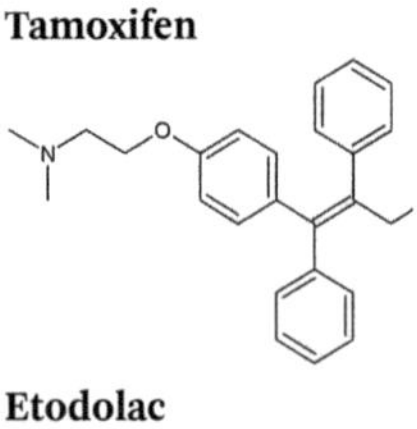	Photoisomerization Photocyclyzation Photooxigenation
6-Monoacetylmorphine 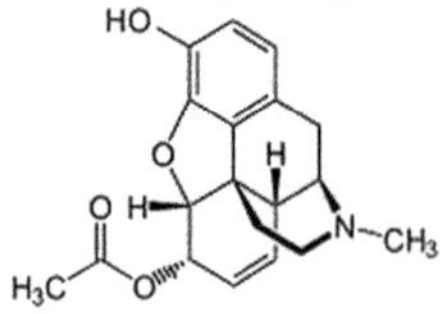	Photooxidation Photoaddition	**Etodolac** 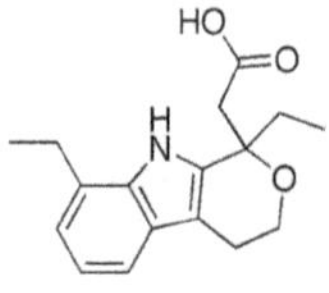	Ring contraction Cyclization

13.3.1. Illicit Drugs

Cocaine, opiates and ketamine are among the most frequently detected illicit drugs in surface waters. The photoinduced transformation of cocaine (COC) and methadone (MET) have been studied by Postigo *et al.*[62,63] in distilled water and sewage treated water, using higher concentrations than usually found in natural surface waters or effluents from municipal wastewater treatment plants. This approach allowed an easier identification of the phototransformation products. Several photo-intermediates were detected in the initial steps of the phototransformation process.

The solar photolysis of COC causes photointermediate levels to continually increase as the COC decreases. Most of the detected photointermediates are readily phototransformed afterwards, and they remain in solution for just a short period of time after COC is completely eliminated. Ecgonine methyl ester (**TP199**), produced *via* cleavage of the benzoyl ester bond of COC, is the major phototransformation product of COC. Benzoylecgonine (**TP289a**), generated *via* cleavage of the methyl ester bond of COC, is the second main solar phototransformation product of COC in aqueous solution. Besides **TP289a**, another isobaric compound is produced in all phototransformation treatments and it is recognized as nor-cocaine, produced *via* *N*-demethylation of COC. Both of these TPs are known COC metabolites as well. Benzoylecgonine has been proposed as a consumption indicator of COC, in order to estimate cocaine consumption at the community level, by means of a sewage epidemiology approach.[64]

Two other COC metabolites, namely ecgonine (**TP185**) and cocaethylene (**TP317**) were also identified as phototransformation products. **TP185** may be generated by further phototransformation of **TP289a** and/or **TP199**, *via* benzoyl ester cleavage and demethylation, respectively. On the other hand, **TP317**, a minor phototransformation product, is produced by COC methylation. Another TP (**TP181**) was detected and attributed to anhydroecgonine methyl ester, which is a pyrolytic product of COC.[65] Its occurrence is consistent with the further transformation of **TP199** by the loss of a water molecule. The applied solar photolytic transformation treatments also generate various monohydroxylated (**TP319**) derivatives of COC. Besides COC phototransformation, results of hydrolysis studies confirm that COC is highly hydrolyzed in benzoylecgonine and ecgonine methyl ester in aqueous solutions with high organic content and high pH. The main identified phototransformation routes, *i.e.*, ester bond cleavage, hydroxylation and *N*-demethylation, are also believed to take place at environmental COC levels due to photolysis and in real wastewater effluent samples treated with the investigated photocatalytic processes.[62]

MET in aqueous solutions does not experience hydrolysis, but it was significantly transformed in the presence of sunlight. Direct solar irradiation resulted in almost complete transformation of MET after 20 h of normalized irradiation time.[63] The main MET phototransformation pathways were observed to be hydroxylation, fragmentation and cyclization. A phototransformation compound **TP277** was identified as 2-ethylidene-1,5-dimethyl-3,3-diphenylpyrrolidine (EDDP), which is also one of the main urinary metabolites of MET[66,67] and is formed through cyclization. EDDP is further transformed into four hydroxylated compounds (**TP293**). Additionally, several monohydroxylated (**TP325**) and dihydroxylated (**TP341**) transformation products of MET were detected.

In the case of ketamine (KET), the metabolic pathways that produce the metabolites norketamine (NK), hydroxy-norketamine (HNK) and dehydronorketamine (DNK) in humans and animals[68–70] have been intensely studied. In contrast, the knowledge of KET environmental occurrence and fate is limited. Dark incubation experiments have shown that KET is not susceptible to microbial degradation or hydrolysis. Conversely, phototransformation was demonstrated to significantly reduce the concentration of KET and NK in river water, and to yield by-products that are similar to human metabolites. KET and NK were consistently detected at a ratio similar to that found in urine samples.[71] Both direct and indirect photolysis led to the *N*-demethylation of KET to form NK and other by-products, including HNK, DNK, hydroxy-ketamine (HK) and isomer forms of KET and NK.

Among opiates, heroin is of great concern. It is rapidly hydrolyzed to 6-monoacetylmorphine (6-MAM) and then more slowly metabolized to morphine, which is the major active metabolite. The occurrence of morphine in surface water is not only due to heroin consumption, but it could also derive from its consumption as a licit pharmaceutical drug. Morphine is in fact widely used as a potent analgesic and it is a metabolite of codeine, ethylmorphine, nicomorphine and pholcodine.

Photodegradation of 6-MAM and morphine was studied in water, methanol and in the solid state.[72] Morphine and 6-MAM are quite stable under UVA light, but they are very sensitive to UVB irradiation. The effect of UV light on the stability of 6-MAM and morphine strongly depends on their physical state and on the medium where they are dissolved. In the solid state, the yield of photodegradation is lower than in solution and, in particular, the photolysis is faster in organic solvent than in aqueous milieu.

The detected UVB-induced degradation products involved photoaddition of the solvent and/or oxygen as the main mechanism of photodegradation and photooxidation. In the absence of the solvent, the only observed path was oxidation, through addition of an oxygen atom to 6-MAM or even more oxygen atoms in the case of morphine. The strong electrophile singlet oxygen is generated by both compounds under irradiation, and this reactive oxygen species reacts with the 7,8 double bond and/or aromatic ring and/or the alpha carbon of the epoxide ring. On the other hand, the expected *N*-oxidation of morphine was not observed. It can be envisaged that some photoproducts formed in water could also be late markers for the presence of the parent drugs (heroin, morphine) in the environment or in sewage waters.

13.3.2. Pharmaceuticals of Particular Concern

Particular concerns have arisen for pharmaceuticals that are widely diffuse and/or characterized by intrinsic toxicity. The latter group comprises antineoplastic drugs, for which studies on photoinduced transformation are limited.[73,74] Based on several OECD guidelines, the persistence assessment of cyclophosphamide (CPA), one of the anti-cancer drugs most frequently found in the environment, revealed that it is released unchanged in the environment and it is rather biorecalcitrant.[75,76] Laboratory simulation tests in lake water showed that CPA had a half-life of 80 days in the dark, while under irradiation the degradation progressed at a faster rate with a half-life of 44 days.[76] Given its persistence and continuous introduction into the aquatic environment, it is not surprising that CPA is frequently detected in the aquatic environment. The formation of six photoinduced transformation products was assessed through a photocatalytic process.[74] The CPA degradation study showed the formation of three previously known human metabolites[77] and the generation of three new compounds. The main known metabolite 4-keto-cyclophosphamide was detected, whereas the quantitatively most abundant compound was NOR-nitrogen mustard. The newly detected transformation products involved hydroxylation, de-chlorination and oxidation.

Among NSAIDs, etodolac (ETD) was detected in river waters at concentration up to 0.3 ng L^{-1};[78] ETD contains an indole system (see Table 13.3), a chromophore present in many molecules of biological interest that show photoreactivity,[79] such as indomethacin[80,81] and vinblastine sulfate.[82] ETD is metabolized in the liver into several metabolites, such as hydroxylated and glucuronide derivatives, all identified in human plasma and urine.[83] Conversely, the drug under irradiation undergoes photooxidative processes only and two major photoproducts were isolated and characterized. They are formed following an initial oxygen addition to the double bond of the five member ring that yields a spiro compound (**TP304**) and a macrolactam (**TP320**). Under UVA both TPs were generated, while under UVB radiation the genotoxic spiro compound was mainly observed. Remarkably, the spiro compound was photostable under sunlight exposure for up to two weeks, which suggests its possible accumulation in aquatic media.

Gabapentine (GAB) is classified as not readily biodegradable[29] but several TPs were identified under polychromatic UV irradiation. Several TPs structures were elucidated and some of them could also be identified at low initial GAB concentrations. The main phototransformation routes are hydroxylation, with formation of monohydroxylated (**TP188**) and dihydroxylated (**TP204**) derivatives, monohydroxylation/reduction (**TP186**), decarboxylation (**TP128** and **TP160**), formation of cyano group (**TP168**) and ring closure (**TP154**); these TPs could also be identified at low initial treatment concentrations. The photo-treatment did not cause an increase in biodegradability.

Tamoxifen, another widely diffuse drug, has been detected in surface water samples at concentrations of up to 71 ng l^{-1}.[84,85] The photoinduced transformation of tamoxifen was studied in water under prolonged sunlight irradiation. The main observed phototransformation pathways involved: (1) photoisomerization to the *cis*-isomer (**TP371**); (2) photocyclization, occurring *via* a dihydrophenanthrene intermediate, with formation of two phenanthrenes (**TP369a** and **TP369b**) and; (3) photooxygenation, a minor reaction leading to a ketone derivative (**TP269**). The formation of benzoic acid was assessed as well.[86]

13.4. Conclusions

Photolysis significantly reduces the predicted environmental concentrations of many recalcitrant drugs through both direct and indirect photochemistry and involves the formation of several metabolites and phototransformation products. The human metabolites of pharmaceuticals are predicted, in most cases, to be less toxic than the parent compounds to common aquatic species. However, there is little available data to demonstrate whether this is generally the case. In other cases, as for instance with anti-cancer drugs, the risk to the environment, and to humans because of the drug occurrence in the environment, is not very clear. Several strategies have been developed for the risk assessment of pharmaceuticals and their related TPs and, although they are mostly directed towards lab-scale simulation tests, they represent the only currently available methods to characterize the risk associated with these compounds.

References

1. Her Majesty's Stationery Office, *British Pharmacopoeia*, Her Majesty's Stationery Office, London, UK, 2013.
2. United States Pharmacopeial Convention, *United States Pharmacopeia 29*, United States Pharmacopeial Convention, Rockville, Md, USA, 2012.
3. A. C. Templeton, H. Xu, J. Placek and R. A. Reed, *Pharm. Technol.*, 2005, **3**, 68.
4. J. K. Sugden, *Int. J. Cosmet. Sci.*, 1985, **7**(4), 165.
5. M. A. Sheraz, S. H. Kazi, S. Ahmed, Z. Anwar and I. Ahmad, *Beilstein J. Org. Chem.*, 2014, **10**, 1999.
6. J. Moan, Benefits and adverse effects from the combination of drugs and light, in *Photostability of Drugs and Drug Formulations*, ed. H. H. Tonnesen, Taylor & Francis, London, UK, 1996, pp. 173–188.
7. *The Biological Effects of Ultraviolet Radiations*, ed. F. Urbach, Pergamon, Oxford, UK, 1969.
8. L. C. Harber, I. E. Kochevar and A. R. Shalita, Mechanism of photosensitization to drugs in humans, in *Science of Photomedicine*, ed. J. D. Regan and J. A. Parrish, Plenum Press, New York, NY, USA, 1982, pp. 323–347.
9. J. H. Epstein and B. U. Wintroub, *Drugs*, 1985, **30**(1), 42.
10. D. A. Lambropoulou and L. M. L. Nollet, *Transformation Products of Emerging Contaminants in the Environment*, Wiley & Sons Ltd., UK, 2014.
11. A. H. Neilson and A. S. Allard, *Environmental Degradation and Transformation of Organic Chemicals*, CRC Press, Boca Raton, Fl, USA, 2nd edn, 2008.

12. C. J. Sinclair and A. B. Boxall, *Environ. Sci. Technol.*, 2003, **37**, 4617.
13. S. D. Richardson, *Anal. Chem.*, 2009, **81**, 4645.
14. D. Fatta-Kassinos, S. Meric and A. Nikolaou, *Anal. Bioanal. Chem.*, 2011, **399**, 251.
15. P. K. Jemba, *Ecotoxicol. Environ. Saf.*, 2006, **63**, 113.
16. M. Isidori, A. Parrella, P. Pistillo and F. Temussi, *Environ. Int.*, 2009, **35**, 821.
17. M. Petrovic, M. J. López De Alda, S. Diaz-Cruz, C. Postigo, J. Radjenovic, M. Gros and D. Barceló, *Philos. Trans. R. Soc., A*, 2009, **367**, 3979.
18. A. L. N. Van Nuijs, B. Pecceu, L. Theunis, N. Dubois, C. Charlier, P. G. Jorens, L. Bervoets, R. Blust, H. Neels and A. Covaci, *Water Res.*, 2009, **43**, 1341.
19. M. Petrovic, C. Postigo, M. J. López De Alda, A. Ginebreda, M. Gros, J. Radjenovic and D. Barceló, in *Water Scarcity in the Mediterranean: Perspectives Under Global Change*, ed. S. Sabater and D. Barceló, Springer-Verlag, Berlin Heidelberg, 2010, pp. 197–228.
20. S. Terzic, I. Senta and M. Ahel, *Environ. Pollut.*, 2010, **158**(8), 2686.
21. M. R. Boleda, M. T. Galceran and F. Ventura, *Water Res.*, 2009, **43**(4), 1126.
22. A. Y. C. Lin, X. H. Wang and C. F. Lin, *Chemosphere*, 2010, **81**(5), 562.
23. International Agency for Research on Cancer (IARC), *IARC Monographs on the Evaluation of Carcinogenic Risk of Chemicals to Humans: Some Antineoplastic and Immunosuppressive Agents*, IARC, Lyon, 1981, vol. 26.
24. M. Allwood, A. Stanley and P. Wright, *The Cytotoxics Handbook*, Oxford Radcliffe Medical Press, 4th edn, 2002.
25. K. Kummerer, T. Steger-Hartmann and M. Meyer, *Water Res.*, 1997, **31**(11), 2705.
26. K. Kummerer and A. Al-Ahmad, *Acta Hydrochim. Hydrobiol.*, 1997, **25**(4), 166.
27. T. Steger-Hartmann, K. Kummerer and A. Hartmann, *Ecotoxicol. Environ. Saf.*, 1997, **36**(2), 174.
28. J. P. Besse, J. F. Latour and J. Garric, *Environ. Int.*, 2012, **39**(1), 73.
29. M. Herrmann, J. Menz, O. Olsson and K. Kümmerer, *Water Res.*, 2015, **85**, 11.
30. F. Bosca, M. L. Marin and M. A. Miranda, *Photochem. Photobiol.*, 2001, **74**, 637.
31. R. Andreozzi, M. Raffaele and P. Nicklas, *Chemosphere*, 2003, **50**, 1319.
32. H. R. Buser, T. Poiger and M. D. Muller, *Environ. Sci. Technol.*, 1998, **32**, 3449.
33. P. F. Robinson, Q. T. Liu, A. M. Riddle and R. Murray-Smith, *Chemosphere*, 2007, **66**, 757.
34. Q. Liu and H. E. Williams, *Environ. Sci. Technol.*, 2007, **41**, 803.
35. Q. T. Liu, R. I. Cumming and A. D. Sharpe, *Photochem. Photobiol. Sci.*, 2009, **8**, 768.
36. *Surface Water Photochemistry*, ed. P. Calza and D. Vione, RSC Publishing, 2016.
37. R. P. Wayne, Basic concepts of photochemical transformations, in *The Handbook of Environmental Chemistry Vol. 2.M – Environmental Photochemistry Part II*, ed. P. Boule, D. W. Bahnemann and P. K. J. Robertson, Springer, Berlin, 2005, pp. 1–47.
38. G. Grabner and C. Richard, Mechanisms of direct photolysis of biocides based on halogenated phenols and anilines, in *The Handbook of Environmental Chemistry Vol. 2.M – Environmental Photochemistry Part II*, ed. P. Boule, D. W. Bahnemann and P. K. J. Robertson, Springer, Berlin, 2005, pp. 161–192.
39. D. Vione, M. Minella, V. Maurino and C. Minero, *Chem.–Eur. J.*, 2014, **20**, 10590.
40. A. L. Boreen, W. A. Arnold and K. McNeill, *Aquat. Sci.*, 2003, **65**, 320.
41. S. A. Loiselle, N. Azza, A. Cozar, L. Bracchini, A. Tognazzi, A. Dattilo and C. Rossi, *Freshwater Biol.*, 2008, **53**, 535.
42. F. L. Rosario-Ortiz and S. Canonica, *Environ. Sci. Technol.*, 2016, **50**, 12532.
43. K. McNeill and S. Canonica, *Environ. Sci.: Processes Impacts*, 2016, **18**, 1381.
44. S. E. Page, W. A. Arnold and K. McNeill, *Environ. Sci. Technol.*, 2011, **45**, 2818.
45. J. Mack and J. R. Bolton, *J. Photochem. Photobiol., A*, 1999, **128**, 1.

46. E. De Laurentiis, M. Minella, M. Sarakha, A. Marrese, C. Minero, G. Mailhot, M. Brigante and D. Vione, *Water Res.*, 2013, **47**, 5943.
47. M. Bodrato and D. Vione, *Environ. Sci.: Processes Impacts*, 2014, **16**, 732.
48. S. Canonica, T. Kohn, M. Mac, F. J. Real, J. Wirz and U. von Gunten, *Environ. Sci. Technol.*, 2005, **39**, 9182.
49. F. Wilkinson and J. Brummer, *J. Phys. Chem. Ref. Data*, 1981, **10**, 809.
50. S. Yan and W. Song, *Environ. Sci.: Processes Impacts*, 2014, **16**, 697.
51. R. Frank and W. Klöpffer, *Chemosphere*, 1988, **17**, 985.
52. K. Fenner, S. Canonica, L. P. Wackett and M. Elsner, *Science*, 2013, **341**, 752.
53. M. Minella, B. Leoni, N. Salmaso, L. Savoye, R. Sommaruga and D. Vione, *Sci. Total Environ.*, 2016, **541**, 247.
54. E. De Laurentiis, M. Minella, V. Maurino, C. Minero and D. Vione, *Environ. Sci. Pollut. Res.*, 2014, **21**, 11770.
55. C. E. Williamson, J. E. Saros and D. W. Schindler, *Science*, 2009, **323**, 5916.
56. C. T. Solomon, S. E. Jones, B. C. Weidel, I. Buffam, M. L. Fork, J. Karlsson, S. Larsen, J. T. Lennon, J. S. Read, S. Sadro and J. E. Saros, *Ecosystems*, 2015, **18**, 376.
57. M. Minella, V. Maurino, C. Minero and D. Vione, *Int. J. Environ. Anal. Chem.*, 2013, **93**, 1698.
58. M. W. Lam, C. J. Young and S. A. Mabury, *Environ. Sci. Technol.*, 2005, **39**(2), 513.
59. A. Y. C. Lin and M. Reinhard, *Environ. Toxicol. Chem.*, 2005, **24**(6), 1303.
60. K. H. Wammer, A. R. Korte, R. A. Lundeen, J. E. Sundberg, K. McNeill and W. A. Arnold, *Water Res.*, 2013, **47**(1), 439.
61. J. Huang and S. A. Mabury, *Environ. Toxicol. Chem.*, 2000, **19**(9), 2181–2188.
62. C. Postigo, C. Sirtori, I. Oller, S. Malato, M. I. Maldonad, M. López de Alda and D. Barceló, *Appl. Catal., B*, 2011, **104**, 37.
63. C. Postigo, C. Sirtori, I. Oller, S. Malato, M. I. Maldonado, M. Lopez de Alda and D. Barcelo, *Water Res.*, 2011, **45**, 4815.
64. A. L. N. Van Nuijs, S. Castiglioni, I. Tarcomnicu, C. Postigo, M. J. López De Alda, H. Neels, E. Zuccato, D. Barceló and A. Covaci, *Sci. Total Environ.*, 2011, **409**(19), 3564.
65. K. Bisceglia, A. Roberts, M. Schantz and K. Lippa, *Anal. Bioanal. Chem.*, 2010, **398**, 2701.
66. A. Goldstein and B. W. Brown, *J. Subst. Abuse Treat.*, 2003, **25**(2), 61.
67. A. Ferrari, C. P. R. Coccia, A. Bertolini and E. Sternieri, *Pharmacol. Res.*, 2004, **50**(6), 551.
68. J. D. Adams, T. A. Baillie, A. J. Trevor and N. Castagnoli, *Biol. Mass Spectrom.*, 1981, **8**(11), 527.
69. K. A. Moore, J. Sklerov, B. Levine and A. J. Jacobs, *J. Anal. Toxicol.*, 2001, **25**(7), 583.
70. J. Wieber, R. Gugler, J. H. Hengstmann and H. J. Dengler, *Anaesthesist*, 1975, **24**(6), 260.
71. A. Yu-Chen Lin, W. N. Lee and X. H. Wang, *Water Res.*, 2014, **53**, 351.
72. G. Miolo, M. Tuccib, A. Mazzoli, S. D. Ferrara and D. Favretto, *J. Pharm. Biomed. Anal.*, 2016, **126**, 48.
73. P. Calza, C. Medana, M. Sarro, V. Rosato, R. Aigotti, C. Baiocchi and C. Minero, *J. Chromatogr. A*, 2014, **1362**, 135.
74. C. Medana, P. Calza, F. Dal Bello and C. Baiocchi, *Curr. Trends Mass Spectrom.*, 2013, **1**, 30.
75. T. Steger-Hartmann, K. Kummerer and J. Schecker, *J. Chromatogr. A*, 1996, **726**(1–2), 179.
76. I. J. Buerge, H.-R. Buser, T. Poiger and M. D. Muller, *Environ. Sci. Technol.*, 2006, **40**(23), 7242.

77. L. A. Vanaerts, J. H. Copius Peereboom and J. Stegeman Noordhoek, *Toxicol. In Vitro*, 1995, **9**, 677.
78. K. Hoshina, S. Horiyama, H. Matsunaga and J. Haginaka, *J. Pharm. Biomed. Anal.*, 2011, **55**, 916.
79. M. Passananti, M. Lavorgna, M. R. Iesce, M. DellaGreca, M. Brigante, E. Criscuolo, F. Cermola and M. Isidori, *Sci. Total Environ.*, 2015, **518–519**, 258.
80. A. B. Wu, H. W. Cheng, C. M. Hu, F. A. Chen, T. C. Chou and C. Y. Chen, *Tetrahedron Lett.*, 1997, **38**, 621.
81. F. Temussi, F. Cermola, M. DellaGreca, M. R. Iesce, P. Passananti, L. Previtera and A. Zarrelli, *J. Pharm. Biomed. Anal.*, 2011, **56**, 678.
82. J. Black, D. D. Buechter, J. W. Chinn, J. Gard and D. E. Thurston, *J. Pharm. Sci.*, 1988, 77, 630–634.
83. D. B. Strickmann and G. Blaschke, *J. Pharm. Biomed. Anal.*, 2001, **25**, 977.
84. K. V. Thomas and M. J. Hilton, *Mar. Pollut. Bull.*, 2004, **49**, 43.
85. D. Ashton, M. Hilton and K. V. Thomas, *Sci. Total Environ.*, 2004, **333**, 167.
86. M. Della Greca, M. R. Iesce, M. Isidori, A. Nardelli, L. Previtera and M. Rubino, *Chemosphere*, 2007, **67**, 1933.

Chapter 14

Photodegradation of Drugs of Abuse in Hair

Giorgia Miolo*[a] and Donata Favretto[b]

[a]Department of Pharmaceutical and Pharmacological Sciences, University of Padova, Italy; [b]Department of Cardiac, Thoracic and Vascular Sciences, University of Padova, Italy

*E-mail: giorgia.miolo@unipd.it

Table of Contents

Light in Forensic Science: Issues and Applications
Edited by Giorgia Miolo, Jacqueline L. Stair and Mire Zloh

Published by the Royal Society of Chemistry, www.rsc.org

14.1. Introduction

14.1.1. Hair Analysis in Crime Investigation

The use and abuse of illegal drugs affects all modern societies, and therefore the assessment of drug exposure and identification of cases of problematic substance use is an important task that needs to be addressed. In recent years, hair has become a fundamental biological specimen, alternative or complementary to the more usual biosamples (blood and urine), for drug testing in the fields of forensic and clinical toxicology.[1]

The use of hair as a biological matrix to assess drug exposure was first described in 1979 by Baumgartner *et al.*,[2] with the determination of morphine in the hair of opiate abusers by using radioimmunoassay (RIA). One of the advantages of this matrix is the extended window for drug detection, from weeks to months or even years after the drug exposure compared to the temporal appearance in blood; moreover, the incorporated drugs and metabolites do not undergo metabolism in hair and are much more stable in keratin than in liquid matrices. Thereafter, the retrospective investigation of chronic or past consumption is possible: the distinction between an occasional dose intake and a repeated or chronic abuse of a drug can be achieved by segmenting the hair shaft, analysing segments and "reconstructing" the history of drug consumption over a period of time that depends on the length of the hair.[3]

Hair matrix is also an important complement to blood and urine analysis giving considerable information about drug facilitated crimes (DFC, rape, sexual assault, robbery, sedation of elderly persons). Moreover, hair analysis may be a body of evidence for a committed crime when there is a long delay between the illegal event and the report to the police occurs; again, hair is a specimen useful in post-mortem cases especially when other bodily fluids are unavailable.[4]

Hair is also an attractive matrix for monitoring drug recidivism, in persons suspected of driving under the influence of drugs and to demonstrate drug exposure when cases of chronic intoxication, use, abuse, or single dose consumption need to be diagnosed in the context of withdrawal controls, doping controls, or workplace drug testing.

14.1.2. Hair Matrix

Human hair, the product of differentiated organs in the skin of mammals, is composed of protein, keratin essentially, a group of insoluble cystine-containing helicoidal protein complexes which form 65–95% of the hair by weight, of water (15–35%), lipids (1–9%), and minerals (<1%).[5]

Keratins form long fibers, which bond together through the formation of disulfide bonds (–S–S–) and through cross-linking with other proteins. Thus, a hair fiber consists of keratinized cells stuck by the cell membrane complex that together form three concentric structures: an outer cuticle that surrounds a cortex cuticle, cortex and medulla (Figure 14.1).

About the one million hair follicles of the adult scalp, approximately 85% of the hair, is in the growing phase (anagen and catagen), the remaining 15% is in a quiescent stage (telogen).[6]

Melanin is synthesized in the melanocytes, specialized cells located in the hair bulb; the melanin granules are located inside the cortex (about 3% by weight). There are two types of melanin, the brown-black insoluble pigments (eumelanins) and the less prevalent, alkaline soluble, red pigments (pheomelanins).[7]

It is generally accepted that eumelanin is photoprotective, while pheomelanin is phototoxic to pigmented tissues.[8,9] The amount, the density and the type of melanin in melanocytes determines the color of hair.

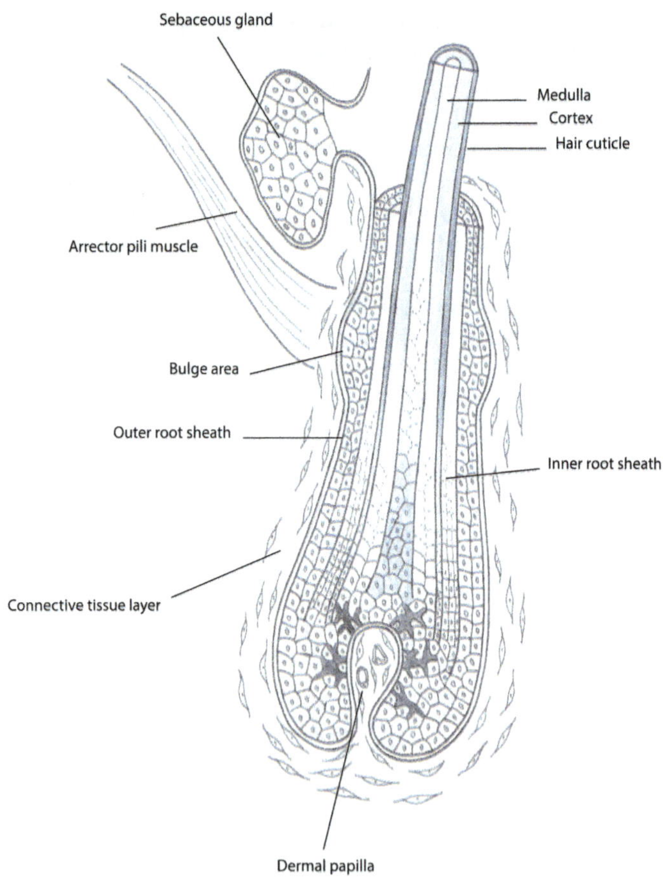

Figure 14.1. The structure of the human hair. Reproduced with permission from *International Journal of Dermatology*, B. Buffoli *et al.*, The human hair: from anatomy to physiology, 2013. Copyright © 2013 WILEY-VCH Verlag GmbH & Co. KGaA, Weinheim.

14.1.3. Drug Incorporation into Hair

Indeed, the capillary blood around the follicle provides not only the nutrients during the hair growth cycle but also delivers trace metals, drugs and any extraneous substances present in the blood stream. Along with passive diffusion from blood capillaries, chemicals are incorporated into hair fiber from surrounding tissues or intercellular fluids by diffusion from sweat or sebum secretions at the level of keratinogenous zone; moreover, drug deposition occurs from the external environment.[10] Hence, xenobiotics can be incorporated into the hair shaft over a length of 1.2 to 1.5 mm between the level of matrix cells and the end of the keratinization zone of the hair follicle and they are carried along as hair grows. In particular the non-keratinous material allows the movement of drug substances into and out of the hair fiber and then, during subsequent keratogenesis, drugs become tightly bound in the interior of the hair shaft. There are many drug binding sites in human hair, such as ionic, amide, disulfide, hydrogen binding sites and Van deer Waal's attractions and, drug binding to melanin-related sites could explain the inter-individual differences in drug uptake into hair of different colors, depending on the differences in melanin levels.[11]

For example, the melanin content in hair plays a role for some weak basic drugs such as codeine, amphetamine, and particularly cocaine and ketamine are known to have a high affinity for melanin since the pH of melanocytes is between 3 and 5. Moreover, pigmented hair contains much higher concentrations of a drug than blond or grey hair taken at the same dosage.[12,13]

Also the polarity of a drug or its metabolite affects their incorporation in hair: more polar metabolites benzoylecgonine, morphine or amphetamine enter the hair in a lesser extent than their more lipophilic precursors cocaine or 6-monoacetylmorphine or methamphetamine.[14–18]

Overall it has been recognized that the melanin content of hair, the lipophilicity through intradermal transfer of lipid-soluble drugs and their basicity are crucial features that influence drug incorporation into hair.[19]

Of course variability in sweat or sebum secretions could also influence the amount of a drug that can be incorporated into the hair shaft by diffusion.

Therefore, drug concentration may vary intra- and inter-individually, even for the same exposure. Some expected ranges of drugs in hair are described in Table 14.1[19] reporting the observed concentrations, in the range of picograms per milligram of hair, in a particular population of drug users (*e.g.* drivers asking for driving licence re-granting after driver under the influence (DUI), offenses).

With regard to drug concentrations in hair, several studies have shown a positive relationship between dose and drug levels[16,17,20] but this is not true for all drugs; indeed, a strong inter-individual variability in subjects receiving the same doses has also been documented. As a result, at the current state of the art, the drug levels in hair cannot be directly correlated to the dosages ingested.[11,21]

Table 14.1. Expected concentration ranges in hair of drug users.

Compound	ng mg^{-1} hair
Cocaine	0.1–28.9
Cocaethylene	0–2.6
Benzoylecgonine	0–4.4
Ecgonine methylester	0–4.4
Heroin	0–16
6-Monoacetylmorphine	0.1–67
Morphine	0.1–10
Codeine	0–4.2
Methamphetamine	3.1–126
Amphetamine	0.8–12

Once incorporated, the concentrations of the drug in hair may be affected by washing, chemical or thermal treatments, the use of cosmetics, deposition from environmental drug smoke, exposure to the environment and in particular to light.[22–27]

Since scalp hair is assumed to be growing at a constant rate of approximately 0.22–0.52 mm per day or 0.8–1.42 cm per month, segmental hair analysis can be used for defining the time-course of drug contact as the position of the drugs along the hair fiber can be correlated with the time the drugs were present in blood.

14.1.4. *Influence of Light on Hair Structure*

Exposure of hair to sunlight damages hair fiber. Protein loss, color change, being more pronounced for light colored hairs, dryness and reduced strength, are some of the observed effects of light on hair. It has been shown that the damaging effect of UVB is about 2–5 times higher than that of UVA plus visible radiation, depending on the hair type.[28]

Photochemical degradation of hair results in the attack on both hair proteins and melanins. Melanin absorbs UVB, UVA and visible radiations. At visible and long ultraviolet wavelengths, melanin releases most of its energy as heat, whereas at short UV wavelengths, reactive intermediates may be formed.[29]

Keratin absorbs mainly UVB radiation but with a very low extinction coefficient ($0.50(mg\ ml^{-1})^{-1}\ cm^{-1}$), at least 70 times smaller than melanin, and the UVA radiation at much lower extent ($0.13\ (mg\ ml^{-1})^{-1}\ cm^{-1}$), while it does not absorb Vis light.

Sunlight on human hair causes melanin oxidation through free radicals, and induces keratin impairment, thus bleaching the hair shaft. Since the outer layers of the fiber receive higher intensities of radiation, in most cases the amino acids of the cuticle are altered to a greater extent than those of the cortex. Cystine, methionine, tyrosine and tryptophan are the most degraded amino acids by UV radiation.[30,31]

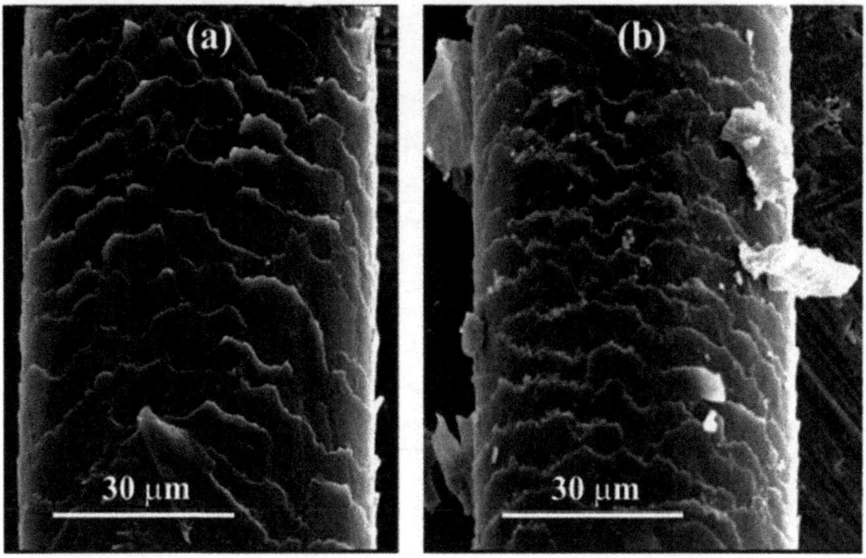

Figure 14.2. SEM images of cuticle of (a) control and (b) irradiated hair shafts. Cuticle damage with partial exfoliation due to irradiation is easily observable. Reprinted from *Journal of Photochemistry and Photobiology B: Biology*, **164**, M. V. Fedorkova, N. N. Brandt, A. Yu. Chikishev, N. V. Smolina, N. G. Balabushevich, S. A. Gusev, V. A. Lipatova, V. M. Botchey, G. E. Dobretsov and E. V. Mikhalchik, Photoinduced formation of thiols in human hair, 43–48, Copyright 2016, with permission from Elsevier.

One of the worst effects of sunlight and long UV exposure on hair is the breakage of the indole ring of tryptophan with the formation of *N*-formylkynurenine *via* a photooxidation mechanism.[28,32,33]

Hair weathering caused by sunlight mostly depends on the UV-induced cleavage of disulfide bonds and the consequent destruction of cystines in the cuticle; hereby an increase in the number of thiols groups occurs in keratins due to irradiation, especially in peripheral regions of the hair shaft, with cuticle damage and partial exfoliation (Figure 14.2).[34]

Oxidation of the amide carbon of polypeptide chains also occurs, producing carbonyl groups,[35] the most common products of protein oxidation and photooxidation.[36]

On the other hand, hair melanins give some photochemical protection to hair proteins, particularly at lower wavelengths, where both the pigments and the proteins absorb light, by absorbing a large fraction of the radiation through the conjugate carbonyl groups and double bonds system and by filtering the incoming radiation.

Moreover, melanin has the ability to scavenge radical species thus immobilizing many of the free radicals formed by photo-sensitive amino acids thus avoiding their transport into the keratin matrix.

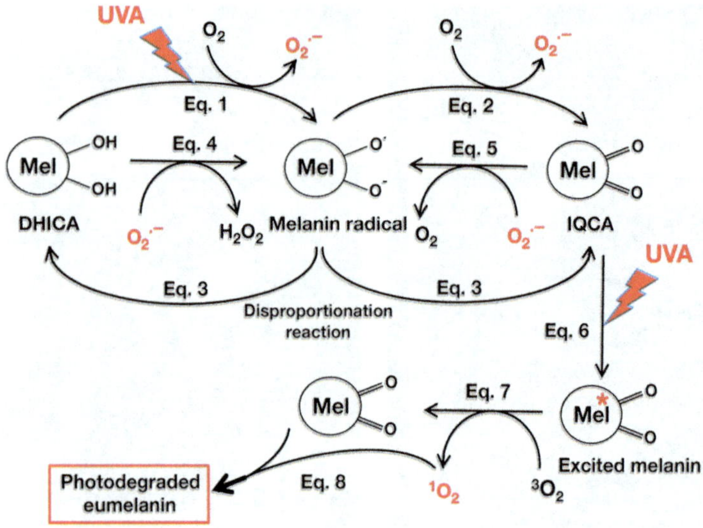

Figure 14.3. Scheme illustrating how DHICA-melanin produces superoxide radical and singlet oxygen and interacts with them to produce photodegraded eumelanin. Reproduced with permission from *Pigment Cell and Melonoma Research*, S. Ito, M. Kikuta, S. Koike, G. Szewczyk, M. Sarna, A. Zadlo, T. Sarna and K. Wakamatsu, Roles of reactive oxygen species in UVA-induced oxidation of 5,6-dihydroxyindole-2-carboxylic acid-melanin as studied by differential spectrophotometric method, 2016. Copyright © 2016 WILEY-VCH Verlag GmbH & Co. KGaA, Weinheim.

Indeed, besides a simple non-radiative decay process following conversion of light energy into heat, the photoprotective action of melanin seems to involve an electron-transfer mechanism that produces free radicals originating from quinone-hydroquinone complex having a long-lived triplet state precursor.[37] At the same time, in the process of protecting the hair proteins from light, the pigments are degraded or bleached (Figure 14.3).[38]

14.1.5. *Exposure of Hair Drugs to Light: Sunlight and Artificial Lamps*

Scalp hair is exposed to sunlight and/or artificial light for many hours per day; hence, the action of light on hair is not only able to damage the hair shaft but it may alter the content of drugs and/or metabolites so that xenobiotics can gradually disappear from the hair shaft. This photodegradation process not only decreases the concentration of the original drugs but can produce new compounds, which, sometimes, have the same structure as natural metabolites.[39,40] Thus, light exposure should be considered as a potential confounder in studies investigating hair xenobiotics. Indeed, variations in hair concentration of the abuse substances due to the effect of light may alter the toxicological investigation measurements giving rise to incorrect drug levels or even false negative results, thus misunderstanding the drug addiction of the patient and her/his substance-use history. On the other hand, the formation of new photodegradation products from illicit drugs in hair under the effect of sunlight could lead to the identification of new useful markers of abuse in forensic medicine.

Therefore, knowledge of the effect of natural or artificial light on drug content in hair can be helpful to the forensic toxicologist, in particular when the examination of drug concentrations above or below pre-determined cut-offs is required in legal cases; meanwhile new markers of abuse, *i.e.* the drug photoproduct, might been identified which could be helpful in forensic evaluations.[41]

Photomodification of xenobiotics incorporated in the hair keratin may derive from the direct action of light on the drug molecules or it may be mediated by intermediate radicals and/or reactive oxygen species produced by the irradiated/excited drug itself or formed by natural component of hair, such as eumelanin and pheomelanin, under the effect of irradiation. UVB is able to breakdown the disulphide bonds inside the hair fibre and on the surface of the cuticle, while UVA mainly produces free radical and reactive oxygen species (ROS) through the interactions with the endogenous photosensitizers. Both eumelanin and pheomelanin can generate ROS, such as superoxide anion ($\cdot O_2^-$), hydroxyl radicals ($HO\cdot$) and H_2O_2, in various quantities.[24,42,43] Moreover, absorption of radiation in photosensitive aminoacids of the hair and their photochemical degradation produces free radicals having an adverse impact on hair proteins, especially keratin, and possibly on incorporated xenobiotics. Melanin can partially immobilize free radicals and block their entrance into the keratin matrix but oxidative stress occurs in any case. Moreover, the production of porphyrins by *Propionibacterium* sp. in the pilosebaceous duct, with photoactivation of porphyrins, demonstrates a failure in the defence mechanisms mediated by melanin.[43] A decrease in the defensive function against oxidative stress is also caused by bleach treatment, since hydrogen oxide can be easily decomposed by light to form the reactive oxygen species such as hydroxyl radicals with high oxidation ability.

Since drug of abuse testing in hair has become routine practice in forensic toxicological laboratories, the study of the behaviour of illicit drugs present in the hair and exposed to sunlight is a relevant aspect in the frame of forensic or clinical toxicology ascertainment for human drug abuse. However, until now only a few studies on the sunlight sensitivity of drugs of abuse in hair samples have been published, as will be presented below.

14.2. Target Drug Analytes in Hair

The most abused drugs worldwide are cannabis, cocaine, heroin, amphetamines, though with different patterns in various regions, *i.e.* in North Europe amphetamines are more prevalent than in South Europe where cannabis and cocaine dominate the illicit market. More recently, starting approximately in 2008, "new" psychoactive substances came to the drug scene, such as synthetic cannabinoids and cathinone derivatives. The target analytes of hair analysis have been therefore, for years, cocaine, opiates and their substitutes (methadone, buprenorphine), cannabinoids and amphetamines, even though the hair matrix has been used as a tool for the detection of a much larger variety of pharmaceuticals, most of them of forensic or clinical interest such as benzodiazepines, antidepressants and other CNS drugs.

As a consequence, a large literature exists on the most prevalent drugs that allows the expected range of their concentrations in hair to be known. The same data are useful to understand the pattern of incorporation and to distinguish between external contamination and systemic exposure. As a matter of fact, together with the active principles, metabolites are determined in hair: for cocaine (COC) the metabolites that may be present are benzoylecgonine (BZE), cocaethylene (CE), nor-cocaine (NCOC), and hydroxylcocaine; for opium derivatives, together with heroin (diaxcetylmorphine), 6-monoacetylmorphine (6-MAM), morphine (MOR), codeine (COD), and acetylcodeine are generally detected; for the opioid-agonist methadone (MTD) its main metabolite product 2-ethylidene-1,5-dimethyl-3,3-diphenylpyrrolidine (EDDP) is found; the group of amphetamine-like compounds investigated generally include amphetamine (AM), methamphetamine (MA), 3,4-methylenedioxymethamphetamine (MDMA or ecstasy), 3,4-methylenedioxyethylamphetamine (MDAEA); for cannabis, the target cannabinoids are the psychoactive component of the cannabis plant, Δ9-tetrahydrocannabinol (THC), cannabinol (CBN), cannabidiol (CBD), and the two metabolic byproducts 11-nor-9-carboxy-Δ9-tetrahydrocannabinol (THC-COOH) and 11-hydroxy-Δ9-tetrahydrocannabinol (OH-THC).

The expected levels, according to literature data,[22] are in the low ng mg^{-1} to hundreds of ng mg^{-1} range and vary among the different drugs and the mode of assumption, depending also on the limits of identification/quantification of the analytical methods employed. For cocaine, concentrations in hair may vary from 0.1 ng mg^{-1} to hundreds of ng mg^{-1}, for 6-MAM from 0.05 to decades of ng mg^{-1}, for methadone and opiates the range is analogous, while for THC the range extends from 0.02 pg mg^{-1} to decades of pg mg^{-1}. Interpretative cut-offs (*i.e.* concentration values that help in clarification) have been proposed by different scientific societies or by jurisdictional authorities. In Table 14.2 the cut-offs proposed by the Society of Hair Testing are presented for the most common drugs of abuse and allow to differentiate single/occasional/non-frequent use from repeated, chronic use or abuse.[44]

With respect to the metabolite/parent drug ratios observed in urine, where the metabolites largely prevail, in hair the parent drug is always the most prominent compound, and its metabolites are much less abundant. This has been interpreted by the rapid incorporation of drugs directly from the blood stream into the growing hair follicle. An exception to this rule is represented by heroin. In fact, following administration (generally *via* intravenous or

Table 14.2. Recommended cut-offs for substances and metabolites in hair to identify use.

	Screening		Confirmation
Group	**Cut-off (ng mg^{-1})**	**Target analyte**	**Cut-off (ng mg^{-1})**
Amphetamines	0.2	Amphetamine	0.2
		Methamphetamine	0.2
		MDA	0.2
		MDMA	0.2
Cannabinoids	0.1	THC	0.05
		THC-COOH	0.0002
Cocaine	0.5	Cocaine	0.5
		BZE, EME, CE, NC	0.05
Opiates	0.2	Morphine	0.2
		Codeine	0.2
		6-Acetylmorphine	0.2
Methadone	0.2	Methadone	0.2
		EDDP	0.05
Buprenorphine	0.01	Buprenorphine	0.01
		Norbuprenorphine	0.01

intramuscular injection, but it can also be snorted or smoked), heroin is readily hydrolysed to 6-MAM by blood esterases which is subsequently hydrolysed to MOR, with an approximate half-life of 45 min. Thus, heroin disappears after a few minutes, while 6-AM and MOR concentrations increase with time. As a consequence, heroin is often absent in hair form heroin users, whereas 6-MAM and MOR predominates; 6-MAM is indeed considered the specific marker of heroin consumption, accompanied by MOR, but in heavy consumers heroin itself can be found in hair. When MOR is found in hair alone, it corresponds to the use/abuse of the pharmaceutical grade narcotic drug MOR. In contrast, when opium or its derivatives other than heroin are consumed, all the cyclopentanoperhydrophenanthrene alkaloids present in the *Papaverum somniferum* latex can be found together with MOR, though at a lower amount. COD can be found alone, when it reflects the use of pharmaceutical grade COD, or together with 6-MAM and MOR due to its natural presence in opium, and 6-acetylcodeine can be found as a by-product of heroin synthesis from opium.

The synthetic opioid-agonist methadone (MTD) is primarily administered for chronic maintenance treatment of heroin addiction and has pharmacological properties similar to morphine. Detection and accurate quantitation of both methadone and 2-ethylidine-1,5-dimethyl-3,3-diphenylpyrrolidine (EDDP) in hair have been determined to be important factors in verifying the use or the abuse of this drug. Some significant pharmacological data of the illicit drugs discussed in this chapter are reported below.

Cannabis products include marijuana, hashish, hemp oil, derived by harvesting Cannabis Indica flowers, leaves, or resin. Tetrahydrocannabinol (THC) is the primary active constituent when Cannabis products are smoked, converting the precursor Tetrahydrocannabinolinic acid into THC. Among more than 100 compounds found in Cannabis, two other cannabinoids often accompany THC in Cannabis derived products: cannabidiol (CBD) and cannabinol (CBN) that do not exhibit any psychoactive response. When THC enters the blood stream and reaches the brain, it binds to cannabinoid receptors, resulting in changes in the levels of dopamine and norepinephrine; which are closely associated with the acute effects of cannabis ingestion, such as euphoria and anxiety. Effects include alteration of perceptions, euphoria, feelings of well-being, relaxation, stress reduction, increased libido. Anxiety and panic attacks are the most commonly reported side effect of smoking marijuana. THC is then excreted in urine mainly as the carboxyl derivative, 11-nor-9-carboxy-delta9-tetrahydrocannabinol (THC-COOH). In hair, THC, CBD and CBN can all be found, at variable ratios, while the presence of the acidic metabolite THC-COOH at very low amount pg mg^{-1} has been taken as a proof of consumption.

Opiates are derived from opium, a naturally occurring product obtained from the plant *Papaver somniferum.* (opium poppy). Opium is the raw material for the production of heroin by acetylating morphine to diacetylmorphine (heroin). Street heroin is, however, a complex mixture of a number of opium alkaloids and synthetic by-products, including morphine (MOR), 6-acetylmorphine (6-AM), codeine (COD), 6-acetylcodeine (6-AC), noscapine and papaverine. In addition, a number of adulterants and cutting agents are also present. Analogously, all these adulterants can be determined in hair. Opiates' effects are sedative, hypnotic and analgesic: their use can depress breathing acting at the brain stem, where automatic body functions are controlled, can increase feelings of pleasure acting at the limbic system, and can block pain messages transmitted through the spinal cord.

Heroin (diacetylmorphine) is in general self-administered *via* intravenous or intramuscular injection, but it can also be snorted or smoked. Less usual administration routes include oral, subcutaneous, sublingual and rectal. Following administration, heroin is readily hydrolysed to 6-AM by blood esterases, which is subsequently hydrolysed to MOR with approximate half-lives of 5 and 45 min for heroin and 6-AM, respectively. Heroin has a high affinity for brain tissue, but it disappears after a few minutes, with 6-AM and MOR concentrations increasing and lasting for up to 30 min; it is believed that 6-AM and MOR are responsible for the narcotic activity of heroin. Heroin is eliminated mainly in the urine, predominantly as free and conjugated MOR although heroin itself and 6-AM may also be present. After heroin injection, abusers typically report feeling pleasurable sensations defined as a "rush." The intensity of the rush is a function of how much drug is taken and how rapidly the drug enters the brain and binds to the opioid receptors. The rush is usually accompanied by a warm flushing of the skin, dry mouth, and a heavy feeling in the extremities, which may be accompanied by nausea, vomiting, and severe itching. After the initial effects, users usually will be drowsy for several hours; mental function is clouded; heart function slows; and breathing is also severely slowed, sometimes enough to be life-threatening. Slowed breathing can also lead to coma and permanent brain damage.[45]

Cocaine is an alkaloid present in leaves from *Erithroxylon Coca* leaves, originating in Peru and Bolivia, now widely cultivated in South America (Columbia) and Asia (India, Java, Ceylon, Indonesia). Coca leaves contain 0.4–2.5% of alkaloids; 50–60% of this is represented by Cocaine. Cocaine is generally assumed by sniffing (inhalation) or snorting; intravenous injection is used especially in the U.S., using the CRACK, obtained in the form of crystals by heating an aqueous solution of Cocaine hydrochloride in the presence of sodium bicarbonate. As for the Heroin, the "street cocaine" is associated with cutting agents, the most frequent adulterants are lidocaine and procaine, while diluents are sucrose, glucose, mannitol, lactose. Cocaine is a strong S.N.C stimulant. and, immediately after use, it causes the consumer a significant mood elevation, with excitement, sometimes with euphoria, loquacity and disinhibition. Enhancement of the mental activity and a significant reduction in the sense of fatigue or hunger have been described.

Acute intoxication by cocaine (in humans the toxic dose is 1–2 grams) mainly affects the S.N.C. and immediately after taking the drug the person becomes excited, talkative and confused and you can observe a variety of clinical symptoms such as tachycardia, irregular breathing, mydriasis and fever accompanied by nausea, vomiting and abdominal pain. After acute intoxication delirium, convulsions and coma occur, and in very short time death occurs by respiratory arrest. Chronic intoxication with Cocaine is characterized, essentially, with a particularly strong psychic dependence.

Amphetamines: the molecule of phenylethylamine is the basic chemical structure for amphetamine-like compounds. Variations of this structure lead to a vast number of compounds whose pharmacological action is characterized by the exaltation or by the reduction of some of the peculiar properties of the base molecule. The central and peripheral stimulating effects can be attributed to increased release of norepinephrine, serotonin and dopamine from the synapses of the sympathetic system, with consequent effects of increased alertness, decreased fatigue, appetite depression, insomnia, agitation and motor incoordination, tachycardia and hypertension, *etc.* The historically most known compounds, and yet abundantly abused in many countries, are the amphetamine and methamphetamine known in the black market as "speed," "crank," "meth," "crystal meth", both strong CNS stimulants. Most of the amphetamine-like compounds was synthesized starting from the '60s and more than a hundred of them proved to be pharmacologically active in humans. The most known molecules refer to the methylendioxyphenylethylamine derivatives such as 3,4-methylendioxyamphetamine (MDA, known as "love pill", "speed for love"), 3,4-methylenedioxymethamphetamine (MDMA, known as "ecstasy", "adam", "xTC"), 3,4-methylenedioxyethylamphetamine (MDEA, known as "eve"). All of these derivatives are defined as "entactogens" or "empathetic", *i.e.* substances able to induce a favourable state to meet other people in a club setting. The amphetamine-like compounds are normally sold in the form of oral tablets; they are readily absorbed from the gastrointestinal tract and reach maximum plasma concentrations after 1–3 hours. They are rapidly distributed to all tissues and are readily detectable in urine, where average persist up to 72 hours after ingestion.

Methadone is a synthetic analgesic and narcotic, having pharmacological activities similar to those of morphine. It is an agonist to the mu1, mu2 and kappa receptors for opiates. Methadone is rapidly absorbed from the gastrointestinal tract and is found in plasma within 30 minutes after oral administration, reaching maximum blood concentrations in 3–4 hours; it is metabolized primarily in the liver: the main metabolites are excreted in the urine and bile along with small amounts of unchanged drug. The two major metabolites are 2-ethylidene-1,5-dimethyl-3,3-diphenypirroline (EDDP) and 2-ethyl-5-methyl-3,3-diphenyl-1-pyrroline (EMDP) none of which is active. Due to its pharmacokinetic properties (biphasic disappearance from the plasma, $t½$ of the first phase from 1.9 to 4.2 h; $t½$ of the second phase from 8.5 to 47 h), methadone is used to treat withdrawal syndromes and addiction to narcotic drugs under medical supervision. However, its use can be diverted to abuse.[46–49]

14.3. Photodegradation of Drugs of Abuse in Hair Samples: A Few Existing Studies

The photodegradation of drugs of abuse in hair has been the object of very few studies, focused on the main classes of drugs.[27,40,50] Skopp *et al.* published a study in which tetrahydrocannabinol (THC), cannabinol (CBN), and cannabidiol (CBD), exhibited a marked instability when exposed in hair samples for ten weeks to natural solar radiation.[27]

Samples were prepared as reported: clipped hair bundles supposed to be cannabis positive were washed and divided into four strands. Two of the strands were kept in the dark at ambient temperature, while the other two strands were stored outside in quartz glass vials and exposed to natural sunlight for 10 weeks at 114 m above sea level, 40°, 300 degrees of latitude. The cannabinoid content in the original and exposed hair samples was determined by gas chromatography–mass spectrometry after hair pulverization, basic hydrolysis, and liquid/liquid extraction in organic solvent, according to Cirimele *et al.*[51] Detection was by ionization in the electron impact mode (70 eV), in multiple-ion monitoring (THC, m/z 299, 314, 271; THC-d3, m/z 302, 317, 274; CBN, m/z 295, 310, 238; CBD, m/z 231, 246, 314). Measurements were taken twice from each strand. The limit of detection was 0.05 ng mg^{-1} of hair for all analytes. The results are reported in Table 14.3.[27]

As may be inferred, drastic changes are detected after exposure to light, that hamper the identification of THC, CBN and CBD in 80–90% of cases; the decrease in cannabinoid content was not connected to the hair color. According to the authors, THC, most likely through a light-induced radical reaction mechanism,[52] might have been converted to CBN in keratinized hair exposed to sunlight. CBN is indeed a chemical degradation product of THC,

Table 14.3. Changes in cannabinoid content in hair strands exposed to global solar radiation for 10 weeks.

		Original concentration			**Findings after exposure to light**		
Sample	**Color**	**THC**[a]	**CBN**[b]	**CBD**[b]	**THC**[a]	**CBN**[b]	**CBD**[b]
1	Brown	0.11	-	+	-	-	-
2	Dark brown	1.72	+	+	0.35	+	+
3	Brown	0.18	+	+	-	-	-
4	Light brown	0.10	+	-	-	-	-
5	Blonde	0.30	+	-	0.11	-	-
6	Dark blonde	0.20	-	-	-	-	-
7	Light blonde	1.20	+	-	0.10	-	-
8	Blonde	0.76	+	-	-	-	-
9	Middle blonde	0.95	+	-	-	+	-
10	Brown/grey	1.18	+	-	-	-	-
11	Red	0.87	+	-	-	+	-

[a]In ng mg^{-1} of hair, mean value (n = 4).
[b]+ , above the limit of detection (0.05 ng mg^{-1} of hair); -, not detectable.

appearing during storage in plant materials or in solution. However, in keratinized hair exposed to sunlight, CBN was further degraded (see Table 14.3) as a consequence of the reactive intermediates and oxygen species generated by both eumelanin and pheomelanin.[29,42,43] An *in vitro* test, where THC stability was studied in ethanol in the presence of hydrogen peroxide, supported this hypothesis.

Those findings may have implications on sample collection and the interpretation of analytical results in hair analysis. In general, only a sample below the top hair, not exposed directly to sunlight, should be collected, and information should be acquired on the outdoor activities of the hair donor.

The same authors examined the stability of cocaine, benzoylecgonine (BZE), 6-monoacetylmorphine (6-MAM), morphine, and dihydrocodeine showing different behaviours.[50]

Clipped hair bundles, supposed to be opiate and cocaine positive, were washed and divided into six strands longitudinally. Two of the strands were kept in the dark in aluminum foil at ambient temperature, while two strands were stored outside in quartz glass vials and exposed to natural sunlight for three months (at 114 m above sea level, 49°, 300 degrees of latitude); two strands were exposed in the open to natural weather (global radiation, rain, wind) for the same period. The results of this experiment are shown in Table 14.4.

It is evident that 6-monoacetylmorphine (6-MAM) and morphine considerably decreased their concentrations under sunlight and even more after global weather exposure. The levels of cocaine and benzoylecgonine decreased to varying degrees; exposure to natural light more strongly affected the cocaine levels than global natural weather. Benzoylecgonine could not be found under weather conditions in four of six originally positive hair segments and was detected only in hair sample D with higher initial concentrations.

Table 14.4. Changes in the drug content of authentic hair samples after three months exposure under global radiation or under natural weather conditions (global radiation, rain, wind) outdoors.

Sample number	Conc. before exposure	Conc. after daylight exposure	Conc. after outdoors exposure
6-Acetylmorphin/Morphin (ng mg^{-1} hair)			
A1	0.7/pos.[b]	0.7/pos.	pos./pos.
A2	1.0/0.9	0.6/0.6	0.6/0.6
B1	17.2/2.9	14.2/0.9	3.3/wr[a]
B2	19.8/3.2	15.8/1.2	5.12/wr[a]
E1	1.2/wr[a,c]	0.4/wr[a]	o.B[a]/wr[a]
E2	1.3/wr[a]	0.5/wr[a]	o.B[a]/wr[a]
Dihydrocodein (ng mg^{-1} hair)			
B1	6.7	8.7	8.6
B2	3.2	7.6	5.1
E1	2.0	2.6	2.6
E2	5.3	5.9	4.1
Cocain/Benzoylecgonin (ng mg^{-1} hair)			
B1	4.9/0.7	1.2/wr[a]	3.8/wr[a]
B2	5.2/0.6	1.3/wr[a]	4.8/wr[a]
C1	1.1/0.9	1.0/0.7	1.0/wr[a]
C2	1.7/0.9	1.2/1.0	0.9/wr[a]
D1	3.3/2.2	1.4/1.0	1.9/1.9
D2	3.2/3.3	1.1/1.3	1.6/1.6
E1	1.1/wr[a]	1.0/wr[a]	0.9/wr[a]
E2	0.8/wr[a]	1.2/wr[a]	0.7/wr[a]

[a]wr = without results.
[b]positive detection below the quantification limit.
[c]findings below the detection limit.

The concentrations of dihydrocodeine were clearly higher after exposure to weathering and even higher under sunlight.

One reason of the lower levels of 6-acethylmorphine and morphine under both conditions was found in the partial destruction of their binding sites on melanin and damage to individual components of the hair fiber, along with rapid hydrolysis and oxidation of the drug molecules. In contrast, cocaine and benzoylecgonine seem to be more degraded under sunlight than after weathering, a behaviour that was explained by access to previously inaccessible melanin domains facilitated by weathering (*i.e.* water) and resulting in a better recovery of drugs with a high melanin affinity.

Dihydrocodeine concentration levels increased, according to the authors, after both sun exposure and global weathering possibly due to the damaging effects of light and water on hair pigment with a better recovery of the drug having a high melanin affinity and a high photostability.

In conclusion, two opposite processes underlie changes in drug levels when hair are exposed to light and/or ambient: hydrolytic and oxidative degradation of drugs or an improved accessibility of the binding sites in the hair.

In the paper published by Favretto *et al.*[40] eight positive samples of hair for three drug classes (methadone, opiates, and cocaine) were analysed to study the photodegradation under controlled UVB radiation; their characteristics are outlined in Table 14.5.

Table 14.5. Hair samples: physical characteristics and drug concentrations determined by HPLC-HRMS in aliquots kept in the dark.

Sample	Color	Thickness	Straight	Curl	Methadone ng mg^{-1}	EDDP ng mg^{-1}	6-MAM ng mg^{-1}	Morphine ng mg^{-1}	Cocaine ng mg^{-1}	BZE ng mg^{-1}
#1 Fair	Grey and white	Very thin	x		—	—	0.23	0.25	—	—
#2 Fair	Light brown	Thin	x		15.29	0.56	0.58	0.28	—	—
#3 Fair	Dark blond	Thin	x		1.90	0.18	—	—	—	—
#4 Fair	Dark blond	Thin	x		6.20	1.20	—	—	2.26	0.48
#5 Dark	Black	Thick			3.25	1.57	0.30	0.50	2.23	0.08
#6 Dark	Brown	Thick	x		4.30	–	0.30	0.30	1.45	0.18
#7 Dark	Dark brown	Thin		x	5.50	0.79	—	—	0.24	0.12
#8 Dark	Dark brown	Thick	x		4.50	1.29	—	—	—	—
#9 Dark	Brown	Thick	x		—	—	—	—	11.98	2.33
#10 Dark	Dark brown	Thin	x		—	—	—	—	31.26	1.82
#11 Dark	Brown	Thin	x		—	—	—	—	8.26	4.04
#12 Dark	Black	Thick	x		—	—	—	—	5.64	0.51
#13 Dark	Brown	Thin	x		2.50	0.90	0.50	0.30	—	—
#14 Dark	Brown	Thin	x		—	—	0.70	0.30	—	—
#15 Dark	Brown	Thin	x		—	—	0.60	0.28	—	—
#16 Dark	Brown	Thin	x		—	—	0.50	0.35	—	—

The hair samples were treated and irradiated as reported: hairs, 5–7 cm long, were divided into two approximately identical strands: one was kept in the dark and the other was put between two optical glasses 50 × 50 × 1.15 mm and exposed to 300 J cm^{-2} of UVB radiation while kept at room temperature. The intensity of the UVB light was 0.16 J cm^{-2} min. The hair samples were irradiated for 32 h, corresponding to approximately 300 J cm^{-2}. The glasses used were of a type able to cut wavelengths below 300 nm which could interfere with and/or alter the results, thereby avoiding those wavelengths emitted by the lamp but not present in the sun's spectrum.

Hair samples (10 ± 0.3 mg) were decontaminated with CH_2Cl_2 and acetone, placed in a vial with one stainless-steel bullet and a mixture of acetonitrile, TFA and water in the presence of an internal standard (IS), then shaken by an automatic pulveriser, following the micropulverized method proposed by Favretto *et al.*[53,54] After centrifugation and filtration, the clear extracts were directly injected into the HPLC-HRMS with an ESI source. Profile full scan mass spectra were acquired in the Orbitrap in the *m/z* range 120–700 with a target mass resolution of 100.000. Detection of the analytes and the IS was based on retention time, accurate mass measurements of MH^+ ions, and correspondence of the observed isotopic pattern to the calculated one. Drug concentrations were determined from peak area ratios of analyte to its IS compared to calibrator curves of peak area ratios to concentrations.

In Table 14.6, the percentage of photodegradation yields of methadone and EDDP, 6-MAM and morphine, cocaine and BZE are outlined, calculated by using the concentration of the sample kept in the dark as 100% reference [% photodegradation = (drug conc. in dark − drug conc. after irradiation)/ drug conc. in dark * 100].

Table 14.6 also shows the yields of degradation for each drug class and for all eight hair samples. These yields are reported as range, average, and standard deviation. A clearly high degradation of methadone (55%) is evidenced when compared to all the other drugs/metabolites (averages from 10% to 21%). Interestingly, a color effect is suggested when considering the

Table 14.6. Hair samples: relative degradation of drugs in aliquots irradiated at 300 J cm^{-2} with UVB light.[a]

	% photodegradation								Average (n = 8)	Range	SD
Hair sample	*#2*	*#3*	*#4*	*#5*	*#6*	*#7*	*#8*	*#13*			
Methadone	85	82	80	53	38	28	34	36	55	28–85	24
EDDP	32	33	32	0	—	10	5	7	17	0–33	15
Hair sample	*#1*	*#2*	*#5*	*#6*	*#13*	*#14*	*#15*	*#16*			
6-MAM	0	29	7	43	22	33	26	11	21	0–43	14
Morphine	0	25	0	43	19	28	15	9	17	0–43	15
Hair sample	*#4*	*#5*	*#6*	*#8*	*#9*	*#10*	*#11*	*#12*			
Cocaine	0	4	26	16	15	0	22	0	10	0–26	11
Benzoylecgonine	24	0	63	0	0	27	17	30	20	0–63	21

[a]% photodegradation = (drug conc.$_{dark}$ − drug conc.$_{irradiatied}$)/drug conc.$_{dark}$*100.

photodegradation of methadone and EDDP whose concentrations fall in the three fair-hair samples. The color of hair depends on the relative amounts of pheomelanin (red) and eumelanin (black) that are thought to represent the first defence against UV in human hair and skin, as mentioned above.[33,55,56]

Generally, a part of the light is absorbed by the hair matrix itself without any photochemical effect. In dark hair, the eumelanin could protect the drugs/metabolites to a higher degree than in fair hair. However, melanin may also react with oxygen under irradiation, producing particularly reactive species such as superoxide anions that can induce the photolysis of melanin itself,[42] thus weakening the photoprotective effect of the melanin.

Regarding opiates, on the basis of previous *in vivo* and *in vitro* experiments on photodegradation[27,57] they were expected to be highly degraded by exposure to light. Unexpectedly, the *in vivo* experiments exhibited a relatively low degradation with an average of 21 and 17% for 6-MAM and morphine, respectively, though a large individual variability was observed.

Figure 14.4 reports, to visually compare, the average photodegradation yields of the target compounds when irradiated in hair and in methanol solution.

In these experiments, cocaine was the most resistant molecule (average photodegradation of 10% in hair, range 0–26), whereas BZE degrades more efficiently than its precursor in both experiments (average photodegradation of 20% in hair, range 0–63). Interestingly, higher degradation yields were observed for cocaine (average of 45%) and BZE (average of 55%) by Skopp[50] when hairs were exposed to natural light (direct and diffuse sunlight) for three months.

To further investigate the role of hair structure, *i.e.* thickness and the integrity of the keratin/melanin organization within the shaft, the authors

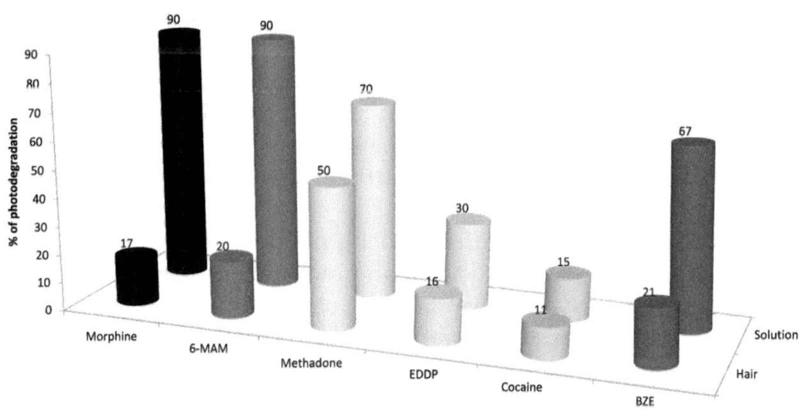

Figure 14.4. Relative photodegradation of methadone, EDDP, 6.MAM, morphine, cocaine and BZE in authentic positive hair irradiated by UVB light (n = 8 for each drug class) and comparison with the maximum relative degradation observed for the methanol solutions of the same drugs.

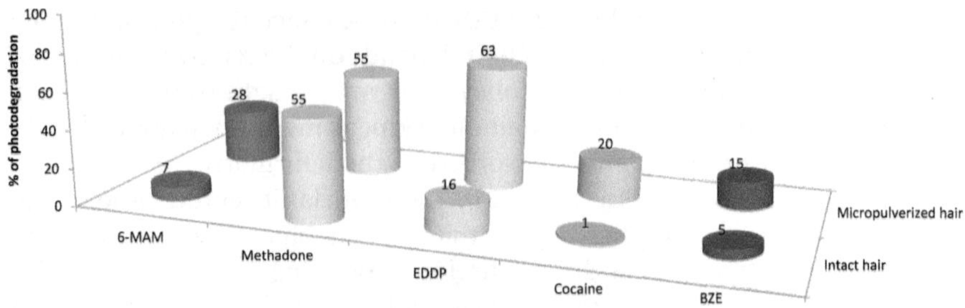

Figure 14.5. Comparison of the relative photodegradation of drugs in authentic positive hair samples as integral hair strands or powder when irradiated by UVB light.

compared the results for hair irradiated as a whole hair and after pulverization. Individual human hair diameter varies from 15 to 120 μm depending upon the hair type and the body region the follicle is located, and this could reflect on the different protective effects of the cortex with respect to the medulla and incorporated drugs.

Pulverization is known to enable drug extraction, increasing the surface in contact with the solvents; moreover, desegregation could sensibly reduce hair thickness and its protective effect, thus exposing the drugs more efficiently to the incoming radiation. Figure 14.5 shows the degradation yields of the supplementary samples that tested positive for the target molecules.

The samples were powdered in a ball mill and put between two optical glasses to be irradiated for 32 h. In the same figure, percentages for photodegradation are compared to those obtained for the corresponding whole hair irradiated for the same time period; the concentration of drugs determined in the aliquots kept in the dark again represented the control. As a general trend, drugs in pulverized hair exhibited higher percent degradations than intact hair (15–63% *versus* 1–55%), demonstrating that the structure of the hair shaft plays an important role in protecting drugs from UVB degradation. Interestingly, methadone, which was the most labile drug in the whole hair tests, did not degrade further in the powdered hair analysis (55% in both cases).

Therefore, the significantly different yields observed for the same drugs in different hair samples (see *e.g.* 6-MAM and morphine in samples #1 and #5 *versus* all the other opiate positive samples) could be ascribed to even slight individual differences in the hair's structure (melanin, keratin, fibres, *etc.*) or to its ultrastructure.[58]

In conclusion, the hair photodegradation studies performed to elucidate how light exposure may decrease concentrations of the most frequent drugs of abuse in keratin matrices, revealed that UVB irradiation is able to modify the levels of methadone and EDDP incorporated in the hair with an important role of the whole hair structure in protecting drugs from light. In contrast, opiates turned out to be slightly photolabile.

In comparing fair hair and dark hair, a much larger photodegradation of drugs was observed in dark hair, supporting the hypothesis of the protective effect of eumelanin.

These findings point out that light influences the stability of entrapped drugs in hair, and reconfirms that it is very important to consider the light exposure history of a hair sample in the interpretation of hair analysis results.

14.4. Conclusions

At the current state of the art, little is known about the photostability of illicit drugs inside the hair shaft when long term exposures to sunlight or artificial light occur. From the few experimental data collected, however, most of the drugs appeared photolabile.

Since the detection of drugs of abuse is often used in courts as evidence of illegal acts, with consequences on the freedom of persons, the UV-Vis effects on the structure integrity of the drugs and their metabolites should be taken in account when single exposures need to be evidenced.

When drugs are covertly administered in drug-facilitated crimes and single drug exposure needs to be ascertained, the expected concentrations in hair are in the range of pg mg^{-1} of hair for sedative and hypnotic drugs.[59] Those low levels allow us to infer how critical the investigation on single doses may be when hair may have been altered by long term exposure to ambient.

On the other hand, when decisional cut-offs are applied to hair analysis results and are used to decide on probation, to grant a driving licence, a job, or a child custody, it must be understood that exposing hair to cosmetic treatments and/or to light may alter the final results producing "false negatives."

The knowledge of the photodegradation of drugs of abuse in hair described in this chapter suggests the need for further studies. In particular, the structural identity of new stable species arising from drug photolysis could provide new markers of drug consumption, incorporation and degradation useful in the forensic investigations.

References

1. V. A. Boumba, K. S. Ziavrou and T. Vougiouklakis, *Int. J. Toxicol.*, 2006, **25**, 143.
2. A. M. Baumgartner, P. F. Jones, W. A. Baumgartner and C. T. Blank, *J. Nucl. Med.*, 1979, **20**, 748.
3. F. Pragst and M. A. Balikova, *Clin. Chim. Acta*, 2006, **370**, 17.
4. P. Kintz, *Anal. Bioanal. Chem.*, 2007, **388**, 1467.
5. C. R. Robbins, *Chemical and Physical Behavior of Human Hair*, Springer, New York, 4th edn, 2002.
6. M. Saitoh, M. Uzaka, M. Sakamoto and T. Kobori, Rate of hair growth, in *Advances in Biology of Skin: Hair Growth*, ed. A. Montana and M. Dobson, Oxford, 1969, pp. 183–194.
7. M. d'Ischia, K. Wakamatsu, A. Napolitano, S. Briganti, J. C. Garcia-Borron, D. Kovacs, P. Meredith, A. Pezzella, M. Picardo, T. Sarna, J. D. Simon and S. Ito, *Pigm. Cell Melanoma Res.*, 2013, **26**, 616.

8. M. Brenner and V. J. Hearing, *Photochem. Photobiol.*, 2008, **84**, 539.
9. J. D. Simon, D. Peles, K. Wakamatsu and S. Ito, *Pigm. Cell Melanoma Res.*, 2009, **22**, 563.
10. L. Potsch and M. R. Moeller, *J. Forensic Sci.*, 1996, **41**, 121.
11. G. L. Henderson, *Forensic Sci. Int.*, 1993, **63**, 19.
12. G. L. Henderson, M. R. Harkey and C. Zhou, *J. Anal. Toxicol.*, 1998, **22**, 156.
13. M. Rothe, F. Pragst, S. Thor and J. Hungen, *Forensic Sci. Int.*, 1997, **84**, 53.
14. F. Pragst, M. Rothe, K. Spiegel and F. Sporkert, *Forensic Sci. Rev.*, 1998, **10**, 81.
15. R. Bajewa, T. D. Sokoloski and P. N. Patil, *J. Pharm. Sci.*, 1977, **66**, 1544.
16. Y. Nakahara, K. Takahashi and R. Kikura, *Biol. Pharm. Bull.*, 1995, **18**, 1223.
17. Y. Nakahara and R. Kikura, *Arch. Toxicol.*, 1996, **70**, 841.
18. D. Favretto, S. Vogliardi, M. Tucci, I. Simoncello, R. El Mazloum and R. Snenghi, *Forensic Sci. Int.*, 2016, **265**, 193.
19. M. Balíková, *Biomed. Pap. Med. Fac. Univ. Palacky Olomouc Czech. Repub.*, 2005, **149**, 199.
20. W. A. Baumgartner, V. A. Hill and W. H. Blahd, *J. Forensic Sci.*, 1989, **34**, 1433.
21. G. L. Henderson and B. K. Wilson, *Res. Commun. Chem. Pathol. Pharmacol.*, 1973, **5**, 1.
22. C. Jurado, P. Kintz, M. Menéndez and M. Repetto, *Int. J. Leg. Med.*, 1997, **110**, 159.
23. G. Skopp, L. Pötsch and M. R. Moeller, *Forensic Sci. Int.*, 1997, **84**, 43.
24. R. M. Trüeb, *Int. J. Trichol.*, 2009, **1**, 6.
25. G. Skopp and L. Pötsch, *In vitro* contamination of hair by marijuana smoke, *Clin. Chem.*, 2004, **50**, 596.
26. L. Pötsch and G. Skopp, Inkorporation von Fremdsubstanzen in Haare, in *Haaranalytik-Technik und Interpretation in Medizin und Recht*, ed. B. Madea and F. Mußhoff, Deutscher Ärzte-Verlag, Köln, Germany, 2004, p. 29.
27. G. Skopp, L. Potsch and M. Mauden, *Clin. Chem.*, 2000, **46**, 1846.
28. C. Santos Nogueira and I. J. Joekes, *J. Photochem. Photobiol., B*, 2004, **74**, 109.
29. S. E. Forest and J. D. Simon, *Photochem. Photobiol.*, 1998, **68**, 296.
30. S. B. Ruetsch, Y. Kamath and H. D. Weigmann, *J. Cosmet. Sci.*, 2000, **51**, 103.
31. C. M. Pande and J. Jachowicz, *J. Soc. Cosmet. Chem.*, 1993, **44**, 109.
32. A. R. Rook and R. P. R. Dawber, *Haarkrankheiten*, Blackwell, Berlin Wien, 1995.
33. T. Herrling, K. Jung and J. Fuchs, *Spectrochim. Acta, Part A*, 2007, **69**, 1429.
34. M. V. Fedorkova, N. N. Brandt, A. Y. Chikishev, N. V. Smolina, N. G. Balabushevich, S. A. Gusev, V. A. Lipatova, V. M. Botchey, G. E. Dobretsov and E. V. Mikhalchik, *J. Photochem. Photobiol., B*, 2016, **164**, 43–48.
35. Z. D. Draelos, *Dermatol. Clin.*, 2000, **18**, 651.
36. L. R. Levine, D. Garland, C. N. Oliver, A. Amici, I. Climent, A. G. Lenz, B. G. Ahn, V. Shaltiel and E. R. Stadtaman, *Methods Enzymol.*, 1990, **186**, 464.
37. T. J. Sarna, *J. Photochem. Photobiol., B*, 1992, **12**, 215.
38. S. Ito, M. Kikuta, S. Koike, G. Szewczyk, M. Sarna, A. Zadlo, T. Sarna and K. Wakamatsu, *Pigm. Cell Melanoma*, 2016, **29**, 340.
39. H. Raoof, P. Mielczarek, K. A. Michalow, M. Rekas and J. Silberring, *J. Photochem. Photobiol., B*, 2013, **118**, 49.
40. D. Favretto, M. Tucci, A. Monaldi, S. D. Ferrara and G. Miolo, *Drug Test. Anal.*, 2014, **6**, 78.
41. V. L. Westera, N. R. P. van der Wulp, J. W. Koper, Y. B. de Rijke and E. F. C. van Rossum, *Psychoneuroendocrinology*, 2016, **72**, 94.
42. M. R. Chedekelt, S. K. Smitht, P. W. Postf, A. Pokorat and D. L. Vessellt, *Proc. Natl. Acad. Sci.*, 1978, **75**, 5395.

43. S. Persad, I. A. Menon and H. F. Habermann, *Photochem. Photobiol.*, 1983, **37**, 63.
44. G. A. A. Cooper, R. Kronstrand and P. Kintz, *Forensic Sci. Int.*, 2012, **218**, 20.
45. NIDA, *Heroin*, 2014, retrieved February 1, 2017, from: https://www.drugabuse.gov/publications/research-reports/heroin.
46. R. C. Baselt, *Disposition of Toxic Drugs and Chemicals in Man*, Biomedical Publications, Foster City, 7th edn, 2004.
47. European Monitoring Centre for Drugs and Drug Addiction, *2009 annual Report on the State of Drugs Problem in Europe*, 2009, http://www.emcdda.europa.eu/publications/annual-report/2009.
48. D. Garside and B. A. Goldberger, in *Drug Testing in Hair*, ed. P. Kintz, CRC, Boca Raton, 1996.
49. T. Reisine and G. Pasternak, in *Goodman and Gilman's the Pharmacological Basis of Therapeutics*, ed. J. G. Hardman, L. E. Limbird and A. G. Gilman, McGraw-Hill, New York, 9th edn, 1996.
50. G. Skopp, L. Pötsch and M. R. Möller, *Rechtsmedizin*, 1997, **7**, 176.
51. V. Cirimele, H. Sachs, P. Kintz and P. Mangin, *J. Anal. Toxicol.*, 1996, **20**, 13.
52. C. M. Bonucelli, Stable solutions for marijuana analysis, *J. Pharm. Sci.*, 1979, **68**, 262.
53. D. Favretto, S. Vogliardi, G. Stocchero, A. Nalesso, M. Tucci and S. D. Ferrara, *J. Chromatogr. A.*, 2011, **1218**, 6583.
54. S. Vogliardi, D. Favretto, M. Tucci, G. Stocchero and S. D. Ferrara, *Anal. Bioanal. Chem.*, 2011, **400**, 51.
55. E. Hoting, M. Zimmermann and S. Hilterhaus-Bong, *J. Soc. Cosmet. Chem.*, 1995, **46**, 85.
56. E. Hoting, M. Zimmermann and H. Hocker, *J. Soc. Cosmet. Chem.*, 1995, **46**, 181.
57. G. Miolo, M. Tucci, A. Mazzoli, S. D. Ferrara and D. Favretto, *J. Pharm. Biomed. Anal.*, 2016, **126**, 48.
58. L. Potsch, G. Skopp and M. R. Moller, *Int. J. Leg. Med.*, 1995, **107**, 301.
59. *Toxicological Aspects of Drug-facilitated Crimes*, ed. P. Kintz, Elsevier Inc, 2014.

Subject Index